The Heart of Leonardo

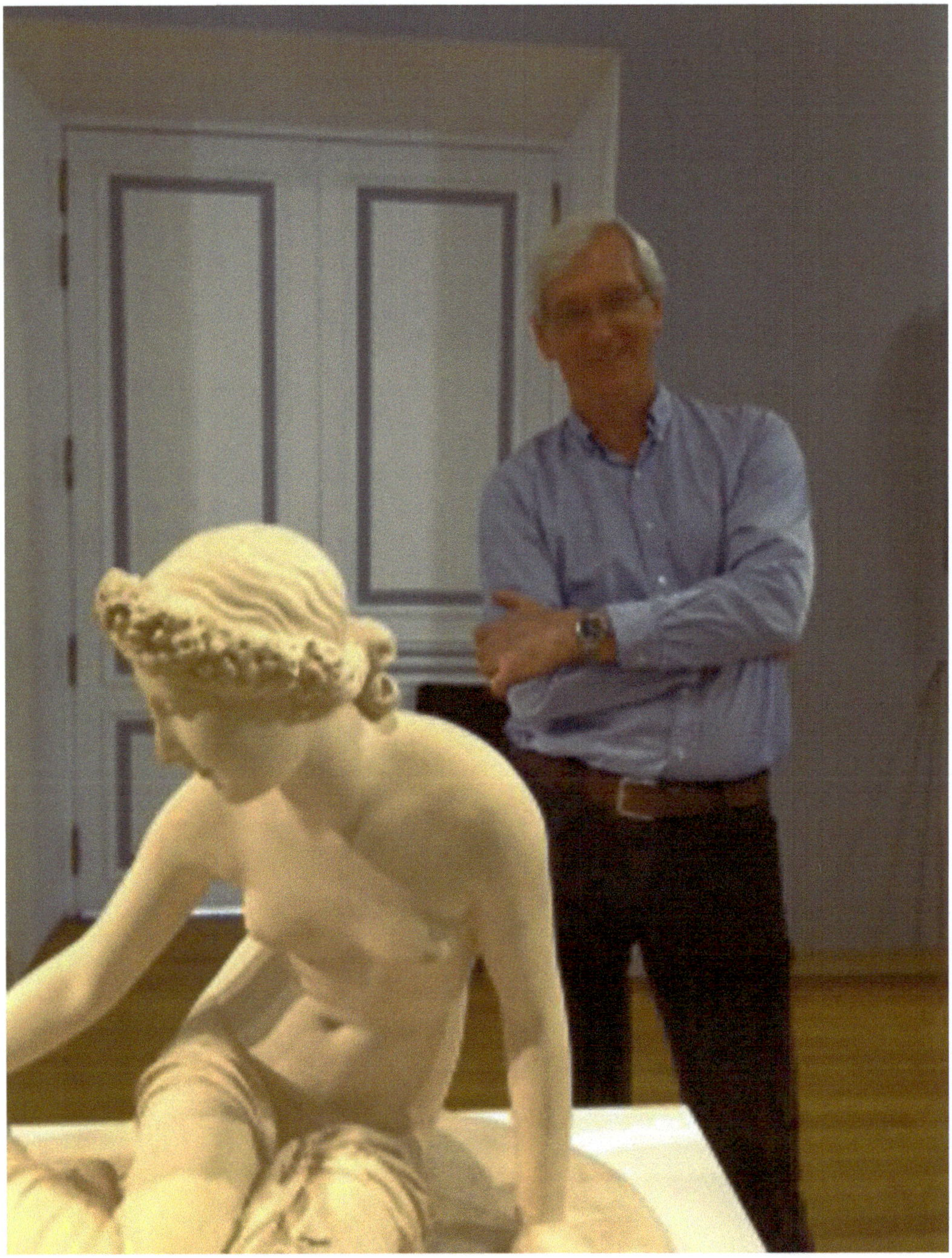

Francis C. Wells

The Heart of Leonardo

Foreword by HRH Prince Charles, The Prince of Wales

Springer

Francis C. Wells
Department of Cardio-Thoracic Surgery
Papworth Hospital NHS Foundation Trust
Cambridge
United Kingdom

ISBN 978-1-4471-4530-1 ISBN 978-1-4471-4531-8 (eBook)
DOI 10.1007/978-1-4471-4531-8
Springer London Heidelberg New York Dordrecht

Library of Congress Control Number: 2013937956

Printed on acid-free paper

Springer is part of Springer Science+Business Media (www.springer.com)

For Joanna, Nicholas and Olivia
semper proxime ad cor meum........

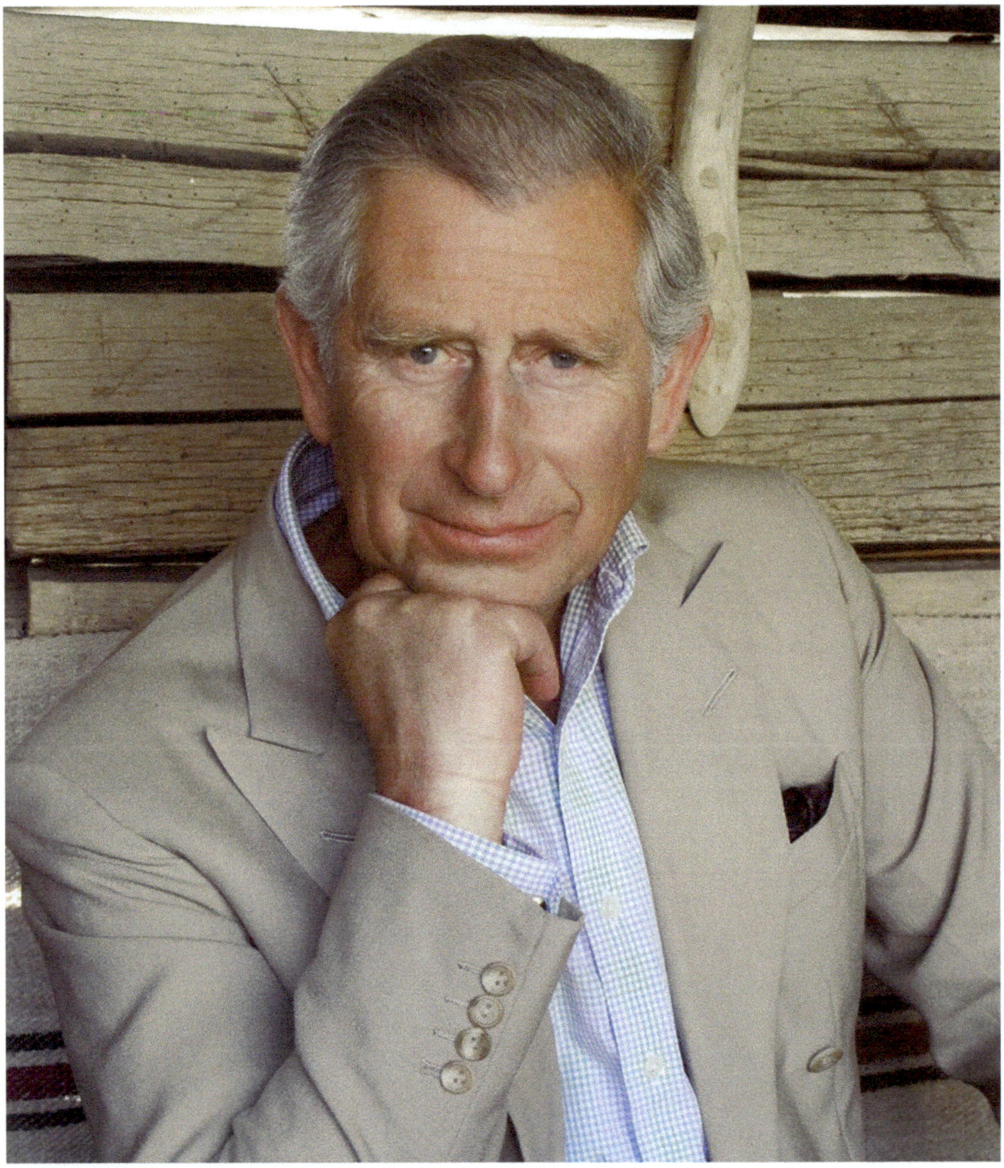

Foreword by HRH The Prince of Wales

CLARENCE HOUSE

Almost half a millennium after his death, Leonardo da Vinci continues to excite universal fascination as the archetypal 'Renaissance Man', equally gifted in the arts and the sciences. The Royal Collection is immensely fortunate to hold the finest surviving group of drawings by Leonardo – over five hundred sheets, which have been together since Leonardo's death and were probably acquired by King Charles II in the seventeenth century.

Two hundred of those sheets are devoted to the field that fascinated Leonardo more than any other: his study of the body. Through these drawings we can follow the blossoming of Leonardo's anatomical knowledge, from his first attempts to come to terms with ancient tradition, using the corpses of animals as his guide, to the studies carried out late in his life, in which he attained a formidable understanding of the structure of the body.

Among these late studies, it is Leonardo's investigations of the heart that stand out as the most detailed and clear-sighted of all of his scientific endeavours. As Francis Wells explains, Leonardo's intuitions allowed him to describe the functioning of the chambers and the operation of the valves with astonishing precision, a hundred years before William Harvey first posited the circulation of the blood. Had Leonardo published his work (as he intended), he would truly have revolutionised the study of anatomy in Europe. But it was only in the twentieth century that these drawings finally saw the light of day; and such was the sophistication of Leonardo's analysis that only now are cardiologists attaining a full appreciation of what he accomplished.

Even more remarkably, Francis Wells demonstrates that Leonardo's studies are not merely of historical interest and that his insights can still inform the work of modern practitioners. It is common to speak of a 'meeting of minds' across the ages; but rarely can the work of a scientist of centuries past be shown to be so relevant to contemporary concerns. And that, I think, is the greatest possible tribute to Leonardo's achievements, teasing apart the hearts of oxen in a villa in Lombardy, exactly five hundred years ago.

Charles

Foreword by Martin Kemp

The twin peaks of Leonardo's science both involved bodies – the human body and "the body of the earth", as he called it. The wonderful drawings of the skeleton and muscles from around 1510, when the artist was in his late 50s and his studies of the heart from 1507 onwards yield to no-one in their acuteness of observation, alertness to function and beauty of representation. His hugely innovatory researches into the vast changes undergone by the body of the earth, culminating in the Codex Leicester of the same period, seem to us to belong to a different scientific discipline. However, as Francis Wells deftly demonstrates, Leonardo was never constrained by what we might think of as separate fields of visual enquiry. His dissection of the old man – who claimed to be 100 years old – indicated that the centenarian's "sweet death" was caused by the enfeebling of his blood supply, which had become tortuous and silted up like meandering rivers. The insights are those of a "master of water" as Leonardo was called, that is to say a hydraulic engineer, and came not least from his work on a great canal to bypass the non-navigable reaches of the Arno west of Florence. It was his observations as he traversed the Arno valley that convinced him that it was once the site of two ancient lakes, one high and one low, just as we see in the *Mona Lisa*. Via his studies of water courses, overground and underground, he comes to define a cycle of "circulation" in the "veins of water" within the earth's body that was more radical than his views on the "circulation" of blood in the human body.

As we discover, on the course of our journey with the author through the human body as envisaged by Leonardo, anatomy is never just a matter of anatomy as we conceive it. Every drawing by Leonardo is simultaneously an act of analysis as well as description. He cannot draw a form without intuitions about its function. To ascertain how a form works – with no deficiency or redundancy – he brings his knowledge of dynamics to bear on his accounts of structures in motion. He knows about weights, levers and pulleys, which he can apply to the functioning of the body. The skeleton and muscles serve as a compound machine, the mathematical trajectory of which enables humans to move their limbs with complete freedom in space – across what he calls a "continuous quality". The complex motion of fluids, above all water, to which he devoted minute and exhaustive attention, becomes vital to his reading of the actions of the heart valves, which he shows to be miracles of geometrical engineering. He cannot see what blood does in the valves, but he can transfer his knowledge of water vortices. He is a natural bio-engineer. Indeed, his plan to make a glass model of the neck of an aorta to test his notion of the internal currents of blood is entirely novel. He is determined to invent means to determine the "impetus" and "percussion" of the surging torrents of blood, even if he cannot see them first-hand within the heart.

We do not *have to* validate Leonardo's science in terms of contemporary knowledge, nor do we *have to* evaluate it by adducing its influence on the history of medicine. His visions of the human body and the body of the earth possess a beauty of insight that are uplifting in their own right, to no lesser degree than his works of art. And they stand as an enduring monument to an unconstrained quest for understanding across disciplines. However, it is thrilling to find that he can still conduct a creative dialogue with a major heart surgeon over the span of 500 years. Francis Wells brings his clinical eye to bear upon drawings we have long admired, and points out major features that we little understood and telling details that we had entirely missed.

He shows that no organ had ever been subjected to such remorseless visual enquiry and functional interrogation as Leonardo devoted to the heart.

The fruits of the dialogue across half a millennium go beyond the clarification of Leonardo's historical eye. How Leonardo looked and what he saw can still hold surprises for the most professional of specialists. His description of the mitral valve encouraged Francis Wells to look again, to re-ask a question that seemed long since to have been put to bed. The results are wonderful and surprising – even to someone like me who is used to being surprised by Leonardo. It is a privilege to realise that a man who lived so long ago and never brought his scientific works to a published state can still live creatively today in the eyes, mind and hands of someone who really knows the geography of the human body.

Martin Kemp
Emeritus Professor of the History of Art
Trinity College, Oxford, UK

Preface

The purpose of this book is to bring to a wider audience the astonishing work that Leonardo did in trying to understand the anatomy and physiology of the heart. This work was carried out at the end of his career as an investigator of anatomy. Note that I have referred to him not as an anatomist, but as an investigator of the human form, for his purpose was very different from that of the usual practitioners of this science. Unlike Vesalius and the many great anatomists who followed, Leonardo was a pupil of natural philosophy, interested in all of the natural world and in particular, in man's place in the microcosm–macrocosm continuum that had exercised the minds of philosophers from Aristotle and Ptolemy to the Renaissance. Like Vesalius, however, he broke new ground in the positioning of the student of anatomy, from one who simply regurgitated the works of Galen to one who thought for himself using the evidence in front of him to describe the human form in new light. Until Vesalius—and here we have to discount Leonardo's contribution, as he did not publish any of his work—the anatomy professors read from the works of Galen; once a year a body was dissected by a prosector and the parts were demonstrated to the students of medicine by the Demonstrator. Needless to say, the marriage of dictat and observations was not a happy one. Vesalius broke new ground in becoming the dissector, the recorder, and the demonstrator all rolled into one, turning the art into a science. Leonardo did the same thing, but his was work done in private and never released to the academic fraternity at large through publication. We may suppose that even if the work had been published, it would hardly have been accepted from one who lacked a formal education and was viewed as an artist and engineer.

It is even more startling, then, to discover in Leonardo's works descriptions of form and function of this immensely complex organ that continue to speak to those of us who work with it every day. Many of his insights, such as the description of the closure mechanism of the arterial valves, pulmonary and aortic, hold true today and indeed were unravelled only in the modern era by two Oxford engineers who published their findings in *Nature* with only one reference—to Leonardo da Vinci!

This book includes chapters on Leonardo's life and times and the state of anatomy in his time. I have included a chapter on his use of drawing, as it is important for the newcomer to Leonardo to realise that he used drawing in many different ways.

In addition to having a large section on the heart anatomy and its relevance to us today, this book has reproductions of contemporary dissections that I have carried out to test the veracity of Leonardo's work. The results are quite astonishing.

This is the first time that all of Leonardo's drawings and writings on the heart have been brought together in one volume, and I am immensely grateful to the Royal Collection for allowing their reproduction along with translations of all of the accompanying text. The text has been transcribed from the mirror image in which Leonardo wrote and has been translated in the hope that this will be a useful resource for other academics who wish to study Leonardo's thinking through his writing and his drawing. Although the audience is likely to be mainly academic I hope that enough general information is included to interest the more general reader.

Acknowledgements

This book has been many years in gestation. Its realisation would not have been possible without the help and support of several people. First I would like to thank His Royal Highness The Prince of Wales for so kindly agreeing to contribute a foreword to this book. I greatly admire the work that he has done through creating the Prince's Teaching Institute for so many state school teachers and their pupils, the next generation upon whom the future of this country depends. It has been my great privilege to contribute, albeit in a small way, to this admirable organization, most ably shepherded by Christopher Pope and Bernice McCabe, the co-directors of the Institute. My lectures on the subject of this book to this inspirational group did much to encourage me to disseminate the information further.

The Honourable Lady Jane Roberts, Librarian and Curator of the Print Room, and Martin Clayton, Senior Curator of Prints and Drawings at the Royal Library in Windsor Castle, graciously allowed me access to the drawings in the Royal Library and then encouraged and supported me in the project. To both I extend my most sincere thanks. Being able to examine the drawings in hand was both revealing and inspirational. Permission by the Royal Collection to reproduce all of the heart drawings, their related transcriptions and translations, and several other manuscript pages and drawings has been essential to the project. Karen Lawson, Picture Library Supervisor in the Royal Library, gave her time to gather the images, and for that I am also grateful. Special thanks must go to Prof. Carlo Pedretti and the family of Dr. Kenneth Keele for allowing the use of their transliterations and translations of Leonardo's original notes; theirs was seminal work and remains the gold standard for the understanding of Leonardo's writing. Regardless of my commentary, I hope that the apposition of these pages against the actual drawings of the heart will render this a potentially valuable source for further research.

First among those who have educated me in this world is Martin Kemp, Emeritus Professor of the History of Art at Trinity College Oxford. His vast knowledge and the rigorous academic approach exhibited in his written work and personally transmitted has both informed me and helped to instill discipline in my interpretations. His persistent encouragement to publish this work has been the main stimulus for its emergence. For all I am extremely grateful.

Dr. Domenico Laurenza and Dr. Matthew Landrus have supported and informed me along the way. Their friendship I treasure and their knowledge I respect. My thanks go to both for their enduring belief in the project. Professor Francis Ames-Lewis's kindness and support is greatly appreciated.

My medical training and surgical experience as a heart surgeon have enabled informed interpretation of the drawings. My entire consultant practice has been based at Papworth Hospital, part of the University of Cambridge. I have been fortunate to enjoy great support from many people, particularly our long-serving chief executive, Stephen Bridge, to whom I owe a sincere debt of gratitude.

Ken Arnold, Head of Public Programmes at the Wellcome Trust, has been enormously supportive in this and other projects. I have thoroughly enjoyed our ongoing conversation and greatly value his practical and intellectual support. It has been a privilege to work with this world-class organization on various projects over the past 10 years. Their support has been invaluable.

A constant support and practical help throughout the years has been my loyal secretary, Mrs. Julie Mazzone. Without her organizational skills, I would not have been able to make the time in a very busy clinical practice to make this book happen.

More recently, MariaClaudia Bautista Gaitan has given me loyal and unstinting practical and emotional support, for which I do not have adequate words of gratitude. Victoria John, Associate Editor of *Clinical Medicine* at Springer-Verlag, has made the final stages of submission a pleasure. I owe a great debt of gratitude to her and all of her colleagues in the publishing house for taking on this challenging Arts/Science project.

Finally, without the support of my immediate family over my entire working life, none of this would have been possible.

Contents

Introduction: The Heart of Leonardo da Vinci

Leonardo da Vinci (1452–1519) is the archetypal Renaissance man.[1] Even in our fast-moving and advanced technological age, Leonardo remains an icon of artistic, scientific, and humanistic achievement (Fig. RL 12726). A true polymath, he left behind a treasure trove of thoughts and ideas expressed on paper unlike any other artist or scientist before or since. These extant drawings and accompanying manuscript notes are the key to his legacy. Despite the fact that much has very probably been lost, more than 4,000 sheets or fragments of paper remain. This body of work is about four times that of even the most prolific sixteenth century draughtsman. In contrast, even though Leonardo is known generally throughout the world as one of the great artists of all time, there are at most 15 paintings that are widely accepted as bearing his hand.[2] Fortunately for us, however, his sublime skills as a draughtsman and the sheer beauty of his drawings caused his work to be sought after by collectors both during his life and after his death. We have to thank his long-term pupil Francesco Melzi for their initial preservation. Leonardo bequeathed to the faithful Francesco his entire collection of books and much else besides.[3] In addition, Giorgio Vasari, regarded by many as the first art historian, wrote in great appreciation of the importance of Leonardo's work in his book *The Lives of the Great Artists and Sculptors*, published first in 1550.[4]

A special mention must be made of the Spanish sculptor, Pompeo Leoni, who by 1590 had bought the majority of Leonardo's manuscripts from Francesco Melzi's brother following Francesco's death around 1570. Leoni preserved them, mounting many of the loose sheets into several large albums (Fig. RL 33320). Amongst them were the anatomical works now in the Royal Collection at Windsor.

Leonardo's range and depth of interests were impressive by any standard. He used his encyclopaedic knowledge most effectively in extensive integrative reasoning, which led him to new and insightful ideas, as particularly well demonstrated in his work on the heart, the subject of this book.

A study of the full range of his notes and drawings reveals vastly differing subjects, often juxtaposed on a single page. Because paper was relatively rare and costly, entries on a single page are sometimes separated by years. To me, these notes have a personal intimacy almost in the character of an occupational diary, and sadly, this is where most of his potential to influence the world remained, as only some geometrical drawings of regular solids that he made for a work on mathematics by Luca Pacioli (*De Divina Proportione*) were published in his lifetime.[5]

---

[1]The term "Renaissance" is derived from the rediscovery (rebirth) of classical Greco-Roman academia and the overt explosion of interest in what we now refer to in secular terms as the arts and sciences. This separation of disciplines was not explicit in the quartocento and cinquecento.

[2]Carmen Brambach (ed.). *Leonardo da Vinci, Master Draftsman.* (New Haven: Yale University Press, 2003), 4–5.

[3]Jean-Paul Richter. *The Notebooks of Leonardo, Vol II.* London 1883. Cited by numbered extract R1-1566.

[4]G. Vasari. *Delle vite de' più eccellenti pittori, scultori et architettori*, 2nd Edition, Florence 1568, III.

[5]Luca Pacioli was a mathematician with whom Leonardo lived and from whom he took lessons in mathematics whilst living in Milan. *De Divina Proportione* was published in Venice in 1498.

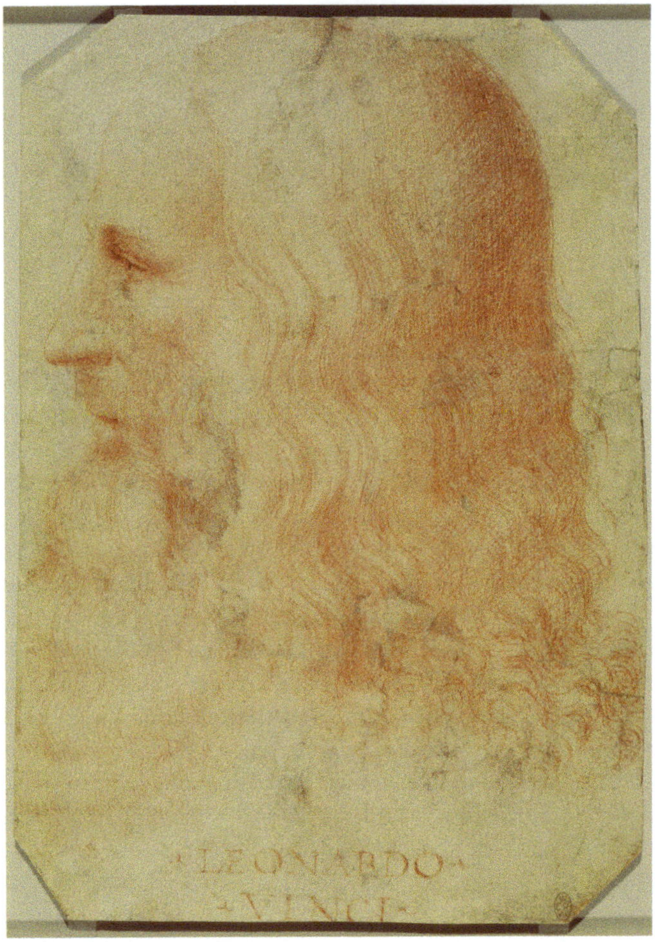

Fig. RL 12726 A portrait of Leonardo attributed to Francesco Melzi, c. 1515. Red chalk (Lent by Her Majesty The Queen. Royal Collection © Her Majesty Queen Elizabeth II)

In the time of Leonardo, much academic thought was focused on the position of man in the natural world, especially on man's place in what was thought of as a continuous universe with man at its centre. The idea of macrocosm/microcosm continuity was not new and had found expression in Ptolemy's "The Geography" (also known as *Geographia* or *Cosmographia*), produced in the second century A.D. The renaissance/rediscovery of this profound work was in the 1400s, when it came to Italy from Constantinople; initially printed in 1477, it was the first book to include engraved illustrations.[6]

Leonardo borrowed from this and described Man as a microcosm or lesser world, the structure of whom reflected the greater scheme of things in the universe.[7] This metaphor was integrated with the profound ideas of the proportionality of man, which in the first century B.C. was reproduced in the design of dwellings and churches by the Roman architect Vitruvius (with whose work Leonardo was also very familiar). The idea of the microcosm stretches back even further to the thoughts of Timaeus in the fourth century B.C., who wrote, "The God, wishing to make this world most nearly like that intelligible thing which is best and in every

[6]Ptolemy was born in 96 A.D. and died in 168 A.D. The whole meaning of the word "Renaissance" is the rebirth or rediscovery of classical Greco-Roman works, be they sculpture, architecture, or philosophy, which was written and passed down through the Arabic scribes and retranslated into Latin in the Renaissance.

[7]RL 19061: In 15 entire figures there shall be revealed to you the microcosm on the same plan as before me was adopted by Ptolemy in his cosmography; and I shall divide them into limbs as he divided the macrocosm into provinces; and I shall then define the functions of the parts in every direction, placing before your eyes the representation of the whole figure of man and his capacity of movements by means of his parts. And would that it might please our Creator that I were able to reveal the nature of man and his customs even as I describe.

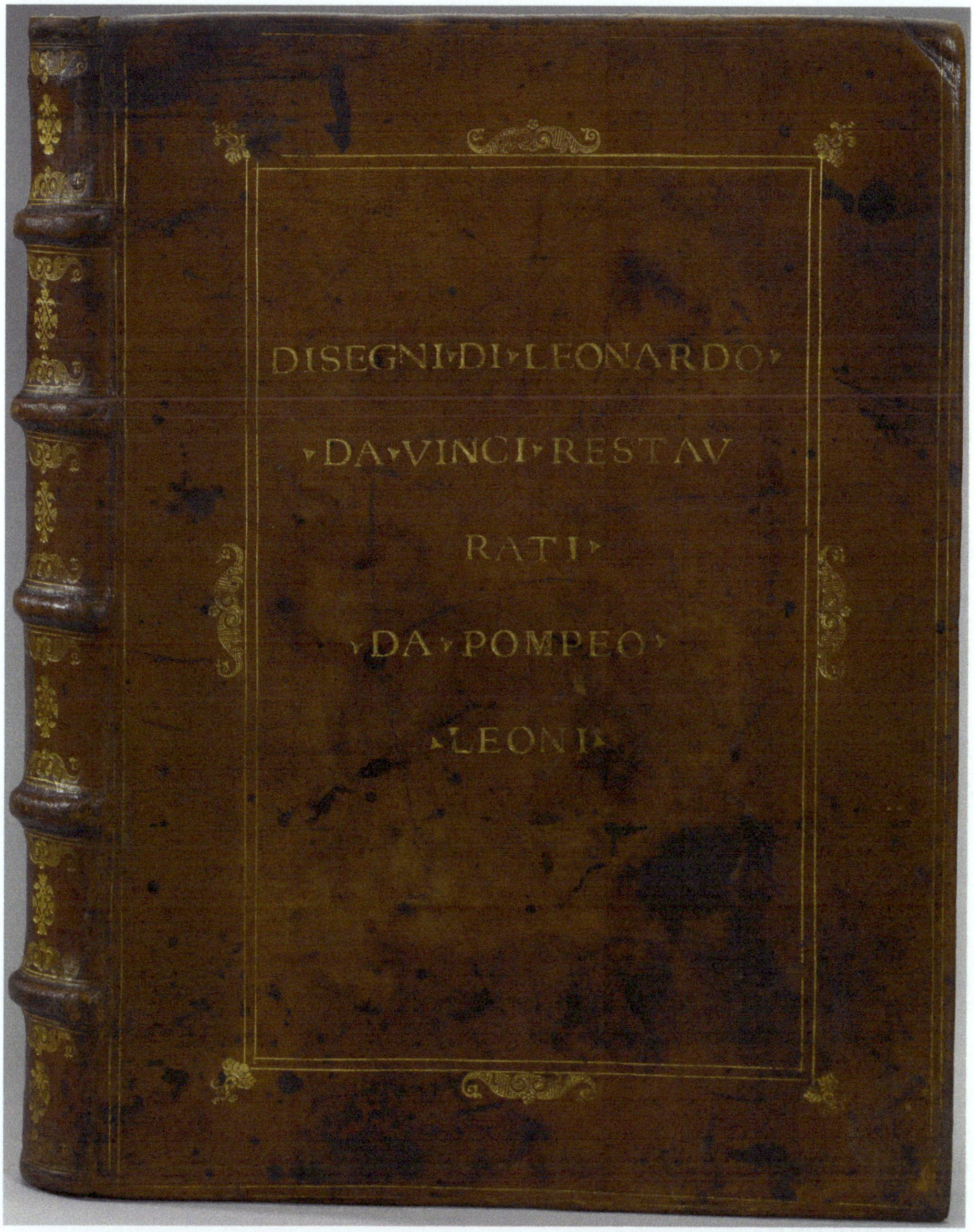

Fig. RL 33320 The Leoni binding, c.1580–1600. A large leather-bound album formerly containing the Leonardo drawings in the Royal Collection. Leather 47×33×6.5 cm (Lent by Her Majesty The Queen. Royal Collection © Her Majesty Queen Elizabeth II)

way most complete, fashioned it as a single, visible, living creature, containing within itself all the living things whose nature is of the same order."[8] This concept of "natural proportionality" extended beyond the anatomical and architectural order of things and into the functional and physiological world. Leonardo gave this account of man as a "lesser world" of natural flux: "By the ancients man was termed a lesser world and certainly the use of this name is well bestowed, because in that man is composed of water, earth, air and fire, his body is an analogue for the world." Leonardo continues in this passage to compare the ebb and flow of the tides

[8]Martin Kemp, *Leonardo da Vinci. The Marvellous Works of Nature and Man, Revised edition.* (Oxford: Oxford University Press, 2006), 95.

with the inspiration and expiration of breathing, and the flow of the blood with the flow of water throughout the world.[9]

These ideas in conjunction with Leonardo's appreciation of beauty in every form in nature made it inevitable that he would be drawn to the inner workings of humans, animals, and plants. Therefore, it can be argued that his attempts to understand Man's place in the cosmic architectural expanse from microcosm to macrocosm were really within the confines of natural philosophy and were not anatomical studies in the way that we may naturally assume, to contribute to medicine and surgery. In that sense, he was very much a man of his time, reflecting the rebirth of classical thought and interpretation of the natural world. The body of man had been forged by the forces of the universe acting upon it, and thereby rendered fit for purpose with "nothing either wanting or superfluous".[10] Modern science describes these forces, which forge natural form and function, in the terms of Darwinian natural selection. Leonardo's desire to achieve a full understanding of the workings of the body was matched by his determination and seemingly limitless energy to complete the work: "Truly it is impatience, mother of folly, who praises brevity" (RL 19084 recto). There can be no shortcuts to true knowledge.

This quest for an all-encompassing microcosmic/macrocosmic explanation of the natural world continues today as physicists search for a unifying theory for quantum and Newtonian physics to marry subatomic and cosmic observations. This inner vision of the continuity of the physical world is a very humanist phenomenon. Recently, I found a quote from Edgar Mitchell, who flew with the Apollo 14 mission and was the sixth man to walk on the moon. He spoke of his epiphany moment on his return to earth: "Every two minutes as you looked out you would see the earth, the moon and the sun, and a 360 degree panorama of the heavens out of the window. You could see the whole solar system. And the star systems are ten times as bright and ten times as numerous in space. It was the recognition that the molecules in my body and the molecules of the spacecraft and in my partners had been prototyped or maybe even manufactured in some ancient generation of stars. This suddenly became damn personal. It wasn't intellectual; it was visceral. And it was accompanied by a sense of ecstasy that I had never experienced before."[11]

Reading Leonardo, there is a sense of this visceral involvement in all things "natural", an urgent sense of his definition of man cosmographically. This abstract ideology became grounded in many areas of Leonardo's work. In his muscular and skeletal anatomy, he is able to define man's emotions and interactions with the physical world. Through his studies of the eye, which is used to interrogate the visible world, he is led into the internal world of vision and the gift of received and developed wisdom through the cerebral integration of sensory information. Some regard his work on optics and vision as the entrée to his desire to understand the workings of the human body. Here we enter the world of physiology through function defining form. This move from a mechanistic description of the workings of the human body to a more physiological one characterizes his later work, in particular his work on the heart and other internal viscera.

The challenge of "knowing" this vital piece of flesh, the heart, must have stimulated his mind in the most invigorating way. Perhaps he could sense the acceleration of his own heart and its increase in power as he made his own discoveries of its mysteries! Unlike breathing, where the movements of respiration can be seen, those of the heart have to be "sensed". It is this distinction that associates the heart and its vagaries with the emotions so much. We "catch our breath" in surprise and can be seen to do so, but our heart "leaps" with emotion and only we can feel it. The very central nature of breathing and the presence of a palpably beating heart

[9]Paris manuscript A 55 verso.

[10]Although human ingenuity makes various inventions, corresponding by various machines to the same end, it will never discover any inventions more beautiful, more appropriate, or more direct than nature, because in her inventions nothing is lacking and nothing is superfluous RL 19115 recto.

[11]Mick Brown, "Down to earth. The men who walked on the moon". *Telegraph* magazine, 3 November 2007, 40.

have for long been known to be essential for life. The association of the heart and lungs with these life-sustaining phenomena is recent, and for centuries it was not appreciated or at best was ill-understood and misrepresented. No one understood the composition of the air we breathe, nor did anyone understand the concept of metabolism and body chemistry, all things that we now take for granted. All that was known in Leonardo's time was that air (pneuma) and food and water were necessary to survive. What happened to these essential ingredients was a matter for speculation. The science that could begin to comprehend these things had not begun. With these facts in mind, one can see the scale of the problem that confronted those in Leonardo's time who were challenged by the fact of the living body. Even knowing the correct questions to ask was a major advance. It was in this as much as in his conclusions that Leonardo revealed his genius. It is in these dark ages of knowledge that Leonardo began to shine the light of revelation in observation, questions, and deductions with embryonic scientific experimentation. To begin to understand the complexities of the heart, this ever-contracting piece of muscle, was a massive challenge, and we should not be surprised at Leonardo's now-obvious mistakes and the questions that he left unresolved. Bearing in mind the vast amount of work that has been necessary over the centuries for us to reach our present (and still incomplete) knowledge of the heart and considering that Leonardo's missing manuscripts may well have contained further, even more insightful information, what we have left of his work remains truly remarkable.

A profound anatomical knowledge, which these days extends to the microcosm of molecular biology, has always been a fundamental necessity for understanding a human, the intelligent, emotional animal. Structure and function are intimately entwined. A grasp of anatomy leads to a desire for physiological explanation. The inevitability of this beautiful, intimate duality was succinctly expressed by D'Arcy Wentworth Thompson in his extraordinary book, *On Growth and Form*. In the introduction, he states, "… the form of an object is a diagram of the forces … acting upon it."[12] Over time, these forces shape and manipulate both form and function, leading to health or disease. Leonardo, in a letter to the cathedral authorities in Milan, when he was trying to win the contract for the crossing dome, wrote:

It is necessary for doctors who are the guardians of the sick to understand what man is, what life is, and what health is, and in what way a balance and harmony of these elements maintains it, and how similarly when they are out of harmony it is ruined and destroyed, and whoever has a good knowledge of the aforesaid characteristics will be better able to heal than he who is lacking in it …. The very same is required by an ailing cathedral—that is, a doctor-architect who well understands what a building is, and from what rules correct building derives, and from where such rules are drawn and into what number of parts they are divided …. I shall endeavour … to satisfy you partly with theory and partly with practice, sometimes showing effects from causes, sometimes affirming principles with experiments, making use, as is convenient, of the authority of the ancient architects.[13]

In this letter, which he probably sent to the cathedral authorities,[14] Leonardo uses human form and function as a metaphor for an ailing architectural body, and he demonstrates the natural lateral thinking that recurs throughout his notebooks.

It is in his anatomy of the heart that Leonardo reaches the peak of his intuitive and deductive powers in abstract anatomy. I call it "abstract" because knowledge of this organ could have no possible application in his time other than knowledge for its own sake—in modern terms, "blue sky" research! Why would anyone want to know how the aortic valve of the heart functions, especially at a time when there was no appreciation of the "circulation" of the blood?

---

[12]D'Arcy Thompson. *On Growth and Form.* Abridged/Edited by J.T. Bonner. (Cambridge: Cambridge University Press, 1961; reprint, Cambridge: Cambridge University Press, 1966), 11.

[13]Codex Atlanticus 730 recto.

[14]Martin Kemp. *Leonardo da Vinci. The Marvellous Works of Nature and Man, Revised edition.* (Oxford: Oxford University Press, 2006), 88. Other authors of the time (such as Alberti and Francesco do Giorgio) also used this human-to-architectural analogy.

Whilst it may have been a desire to understand sight and vision that drew Leonardo into a detailed study of anatomy, it appears that at the height of his powers, it was the heart that captured his imagination and to which he devoted much of his last documented period of anatomical study. It is in his cardiac anatomy that he begins to show signs of shedding some of the manacles of Galenic dogma, replacing it with his own ideas. In his first general comment upon the heart, he wrote, "Of the heart. This moves of itself and does not stop unless forever …. Marvellous instrument, invented by the supreme Master."[15] In a second quotation, his powers of observation and deductive reasoning come to the fore: "The heart in itself is not the beginning of life, but it is a vessel made of muscle, vivified and nourished by the artery and vein, as are the other muscles."[16] In other words, although it is awe-inspiring, this thing we call the heart is made of muscle and like other muscles it needs its own blood supply and is a response to the needs of life and not the magical progenitor of it—a practical, down-to-earth observation that is correct.

This book is not the first attempt to discuss Leonardo's work on the heart. In 1952, Kenneth Keele published *Leonardo da Vinci on Movement of the Heart and Blood*.[17] At the time of writing, Dr. Keele admitted that there remained questions about Leonardo's drawings and interpretation of heart function that "… still baffle interpretation".[18] Now, with the passage of another half century, knowledge has moved on. Amazingly, the new information has given some of Leonardo's descriptions an even greater ring of truth. An example is the importance of the energy-absorbing capacity of the aortic root on aortic valve closure. This capacity had no impact or importance in 1952, a time when open-heart surgery was yet to be developed. With the advent of valve replacement surgery, an appreciation of the impact of the energy released into the tissues surrounding the valve by its sudden closure has been used in the design of replacement heart valves, and the fact of its existence has been used to explain the early failure of some surgical solutions.

This book will acknowledge Leonardo's shortcomings in continuing to accept untrue Galenic dogma, particularly with regard to the circulation, but whilst relating the entirety of his comments on the heart and the movement of the blood, it will focus upon areas where his observations were prescient and indeed instructive for us today.

The work described herein is the result of a long-time personal interest (and indeed, fascination) with Leonardo da Vinci and a desire to make a contribution to the appreciation of his depth and subtlety of thought that can be appreciated in the modern era of cardiology and cardiac surgery. It is an opportunity to reproduce for the first time in a single volume and in high definition all of Leonardo's existing cardiac drawings in full color, accompanied by a full transcription and translation of all of the notes. This format will allow others interested in this fascinating subject to begin their own review and examination of the manuscript sheets and notebook pages in one volume.

That a significant amount of material is lost will become apparent as the book unfolds. We may never know how much further Leonardo's knowledge and understanding of the heart and the circulation had progressed beyond the extant pages, but it is my hope that this book will encourage others to study his magnificent achievements in greater detail.

[15] An. B 12r (Dell' Anatomia B, Torino, Roux e Viarengo, 1901).

[16] An. B 33 (Dell'Anatomia B, Torino. Roux e Viarengo, 1901).

[17] Kenneth Keele. *Leonardo da Vinci on Movement of the Heart and Blood.* (London: Harvey and Blythe Ltd., 1952). Limited edition print run of 1,000 books.

[18] Kenneth Keele. *Leonardo da Vinci on Movement of the Heart and Blood.* (London: Harvey and Blythe Ltd., 1952), xvii.

The Making of a Master: Leonardo, His Life and Times

1

Wisdom is the daughter of experience

(Forster III; 14r.)

Late one evening, in the springtime of 1452, in the low foothills of Montalbano, northwestern Tuscany, an illegitimate baby boy was born to a local farmer's daughter called Caterina. The child was Leonardo da Vinci. He was conceived, perhaps, as the result of a brief romantic entanglement between the relatively affluent son of an old Vinci family of notaries (Ser Piero da Vinci) and a woman of lower birth, about whom very little is really known. He would grow into the original Renaissance man, whose good looks, charm, and charisma became a legend in his own time. Furthermore and pertinent to our story, he would become a contemporary icon for both science and the arts.

Although very little is known about his early life, we do have the exact date and time of his birth. A German research worker, Dr. Emile Möller, discovered it in the state archives in Florence in the 1930s. Möller discovered the details of the birth on the back of an old notary protocol book. Leonardo's grandfather, Ser Antonio da Vinci, had recorded them there, along with the births and baptisms of his own four children. The notebook was clearly of some significance to the family, as Leonardo's great-grandfather, Ser Piero di Guido da Vinci, had previously owned it. He wrote: "There was born to me a grandson, the son of Ser Piero my son, on the 15th day of April, a Saturday, at the third hour of the night (10:30 p.m.). He bears the name Lionardo; he was baptized with the name Lionardo by the priest Piero di Bartolomeo da Vinci".

The baptism of Leonardo in front of the family and ten witnesses suggests that despite his illegitimacy there was no attempt to conceal his arrival. Antonio's note continues with the names of those witnesses. The parish priest in Vinci was Piero di Bartolomeo, a near neighbour of the da Vinci family home. Therefore, although the birth may have been in Anchiano, it is likely that the baptism took place in Santa Croce, the Vinci parish church. The font around which this event probably took place can still be seen, and is celebrated by a stone plaque on the wall of the small baptistery.

The Anchiano house stands on a promontory overlooking the Arno valley, with a view that finds reprise in Leonardo's first known drawing (Fig. 1.1). Above the door of the house is the "da Vinci" family crest.[1] It is known that Leonardo's grandfather was involved in the sale of that house, as he is recorded as being called upon to draw up the sale contract in October of 1499.

At that time, illegitimacy bestowed some penalties in education and potential social position, with limitations on access to some professions. For Leonardo, these limitations seem to have affected the breadth of his education, as later in life he records his struggle to learn Latin and Greek, but it seems that his illegitimacy did not stand in the way of a constructive childhood. It is very likely that his birth into a family of notaries and lawyers meant that he was nurtured in an environment of some scholastic significance; his beautifully neat handwriting with notarial flourishes attests to a significant degree of guidance and support in the basics of his education. Vasari reported that he was gifted in his abilities in mathematics, and his skills in geometry were certainly honed later through his association with Luca Pacioli. He was very well aware of the classics in his later years, as his own catalogue of his library attests. He is described as being a competent musician, performing with the lyre[2] to a standard that brought him an appreciative audience.

The extent of the influence of his father in all of this is impossible to know. There is significant evidence to prove that Ser Piero no longer resided in Vinci at that time. Records show that in 1450, following a brief period in Pisa, he had already taken up a post as a notary in Florence. Therefore by that time Piero must have been a visitor to his home, perhaps drawn back by a beautiful girl with whom there could actually be no future. No more than 18 months after Leonardo's birth, Piero married Albiera degli Amadori, the daughter of a wealthy Florentine notary. Being of higher birth, Albiera

[1]A reproduction of the crest can be seen there now, but the original has been taken down for conservation reasons. It is to be found in the da Vinci museum in Vinci.

[2]*The lira da braccio* was a forerunner of the violin. It was often made with a very fanciful sounding box, an attribute that Leonardo would later exploit.

F.C. Wells, *The Heart of Leonardo*,
DOI 10.1007/978-1-4471-4531-8_1,

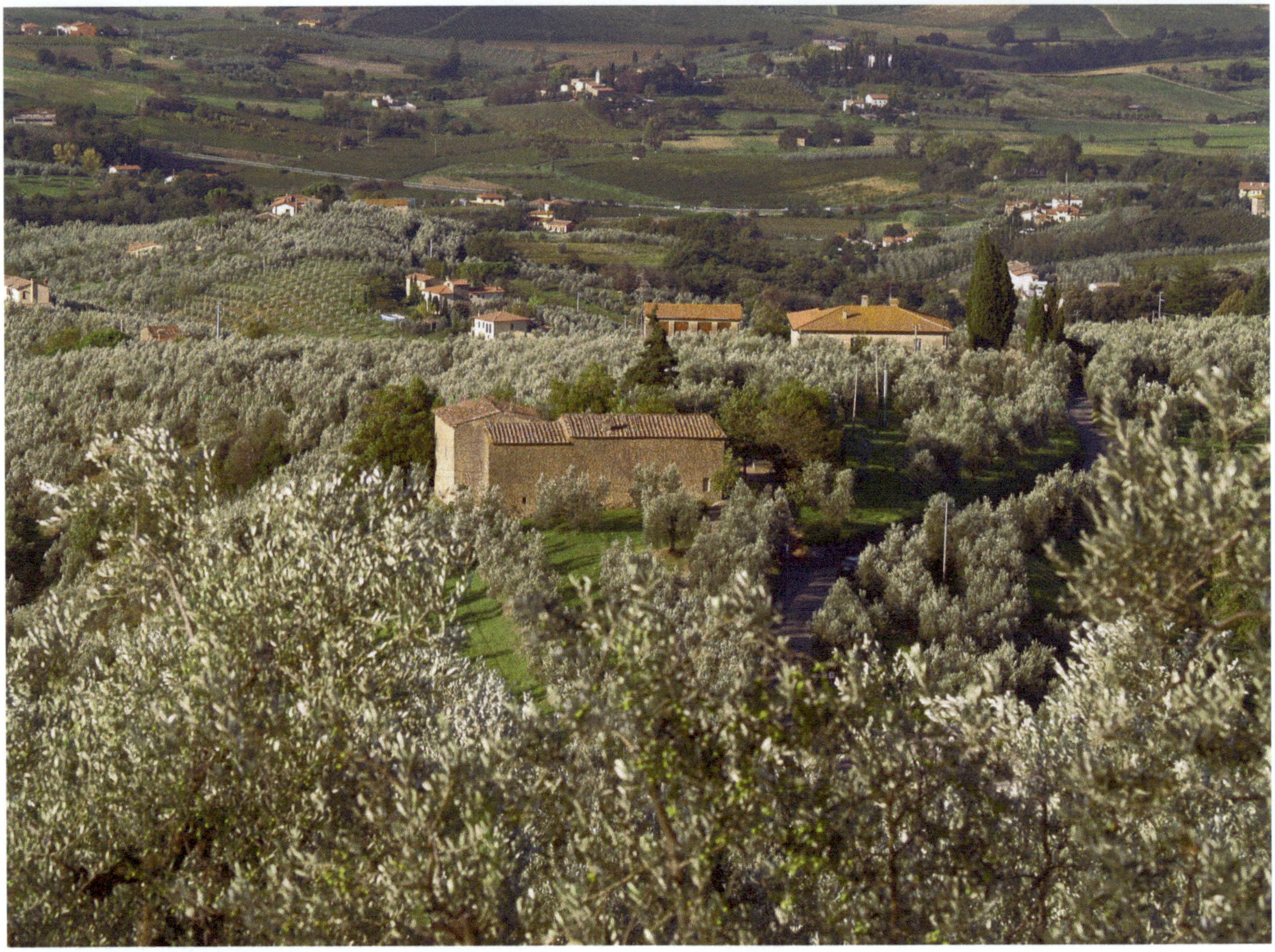

Fig. 1.1 The house in Anchiano thought to be Leonardo's first home

would have been much more in keeping with his intended station in life. It is important to note that following Leonardo, Ser Piero had no children by Albiera, who died in childbirth, and no further children until Leonardo was 20 years old.

It would appear that Leonardo grew up with his grandparents as guardians. In his 1457 tax return, Leonardo's grandfather, Antonio, listed the 5-year-old Leonardo as being part of his household.[3] His grandmother Lucia and his young uncle Francesco, with whom he had a lifelong relationship, were all together at that time. It is likely that Leonardo later spent a significant amount of time with his father in Florence, especially when apprenticed to the studio of Master Andrea del Verrocchio.

Meanwhile, his uncle Francesco remained childless and worked locally on the family properties around Vinci. Francesco owned a mill, and he and his brother Piero had rebuilt and managed a local kiln at the convent of San Pier Martire.[4] His paternal grandmother had come from a family famous for their majolica pottery. Hence, Leonardo would have been exposed to artisan activities as well as being surrounded by olive groves and vineyards that are not much changed to the present day. As Leonardo's uncle Francesco was closer in age to Leonardo than his brother Ser Piero, it is probable that he would have been a companion for Leonardo, as well as a substitute parent.

Leonardo's father, Ser Piero, went on to have 4 wives and 11 legitimate children. It was perhaps fortunate for Leonardo's emotional development that these stepsiblings all arrived after he had reached manhood and independence. When his father died, he left nothing to Leonardo, his first-born son. Yet when Francesco died, he left his entire estate to Leonardo. The result was a bitter feud between Leonardo and his stepsiblings.

[3]This is only circumstantial evidence, as the presence of another mouth to feed was tax deductible.

[4]Matthew Landrus, *Leonardo da Vinci.* (London: Carlton Books, 2006): 6.

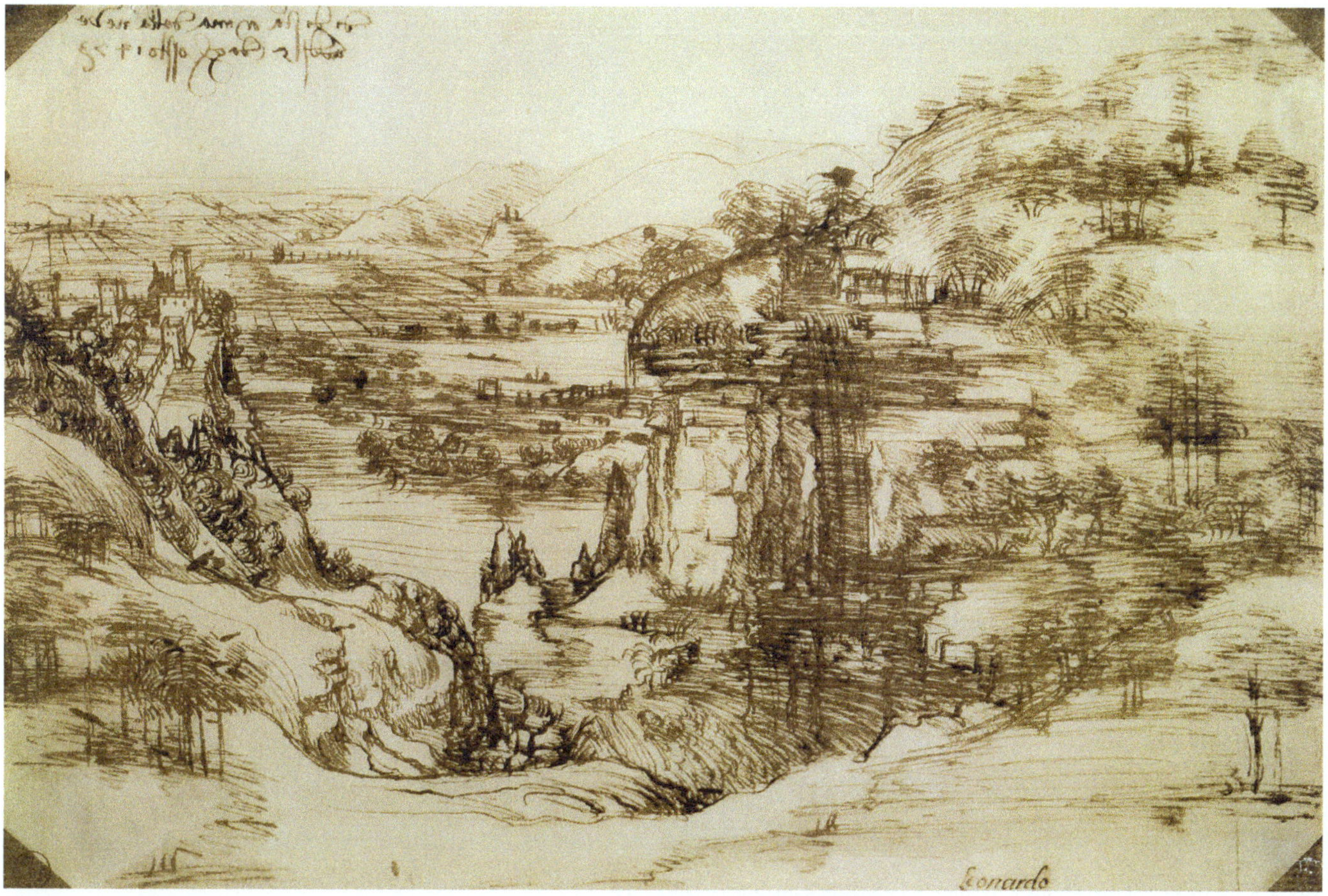

Fig. 1.2 The first known drawing by Leonardo, depicting an imaginary landscape. ALG 152192. Arno Landscape, 5th August, 1473 (pen and ink on paper), Leonardo da Vinci/Galleria degli Uffizi, Florence, Italy/Alinari/The Bridgeman Art Library

One of the few facts known about Leonardo's mother is that shortly after Leonardo was born, she married a local man, Antonio Buti, nicknamed "Acattabrigha" (the brawler). In local tax returns, Caterina is recorded as living with her family in Campo Zepo, close to Vinci. Caterina produced six children during the next 11 years, five girls and one boy. Her only legitimate son, Francesco, was killed by a shot from a military catapult.

Wherever Leonardo laid his head at night in those most formative of years, it seems clear that the family arrangements for the young Leonardo were not simple, and he must have experienced emotional tensions from many directions. Caterina's growing family needed a lot of her attention, and Ser Piero's busy life with a new wife in Florence kept them apart, Leonardo's emotional and intellectual centre of gravity was probably with his grandfather, grandmother, and uncle. In his later notes, Leonardo refers to the funeral of a Caterina in an unemotional way.

His keen observation of nature and his interest in the landscape can be seen in his earliest surviving known drawing (Fig. 1.2). This work by the young Leonardo can be dated quite precisely from the inscription upon it, written in his hand as 5 August 1473. The influence of his notary family can be seen in the flourish of his handwriting. At this time, he was 21 years of age and this was 1 year after his admission to the guild of painters in Florence. The beauty of the local countryside and the bustle of the small but strategically important regional town of Vinci would have had a very significant impact on the development of the young Leonardo. Certainly his interest in the natural world and man's place within the macrocosm were to dominate his thinking throughout his life. Although this first landscape is more of a fantasia of his memory of home, there are identifiable landmarks. The contemporary landscape reveals features that are identifiable in the drawing (Fig. 1.3).

The enormous gifts of Leonardo are eulogized by Vasari in the first recognized reference text in art history, *The Lives of the Great Artists and Sculptors*.[5] Among the many talents Vasari described, Leonardo was said to be a musician, "being gifted in music with a fine voice and a competent player of the Lira da Braccio," but Vasari went on to give an impres-

[5]Giorgio Vasari, *Lives of the Artists*. Volumes I & II. London: Penguin Classics, 1987.

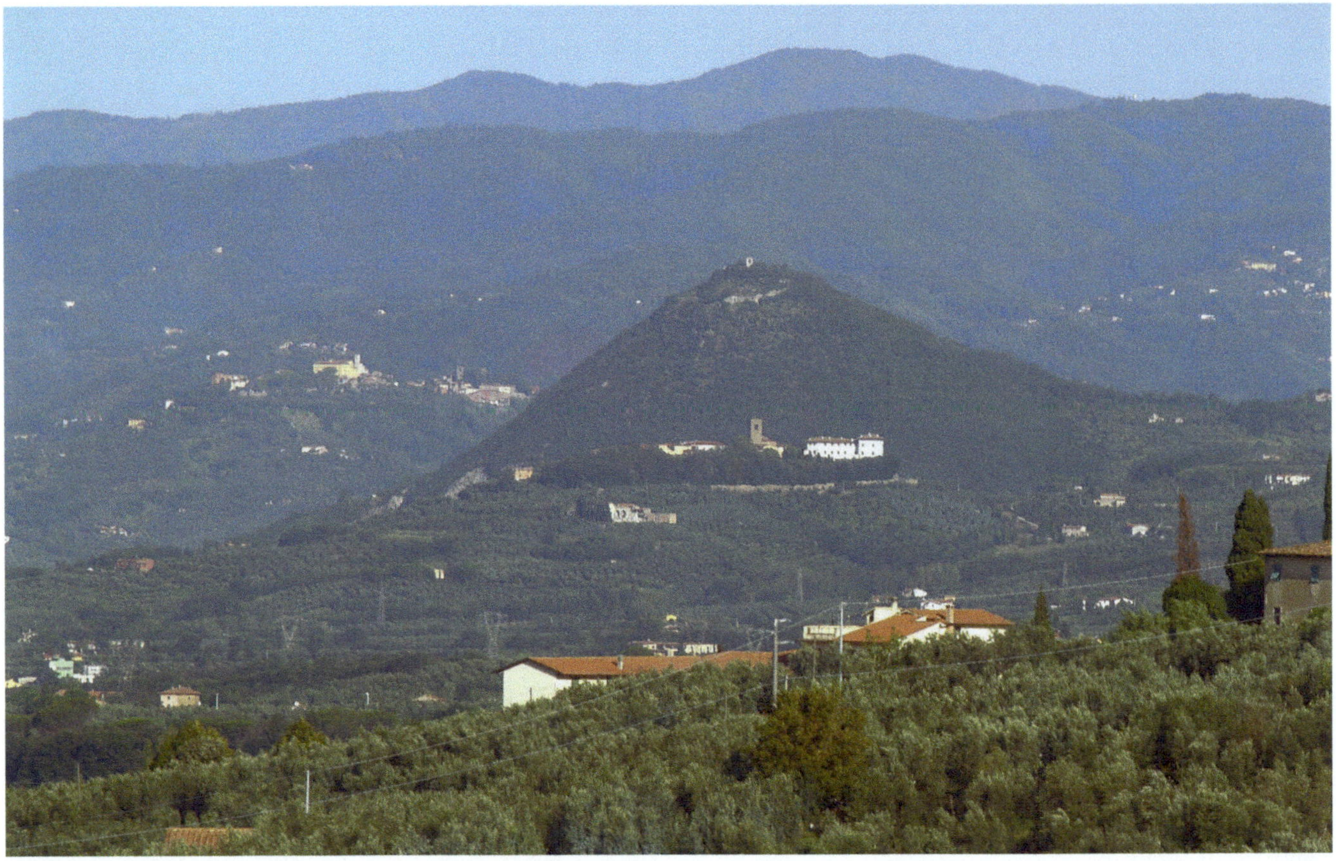

Fig. 1.3 A small hill that seems to feature in Leonardo's imaginary landscape

sion of an inconstant Leonardo, who undermined his education with whimsical application, flitting from subject to subject (in which, despite his abbreviated attention, he often excelled). Vasari reported that soon after Leonardo started arithmetic, he could baffle his masters with his ability, but then he moved on to something else.

With regard to his artistic skills, Vasari reported that throughout his youth Leonardo never ceased to draw. Leonardo's father recognized his son's potential in the arts. Perhaps he saw in this skill an easy solution to a career for his illegitimate, capricious offspring. He introduced Leonardo to Andrea del Verrocchio, a friend and master of one of the most significant studios in Florence at that time.

The year 1464 was very difficult for Leonardo's father. He had lost both his first wife and his own father. It would seem natural to want to move his son from Vinci into his household at that time. A common age for a young man to begin an apprenticeship was 14. It is not known exactly when Leonardo entered his apprenticeship but these traumatic events could have had some bearing on Ser Piero's future plans for his son. At some time between 1464 and 1470, his father displayed some of Leonardo's drawings to Master Verrocchio, leading to his acceptance as a pupil in this distinguished Botega. Verrochio was about 30 years old when Leonardo joined him, still a young man and not too dissimilar in age from Leonardo's father. Verrochio's studio was close to Ser Piero's offices in Palazzo del Podestrè, opposite to the Bargello. In addition to musical and mathematical skills, Vasari attributed to Verrochio a very restless and enquiring mind. That being the case, he would have seemed an excellent tutor for Leonardo. It is likely that Leonardo's father perceived him to be a good role model for his son. In Verrocchio's studio, Leonardo was surrounded by much of the greatest artistic talent of the day in Florence. Some of Leonardo's fellow pupils included Perugino, Lorenzo de Credi, and Ghirlandaio. The apprenticeship served by Leonardo in such a high-profile and prolific workshop must have fanned the flames of his creativity, but perhaps also fed into his capricious nature. We can find examples in his notebooks. On the cover of manuscript F (Paris), he wrote a stream of unrelated questions that were clearly in his mind at the same time: "inflate the lungs of a pig. Avicenna on fluids. Map of Elefan of India, which Antonello Merciaio has. Enquire at the stationers for vetruvius. Ask Maestro Mafeo why the Adige rises for 7 years and falls for 7 years. Go every Saturday to the hot baths and you will see naked men".

The artist's studio of those days was called a "bottega," a workshop or workplace. As the word implies, this was no simple studio designed only for painting or sculpture. These were places of great endeavour open to commissions in many disciplines. The work of the bottega would have ranged from engineering and architecture to the fine arts. Although the inventory of Verrochio's studio's skills was complete, he was best known as a goldsmith and a sculptor. He was described by Vasari as a wooden painter, gifted in effort and application rather than having innate talent. An example of his work, the *Baptism of Christ*, can be seen in the National Gallery in London. Although ascribed to the head of the studio, often many hands were involved in a commission such as this. Indeed, the studio method of the day meant that the production of such a painting would be a team effort by the master and his pupils.

A product of Verrocchio's studio, utilizing the dual skills of engineer and metal worker and still visible today, is the gilded copper orb that sits on the top of the lantern of Brunelleschi's dome of the cathedral in Florence. This was completed by May 1471 (Fig. 1.4). Leonardo witnessed its construction and may indeed have had a hand in it. He referred to it years later in Paris manuscript G (G.84v), written in 1515, whilst in Rome.

Other than putative portraits of Leonardo as an aged man, there are no definitely attributable images of him. As a young man, he was described as strikingly good looking and graceful, with great intelligence that charmed all who came in contact with him. Vasari wrote that "He was a man of outstanding beauty and infinite grace. He was striking and handsome and his great presence brought comfort to the most troubled soul." Vasari went on to say that "he owned nothing…and worked very little, yet he always kept servants and horses." The Anonimo Gaddiano said of him that "He was very attractive, well-proportioned, graceful and good looking," with beautiful hair, arranged in ringlets "down to the middle of his chest." He was said to have dressed flamboyantly, and one reference describes him in a pink tunic. He was reportedly exceptionally strong, and Vasari said that he had been seen to bend an iron horseshoe with his bare hands. How much of this is true is not known, but the mythology is so strong and reported from so many different sources that some of it is likely to be true. Perhaps Leonardo's youthful appearance can be guessed at through a possible likeness captured by his master Verrocchio in his beautiful bronze sculpture of David (now held in the Bargello in Florence) and a possible self-portrait to the right-hand side of the unfinished *Adoration of the Magi* owned by the Uffizzi gallery, also in Florence (Fig. 1.5).

The image of an athletic, graceful youth with long hair in ringlets is realised in several images in Leonardo's notebooks. Whether all of these are merely stereotypes of physiognomy that Leonardo found attractive or whether they are more personal is interesting to speculate. He certainly cultivated an impression of poise and individuality. His description of the refinement of his profession is well drawn in this quotation by Leonardo: "The painter sits in front of his work at perfect ease. He is well dressed and wields a very light brush dipped in delicate colour. He adorns himself with the clothes he fancies; his home is clean and filled with delightful pictures and he is often accompanied by music or by the reading of various beautiful works"[6] All surviving biographies from the time testify to his aura of gracious brilliance.[7]

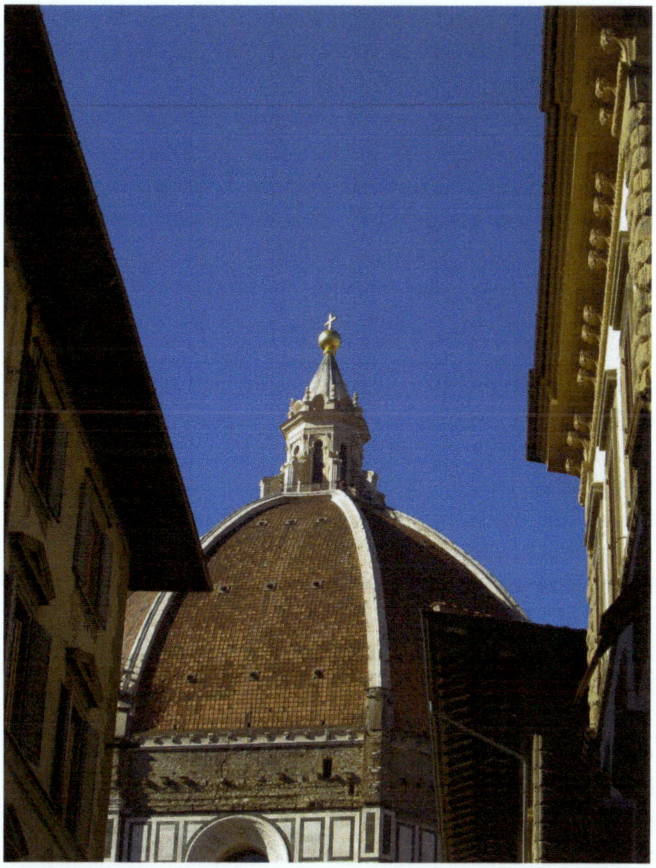

Fig. 1.4 The gilded orb atop Brunelleschi's dome of the cathedral in Florence

The studio system of the time relied upon commissions of many kinds, especially religious paintings. Alongside the hand of the master, these often had the hand of some of the apprentices as they matured in their skills. Perhaps the earliest extant example of Leonardo's hand with the paint brush can be seen in the little dog at the bottom of the painting entitled *Tobias and the Angel*.

The maturation of Leonardo as an artist is revealed in a story recounted by Vasari, who suggested that in one such project, a painting of the baptism of Christ (1473–1476), Verrocchio was so overwhelmed by Leonardo's rendering of

[6]Leonardo da Vinci. Codex Urb. 20r-v.

[7]Martín Kemp, *Leonardo da Vinci. The marvellous works of nature and man*. (Oxford: Oxford University Press, 2006): 165.

the head of the angel at the left side of the picture that he was moved to put down his brush and not to paint again, in deference to Leonardo's superior skill.

In 1472, at the age of 20, Leonardo was admitted to the Florentine confraternity of painters, the Company of St. Luke, which also included the doctors. This signaled the end of his apprenticeship and the beginning of his career as an independent artist. His relationship with his now ex-master, Verrocchio, was clearly a strong one: Despite completing his apprenticeship, he is registered as part of Verrochio's household in 1476.

During Leonardo's apprenticeship and throughout his early years in Florence as a full-fledged artist, the titular head of the Florentine republic was Lorenzo de Medici. He held this position through the power and financial interests of the Medici family. An intelligent, well educated, and articulate man of significant physical stature, Lorenzo must have been an impressive personality to the young Leonardo. During the period of 1475–1482, the Anonimo Gaddiano suggested that Leonardo enjoyed the patronage of Lorenzo de Medici, at times working and studying in his famed sculpture garden, later frequented by Michelangelo. Although little other than the Anonimo's report remains in the historical record to link Leonardo to Lorenzo with any intimacy, they may have had some kind of relationship. Born in 1449, Lorenzo was of similar age to Leonardo. He rose to prominence during Leonardo's formative years in Florence. Both Leonardo's father and his master Verrocchio were likely to have had contacts with the Medici household. Ser Piero would provide legal and notarial services for the Medici, and Verrocchio carried out commissions for them, some of which Leonardo was likely to have had a hand in. Interestingly Vasari does not mention Lorenzo in context with Leonardo.

In 1476, whilst living in Andrea del Verrochio's house, Leonardo was anonymously accused of sodomy. His was one of four names on a list left in an anonymous drop-box with the accusation that those listed had had an illicit relationship with Jacopo Saltarelli, a 17-year-old apprentice in a goldsmith's workshop. Another named was a young man of the Turnabuoni family, a relative of Lorenzo de Medici's mother. Leonardo was cleared of any wrongdoing and little more is known of the event.

The years leading up to Leonardo's departure from Florence were very turbulent. In 1478, Lorenzo de Medici's brother Giuliano was murdered by the Pazzi family, the opposing dominant family in Florence, and Lorenzo himself was targeted in that assassination attempt. Although the coup failed, it left the city in turbulence. A co-conspirator of the Pazzi family, Archbishop Salviati, was hung unceremoniously in his full clerical regalia, and as a result the Papal States declared war on the house of the Medici. Bernardo Bandini, another co-conspirator who had escaped to Constantinople, was extradited to Florence at the request of Lorenzo and also was hanged. Leonardo witnessed these events and is reported to have been commissioned by Lorenzo to complete a fresco on the wall outside the Signoria (begun by Botticelli, a favorite of Lorenzo) to remind the populous

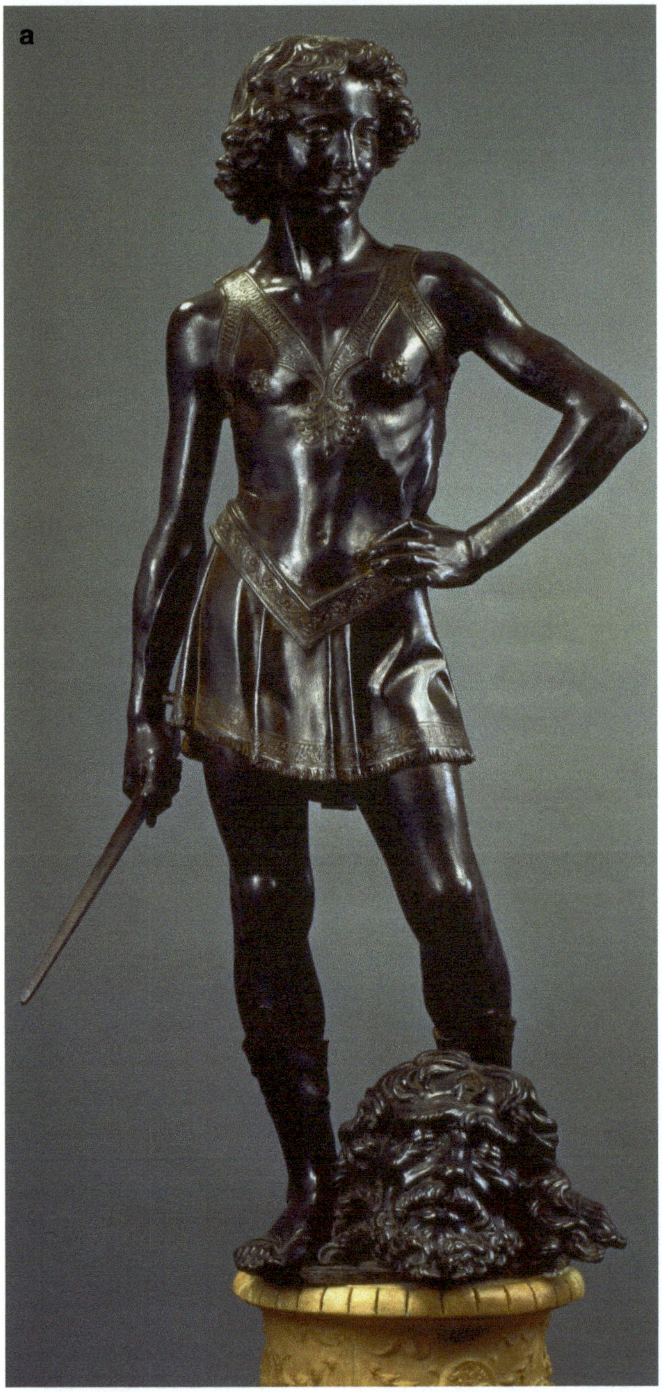

Fig. 1.5 (**a**) Verrocchio's *David*. BEN 93596. David, c.1470 (bronze) (see also 220043), Verrocchio, Andrea del (1435–88)/Museo Nazionale del Bargello, Florence, Italy/The Bridgeman Art Library. (**b**) Detail from *The Adoration*. BAL 33484. The Adoration of the Magi, 1481–2 (oil on panel), Leonardo da Vinci/Galleria degli Uffizi, Florence, Italy/ The Bridgeman Art Library

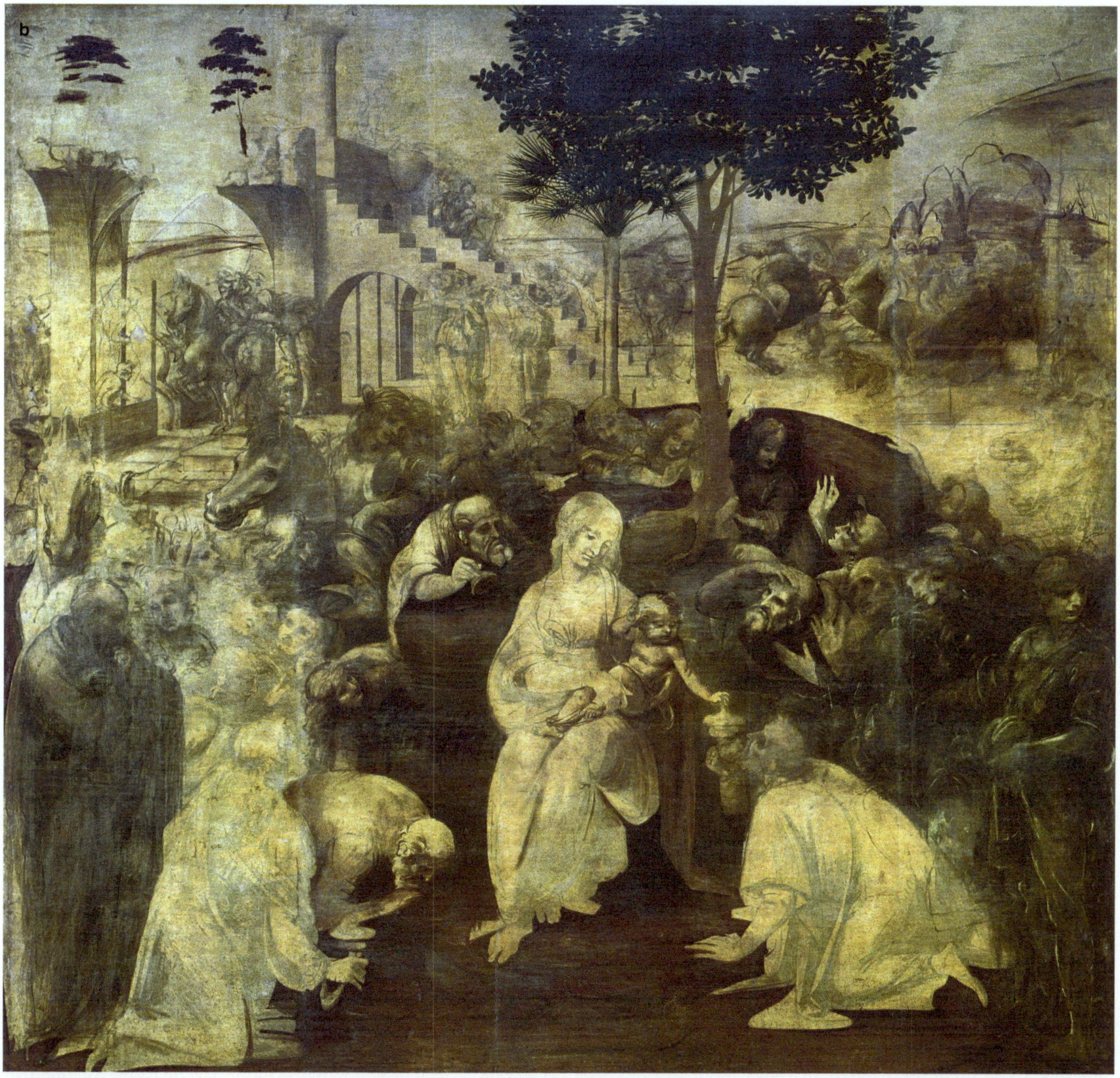

Fig. 1.5 (continued)

of Florence of that terrible day and of the outcome for the perpetrators of the crime.

In 1478, Leonardo began entries into two of his great codices, Arundel and Atlanticus, which ran on until his death. In 1481, Leonardo was commissioned to paint *The Adoration of the Magi* for the Augustinian monastery of San Donato at Scopeto, a village just outside the city walls. Leonardo's father had handled the monastery affairs since 1479, which may have led to the commission. Unfortunately as with so many others, this commission remained unfinished, despite being paid. The last payment record for Leonardo's *Adoration* (paid in wine, not money) is also the last record of him in Florence.

Sometime after 1482, Leonardo left Florence and moved to Milan, where he lived and worked for more than 18 years. The reasons for his departure can only be guessed at. In 1481, Lorenzo sent several Florentine artists to Rome on a very prestigious commission, in the spirit of the newly mended relationship between the two states. Perugino, Botticelli, and Ghirlandaio were among these artists; Leonardo was not, which may have distressed him and added to his desire to move to new pastures. Perhaps he was drawn to Milan as a result of a suggestion that he would be able to pursue his other rapidly developing interests in engineering and the machines of war. Certainly there had been political turmoil, and military turmoil around the Milanese state threatened at that time. Duke Galeazzo Maria had been assassinated in Milan in 1476. His 10-year-old son Gian Galeazzo had succeeded his father in name, but Galeazzo's brother Ludovico Sforza, the boy's uncle, ran the Milanese state as Regent, isolating the boy and his mother, the Duchess Bona of Savoy. Ludovico, also known as "Il Moro" because of his swarthy complexion, was the same age as Leonardo, being born in Vigevano early in 1452. Although there is no evidence that the two men had met, it is likely that Leonardo had seen him, as Ludovico had visited Lorenzo in Florence on state visits.

There is some reason to believe that Lorenzo was instrumental in Leonardo's move to Milan in the guise of a musician. According to the Anonimo, Lorenzo sent him on his way with the musician Atalante Migliorotti and a gift of a silver *lira da braccio*, made in the shape of a horse's head. Fantastical shapes for musical instruments morphing into animal representation were not unusual at the time, and many examples exist in collections today.

The letter of introduction that Leonardo wrote for Ludovico Sforza is recorded in his notebook. In it, Leonardo rated his capabilities as a war engineer above all else. It begins, "Most illustrious lord, having sufficiently seen and considered the works of all those whose are reputed to be masters and contrivers of war machines, and that the invention and operation of the aforesaid instruments are none other than those in common use, I will strive, without disparaging anyone else, to show my intentions to your Excellency, showing my secrets to you, and then offering all of them for your approbation, to work effectively at opportune time on all those things which are briefly noted in part below".[8] He then listed nine categories of military engineering of which he claimed special knowledge: portable bridges, methods of destroying and burning those of the enemy, draining trenches, building bridges and covered ways, ways to destroy castles and fortresses (even embedded in rock), how to construct bombarding machines, methods of tunneling, and new chariots and guns of beautiful and useful form not in common usage, as well as catapults. He also claimed devices for offence and defense at sea. Finally, as though as an afterthought, he reminds Ludovico that he can carry out any sculpture in marble, bronze, and clay, and that he is as capable in painting as any other artist. The reference to his capabilities in bronze sculpture was probably intended to draw attention to the possibility of a commission for a memorial to Francesco Sforza. Interestingly, just before Leonardo left Florence, his former master and now colleague, Verrochio, had been awarded a commission to cast an enormous equestrian bronze as a monument to Bartolomeo Colleoni (Fig. 1.6).

Sometime between 1482 and 1485 (the actual date remains elusive), Leonardo began *St. Jerome Praying in the Wilderness*.[9,10] This painting is interesting from the anatomical point of view. In it, one can see very clearly the musculature of the head, neck and upper limb. The influence of anatomical study is clear. Although reportedly antedating the Windsor skull studies (c.1489) and the dissection of an old man in Florence in Santa Maria Nuova (c.1507–8), the accuracy in depicting the muscles reveals a knowledge that was likely to have been informed by either the inspection of someone with little subcutaneous fat, or detailed anatomical study (Fig. 1.7).

Although this first Milanese period has left few artifacts of his anatomical quest, Leonardo was showing signs of interest in producing some kind of statement on anatomy. On a page of his beautiful skull drawings, he wrote: "On the 2nd day of April, 1489, the book entitled 'On the human figure'".[11] This statement clearly represents some kind of starting point for his project on anatomy, but Leonardo's time for anatomy was probably limited during his first period in Milan. He spent a great deal of time on two famous projects. One was the Sforza horse, which would have been the largest sculpture cast in bronze until that time. In keeping with many other projects undertaken by Leonardo, this project remained unfinished. When the French King Charles VIII invaded Italy, the immense amount of bronze that had been accumulated for casting the horse was sent to Ferrara to make cannons. The invading French bowmen destroyed the life-size model that Leonardo completed in clay, using it for target practice. Perhaps stimulated by knowledge of the preparatory work for this project, Vasari reported in his biography of Leonardo, "Because of Leonardo we have a deeper knowledge of human anatomy and the anatomy of the horse."[12] The only drawings of the horse that remain are of surface detail and perspective.

The second great project was one of his most famous, the *Last Supper* painted on the refectory wall of Santa Maria delle Grazzie. This was begun in the latter part of 1496. As much an adventure in the science of painting as the construction of one of the most recognizable compositions in the western world, it did not survive for long in its original form. Leonardo's desire to work slowly to achieve the detail and subtlety of realism

---

[8]C.A. 1082 recto.

[9]Carmen C. Brambach, Editor, *Leonardo da Vinci, Master Draftsman* (*New York Metropolitan Museum of Art Series*). (New Haven: Yale University Press, 2003): 375.

[10]Kemp (p. 57) suggests that this painting should be dated 1481 but "would not be surprised if evidence emerges that assigns it firmly to his time in Milan."

[11]RL 19059 recto.

[12]Giorgio Vasari, *Lives of the Artists*. Volumes I & II. London: Penguin Classics, 1987.

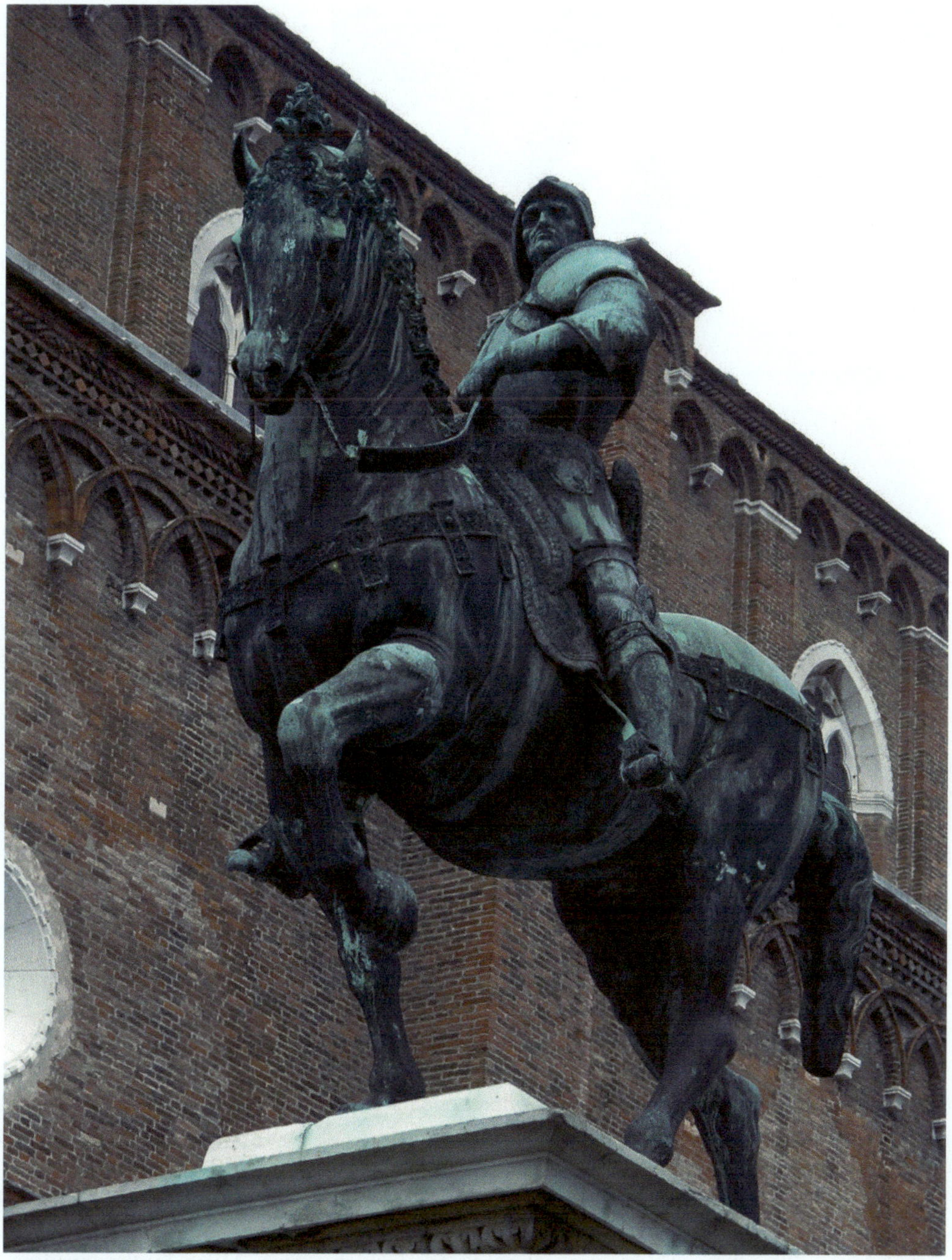

Fig. 1.6 Verrocchio's equestrian bronze monument to Bartolomeo Colleoni in Verona

that he wanted caused him to use oil-based paints instead of the normal fresco techniques. Because of uneven drying, the paint began to flake from the wall, leaving only a shadow of its former self as early as 15 years after its completion. About the time of the completion of the *Last Supper*, Leonardo received a gift from the dukedom of a garden vineyard near Santa Maria del Grazzie, perhaps in part payment or as an extra award in recognition of his genius in its execution.

In 1493, Leonardo records that a "Caterina" came to stay. Whether this was his mother is not known, but within 2 years she was dead and Leonardo records her funeral expenses in a dry and soulless way. The funeral was inexpensive.

During this time in Milan, Leonardo began two very important friendships. The first was with Giacomo Caprotti, a young man from Oreno near Monza. Leonardo soon gave him the nickname of Salai, which means "little devil." It appears from Leonardo's notes that Salai was a very mischievous young man— "thief, liar, obstinate, glutton" (C. 15 verso).[13] Leonardo expressed considerable frustration with his young assistant but also showed great affection for him.

[13]*Les Manuscrits de Leonard de Vinci. Manuscrit C.* ed. C. Ravasson-Mollien, 6 vols, Paris, 1881–91.

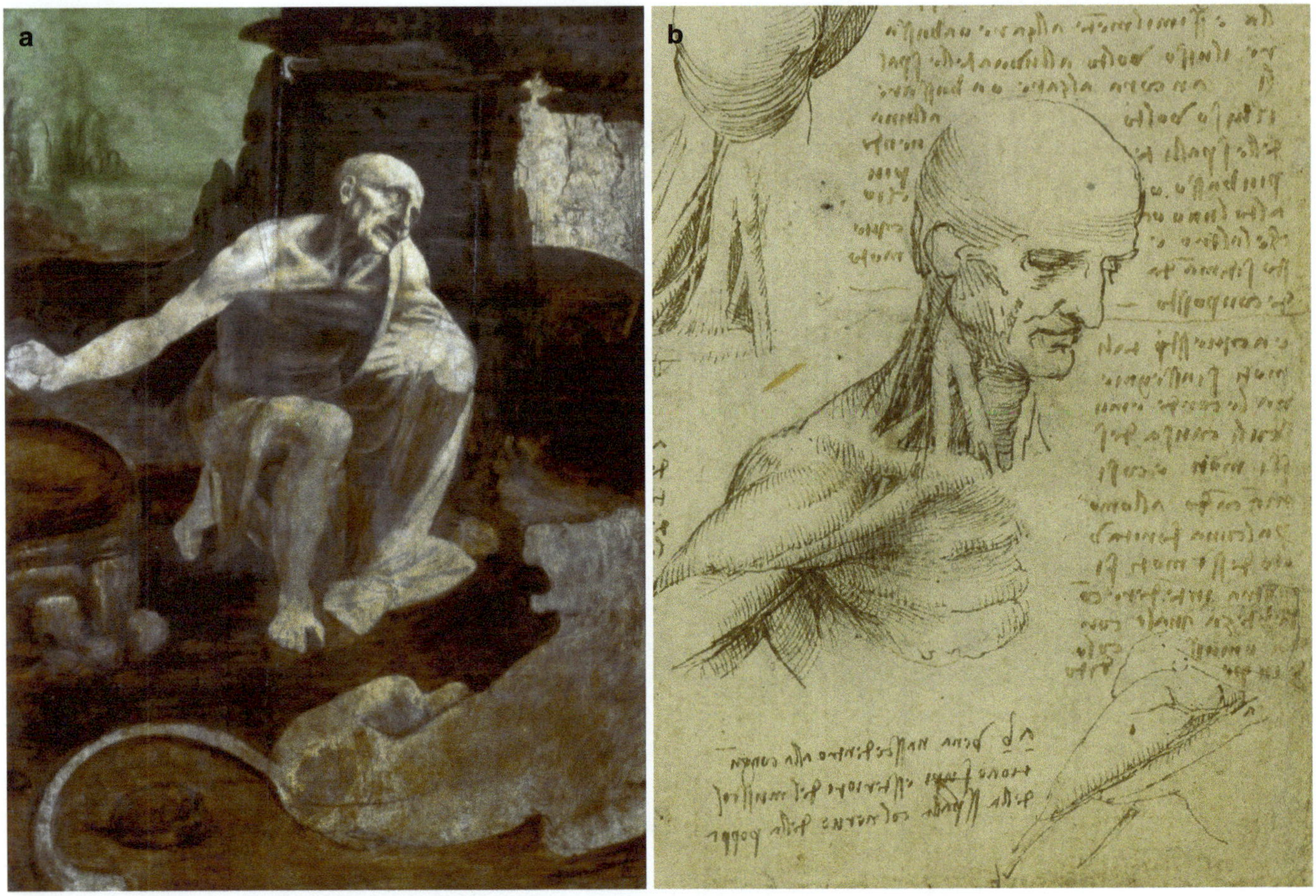

Fig. 1.7 (**a**) St. Jerome: detail of musculature in anterior triangle of the neck. REV 685. St. Jerome, c.1480–82 (oil and tempera on walnut), Leonardo da Vinci/Vatican Museums and Galleries, Vatican City/The Bridgeman Art Library. (**b**) Leonardo drawing of the musculature in the same anatomical region. Detail from RL 19003 recto, The superficial anatomy of the shoulder and neck. Leonardo da Vinci, c.1510–11. Pen and ink with wash, over black chalk (Lent by Her Majesty The Queen. Royal Collection © Her Majesty Queen Elizabeth II)

"Salai" was to stay with Leonardo continuously for 27 years. A possible image of Salai can be seen as what appears to be a doodle in RL 19093 recto.

The second was with Luca Pacioli, a mathematician, who had been invited to lecture in Milan. Pacioli stayed with Leonardo for 4 years, and in that time he completed his geometrical treatise, "De divina proportione," for which Leonardo drew the illustrations of geometrical solids (Fig. 1.8). Leonardo's association with Pacioli was to prove very important to him. For several years, Leonardo was preoccupied with geometry and mathematics. In his notes, he made many impassioned pleas for the necessity of mathematical proof in any scientific endeavor.

During this lengthy period in Milan (circa 17 years), Leonardo lived within the imposing Sforza castle. He contributed to the design of the castle's defenses and latterly created another of his naturalist masterpieces, the wall and ceiling frescoes in the Sala delle Asse. In modern terms, this wall and ceiling painting was in trompe l'oeil form. Although much of it is missing or has been painted over, the magic and beauty of what it once must have been is still discernible. Gualtieri Bascape reported to Lorenzo that Leonardo was working on two rooms, the second being the saletta negra.[14]

Sadly, Leonardo's settled life was about to be disturbed again. From 1497 or so, the French had been threatening to invade. In 1499 it happened. Ludovico was forced to flee from Milan by the invading army of Louis XII, the former duke of Orleans. Five years earlier, Ludovico had laughed off his claim to the duchy of Milan. Unfortunately, this backfired badly on him and he ended his days in captivity in France. Leonardo witnessed the invasion of the French and probably came to an accommodation with them, as there are records of his involvement, not least some cryptic notes in his manuscripts referring to liaisons and latterly commissions such as *Madonna of the Yarnwinder*, for Florimond Robertet, a favorite of King Louis. Some of the anxiety related to Leonardo's departure from the city can be felt in a

[14]Irma A. Richter, Editor, *The notebooks of Leonardo da Vinci.* (Oxford, Oxford World , 1998): 334.

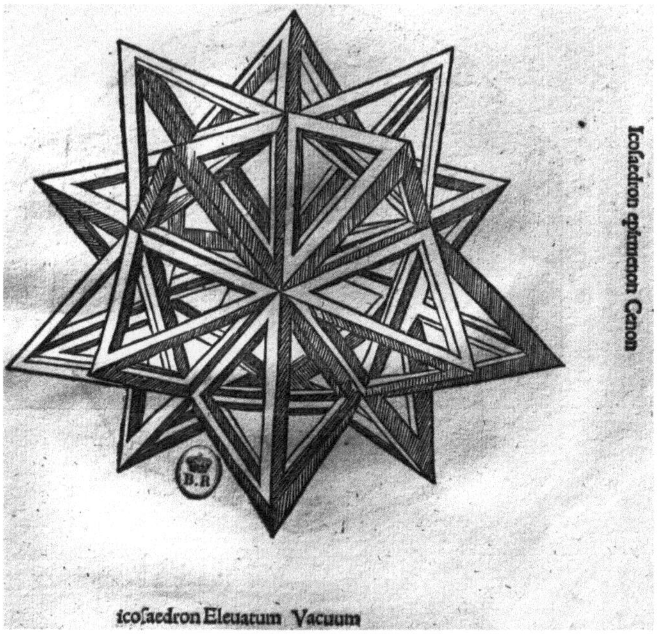

Fig. 1.8 Plate from *La Divine Proportione* by Luca Pacioli, illustrated by Leonardo. XIR 217647. Dodecahedron, from 'De Divina Proportione' by Luca Pacioli, published 1509, Venice (engraving) (b/w photo), Leonardo da Vinci/Bibliotheque Nationale, Paris, France/Giraudon/ The Bridgeman Art Library

phrase written in haste as part of a to-do list: "Have boxes made….sell what you cannot take with you."[15]

Leonardo escaped in the company of Luca Pacioli and traveled to Venice via Mantua. There he met and spent a short time in the court of Isabelle d'Este. A memento of that visit is a small portrait in crayon of Isabelle. It is unlikely that he was there for more than a few weeks before moving on. It is possible that whilst in Venice he witnessed anatomy demonstrations by Alessandro Benedetti, then eminent in the subject. There is no doubt that Leonardo was aware of his work, as he mentions Benedetti's treatise on anatomy in a notebook dated September 1508.[16]

By April of 1500, Leonardo was back in Florence after a brief sojourn in Rome. On his return, some things were unchanged; his father was still practicing as a notary from his offices opposite the Bargello. By August 18, 1502, Leonardo had been appointed family architect and general engineer for Cesare Borgia. He was to look after the fortresses in the conquered provinces and was given power to requisition all that was needed. Cesare spent his time building empire in central Italy on behalf of his father, Pope Alexander VI. This he did brilliantly, and his chief engineer, Leonardo, spent a lot of time traveling around managing projects for him. Cesare had the backing of the French King and had the title of the Duke of Valentinois bestowed upon him. Leonardo referred to him in familiar terms in his notes as Valentino.[17] During this time, he found himself held up with Cesare in Imola as Cesare's generals revolted against him. During that time, he produced a beautiful and accurate map of the city (Fig. 1.9). This map is one of the masterpieces of cartography of the time. It once again reveals Leonardo's wide self-education, as the techniques used to produce it are taken from Alberti's *Descriptio Urbis Romae* and *Ludi Matematici*.[18]

During Leonardo's time in Imola, Florence made offers of help to Cesare via Niccolò Machiavellia, who was dispatched there by the Signoria of the city. It was there in Imola that Leonardo and he became friends. They both would have witnessed the murder by Cesare of his rebellious generals on the evening of 12 December, when they had met to try to resolve their differences. With the campaign over in February 1503, Cesare left for Rome and Leonardo and Machiavelli left for Florence.

Within a few months of his return to Florence, Leonardo was again dispatched on a military engineering mission, this time near Pisa, where he was instructed to form a plan to divert the Arno away from the city to deprive it of its water supply. This plan would make Florence directly accessible from the sea. Niccolò Machiavelli was also involved in this project. Leonardo's drawings showing the rerouting of the river are in the Windsor collection. This task, which would pose massive problems for engineers today, failed, but only after the initial digging began. His work and experimentation over many years in this specialty of hydrodynamics gave him important insights, which he employed to good use in his understanding of the mechanism of valve function within the heart.

On 15 October 1503, Leonardo, back in Florence, was reinstated to the confraternity of St. Luke. His reinstatement seems to have coincided with an extremely prestigious commission to produce a fresco for the Signoria celebrating the Florentine defeat of the Milanese at the battle of Anghiari. This commission may have been arranged through the offices of Niccolò Machiavelli. The project was to place him "head to head" with Michelangelo, who had been commissioned to paint another battle scene on the opposite wall, celebrating the battle of Cascina. This and the battle of Anghiari were classic landmarks in the life of the Florentine republic. Who better to portray these seismic events than the two giants of contemporary Florentine arts?

During the preparatory work for his depiction of the battle of Anghiari, Leonardo was given rooms in Santa Maria Novella, a monastery hospital. It was in that hospital that Leonardo banked his money and stored some of his books, and it is thought that it was there that he spent some of his nights dissecting corpses.

[15] C.A. 247 recto.

[16] Venice 1496 and 1502 and Paris O on cover of manuscript.

[17] Codex Arundel in the British Museum. 202 verso.

[18] Martín Kemp, *Leonardo da Vinci. The marvellous works of nature and man.* (Oxford: Oxford University Press, 2006).

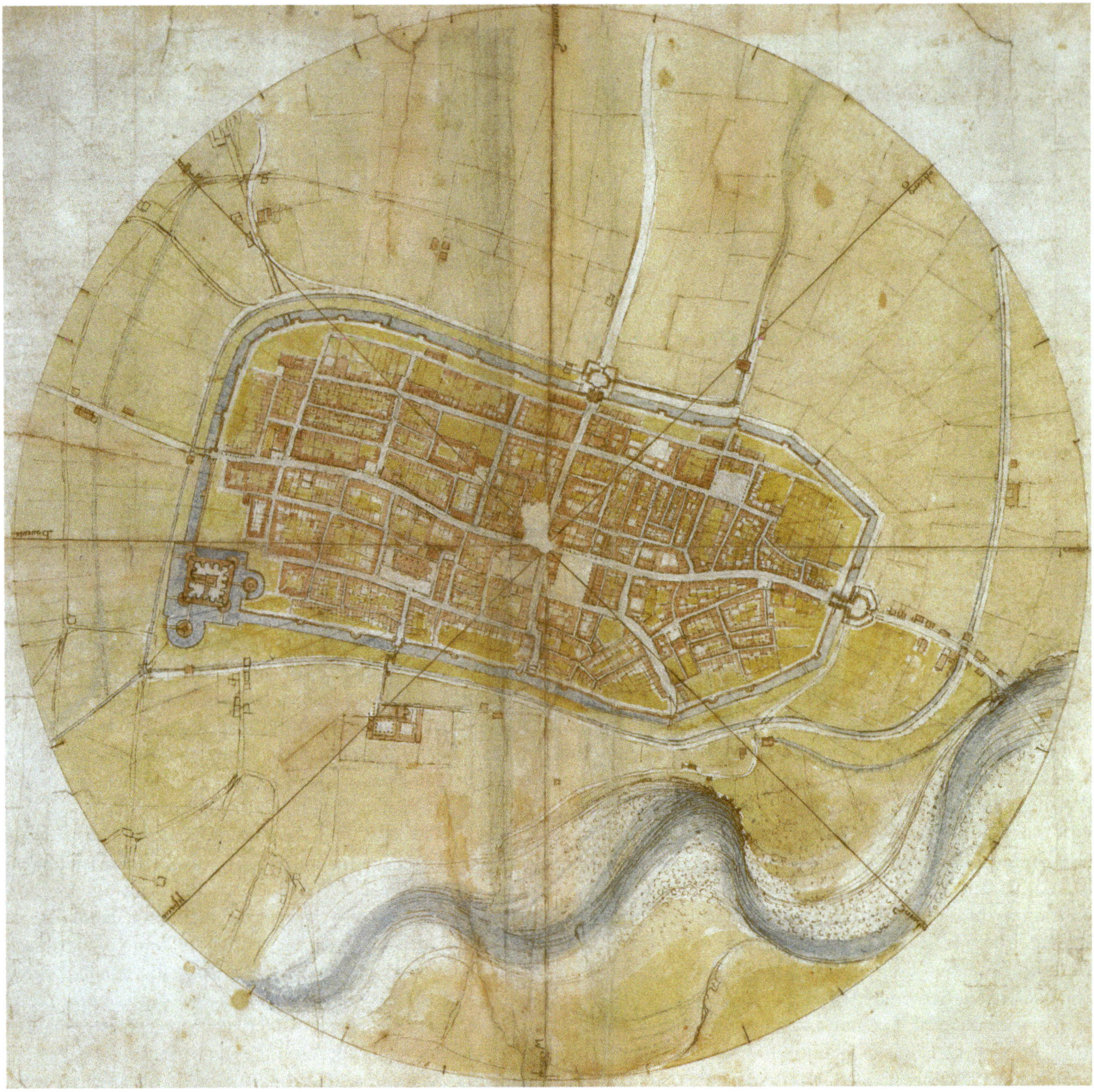

Fig. 1.9 Map of Imola. RL 12284. Leonardo da Vinci (Vinci 1452-Amboise 1519). c. 1502. Pen and ink, with coloured washes, and stylus lines, over black chalk (Lent by Her Majesty The Queen.Royal Collection © Her Majesty Queen Elizabeth II)

In 1503, Leonardo began the *Mona Lisa* and *Leda and the Swan*. He seems to have worked on both of these paintings until near his death. The persistent reworking of the *Mona Lisa* suggests that this project was at one level a piece of research, a scientific endeavor in which Leonardo was employing all of his principles of painting and perspective and developing new ones.

Leonardo's father, Ser Piero, died on 9 July 1504. He was intestate, and thus began a quarrel between Leonardo and his step-siblings that lasted for almost 6 years.

In the second half of 1505, Charles II d'Amboise negotiated Leonardo's release from Florence prior to the completion of his painting of the battle of Anghiari, thereby beginning Leonardo's second Milanese period. In September of 1507, Leonardo met another key person in his life: Francesco Melzi, a young man of approximately 15 years of age who came from a noble family in Vaprio d'Adda near Milan. He became Leonardo's pupil and companion for the remainder of his life. Ultimately, Leonardo repaid Francesco handsomely by making him his principal heir.

In 1508, Leonardo was back in Florence, and at this time he made the important dissection of an old man who claimed to be 100 years old. He reported that he sat on the centenarian's bed in Santa Maria Nuovo and conversed with him until he died, effectively taking a medical history from him. The old man told him that he had suffered no serious illness but was just wasting away. Leonardo tells us that he then carried out a dissection on the man's death. He simply states, "And I made an anatomy!" The results of this dissection are discussed in detail later. It appears that he gained a wealth of anatomical information from this dissection, which he was able to call upon for years to come. It is the only complete dissection of which we have certain knowledge, although near the time of his death he reports having dissected 30 or more corpses.

In 1510, back in Milan, it is likely that Leonardo met and began work with Marcantonio della Torre, a young and brilliant professor of anatomy who had trained in Padua and was appointed professor at a very early age in Pavia. Some have suggested that it was this association that educated Leonardo in the subject. There must be some truth in this idea, but Leonardo had completed a great deal of anatomical study before this relatively brief relationship. There is only a single reference to Marcantonio in Leonardo's notes, referring to a "book on water" (*libro dell' aque*)[19] (possibly a book on urinary-based diagnosis, although this could be mistaken). With Leonardo's extensive interest in hydrology, it is also possible that this reference is to a book on hydrology. The family name della Torre (literally, "of the tower") and the given name Marcantonio were common names (and remain so), but there is significant circumstantial evidence that the Marcantonio referred to by Leonardo in this text was indeed the young physician. Vasari is very clear in this assertion. Paris manuscript K, in which Leonardo refers to Marcantonio, was begun in 1509 and was actively used by him in this period, the same period when Leonardo would have been associated with him.

An article by F. Saxl published by the Warburg Institute[20] gives the following account of Marcantonio della Torre and his father, Girolamo della Torre: "Girolamo was a famous physician, who lectured in Padua and was consulted by the noble families. Piero Valeriano, a devoutly Christian humanist, who was equally interested in astronomy and the antiquities of Belluno, gave the funeral oration for Girolamo. Piero implored Girolamo's sons to follow the Christian values of their father so that they might be reunited in the hereafter". Marcantonio thus would have grown up in the environment of humanist scholarship. Faxl goes on to describe Marcantonio as one who was a bitter enemy of traditional medical lore. In this he would have found an ally in Leonardo, who wrote of physicians, "Endeavour to preserve thy health in which thou wilt succeed the better the more thou guardest thyself from the physicians. For their mixtures are a kind of alchemy on which there are no fewer books than there are remedies!"[21] Marcantonio della Torre is reported by Saxl to have been a keen dissecting anatomist, an observer of the nature of plants, and an impressive teacher and scholar who could quote Greek authorities in debate. He goes on to say that Marcantonio was a man of such winning appearance, serenity, and gentleness that he attracted those with whom he came in contact. His brother Giulio, a jurist and medal maker, represented Marcantonio as a nude figure galloping away on a Pegasus. Descriptions of other family members follow. They all add up to portray a certain uniformity of the younger della Torre family, who were humanists, students of nature and of man's intellect, classical scholars, artists, and men of letters. They were not, however, intent on religious and moral problems.

A friend of Marcantonio, Girolamo Fracastoro, is described as being of outstanding importance in the history of medicine. He was an interesting astronomer and psychologist, an enthusiastic Aristotelian as regards his theory of poetry, and a very elegant Latin poet. It is very easy to imagine and accept that Leonardo would have been with a group of like-minded people with this family and their friends. This association, although undoubtedly important, would only last a short time, as Marcantonio died of the plague about 1511.

At the same time, the French domination of Milan came to an end with aggressive action by the Swiss Guard. Leonardo left the city and moved into the Melzi household in Vaprio. Skirmishes of war continued around him, and Leonardo made a drawing that recorded fires set by the Swiss Guard in Milan. Although he lived outside the city, it is unlikely that he was cut off from life inside the city walls, and he may well have had continued access to human remains as well as ready access to animal carcasses. In January of 1513, Leonardo was at work on some of the drawings now held in Windsor.

On 24 September 1513, he recorded his departure from Milan for Rome, accompanied by Salai, Melzi, and Lorenzo. By December 1st he was listed as part of the household of Guiliano de' Medici, the brother of Pope Leo X. There he set up a workshop in the Belvedere wing of the Vatican. His rooms were refurbished for him at that time. He appeared to travel quite a lot, visiting Parma and Citavecchia on various projects. One in particular is the draining of the Pontine marshes for Leo X.

He was certainly dissecting at that time, as there is an important reference to his being stopped by the intervention of a German mirror maker, Giovanni degli Specchi. It appears

[19]Man. K 48 verso.

[20]F. Saxl, Pagan sacrifice in the Italian Renaissance. *Journal of the Warburg Institute*, Vol. 2, No. 4. (April 1939): 346–367.

[21]This entry by Leonardo is part of a paraphrase from Regimen Sanitas of Arnold of Villanova or one of the numerous modifications of that work such as the tract of Vgr. Brenzo of Milan 1481.

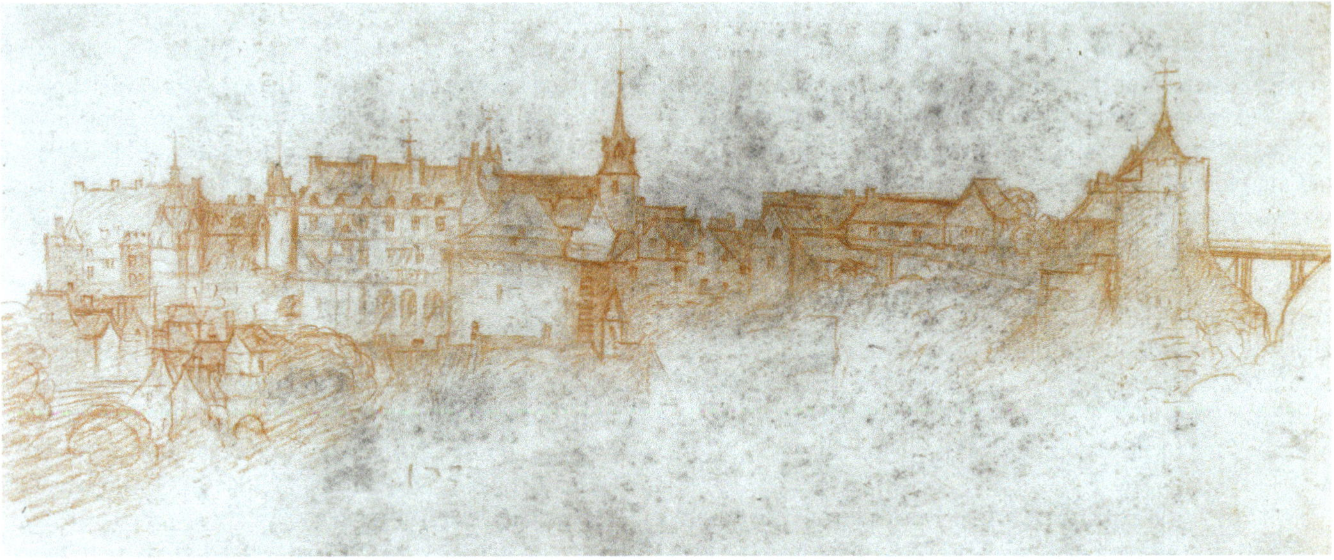

Fig. 1.10 A sketch by Leonardo's pupil Melzi of a view of Chateau Amboise from Cloux. RL 12727. Workshop of Leonardo da Vinci (Vinci 1452-Amboise 1519). c.1517–19. Red chalk (Lent by Her Majesty The Queen.Royal Collection © Her Majesty Queen Elizabeth II)

they fell out over an artisan that Leonardo had billeted and partly employed, who was called Giorgio (also German). It would seem that there were also jealousies between them related to Leonardo's innovative development of mirrors. Leonardo complained to Guiliano that Giovanni was snooping on the projects in his studio and using Giorgio to spy for him. In return, Giovanni reported Leonardo to the Pope and officials of the hospital, San Spirito, where Leonardo was engaged in dissection. Leonardo wrote,"He has hindered me in anatomy, denouncing it before the Pope and also the hospital."[22]

In 1515, Louis XII died and Francis I succeeded him as King of France. Leonardo produced a mechanical lion for presentation to Francis, a gift from Lorenzo di Piero de' Medici, governor of Florence and a nephew of Pope Leo X. Later that year, Leonardo accompanied Guiliano and his brother the Pope on a triumphal return to Florence. Whilst there, he was involved in plans for a new palace for Lorenzo, near to the Medici palace. In December, he traveled to Bologna to meet Francis I and then traveled back to Rome with Guiliano.

In March 1516, Guiliano, his friend and patron, died. At about this time, Francis I gave Leonardo the official position of painter to the King and invited him to France. At age 64, Leonardo traveled with his faithful pupil Melzi and his helper the "devilish" Salai to the Loire Valley, where he was housed in Castle Cloux, near to Amboise (Fig. 1.10).

On Ascension Day, 21 May 1517, Leonardo records that he is in Cloux and Amboise. In October of that year he was visited by the Cardinal Luigi d'Aragona. The Cardinal was a cousin of Isabella of Aragon, whom Leonardo knew well. Antonio de Beatis accompanied the Cardinal in his travels as chaplain and secretary, and a record of what transpired between Leonardo and Luigi is recorded in Antonio's journal.[23]

The beginning of the entry records their pre-amblings to the Chateau at Amboise and on to Castle Cloux. It also mistakenly identifies Leonardo as being over 70 years of age. This mistake may tell us something of the way that Leonardo looked at that time. He had already suffered the first of his two strokes and was said to have residual neurological deficit in his right arm. Antonio reported that Leonardo, instigated by Guiliano de' Medici, showed them pictures "quite perfect" of a "certain" Florentine lady (which may have been La Giaconda, which we now know as the Mona Lisa), a Madonna with a child placed upon the lap of St. Anne, and another of a young John the Baptist. Antonio goes on to report, however, that "We cannot expect any more great work from him, since he is now somewhat paralyzed in his right hand." The report continues, "He has trained a Milanese pupil who works well. And while Master Leonardo can no longer color with such sweetness as he used to he is none the less able to do drawings and to teach others." These comments make sense in that painting requires the use of both hands, especially for the kind of detail and sfumato and chiaroscuro techniques that defined Leonardo's work. One can imagine him explaining this to the Cardinal, who may even have made the visit in an attempt to commission work from this contemporary genius.

[22]C. Atlanticus, fols 500v, 671r, 768r.

[23]Beatis 1979, 131–4. The original journal is in the National Library of Naples, X.F.28.

Of greater pertinence to this commentary, however, are the following words on Leonardo's anatomy. Antonio wrote, "This gentleman has written a great deal about anatomy, with many illustrations of the parts of the body, such as the muscles, nerves, veins, and the coilings of intestines, and this makes it possible to understand the bodies of both men and women in a way that has never been done by anyone before. All this we saw with our own eyes and he told us he had already dissected more than thirty bodies, both men and women of all ages." This contemporaneous record suggests the wealth of Leonardo's material that may have been lost.

The remainder of Leonardo's life was lived out in the Loire Valley in this pretty setting. He was involved in the plans for a palace for the King in Romorantin and in drainage work of the land between the Loire and the Saône rivers. He was involved in the reveries for the wedding between Lorenzo de' Medici and the niece of Francis I, Maddalena de la Tour d'Auvergne. The proceedings of the wedding included performance of a play for which Leonardo had produced mechanical props and scenery in Castella Sforzesco in Milan. Amongst his pursuits, Leonardo remained actively engaged with mathematical and architectural problems, among them the volumetric analysis of Bramante's St. Peter's Basilica.

In the springtime of 1519, on April 23, Leonardo asked the court at Amboise to acknowledge his will. Nine days later, on May 2, he died.

Did Leonardo approach death with equanimity? In his later anatomical period, 1510–1513, he wrote at the heading of a page of wonderful drawings of a dissection of the shoulder the following:

> And thou man, who, by these my labors dost gaze upon the marvelous works of nature, shouldst thou judge mutilation to be an impious thing, reflect that it is of infinitely greater impiety to take the life of a man. Though his structure seem to thee a marvel of artifice, consider that it is nothing compared to the soul which abides within this dwelling, and in truth, whatever it is, it is a divine thing which lets the soul dwell within its works and at its good pleasure and will that not anger and malignity should destroy such a life since, in truth, he who values it not, deserves it not. For it parts from the body so unwillingly and I, indeed, believe that its tears and anguish are not without cause.[24]

In the Codex Atlanticus he wrote: "The soul desires to remain with its body, because without the physical instruments of that body it can do nothing and feel nothing."[25]

What remains without doubt is the fact that he left so much for following generations in his voluminous notebooks. These archives allow us to assess the true worth of the man, as without them we would know little of his innovative genius other than through myth and reportage. Doubtless much is missing. Kemp and others have suggested that as much as three quarters to four fifths has gone. How much more he had discerned in his anatomical studies and the prescient functional deductions that he made from his powers of observation we shall probably never know. That which we do know, taking into account the context of the time in which he lived, is profound in every sense of the word.

The product of Leonardo's unstinting energy and ceaselessly questioning mind, allied to his giant intellect, continue even today to act as a beacon of inspiration for artists and scientists alike—a human compass by which we may still navigate man's place in the world. Few others have contributed so much in so many areas of nature and humanity. In this modern world, we would all be wise to look, and look, and look again; then what might we see?

> Nature is concerned with the production of elementary things. But man from these elementary things produces an infinite number of compounds; although he is unable to create any element except another like himself—that is in his children."[26]

[24] RL 19001 recto.

[25] C Atlanticus 166r/59r-b, R 1142.

[26] RL 19045 verso.

2 Leonardo, Anatomist or Natural Philosopher?

No other subject engaged Leonardo more than his work on the anatomy of the human body. It was the subject in which he argued most forcibly for the importance of independent observation allied to original thought. This process he called "experience"[1] of the world around him "gained through the senses" was his starting point for new understanding. Reason and contemplation brought to bear on the sensed experience allowed the development of a hypothesis. Experimental testing and mathematical proof then gave authority to that hypothesis, the veracity of which could be stated. This approach, proclaimed by Leonardo as essential for rational progress, presages the modern method of science.

This reliance upon dependent observation, throwing off the cloak of didactics, clearly brought him into conflict with some of the academics of the day. He can be heard to rail against his detractors who seem to be criticizing him for original thought; though the context may have been more general, the control on perceived wisdom by the physicians of the day would seem a likely target. Leonardo wrote:

> Many will think that they can with reason blame me, alleging that my proofs are contrary to the authority of certain *men held in great reverence* by their inexperienced judgment, not considering that my works are the issue of simple and plain experience which is the true mistress. These rules enable you to know the true from the false—[The rules that he refers to are those that we would speak of as scientific method.]—and this induces men to look only for things that are possible and with due moderation—and they forbid you to use a cloak of ignorance, (allowing the enquirer to admit that he does not know the answer) which will bring about that you attain to no result and in despair abandon yourself to melancholy.
>
> I am fully aware that the fact of my not being a man of letters may cause certain presumptuous persons to think that they may with reason blame me, alleging that I am a man without learning. Foolish folk! Do they not know that I might retort by saying, as did Marius[2] to the Roman Patricians, 'They who adorn themselves in the labours of others will not permit me my own.' They will say that because I have no book learning, I cannot properly express what I desire to treat of—but they do not know that my subjects require for their exposition experience rather than the words of others. Experience has been the mistress of whoever has written well; and so as mistress I will cite her in all cases.[3]

Perhaps the melancholy he speaks of refers to his own state of mind in dealing with the obdurate academics that clearly were scoffing at his work. He was an original thinker prepared to think outside of the box, and his preparedness to challenge ancient accepted authority can be found in these words, which open the passage quoted above: "Consider now, O reader! What trust can we place in the ancients, who tried to define what the Soul and Life are—which are beyond proof—whereas those things which can at any time be clearly known and proved by experience remained for many centuries unknown or falsely understood."[4]

The dominant academic force in anatomical teaching of the period (and for more than a century before and afterwards) was the work of the Greek physician Galen, who was born in Pergamum, on the seaboard of Asia Minor, in 129 A.D. Galen, who was physician to the gladiators, amongst other things, declaimed Hippocrates' stance on the importance of knowledge of anatomy in the understanding of the human condition. Galen wrote, "He [Hippocrates] thought that one should have a precise understanding of the nature of the body, that this was the source of the whole theory of medicine."[5] As we shall see, Galen did not completely hold to that proclamation himself.

Galen's anatomy was based upon the dissection of animals, principally monkeys; he argued that as a result of their external similarities, their internal form would mimic that of the human being. As a result, there were considerable inaccuracies in his descriptions. Despite maintaining the position

[1]Experience is the term used quite specifically by Leonardo to distance his testable approach to understanding from that of the majority of academics, who were happy to quote the work of others.

[2]Marius's speech to the Quirites in Sallust, *Bellum Iugurthinum*, ch. 85, §25.

[3]C.A. 119v.

[4]*Ibid.*

[5]*Quod optimus medicus sit quoque philosophus*. Khun, Karl Gottleib. Leipzig reference I 54. Claudia Galeni Poera Omnia. Leipzig C. Cnobloch 1821–33 rpt. Hildesheim. Georg Olens 1964–5.

F.C. Wells, *The Heart of Leonardo*,
DOI 10.1007/978-1-4471-4531-8_2,

that anatomy had an important role, Galen's therapeutic approach to disease was based on the balance or imbalance between the four humors—blood, yellow and black bile, and phlegm—so the role of anatomy in the diagnosis of medical conditions was very limited. In both Galen's and Leonardo's time, knowledge of the superficial veins and arteries was important for the barbers who "let blood" from the sick, and knowledge of muscle, bones, and joints was important for the surgeons, who dealt mainly with trauma. Beyond these uses, however, there was little practical application for more advanced anatomy, other than demonstration to medical students of the structures that Galen described.

Galen was much more than just a physician, and he considered it impossible to be a doctor without a full knowledge and use of the canonical branches of philosophy (namely logic, physics, and ethics). This approach brings the disciplines of philosophy and medicine inextricably together. Galen wrote of this in the *De anatomicis administrationobus*. He said that there are applications for anatomy "that are more useful for philosophers than for physicians."[6] He continued, "Of what use to aetiology, diagnosis and treatment [of disease] could a knowledge of the muscles of the tongue or of the muscles, nerves, arteries and veins that run through the heart and viscera if it was impossible to act upon them in any way?"[7] Herein lay the context in which Leonardo was drawn so heavily into the challenge of understanding the human body. Not specifically to derive information that would be of use to the physicians and surgeons of the day but more to answer questions within the confines of Natural Philosophy.

Interestingly, as Andrea Carlino points out in his excellent book, *Books of the Body*,[8] this epistemological approach discussed by Galen affected the reception of Vesalius's *de Fabrica* in the general medical world for several decades, with the continuation of the ritualisation of dissection for didactic ends and as a visual aid to the textbooks of the day, rather than developing the craft of dissection as a research tool, as Vesalius championed.

Leonardo's work and his ideas on "experiencing the truth" through dissection discussed in his manuscripts predated Vesalius's standard-setting *De Humani Corporis Fabrica* (1543) and others such as the *De Re Anatomica* of Realdo Colombo (1559) by a generation. Sadly, as Leonardo did not publish any of this work, there was no way that he could be recognized for the originality of his approach.

With this setting in mind, the observer can only marvel at the insightful questions that Leonardo asked of himself and the outstanding quality of dissection that was necessary to reveal aspects of previously unknown parts of anatomy. These skills, allied to the originality and sheer beauty of the representation to be found in his drawings, make his work quite extraordinary. Also, although his anatomical work was known about by authorities of the time and is referenced with significance in Vasari's *Lives of the Great Artists and Sculptors*, there is no reference to it in the authoritative anatomical texts of his or the next generation. Vesalius produced *de Fabrica* whilst working in Padua.

The comparatively primitive work of the anatomist/physicians de Ketham and Berengario exemplifies the world of anatomy in Leonardo's time. It must be stated, however, that Berengario in his *Commentaria* corrected some aspects of Galen's work and was committed to carrying the specialty forward. He wrote, "We know how to carry out science by adding parts to other parts: and we are as children standing on the shoulders of giants: we are able to see much farther than antiquity could."[9]

Although it took time,[10] it was in this setting that Vesalius's work was able to make such an enormous impact. Yet Leonardo had developed the inquisitive approach to anatomy decades earlier. His progress from a standing academic start, though disappointing to him, was enormous. The picture emerges, however, of a few other farsighted men who appreciated the relevance and importance of original thought. Again, Berengario wrote, "And in this discipline nothing is to be believed that is acquired either through the spoken voice or through writing: since what is required is seeing and touching."[11] In fact, he goes on to stress that Galen should not be followed "where seeing and touching [Leonardo's experience] are in opposition." Leonardo's attitude antedates these sentiments and indeed may have stimulated them. Berengario was an approximate contemporary of Leonardo, being born in 1460 and attending Bologna University. His *Commentaria* was not published until after Leonardo's death (1521), but Leonardo may well have known of Berengario's stance on anatomical exploration. Berengaria was a considerable collector of art and had dealings with Raphael, so he is likely to have been aware of Leonardo's prowess in the arts world.

The importance of the subject to Leonardo is clear from the breadth and depth of his work as well as from the sheer volume of extant material. Bearing in mind that much has been lost, this body of his work is very substantial indeed and suggests a great deal of time spent looking and dissecting, in addition to the many hours of writing and drawing that his manuscripts reflect.

[6] *Andrea* Carlino, *Books of the Body*. (Chicago: University of Chicago Press, 1997), §4.

[7] Galen, *De anatomicis administrationobus*, KII 284 and 287.

[8] Andrea Carlino, *Books of the Body*, §6.

[9] Berengario da Carpi, Commentaria, fol. IIIv. A metaphor used famously by Isaac Newton.

[10] Masterful as Vesalius's work was, the lack of obvious practical application slowed its appreciation outside the major teaching institutions (Andrea Carlino).

[11] Berengario da Carpi, *Commentaria*, fols. VIv and VIIr.

His study of anatomy spans three decades of his life. If, as seems most likely, he was first stimulated to learn anatomy for use in his artistic endeavors, what then drew him into the world of what effectively became a research scientist? His studies of proportion and the muscular and skeletal anatomy fit well with that of an artist struggling to achieve perfection in the representation of the human form in motion and at rest. His detailed analyses of the viscera and the cardiovascular and central nervous systems are something quite different.

A clue to this interest can be found in an interesting note on the earliest page of his extant anatomical notes. Dated as early as 1482 by some influential scholars and headed "Tree of vessels," this drawing of a man standing with legs apart and arms partially outstretched looks as though it may have been drawn to help those who may wish to "let blood," as the major arteries and veins are drawn in.[12] It also has a representation of some of the internal organs, namely the heart, liver, and kidneys. To the left side of the page is a note reminding him to "Cut through the middle of the heart, liver and lung and kidneys that you may entirely figure the tree of vessels." These internal vital organs he refers to as "The Spiritual parts," suggesting perhaps that their investigation linked more to philosophical enquiry than the practical.

Anatomy in the Time of Leonardo

The first records of dissection of the human body arise in Alexandria during the third century B.C. Surprisingly, from that time until the fourteenth century there is no reliable archival record of formal human dissection taking place, including in the work of Galen, whose anatomy (as pointed out earlier) was based on animal dissection.

In Leonardo's time in the late fifteenth and early sixteenth centuries, dissection as practiced amongst physicians was largely in the form of autopsy to establish or confirm the cause of death. Interpretation of the findings was very limited by the teaching of the time, which was based upon the idea that the driving forces in the body were rooted in the relationship between the four elements of nature—earth, air, fire, and water—and the four humors of man—black bile (earth), blood (air), yellow bile (fire), and phlegm (water). The lack of equilibrium resulting from the predominance of one force over another gave rise to what Leonardo called the four dispositions of man: melancholy, choleric, sanguine, and phlegmatic.[13] In this framework, the body was seen to be "opened" for inspection rather than to be systematically dissected. In the writings of Mundinus and Avicenna, the body was seen as three cavities: the abdomen, the thorax, and the skull. The surgeon anatomists considered the limbs separately, with the purpose of structural repair. The first cavity to be opened would have been the abdomen because of the rapid putrefaction of the parts found therein after death. Also, for this reason, dissection tended to be exclusively a wintertime activity. Leonardo's own commentary on dissection powerfully displays the distasteful nature of the task:

> "And though you have love for such things you will perhaps be hindered by your stomach; and if that does not impede you, you will perhaps be impeded by the fear of living through the night hours in the company of quartered and flayed corpses fearful to behold. And if this does not impede you will lack the good draughtsmanship which appertains to such representation; and even if you have the skill in drawing it may not be accompanied by a knowledge of perspective; and if it were so accompanied, you may lack the methods of the geometrical demonstration and methods of calculating the forces and strength of the muscles; or perhaps you will lack patience so that you will not be diligent. Whether all these things were found in me or not, the 120 books composed by me will give their verdict yes or no. In these I have been impeded neither by avarice or negligence but only time. Farewell."[14]

Several professional groups were interested in the body and its workings. At the head of the professional hierarchy were the physicians, who rated themselves on a par with the philosophers as part of the liberal arts. Thus they could not be seen to partake of manual occupations but worked within the academic boundaries of natural philosophy. Below them were the surgeons, barbers, and apothecaries, who worked at the bidding of the physician. In addition to these groups were the midwives, the empiricists, and, of course, the charlatans!

Furthermore, as practitioners of the liberal arts, architects were interested in the perspectival aspects of the human frame. Francesco de Giorgio and Alberti were influenced by the writings of the first century Roman architect, Vitruvius. His famed ten books on architecture were composed in the first century A.D. and were to architecture what Galen was to medicine. Leonardo was familiar with and significantly influenced by Vitruvius and by Alberti's interpretation of Vitruvius. One of the manifestations of this influence was the various forms of imagery of the iconic Vitruvian man.[15] For Vitruvius and his followers, the body was seen as a cultural entity. The standard layout of a church, which is based upon the head as the chancel, the extended arms as the side chapels, and the remainder of the body as the nave, can be found in di Giorgio's manuscripts on architecture, a volume possessed by Leonardo. Leonardo's own iconographic image of the Vitruvian man fits directly into this paradigm (Fig. 2.1).

The use of anatomical knowledge in the development of philosophical ideas about man's place in the cosmos cannot be overstated. The question of how the human body was

[12]RL 12597 recto.

[13]RL 19037 verso.

[14]RL 19070 verso.

[15]There are numerous examples of Vitruvian forms relating man to the geometrical shapes of the square and the circle. The image by Leonardo da Vinci happens to be the most famous.

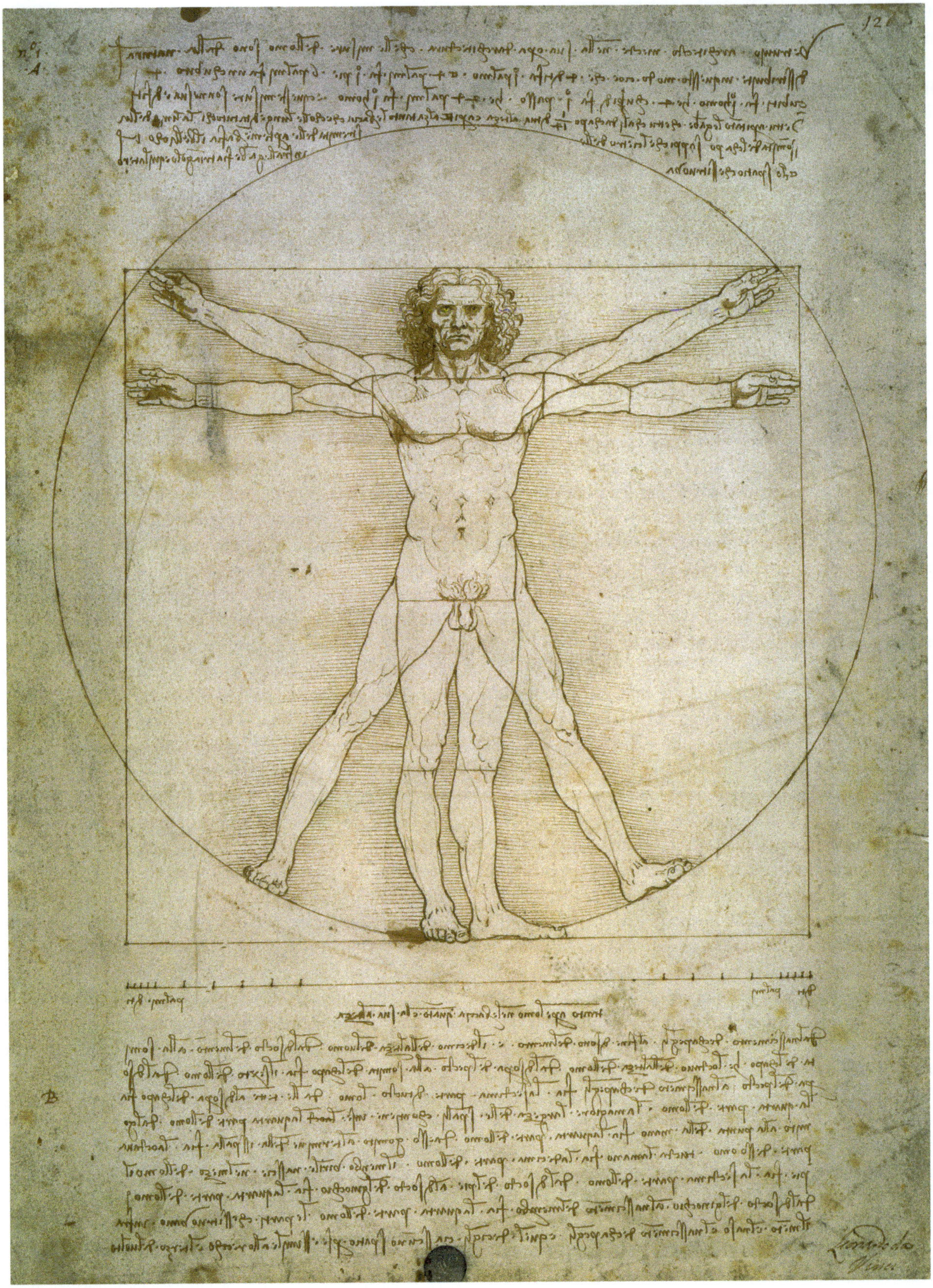

Fig. 2.1 Leonardo's Vitruvian man. BAL 4146. The Proportions of the human figure (after Vitruvius), c.1492 (pen & ink on paper), Vinci, Leonardo da (1452–1519)/Galleria dell' Accademia, Venice, Italy/The Bridgeman Art Library

made pertained as much to natural philosophy as to natural science. This positioning of anatomical knowledge within the sphere of philosophy continued to be influential upon Leonardo, as one would expect from his reading of the texts of the day. Indeed, Vesalius defines anatomical discipline in *de Fabrica* as belonging to natural philosophy.

Strict rules controlled access to human cadavers for dissection. To preserve anonymity, the unfortunate subject had to have been domiciled in an area distant from the place of dissection. He or she had to have been convicted of unpleasant crimes and not be a Catholic, but preferably of Jewish origin. The element of societal control was very important. Public dissections, which went on over several days, were highly ordered. In some cases, they were organized as the end phase of an execution. In such cases, the process was planned as two separate steps with significant ceremony. By such means was the sanctity of the human body managed and maintained.

Where Vesalius, and indeed Leonardo before him, broke with tradition was in their belief that the detailed structure of the human body should be studied for itself, independently of its relevance to the practice of medicine. Thus the study of anatomy as an independent scientific discipline began in the middle years of the sixteenth century. As a result of this new exploration of the anatomical universe of the human frame, there was an inevitable reevaluation of Galen's ideas and an improved record of observations on anatomy. This change was at first more subtle than might have been expected. In Berengario da Carpi's commentaries on the text of Mundinus,[16] and also in Vesalius's *de Fabrica*,[17] neither man was explicitly critical of his forebears, but utilized their experience from dissection as a means to point out differences in their developing knowledge base from that of Galen and Mundinus.

In the same period, philosophers were using anatomical insight to provide indisputable evidence of the supreme authority of the principles that inform man and nature. These questions largely revolved around the female form and the process of conception. This philosophical stance has reverberations in the words of Leonardo in his reference to the marvelous works of nature and man: "And would that it might please our creator that I were able to reveal the nature of man and his customs as I describe his figure!"[18]

The legitimacy of dissection in this period is often questioned. It has been suggested that the Vatican had banned dissection. Pope Bonefacio VIII issued a Papal Bull in 1299, declaring that there should be no separation of the parts of the human body and no boiling of these separated parts. This has been taken as a prohibition of anatomical dissection, but in fact it was nothing of the kind. This Bull was declared in response to the attempts to preserve parts of the bodies of noblemen (and occasionally women) who died on the Crusades and whose relatives and friends did not want them buried in the unconsecrated ground of the Muslim homelands. The separation and boiling was to allow a degree of preservation of the tissues for repatriation of the parts for burial at home. Separation of the body into parts also allowed for several burial sites, allowing multiple opportunities for simultaneous prayer, thereby increasing the chance of the departed to pass from Purgatory and enter Heaven. In 1301, Cardinal Le Momier widened the Bull to "all desecration of the human body," including cremation, but this was nothing more than a commentary and had no legal status. An anatomist at the time of Mundinus comments that as a result of the Bull, dissection of humans should not be practiced because of the revulsion that it would cause. Nevertheless, Mundinus pressed on with his dissections just a few years later.

As Latin translations of the Arabic versions of Galenic texts became more available in the thirteenth and fourteenth centuries, factual verification of anatomical dissection was deemed necessary by the reading of the accepted texts. It was this "reading" by the Professor and the "demonstrating" of anatomy revealed by the "Prosector" that was the method of anatomical education in the time of Leonardo. This relationship is clearly depicted in "La lezione di anatomia" in the *Fasciculus Medicinae* of Johannes de' Ketham (Fig. 2.2).

Despite Berengario's words urging the anatomist to learn from his own experience, the same presentational relationship between Reader, Prosector, and Demonstrator is revealed in the frontispiece to his text *Isagoge Breves*.

It is from this teaching method that Vesalius really broke with tradition. In the frontispiece to his book *de Fabrica Humana Corporis*, he illustrates himself at the side of the corpse, personally carrying out the dissection and the demonstration of the parts. He has his quill pen and paper at his side to allow faithful recording of his findings as the dissection proceeds (Fig. 2.3). Here the paradigm has shifted very significantly: the Professor, or Reader, has become Prosector, Demonstrator, and Recorder, revealing the investigational approach to anatomy that put Vesalius apart from all of those between himself and Galen.

The principal texts of the time were the Canon of the Persian physician and philosopher Abd Allah Ibn Sina, known as Avicenna (c. 980–1037 A.D.) and, closer to home, the Italian professor of medicine, Mondino De' Luzzi (Mundinus). Leonardo was versed in both texts and had Mundinus in his personal library. He was quite familiar with at least parts of Galen's *De Usu Partium* through his later collaboration with Marcantonio della Torre. Avicenna was read and quoted widely at that time, and McMurrich refers to

[16] A brief, early fourteenth century (1316) affirmation of Galenic traditional anatomy, flavored by the description of the dissection of two female bodies between January and March of 1315.

[17] This book was in many ways a commentary upon Galenic anatomy.

[18] Quaderni d'Anatomia I.2, recto. Royal Library Windsor.

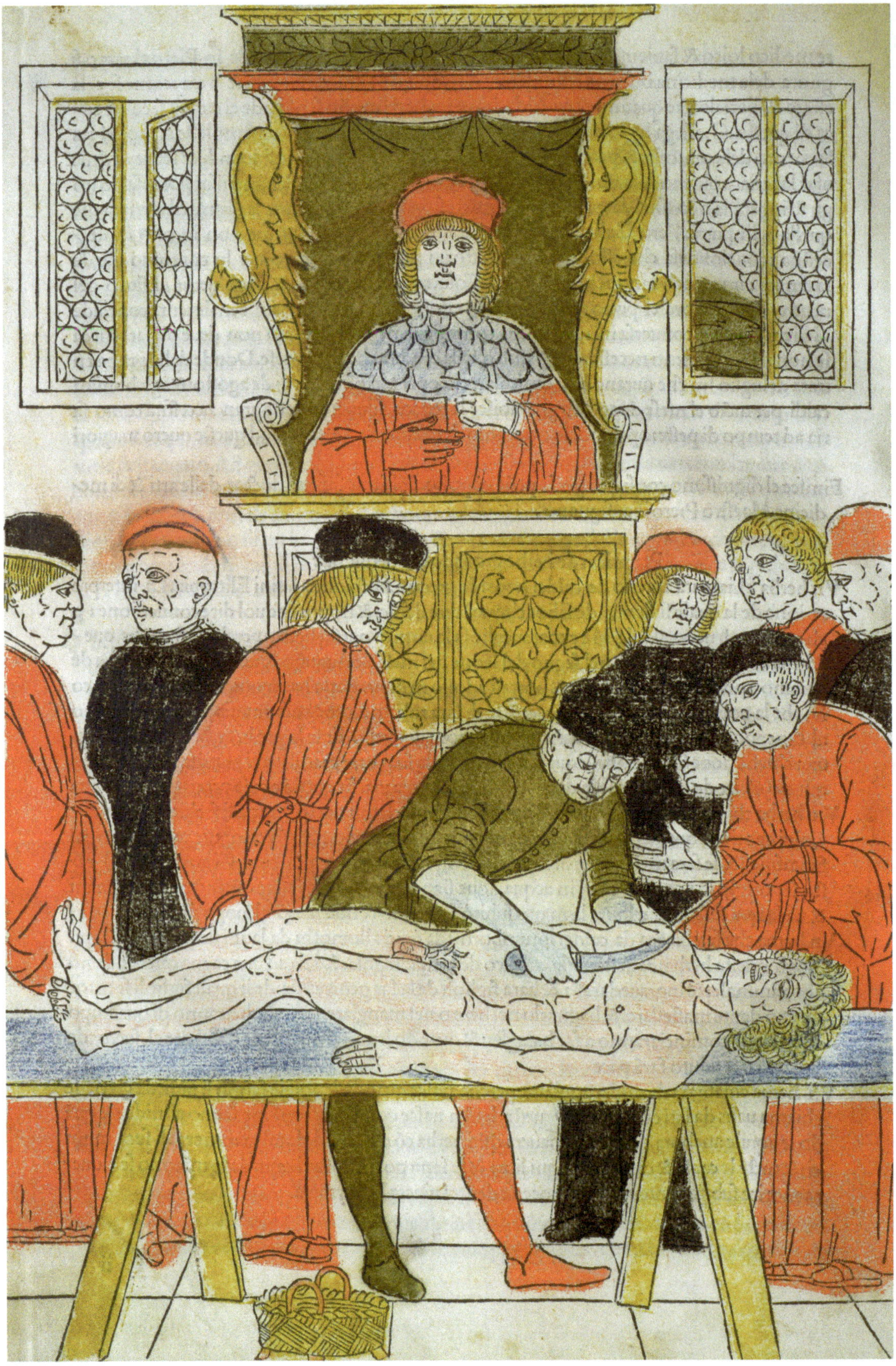

Fig. 2.2 Johannes de' Ketham, The Dissection. ALG 178332. Illustration from *Fasciculus Medicinae* by Johannes de' Ketham (d.c.1490) 1493 (woodcut), Italian School (fifteenth century)/Biblioteca Civica, Padua, Italy/Alinari/The Bridgeman Art Library

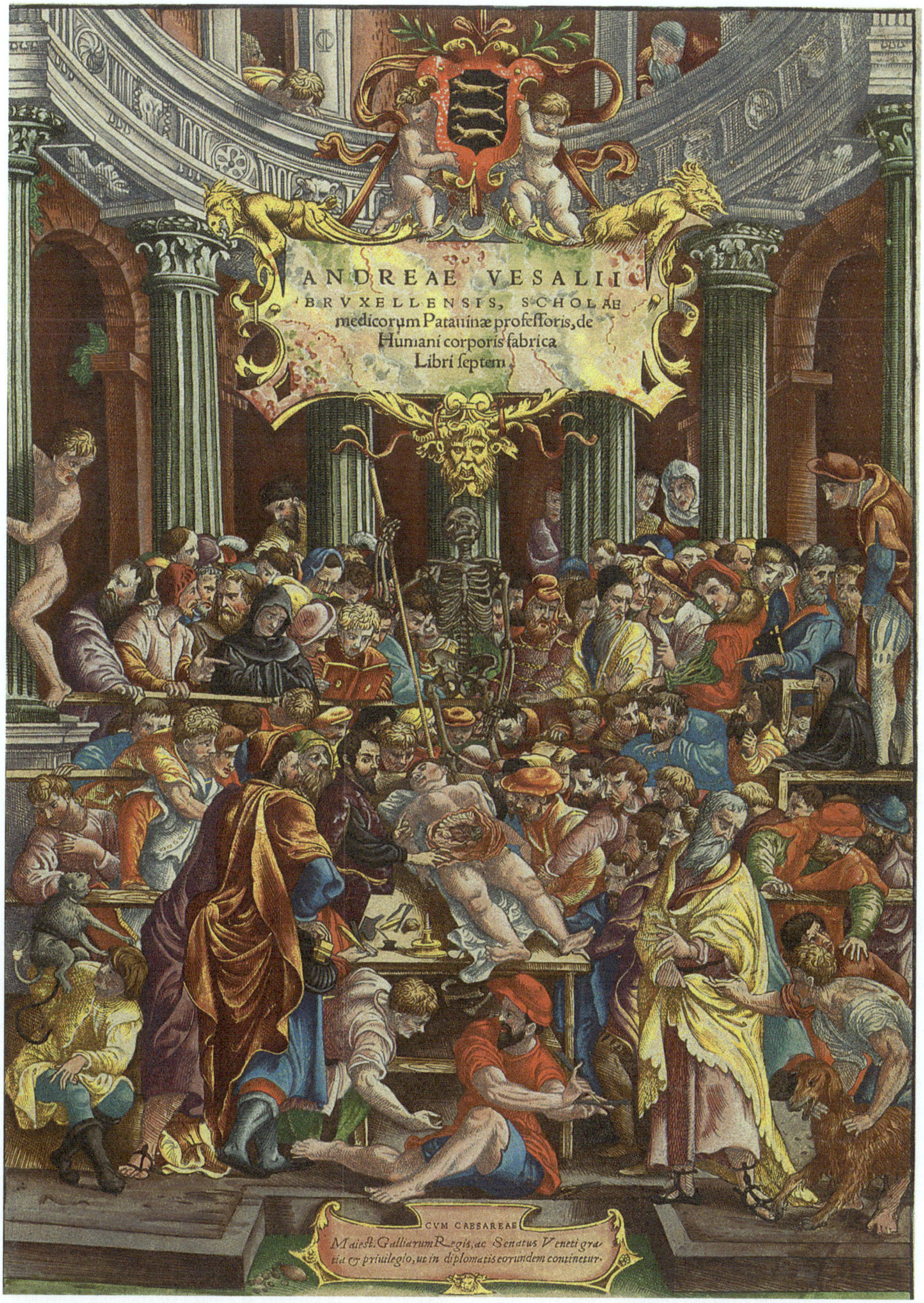

Fig. 2.3 Frontispiece of *de Fabrica*. CH 466697. Frontispiece to *De Humani Corporis Fabrica Libri Septem* by Andreas Vesalius (1514–64), published by Johannes Oporinus, Basel, June 1543 (colour woodcut) (see also 233019), Venetian School, (sixteenth century)/Private Collection/Photo © Christie's Images/The Bridgeman Art Library

this and other Arab contributions as the "keepers of the 'Light' through the dark ages."[19]

That Leonardo was aware of the contents of Mundinus is clear, as he quotes from him in his notes. Leonardo broke new ground by being prosector, demonstrator, and recorder all in one person well in advance of Vesalius by conducting his own dissections and recording his observations, flavoured with his own interpretation. Both Leonardo and Vesalius recorded the sorts of instruments they used (Fig. 2.4). It is possible that later (1510–1512), Leonardo also recorded dissections carried out by Marcantonio della Torre, the young and brilliant professor of anatomy at Pavia University, and his students.

Anatomy for the Artist

The emergence of neoclassicism and humanism in the fifteenth century defined artistic direction in the Italian Renaissance. Bernard Berenson likened it to a human reawakening of the sense of personality, resulting in a boundless curiosity about all natural things.[20] The whole idea of art imitating nature was built upon the Greco-Roman tradition as exemplified in Pliny's *Natural History*. Pliny described with admiration the massive leap forward made by sculptors and painters of that period. The power and grace of the human form hewn from marble and stone as seen in the newly excavated antiquities of Rome made a huge impression on the artists of the quatrocento. Unlike today, however, the antique (admirable as it was) was not seen necessarily as the ultimate in form, but rather as a starting point for betterment. In other words, the style and technique could be studied and copied to perfection, but then it was the challenge for the artist to take this realism to the ultimate. For this reason, Leonardo recommended that the student first draw from the antique before moving on to study nature. In short, says Berenson, "It (the renaissance) looked upon the antique as a short cut to knowledge."[21]

For the artist to achieve the goal of absolute realism, it was perceived that an extensive general education would be necessary. Ghiberti suggested that the artist should be conversant with grammar, geometry, philosophy, medicine, astrology, optics, history, anatomy and arithmetic, as well as being an accomplished draughtsman.[22]

The pursuit of ever more tangibly natural forms in commissioned works of art demanded the study of nature in depth. This development necessitated the use of human and animal dissection, so it is not surprising to discover that commentators upon the Florentine studio curriculum of the fifteenth century often described this activity. The studios of the Pollaiuolo brothers and Andrea del Verrochio were known to have undertaken these activities. Vasari recalls the importance of anatomical knowledge to Raphael and Michelangelo as well as to Leonardo. It is known that Michelangelo was interested in anatomy and worked with the great anatomist Colombo in the mid 1500s,[23] but there is no record of his becoming interested in much other than the musculoskeletal system. Vasari speaks only of Leonardo as an "anatomist" who made great contributions to the fund of knowledge in this subject. He wrote, "Because of Leonardo we have a deeper knowledge of human anatomy and the anatomy of the horse" and he continued, saying that Leonardo's fame would never be extinguished, "though this will rather be by word than by deed!" Thus he acknowledged Leonardo's infuriating tendency not to finish things but also confirmed the massive legacy of his notebooks and manuscripts.[24]

The frequency of dissection is more conjectural. It is known that studios possessed casts of human torsos and limbs, which would have been used on a regular basis. Indeed, Leonardo lists an inventory of artistic "blueprints," amongst which there were likely such casts.[25]

Leonardo's Anatomical Work

The exact dating of all of Leonardo's anatomical studies remains to be completed, but much of it is known with some certainty. From the extant record, it appears that there were periods of quite intense activity separated by periods of indus-

[19]J. Playfair McMurrich, *Leonardo da Vinci, the Anatomist.* (Baltimore, Williams & Wilkins [published for the Carnegie Institution of Washington], 1930), 8.

[20]Hanna Kiel, ed., *The Bernard Berenson Treasury.* (New York, Simon and Schuster, 1962), 75–7.

[21]Ibid., 64.

[22]Martin Kemp, *The Marvellous Works of Nature and Man.* (Oxford: Oxford University Press, 2006), 18.

[23]Colombo had been a surgeon in Padua before taking up the post of "Master of Anatomy and Surgery" in Pisa in 1544. He had been invited there by the Duke Cosimo. Colombo went to Rome to teach at the papal university, the Sapienza, initially in 1547 and permanently 1 year later. In a 1548 letter to Cosimo, he stated that his reason for leaving was "to pursue my dissections and supervise the painters." He also referred to his intended collaboration with "the greatest painter in the world" (Michelangelo) on a proposed book of anatomy. As with Leonardo, however, the planned treatise on anatomy did not materialize. Colombo did produce *De re anatomica Libri XV*, which was published a few months after his death in 1559. Its sole illustration is the frontispiece, which features the presence of Michelangelo at the dissection table. There are further hints that Michelangelo's involvement in anatomy was significant. Passarotti's *Michelangelo Conducting an Anatomy Lesson* illustrates an array of artists gathered around the master. Included in the picture are Andrea del Sarto, sitting on a stool to the left of Michelangelo; Sebastien del Piombo, at his right; and Baccio Bandinelli, with his back to the viewer. Raphael is to be found facing forward, supporting the hand of the flayed corpse. The Accademia del Disegno in Florence, of which Michelangelo had been nominal head, recognized his knowledge of anatomy as a major contributing factor to the perfection of his art.

[24]*Vasari,* Giorgio. *Lives of the Artists.* Volumes I & II. London: Penguin Classics, 1987.

[25]*Kemp, Marvellous Works,* 17.

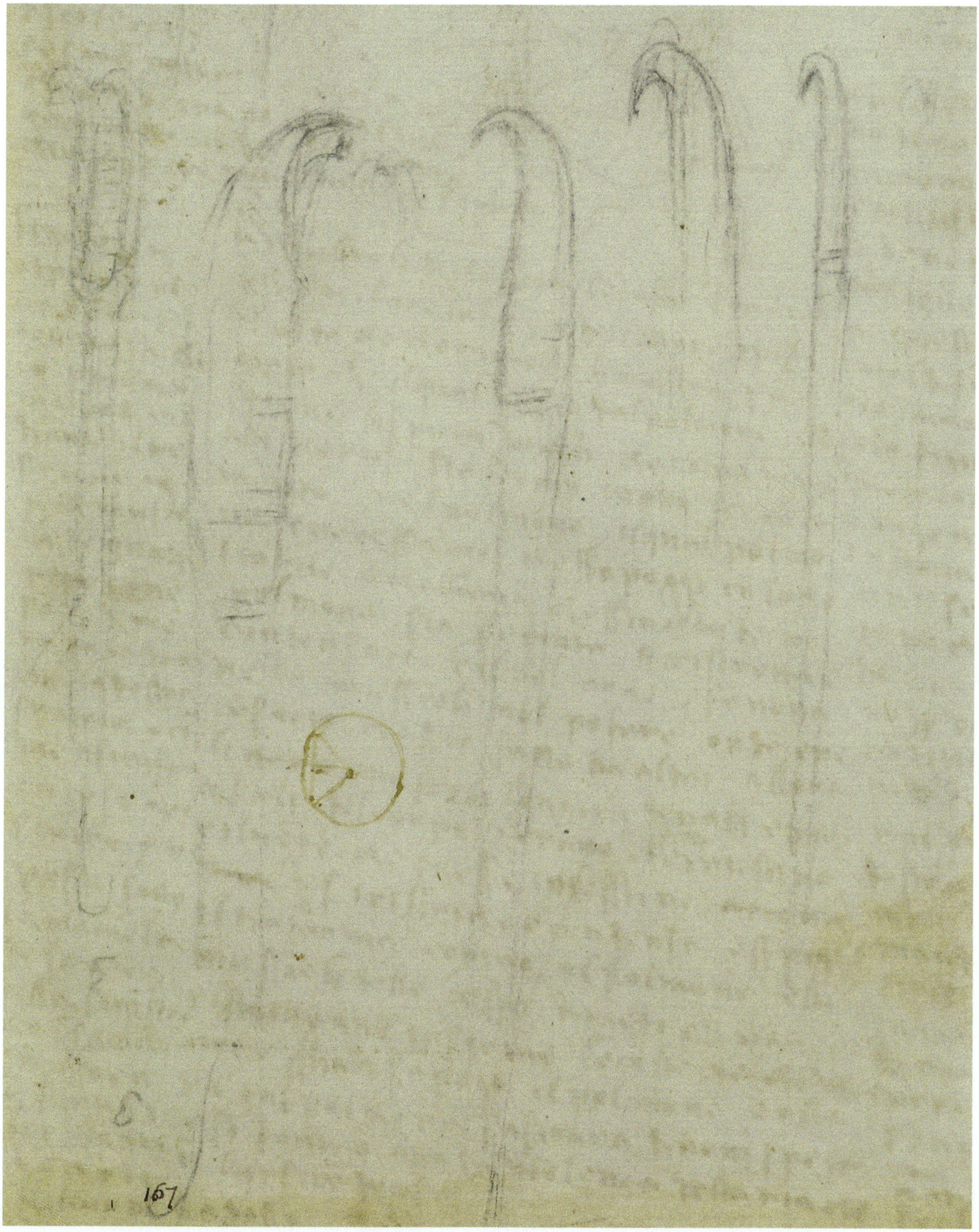

Fig. 2.4 (**a**) Sketches of the intestines, scalpels, and hooks for dissection, and notes. Detail from RL 19070 recto, Leonardo da Vinci, c.1508–10. Pen and ink and black chalk (Lent by Her Majesty The Queen.Royal Collection © Her Majesty Queen Elizabeth II). (**b**) Vesalius's dissecting table with instruments displayed. FAB 45935. Illustration from De Humani Corporis Fabrica (b/w print), Andreas Vesalius (1514–64)/Fratelli Fabbri, Milan, Italy/The Bridgeman Art Library

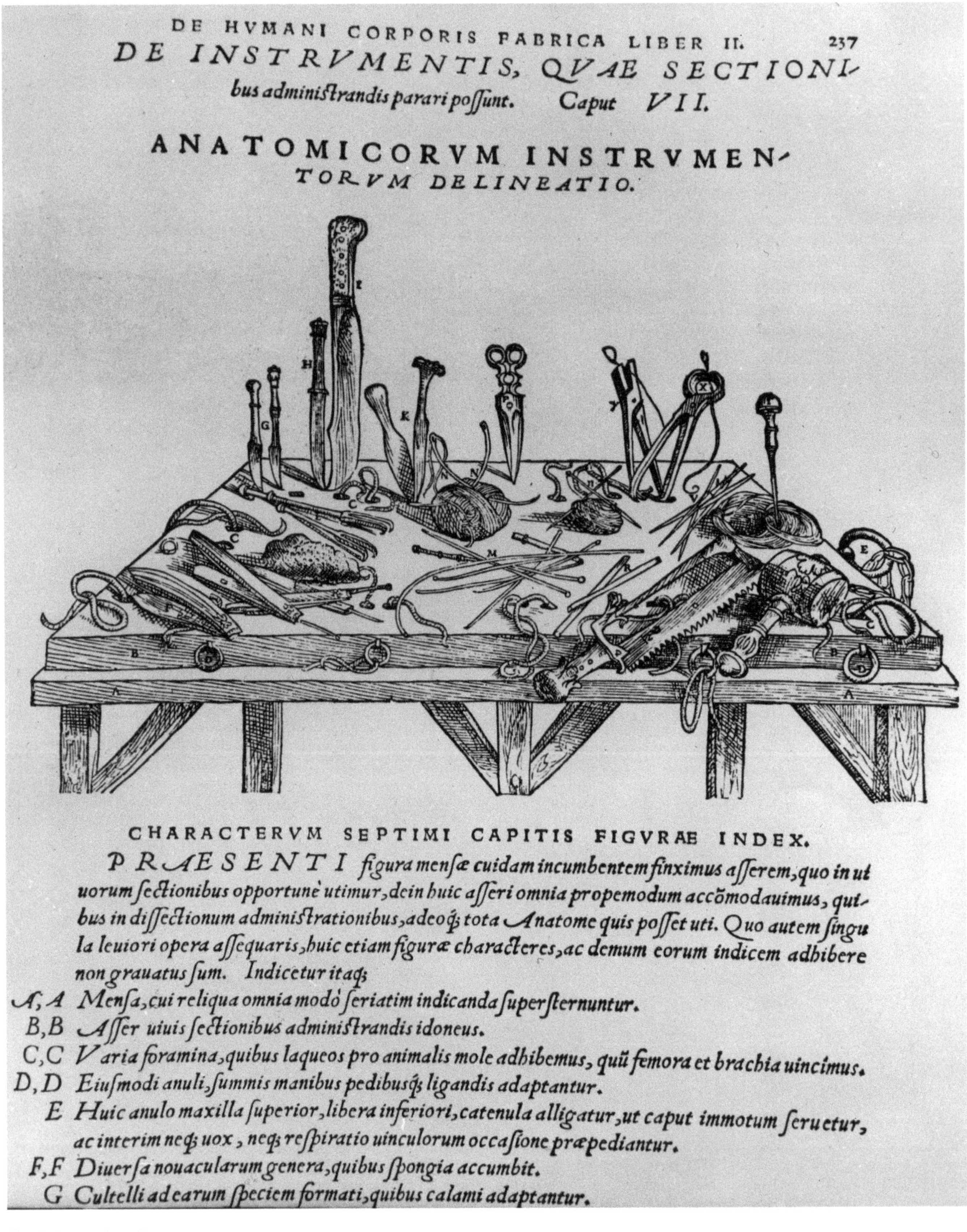

DE HVMANI CORPORIS FABRICA LIBER II. 237

DE INSTRVMENTIS, QVAE SECTIONIbus adminiſtrandis parari poſſunt. Caput VII.

ANATOMICORVM INSTRVMENTORVM DELINEATIO.

CHARACTERVM SEPTIMI CAPITIS FIGVRAE INDEX.

PRAESENTI figura menſæ cuidam incumbentem finximus aſſerem, quo in uiuorum ſectionibus opportunè utimur, dein huic aſſeri omnia propemodum accõmodauimus, quibus in diſſectionum adminiſtrationibus, adeoq; tota Anatome quis poſſet uti. Quo autem ſingula leuiori opera aſſequaris, huic etiam figuræ characteres, ac demum eorum indicem adhibere non grauatus ſum. Indicetur itaq;

A, A Menſa, cui reliqua omnia modò ſeriatim indicanda ſuperſternuntur.
B, B Aſſer uiuis ſectionibus adminiſtrandis idoneus.
C, C Varia foramina, quibus laqueos pro animalis mole adhibemus, quũ femora et brachia uincimus.
D, D Eiuſmodi anuli, ſummis manibus pedibusq; ligandis adaptantur.
E Huic anulo maxilla ſuperior, libera inferiori, catenula alligatur, ut caput immotum ſeruetur, ac interim neq; uox, neq; reſpiratio uinculorum occaſione præpediantur.
F, F Diuerſa nouacularum genera, quibus ſpongia accumbit.
G Cultelli ad earum ſpeciem formati, quibus calami adaptantur.

Fig. 2.4 (continued)

try in other very disparate disciplines. How much of this is real and how much may be an impression created by the loss of material is impossible to know. It seems that the remaining notes can be divided into three fairly distinct periods. The relevance of the consideration of the anatomical studies in chronological episodes is that it allows emphasis upon Leonardo's intellectual and artistic development within this subspecialty of his work. Despite the clear influence of Hippocratic and Galenic thought on the functional aspects of the "soft or *spiritual* parts" throughout most of his work, there is an intensity of original observation and representation allied to exquisite dissection that is obvious from the first drawings of the skull. The wonderful realizations of the upper and lower limb muscles, bones, and joints and the superlative renderings of the complexities of the heart and other viscera attest to his independence of mind in observation of the true structure and relationship of the parts. It is in the later stages that his own interpretations of bodily function come to the fore.

The first period, from the early 1480's to 1495 in Milan, includes a very primitive and Galenic rendering of the circulation seen a drawing from about 1482,[26] and the Hippocratic notion of procreation is seen in Windsor 19097v, where the semen is depicted as being derived in a very fanciful way from the spinal cord. Also from this period is the wholly believable modern interpretation of the bony structure of the skull in drawings from 1489. How much knowledge gleaned from dissection was available to him at that time we do not know, but on the page of the "procreation" drawing, the penis is shown as having two channels, both of which appear to open into the urethral meatus, which is manifestly untrue. The second channel may be the dorsal vein of the penis, which Leonardo misconstrued.

The originality of the skull drawings of 1489 is so fundamentally different and superior to all other extant illustrations of the time that they are completely out of character with the age (Fig. 2.5). The lines of incision in the skull probably represent drawing summaries of the information gained from a series of dissections. They reveal the air spaces (the sinuses) and the complexities of the internal surface of the base of the skull that were not matched for centuries to come. They stand as completed demonstrations of virtuosity in artistic skill and insightful basic scientific questioning, yet they were produced within the early period. Within them are geometric lines placed in search of the sensus communis, a very Galenic idea of human function.

The second period of his anatomical record was created following his return to Milan from Florence, between 1506 and 1509. Whilst being grouped together by previous historians, this is a somewhat fragmentary period of Leonardo's life, with significant travelling between households in Milan and Florence. It is probably during this period, whilst in Florence during the winter of 1507–1508 that he carefully documented a dissection that he carried out in the Santa Maria Novella. Leonardo had returned to Florence from Milan in an attempt to resolve the prolonged dispute between him and his brothers over his father's estate. Carmen Bambach dates this episode as 12 months later, in the winter of 1508.[27] According to Clark, however, quite accurate dating may be possible[28] because an outline of the passage to be found on the reverse side of the manuscript sheet describing the changes in the mesenteric arteries and veins in old age is also found on ff.1, recto of Paris MS F,[29] and that is dated 12 September 1508.

Here, Leonardo describes his observation and conversation with an old man who died peacefully, claiming to have reached 100 years of age:

> And this man, a few hours before his death, told me that he had passed one hundred years, and that he was conscious of no failure of body, except feebleness. And thus sitting upon a bed in the hospital of Santa Maria Nuovo at Florence, without any untoward movement or sign, he passed from this life.
>
> And I made an anatomy to see the cause of a death so sweet, which I found to proceed from debility through lack of blood and deficiency of the artery which nourishes the heart and the lower members.[30]

This very remarkable passage of writing reveals several things. First, it confirms that Leonardo undertook at least one human dissection. Second, it epitomizes his unceasingly enquiring mind; and third, it once again reveals his considerable powers of observation. This human autopsy report contains a description that constitutes the first report of the disease arteriosclerosis (hardening of the arteries). It also documents the intellectual progress that Leonardo is making from a purely didactic "Galenist" to a free-thinking academic. Information gleaned from the autopsy on the centenarian can be seen to inform much of his subsequent work.

This episode then marks the one definite human dissection by Leonardo. He variously claims to have dissected "more than ten human bodies…repeated twice over"[31] including the young and the old! As described earlier, Antonio de Beatis in his diary reported that Leonardo described dissecting more than 30 bodies in his lifetime when the Cardinal of Aragon visited him in Amboise in October 1517.

26RL 12597 recto.

27Carmen C. Bambach, *Leonardo da Vinci, Master Draftsman (New York Metropolitan Museum of Art Series)*. (New Haven: Yale University Press, 2003), 236.

28Kenneth Clark, *The Drawings of Leonardo da Vinci in the Collection of Her Majesty the Queen at Windsor Castle*. (Vol. 3, Anatomical Drawings, Edition 2). (London: Phaidon, 1969), 13.

29Paris Manuscript F folio 1.

30RL 19027 verso.

31RL 19076 verso.

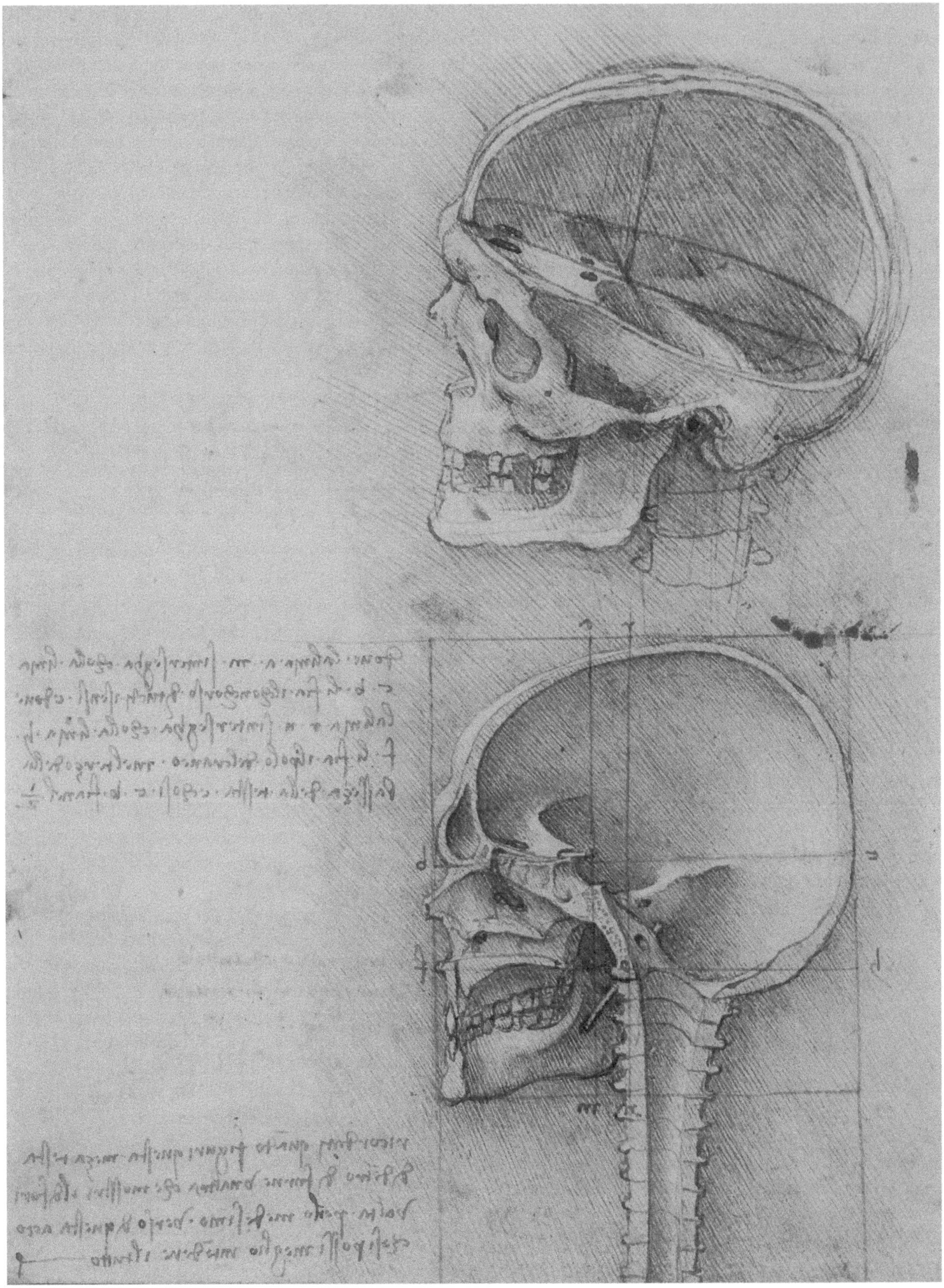

Fig. 2.5 Skull drawings carried out early in Leonardo's career. RL 19057 recto, The skull sectioned. Leonardo da Vinci,. c. 1489. Pen and ink over black chalk (Lent by Her Majesty The Queen.Royal Collection © Her Majesty Queen Elizabeth II)

Places in which Leonardo probably either took part in or witnessed dissection were Santa Croce and Santa Maria Novella in Florence, where he may have participated in public demonstrations of anatomy; the University of Pavia, working alongside Marcantonio della Torre; and later, Santa Spirito in Rome, where he probably performed his last dissections, including his work on the heart. There is no record of this activity after Leonardo traveled to France at the invitation of King Francis I in 1516.

The instruments used for dissection were probably taken from surgical sets as well as being made up by the anatomist's assistants. One of Leonardo's last anatomical notes begins, "Definitions of the instruments." Under this heading, he writes a page of notes on the mechanical functions of the instruments. On the other side of the manuscript sheet on which he described the unpleasant nature of dissection and the skills that were involved, he made a sketch of some of the dissecting instruments that he used (Fig. 2.4a).

In the beginning of the third period, from 1510 to early 1513, Leonardo was in Milan. During this time he is thought to have been working with Marcantonio della Torre in Pavia. This potentially important liaison was short-lived, as Marcantonio died of the plague in 1511.

In 1513, the Swiss Guard invaded Milan. For a time, Leonardo boarded with the family of his long-term pupil, Francesco Melzi, in Vapprio d'Adda; he then moved to Rome at the invitation of Guiliano de Medici, the brother of the Medici Pope.

The heart work was done during the period 1513–1516, covering the late Milanese period to the end of his stay in Rome. Many of the drawings of dissections of the ox heart seem to have been done during his stay in Vapprio, which is logical, as the displacement from his apartments in Milan and the dislocation of his engagement with the Sforza family during these troubled times would probably make it difficult for him to carry on a normal working life. Ox hearts would be easily accessible and allow Leonardo to continue some meaningful work whilst in domestic transition.

Within this period falls the production of his famous rendering of the female body, referred to by some as *The Great Lady* (Fig. 2.6). Close inspection of this drawing reveals an amalgam of observations and ideas. There appears to be ox anatomy, human foetal anatomy, and some adult human anatomy all blended together, giving evidence of the continuing Galenic influence of the cross-referencing of animal anatomy to the human. This drawing was probably produced for some type of formal use, as all of the major lines are pricked through for transfer of the image by the technique of "pouncing."[32]

In his last period of anatomy, during his stay in Rome, Leonardo produced his extraordinary work on the heart and other viscera. Although there are examples of work on the respiratory system in earlier years, some of the most impressive work, in which he discovered the bronchial arteries and correctly defined their function in supplying the bronchi rather than the lung parenchyma, was done in 1513.[33] On this page, he also described the nature of the supporting cartilages in the major airways.[34] He also produced extraordinary work on the muscle of the diaphragm and the intercostal muscles in the same period.[35] In addition he also records detailed work on the larynx, tongue, and palate.[36]

Once Leonardo had left Rome (under what seems to have been difficult circumstances), the extant record and evidence of further anatomy ends. One wonders whether Leonardo may have worked on producing his anatomical treatise whilst the guest of Francis I in Amboise. He certainly had many or all of the manuscripts with him, as recorded by Antonio de Beatis, the secretary to Cardinal Louis of Aragon. By this time, Leonardo was partially paralysed by stroke and close to the end of his life.

The Intellectual Process

It is important to realize that Renaissance academic endeavor did not closely correspond to the methods of the natural sciences of today. Rhetoric and Aristotelian logic were the language of the academics of the day and may be reflected in Leonardo's analytical approach. Some authors have gone further and claim that they find unequivocal evidence of the use of the Renaissance analytical process, the "regressus demonstativus" or "demonstrative regress." This rhetorical analysis utilizes an inference drawn from an observed effect to its proximate (closest, before or after) cause, which is then combined with a further inference about the proximate cause and the observed effect.[37] This analytical method is discussed by Aristotle in *Posterior Analytics*, in which he contrasts two sorts of syllogisms. He gives this example: "Heavenly bodies that are near the earth do not twinkle; the planets are near the earth; hence the planets do not twinkle." In this syllogism, the observed effect is "not twinkling" and the subject is "the planets." The middle term is being "near the earth," which is the proximate cause of the observed

32"Pouncing" is a technique whereby the drawing is dabbed with a muslin or other such cloth filled with carbon or silver dust, which passes through the holes and leaves behind an outline of the "traced" image.

33Kenneth Clark catalogue of Windsor drawings. (RL 19071 recto).

34Ibid. RL 19072 recto.

35Ibid. RL 19108 verso and RL 19109 recto.

36Ibid. RL 19002 recto.

37Zwijnenberg, Robert. *The writings and drawings of Leonardo da Vinci. Order and chaos in early modern thought.* (Cambridge: Cambridge University Press, 1999), 151.

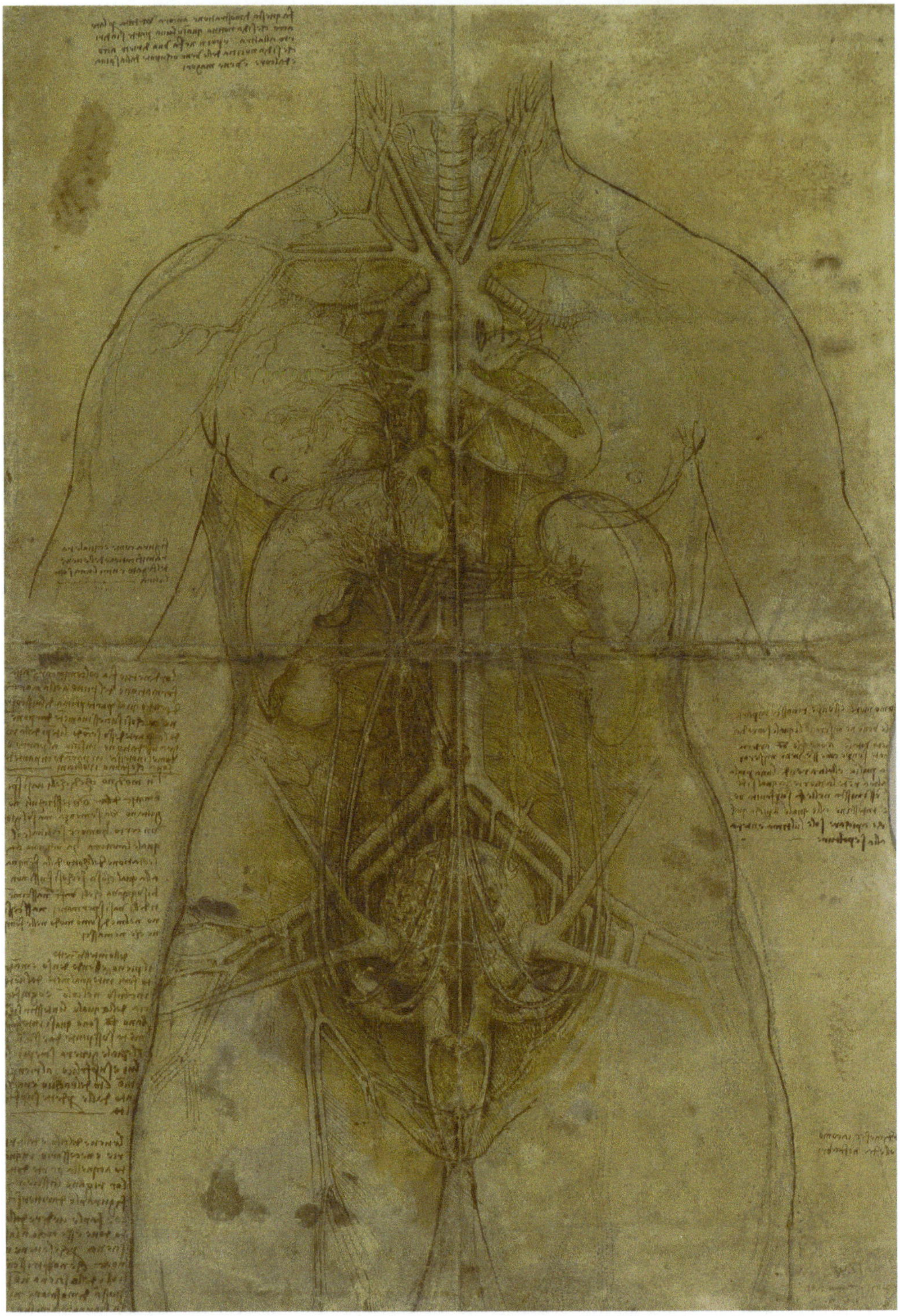

Fig. 2.6 Leonardo's composite drawing of the anatomy of a woman, showing the cardiovascular system and principal organs. RL 12281 recto, Leonardo da Vinci, c.1509–10. *Black* and *red* chalk, ink, *yellow* wash, finely pricked through (Lent by Her Majesty The Queen.Royal Collection © Her Majesty Queen Elizabeth II)

effect. By rearranging the terms, a demonstration of the fact is obtained in which the middle term expresses the effect rather than the cause. Thus, heavenly bodies that do not twinkle are near the earth; the planets do not twinkle; hence the planets are near the earth. When we come to discuss Leonardo's groundbreaking description of the true mechanism of aortic valve closure, this method will become evident to some extent.

Whilst there are signs of the use of these methods in Leonardo's workings on paper, he was clearly most comfortable with his observations of the interaction of proximate cause and effect in the natural world around him. These natural phenomena he adopted as his gold standard. The teleological approach used by Leonardo in his analysis is best seen in his description of the function of the aortic valve. In writing about the flow of blood through the atria and ventricles of the heart, he wrote, "Nothing is superfluous and nothing is lacking in any species of animal or product of nature unless the defect comes from the means which produce it"[38]; and again, "Although human ingenuity makes various inventions, corresponding by means of various machines to the same end, it will never discover any inventions more beautiful, more appropriate or more direct than nature because in her inventions nothing is lacking and nothing is superfluous."[39]

Leonardo based much of his deductive reasoning on this belief that all structural forms were forged by the forces acting upon the structure in question. To him, it was therefore perfectly legitimate to describe the function of something by studying and understanding its form and the forces that shaped it. This argument was by no means new. It was described at length by Aristotle, whose work Leonardo studied, having translations of his writings in his personal library.

Leonardo's Book of Anatomy

> This, my illustration of the human body shall be demonstrated to you, not otherwise than if you had a real man before you.[40]

The Achilles heel for posthumous recognition in all of Leonardo's endeavors was the lack of any publication of his work during or after his lifetime. Leonardo clearly stated his intention to produce a book on anatomy, and indeed refers to the "many books produced by me." The books that he describes must surely reflect his copious manuscript notes, many of which remain available to us. It is possible that more substantive publications did exist, but their existence is very unlikely, as there is no corroborative evidence.

Leonardo's extensive use of notebooks and manuscripts probably arose from his experiences in the workshop of Verrochio. The use of the "common-place book" peaked in the Renaissance. This was a book where passages were collected and often indexed and themed. Tania Kovats described them thus in her opening remarks on a contemporary book on drawing: "They [the common-place books] sanctioned the selection of passages made significant by personal experience; they were personal anthologies, a forest of thinking governed by the objective selection of the compiler and as such represent to some extent what that individual's memory might look like."[41]

Leonardo set out his ambition for his book on anatomy on two separate occasions. On the first occasion, in 1489, he sounds more like the artist and anatomist rather than a scientist. He writes of portraying the conception and birth of a child and continues with descriptions of physiognomy and the emotional states that he wishes to represent. He relates the function of the eye to that of understanding perspective, and the ear to the hearing of music.[42] Twenty years later, he again sets out his aspirations for his book on anatomy and begins with these words: "This my depiction of the human body will be shown to you just as though you had a real man before you. The reason is that if you wish to know thoroughly the parts of an anatomized man, you must either turn him or your eye so as to examine him from different aspects, from below, from above and from the sides, turning the subject around and investigating the origin of each member and in this way you will satisfy yourself as to your knowledge of the actual anatomy."[43]

This description is more revealing of his intended objectives. He writes with the demanding eye of the artist, aspiring to achieve a true and lifelike representation of the human body. Schematic representations will not do. He states that because the process of dissection destroys some parts in the process of revealing others, at least 12 dissections are necessary to appreciate the human form: three for the veins and arteries; three for the ligaments, muscles, and cords; three for the bones and cartilages; and three for the membranes. In addition, he states that another three dissections of a woman are necessary to understand the mystery of the uterus and its foetus. His thorough approach is further underscored by his anger at those who try to take shortcuts to knowledge, and he rails at "abbreviators" of knowledge of any kind. He states that there can be no shortcut to a true appreciation and an understanding of the "marvel of artifice"[44] which is the human body: "The abbreviators of works do injury to knowledge and to love, for love of anything is the offspring of knowledge, love being more fervent in proportion as knowledge is more certain, and this certainty springs from a thorough knowledge of all those parts, which united, compose the whole.... Truly it is impatience, mother of folly, who praises brevity."[45]

The three views that Leonardo speaks of refer to his architectural discipline, but as he progressed through the more complex anatomical forms he realized that these would not be enough and increased the number to six. Indeed, in his representation of the arm he employs an almost cinematographic approach. Superimposition of the drawings into a "flick book" format would almost produce the effect of the

[38]RL 19063 verso.

[39]RL 19115 recto.

[40]RL 19009 recto.

[41]Tania Kovats, ed. *The Drawing Book: A Survey of Drawing: The Primary Means of Expression.* (London: Black Dog Publishing, 2007).

[42]RL 19037 verso.

[43]RL 19061 recto (1509–13).

[44]Kemp, *The Marvellous Works*, 279.

[45]RL 19084 recto.

viewer moving around the arm, or the arm rotating in front of the viewer. Leonardo refers to this effect in this passage: "If you wish to understand all the parts of an anatomised man, you turn him or your eye through all of the various positions…turning him around and searching for the origins of each member."[46]

Hence it can be seen that Leonardo's plans were quite revolutionary when compared with what was available at the time and indeed with what was to later be heralded as the most important contribution to anatomical knowledge and study, namely *De Fabrica de Humana Corporis* by Vesalius. However, the very exhaustive scope of the proposed book almost certainly precluded the possibility of its ever being accomplished. He wrote in 1508:

> Begun at Florence in the house of Piero di Braccio Martelli on the 22nd day of March 1508. This makes a collection without order, taken from many sheets which I have here copied hoping to arrange them later, each in its place according to the matter of which they treat. I believe that before I make an end to this I shall have to repeat the same things many times, for which O reader, do not blame me, for the subjects are many and the memory cannot retain them, and say, "This I do not need to write because I have written it before." If I wished to avoid falling into this error, it would be necessary in every instance when I desired to copy, that I should read over all that had gone before so as not to repeat myself, and especially so since the intervals of time between one writing and the next are long![47]

Thus we find that an order of work is frequently lacking in Leonardo's manuscripts and notebooks. That they were used as a paper laboratory of thought soon becomes very clear. Leonardo worked and reworked ideas incessantly and frequently over many years. Entries can be found on sheets years apart. How he kept track of it all, if he ever did, is a mystery. Perhaps this fact alone suggests that there was more order to it than we can now comprehend, with the separation and loss of material that is so evident. In codex Leicester he wrote, "I will not consider the demonstration here because I will reserve them for the ordered work; my concern now is to find cases and inventions, gathering them as they occur to me; then I shall have them in order, placing those of the same kind together; therefore you will not wonder or laugh at me, Reader, if here I make such great jumps from one subject to the other."

One can find a note of optimism from Leonardo in his late anatomical studies. About 1507–1509, he wrote, "Have your books on anatomy bound,"[48] and later, "This winter of 1510 I believe that I will expedite all this anatomy."[49] Sadly, we have no evidence that this optimism came to fruition. There is no magnum opus from him by which to place him in the history of anatomy. The manuscripts and notebooks that we do have, however, reveal a giant intellect, not in any way daunted by the scale of the problems that presented to him but ready to tackle any and all keenly observed structures and phenomena with vigorous interrogation and abundant lateral thought. For that alone, his notes and deductions bear intense scrutiny through the scientific eyes of today.

[46]RL 19016 recto.

[47]Codex Arundel. B.M. 1r.

[48]RL 19070 verso.

[49]Ibid.

Leonardo's Writing

It is easy to focus on the beauty of Leonardo's drawings, but it is also worth studying his writing style with care. He writes directly and often with beautiful simplicity but also uses imagery and metaphor quite frequently. Irma Richter wrote, "It was not only the beauty of nature but also the spirit at work beneath the world of appearance that fascinated him." This passion for the representation of the truth of the beauty of nature can be heard in this passage on the formation of the child in the womb as the most perfect triumph of nature, something that manual human construction can never achieve:

> Though human ingenuity may make various inventions answering by different machines to the same end, it will never devise an invention more beautiful, more simple, more direct than does Nature; because in her inventions nothing is lacking and nothing is superfluous. She needs no counterpoise when she creates limbs fitted for movement in the bodies of animals, but puts within them the soul of the body which forms them, that is the soul of the mother which first constructs within the womb the shape of man, and in due time awakens the soul that is to be its inhabitant.[50]

His use of metaphor, which translates well into today's world of medicine and surgery, in which new technologies are often introduced before being properly assessed, is beautifully shown in this fable about the butterfly:

> The vain and wandering butterfly, not content with being able to fly at its ease through the air, overcome by the tempting flame of the candle, decided to fly into it; but its sportive impulse was the cause of a sudden fall, for its delicate wings were burnt in the flame. And the hapless butterfly having dropped, all scorched, at the foot of the candlestick, after much lamentation and repentance, dried tears from its swimming eyes, and raising its face exclaimed: O false light! How many must thou have miserably deceived in the past, like me; or if I must indeed see light so near, ought I not to have known the sun from the false glare of dirty tallow?[51]

Much of Leonardo's notes is written with his left hand from right to left and in mirror image. Many reviewers in the past have thought that this was a device to protect the secrecy of his notes, but it is most likely that it was his natural way of writing. Recent study of left-handed people has shown that

[50]RL 19115 recto.

[51]C.A. 67r.

as many as 10 % may be able to write naturally in mirror image. Of course it may have suited him to conceal his thoughts from time to time, but there is no evidence that this was a strategy he adopted. There are sheets where he writes from left to right in a fully legible way. Some of these sheets are clearly for demonstration purposes.

Conclusions

We will probably never know in detail the reasons behind Leonardo's almost lifelong fascination with the structure of the human body. There seems no doubt that it was in the pursuit of the truth about its structure from the point of view of a natural philosopher and not as part of an explanatory process for use in medicine. His opinion of the medical profession can be felt in this passage: "Endeavor to preserve thy health in which thou wilt succeed the better the more thou guardest thyself from the physicians. For their mixtures are a kind of alchemy on which there are no fewer books than there are remedies."[52]

I will leave the last words on this topic to Martin Kemp and Leonardo himself. Kemp writes "To his already extensive range of studies in the late 1480s he added in 1489 his project for a treatise 'On the Human Body'. This was to be not so much an anatomical book as a far-reaching exposition of man's role in the natural order of things."[53]

Leonardo wrote, "The Human race in its marvelous and varied works seems to reveal itself as a second nature in this world."[54]

[52]Part of a paraphrase from *Regimen Sanitatis* of Arnold of Villanova or one of the numerous modifications of that work, such as the tract of Ugo Benzo, Milan, 1481.

[53]*Kemp, The Marvellous Works*, 106.

[54]*B.L.* 151v.

Leonardo's Use of Drawing

3

I am really drunk with intellectual vision whenever I take a pencil or engraver into my hand.

William Blake[1]

The need to draw is deeply rooted in our psyche. The intellectual energy of the mind finds a profoundly important release for human creativity and expression in drawing. The commitment of ideas and mental images to paper is a transition from a wholly personal experience to a form of external self-examination and visual communication. Henri Matisse wrote, "I have always seen drawing not as the exercise of a particular skill, but above all as a means of expression of ultimate feelings and states of mind."

The act of visualizing our thoughts on paper or other media has both constructive and therapeutic elements. It is a pathway of intellectual and emotional expression, which encourages a refinement of ideas. Yet there is often a dissonance between what we see in "the mind's eye" and that which our hand reveals, not always as a result of our drawing and writing deficiencies. We commonly find ourselves frustrated by the manual revelations of our grand thoughts and imaginings. If we have tenacity and determination, the process of refinement of image or sentence begins to define us in a way that is identifiable to outside observers. We develop a "style," which is self-defining. Girolamo Savonarola, the Catholic fanatic and direct contemporary of Leonardo, captured the inescapable truth of this: "Every painter paints himself.... Of course he isn't doing a self-portrait when he is making pictures of lions or horses, or of [other] men and women, yet even so he is painting himself because he is a painter and expresses what is in his mind. However varied his subject, his work bears the stamp of what he is thinking."[2]

Leonardo repeatedly argued that words alone lack the power to convey a clear notion of visible things, but he was also quick to point out how words are superior in conveying the meaning of things of a philosophical and verbal nature. Bernard Berenson was an American art historian who was active in the first half of the twentieth century, the end of the age of connoisseurship. He wrote extensively on the Italian Renaissance and was gifted with a genius for literary description with an extraordinary distillation of words. He described Leonardo's notebook content and style in the following way:

> In his capacity as a draughtsman his activity was prodigious precisely because he was of all artists the farthest removed from being "merely the artist." The simple painter or sculptor makes his sketches, as many as you will, for a work that he has in hand, and then like the rest of us, returns to words. But Leonardo's mind was a universe of which painting and sculpture were scarcely more than outlying continents. His chief interest lay in every form of mechanical and physical science at which a mortal of his day could possibly grasp; and the results of his investigations had to be expressed in words. Now Leonardo, when writing connectedly, was a master of a prose, supple, dignified, even elevated; the prose, in short, of a man of supereminent genius. Nevertheless...he did not believe greatly that words could give a clear notion of visible things. Consequently he rarely wrote without at the same time sketching."[3]

Close scrutiny of the labyrinth of words and images that Leonardo created in his notebooks allows the student a quite exceptionally intimate exchange with their author. Berenson considered that Leonardo's drawing style and execution was subordinate to purpose. They were not meant for aestheticism but for Leonardo's eyes alone, recording the development of his thoughts and observations. When we study his pages of notes, we are sharing with him his transference of thought to paper like no other artist/scientist before or since. Quoting Berenson again, "The quality of qualities, then, in Leonardo's drawing is the feeling it gives of unimpeded, untroubled, unaltered transfer of the object in his vision to the paper, thence to our eye; while, at the same time, this

[1] Letter to his patron William Hayley, 23 October 1804. (Geoffrey Keynes, ed., *The letters of William Blake* [London: Hart-Davis, 1956], § 138.)

[2] This observation is all the more poignant as it emanates from the man who was responsible for the "burning of the vanities" in Florence. Following the expulsion of the Medicis, Savonarola became at once succor to the ordinary Florentines and the scourge of the "godless" upper classes.

[3] Hanna Kiel, *The Bernard Berenson Treasury*. (New York: Simon & Schuster, 1962), 116–8.

F.C. Wells, *The Heart of Leonardo*,
DOI 10.1007/978-1-4471-4531-8_3,

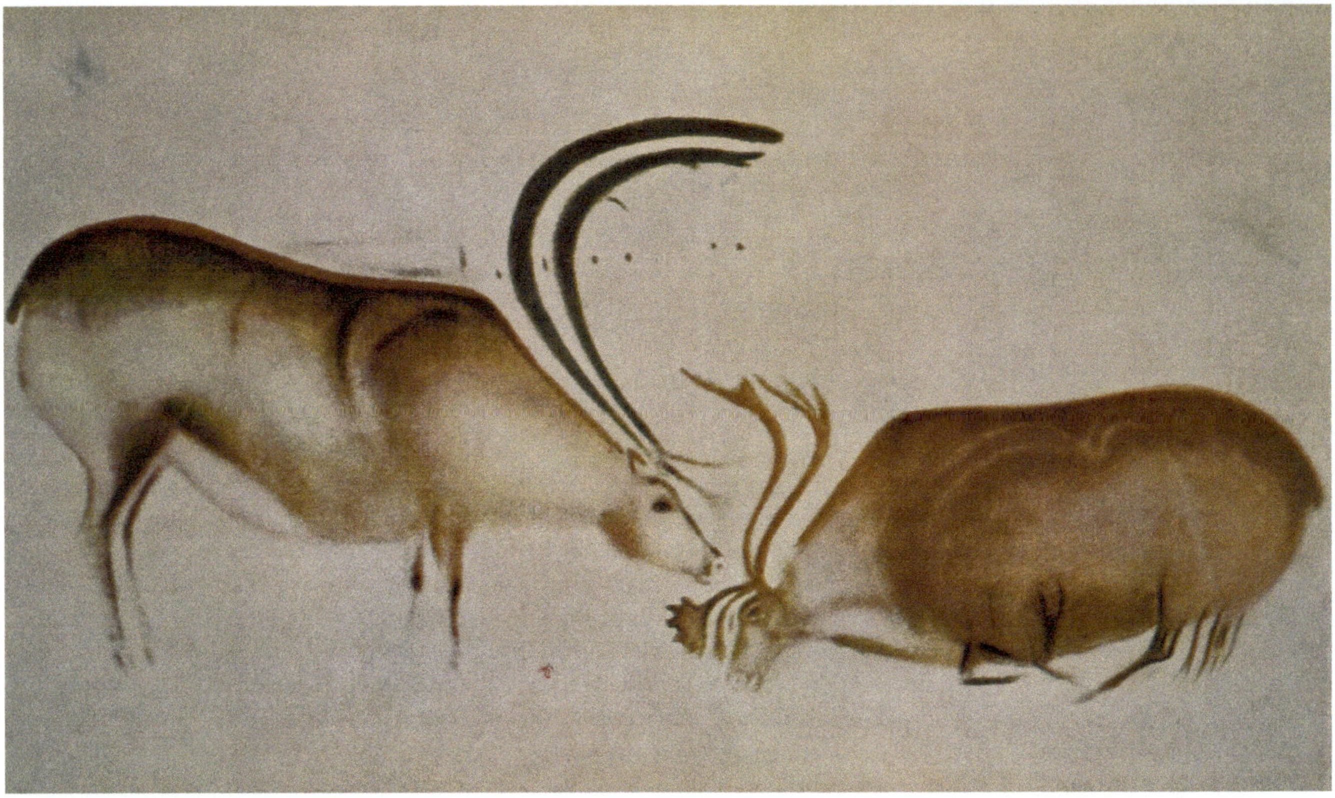

Fig. 3.1 Cave painting. BAL 3703, Male and female deer, Magdalenian school, c.13000 B.C. (cave painting), Paleolithic/Font de Gaume, France/ The Bridgeman Art Library

vision of his has such powers of penetrating, interpreting, even of transfiguring the actual, that no matter how commonplace and indifferent his subject-matter would seem to ourselves, his penetration of it is fascinating and even enchanting…."[4] His spontaneity, precision, readiness, and intimacy of line characterize his work and mature with his age. Thus, in Berenson's words, "…none of his sketches exhibits these qualities better than his late studies of human anatomy".[5]

In this section, I hope to develop these themes in particular relationship to the cardiac studies of Leonardo, which in my view are the apotheosis of his scientific endeavors and are recorded in a style that is the culmination of years of thoughtful recording.

My interest in his heart anatomy caused me to carry out a series of dissections on the heart of an ox. These dissections and the knowledge that I have gained from intimate daily personal experience of the human heart have led me to believe that many of the sheets of Leonardo's notes are visual and literal dissection notes, made at the time of the dissection or soon after.

Our desire to communicate emotionally and intellectually is an original and profound part of the way that we are constituted. Man's earliest known forms of communication involve the figurative use of line. From the French caves of Lascaux and Pindral in northern Spain, to the Egyptian hieroglyph, the means of nonverbal communication was drawing, rather than writing as we would think of it today (Fig. 3.1). It is a natural extension of the visualization of emotions, thoughts, and ideas therefore for them to be recorded by hand in a variety of media as drawings of one form or another. Anthony Gormley wrote of drawing, "For me drawing is a form of thinking. But it is also about medium: using the intrinsic qualities of substances and liquids: a kind of oracular (mysterious or ambiguous) process that requires tuning in to the behaviour of substances as much as to the behaviour of the unconscious, like reading images in tea leaves, trying to make a map of a path of feeling, a trajectory of thought."[6]

This theme of a "trajectory of thought" was explored in some depth by the artist Joseph Beuys in his drawings stimulated by Leonardo's Madrid Codex. Kemp, in an essay included in the Dia publication of this work, wrote, "Leonardo

[4] Ibid., 117.

[5] Ibid., 117.

[6] Tania Kovats, ed, *The Drawing Book: A Survey of Drawing: The Primary Means of Expression.* (London: Black Dog Publishing, 2007), 60.

achieved what Beuys called 'materialized thought' through a brainstorm style of rough sketching, in which forms emerge from a maelstrom of lines. The power of invention to which this drawing style gave vent was prized by Leonardo and stood at the very center of his enterprise to remake nature in his art".[7] For Beuys, the use of words as an art form was as natural as the act of drawing. There is an element of this combination in the pages of Leonardo's notebooks, where words and images blend with each other.

An appreciation of the various uses of drawing in the context of the visualization of thought is important when studying the drawings of Leonardo da Vinci. Here was a draughtsman with divine skills, who produced a line of beauty almost every time his drawing implement touched paper. It is therefore tempting to make the mistake of thinking that each and every drawing by someone who is perhaps known preeminently as an artist is a representation of exactly what he saw. In the context of the anatomical drawings, this is not the case. In this chapter, I hope to demonstrate some of the many ways that drawing may be used as an act of communication, in an attempt to help the observer understand the types of drawings that can be seen within Leonardo's anatomical notebooks. This is important if one is to attempt to explain some of the criticisms of the accuracy of his work. It is also important to recognize that a considerable proportion of the heart drawings are from the ox; perhaps two, if any, are of human origin. Indeed, one very rough sketch may be taken from a small mammal such as a rabbit or a mouse.[8]

Drawing is a highly intellectual process and for best effect requires a considerable degree of insight and higher cerebral function. One only has to watch a professional artist at work to realize the enormous amount of thoughtful observation that goes into each stroke of the pen. Several examples of this process have been captured on film. One of the best is of Matisse at work producing the extraordinarily beautiful line drawings that are so well known from the latter part of his career. He can be seen in front of the model, pen in hand, observing intensely for considerable periods before making a finished line.

Anyone who has practiced the art of drawing will know that a deep appreciation of the subject is essential if a line true to the object can be made without several alterations. Careful observation of so many of the anatomical drawings of Leonardo, many of which are only sketches, reveals the accuracy of his use of line. Some of his earlier work is done using silver point, a technique that allows no rubbing out and requires absolute accuracy of line to produce a satisfactory effect. There is no doubt that Leonardo was the master of this medium, as with all others. In other preparatory drawings for artistic projects, there are multiple lines around the finished object, allowing the viewer to see the way his thoughts drew him to the desired finished image. These "pentimenti" are described in more detail later.

In addition, his mastery of the medium can be witnessed by close examination of the perfect gradation of shading used to clarify very complex forms within the heart. Leonardo's rendering of the opened right ventricle, as shown in RL 19119 verso, allow an immense clarity of depth perspective that is only realized when studied with a hand lens. Often it is as simple a maneuver as a rotation of the wrist or fingers to change the thickness or weight of the line, as seen in the detail in RL 19117 verso (top right), which shows the individual drawing of the aortic valve opening.

Therefore, in addition to a complete mastery of the form in the mind's eye, the neuromuscular control of the drawing hand needs much practice to give the correct and touch required to produce the finished line, giving vivacity through variations of thickness and weight. This is the skill honed with hours of practice, which in harmony with natural talent produces the work of the master. In this respect, the process may be likened to the technical skill of the musician, the sportsman, or (dare it be said?) the surgeon, to name but a few examples. The result may be practical or beautiful—or in the case of Leonardo, both practical *and* beautiful.

Leonardo used a variety of media for his drawing. As already mentioned, some of the earliest anatomical works are in a thin and tentative silverpoint style on blue prepared paper. This exceedingly demanding form of drawing leaves every stroke of the metal tracer bound in the prepared surface of the paper; it cannot be erased or reworked. His study of hands (Fig. 3.2) is one of his most beautiful creations in silverpoint. It is very close in imagery to Verrochio's terracotta bust, *Lady with a Bunch of Flowers*, and may be related to work in Verrocchio's studio.[9] Most of the anatomical heart illustrations are in pen and ink with black chalk. Most are done on blue paper, but he seems to have switched to cream paper during his last phase of heart drawings, probably completed in Rome.

The definitive archaeology of these drawings remains to be completed and we depend upon the scholarship of Kenneth Clark, Kenneth Keele, Carlo Pedretti, and Martin Kemp for our current knowledge.

The Types of Drawing

Drawing underpins so many of our activities. In almost every walk of life, drawing has an important function. This very natural act extends from such everyday things as communi-

[7] Joseph Beuys, *Drawings After the Codices Madrid of Leonardo da Vinci*. (New York: Dia Art Foundation, 1999), 35.

[8] RL 19097 recto. Man B.

[9] Andrea del Verrocchio, *Lady with a bunch of flowers*, marble, Museo Nazionale del Bargello (Inv. No. 472), Florence.

Fig. 3.2 Silverpoint heightened with white, on pink prepared paper; probably before 1490. RL 12558, A study of a woman's hands. Leonardo da Vinci c.1490. Metalpoint with white heightening over charcoal on pale buff prepared paper (Lent by Her Majesty The Queen. Royal Collection © Her Majesty Queen Elizabeth II)

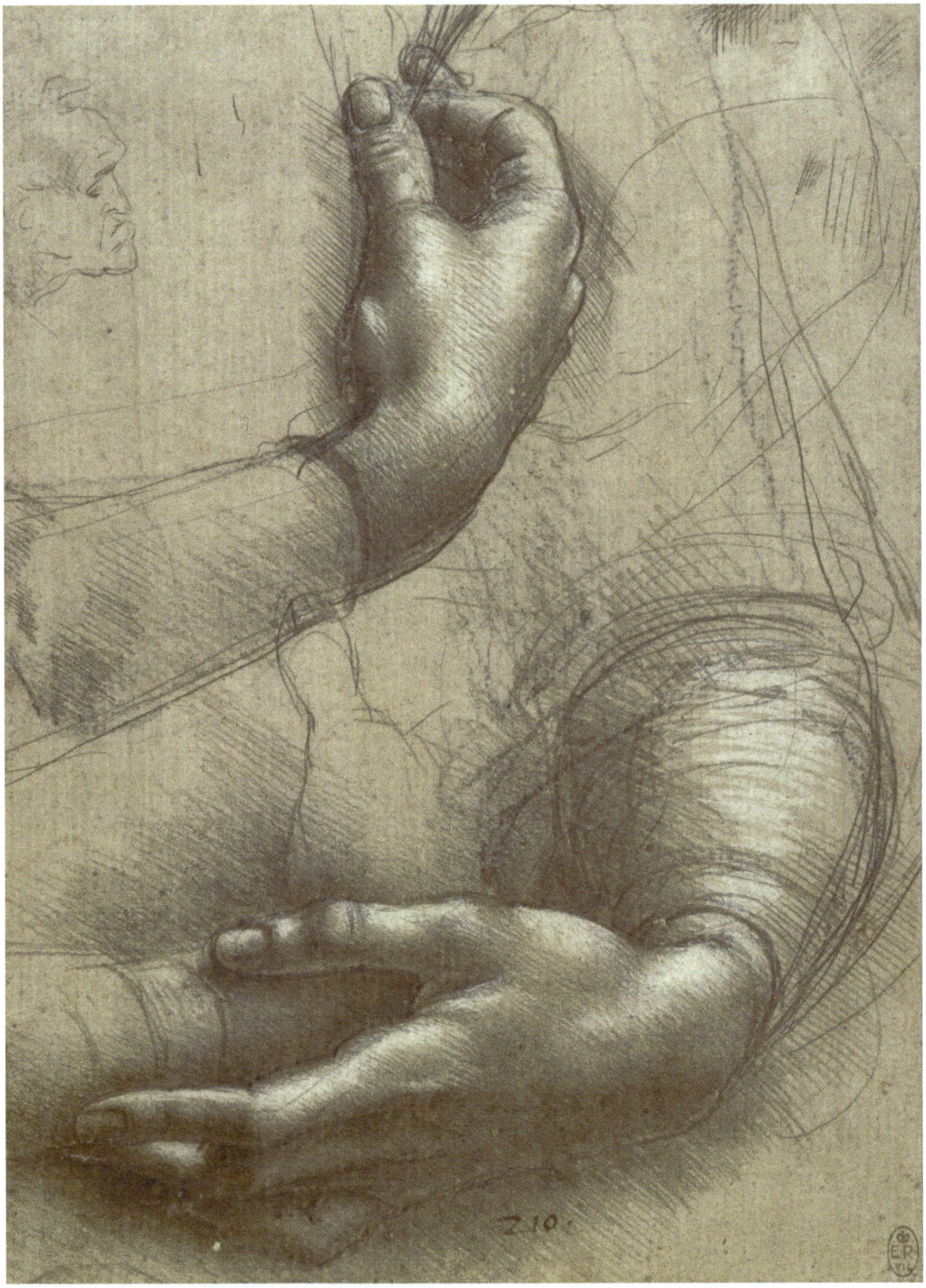

cating travel directions to a friend to the setting down of complex architectural drawings and engineering plans; from the aesthetic beauty of an Old Master drawing to the simple, mindless doodle that we all engage in from time to time. Between these extremes occur a myriad of uses. This part of the chapter reviews with examples some of those used in Leonardo's anatomical notebooks.

But first consider Leonardo's invention for generating ideas simply by studying stains on a wall. He explained how painters can find new *invenzioni* (ideas) from allowing the mind to drift whilst studying stains. He wrote:

> Way to augment and stimulate the mind towards various discoveries. I shall not fail to include among these precepts a new discovery, an aid to reflection, which, although it seems a small thing and almost laughable, nevertheless is very useful in stimulating the mind to various '*invenzionis.*' This is: look at walls splashed with a number of stains or various stones of mixed colours. If you have to invent some scene, you can see the resemblances to a number of landscapes, adorned in various ways with mountains, rivers, rocks,

> trees, great plains, valleys and hills. Moreover, you can see various battles, and rapid actions of figures, strange expressions on faces, costumes, and an infinite number of things, which you can reduce to good integrated form. This happens thus on walls and vari-coloured stones, as in the sound of bells, in those pealing you can find every name and word you can imagine. Do not despise my opinion, when I remind you that it should not be hard for you to stop sometimes and look into stains on walls or the ashes of a fire, or the clouds, or mud, or like things in which if you consider them well you will find marvelous new ideas. The mind of the painter is stimulated to new *invenzioni*, the composition of battles of animals and men, various compositions of landscapes and monstrous things such as devils and similar creations, which may bring you honour, because the mind is stimulated to new *invenzioni* by obscure things.

Although others had described this practice before Leonardo, his characterization of these mental perambulations as being positive and creative is new. The first person that is known to have written about it, albeit in a noninventive way, was Aristotle in *De Somniis*. He wrote "…that we are easily deceived respecting the operation of sense perception when we are excited by the emotions…. This is the reason, too, why persons in the delirium of fever sometimes think that they see animals on their chamber walls, an illusion arising from the faint resemblance to animals of the markings thereon when put together in patterns."

Leon Battista Alberti, whilst not extolling the process as desirable, wrote of it as an explanation for the imitation of natural things by artists. He wrote, "I believe that the Arts of those who attempt to create images and likenesses from bodies produced by nature, originated in the following way. They probably occasionally observed in a tree-trunk or clod of earth and other similar inanimate objects certain outlines in which with slight alterations, something very similar to the real faces in nature was represented. They began, therefore by diligently observing and studying such things to try to see whether they could not add, take away, or otherwise supply whatever seemed lacking to effect and complete the true likeness."[10] It seems fair then to assume that Leonardo was the first to record his thoughts of this method as a constructive and creative process.

A distinctive aspect of Leonardo's drawing style is his regular acceptance or even use of allowing inaccurate lines or "mistakes" to remain on the page. This practice is referred to as *pentimenti*. Instead of erasing imprecise lines, Leonardo left them; the true shape of his creation was to appear within the boundaries of several attempts. This appearance almost gives another life to the drawing, which is lost within the perfection of the perfect line (Fig. 3.3). These pentimenti can be found with infrared spectrographic and x-ray examination of his paintings and is used as a way of establishing the authenticity of a work by Leonardo.

[10] Alberti-Grayson, 1972, 120. De Picture, § 28, also contains a reference to Pliny's description of images made by nature.

Interestingly, pentimenti are not really a feature of his anatomical drawings of the heart, suggesting that most of them are quick representations of ideas and that the image that we see is not as important as the record of the thought. The more finished drawings may have been intended to be seen publicly; they were prepared for transfer to copper plate, Leonardo's preferred method for printing, as it gave a much more refined image. In these drawings, the lines of inquisitive thought have been removed.

To get the most from looking at Leonardo's anatomical manuscripts, it is important to think about the uses that Leonardo makes of drawn representation. Coming from the hand of Leonardo, perhaps the greatest draughtsman of all time, it is easy to understand that the new observer will expect each and every drawing to be a finished masterpiece. What is found, however, is in some ways more interesting. The observer can see Leonardo's various uses of drawing. Drawing is visual thinking as well as the representation of truth and beauty. This range of uses is very well summarized in Leonardo's techniques, which we can group into five categories:

Thinking through drawing
Compilation and summarized memory
Schematic and diagrammatic drawing
Representational drawing
Four-dimensional drawing implying movement

Thinking Through Drawing

Drawing as part of "thought" is something that most of us do at some time, if only to find ourselves "doodling the time away." As our minds are far away from the moment, we find ourselves producing images partly as aesthetic relaxation and partly as if in a trance. The physical act of drawing is linked to the deeper recesses of our minds. An example of a doodle amongst Leonardo's cardiac anatomical notes is the drawing of a beautiful young man with curls in his hair, within the outline of whose chest is a drawing of a section of the heart (Fig. 3.4). This profile of the young man is not dissimilar to Verrocchio's bronze sculpture of David or the outwardly looking young man found at the extreme right side of Leonardo's unfinished painting of the *Adoration of the Magi*. Perhaps it is a self-portrait, perhaps it is his wayward helper Salai, or perhaps it is just a fantasy. Is the preexisting drawing of the heart subliminally involved in the fantasy also? (Be still, my beating heart!). It is clear that the chronology of the drawing of the young man is separate from the sectional drawing of the heart enclosed within it. We may make this assumption, as the surface anatomy of the position of the heart is completely wrong for that of the rest of the drawing. Perhaps the doodle was made around the preexisting heart drawing, as it is likely that the drawings were made at different times, perhaps separated by some years.

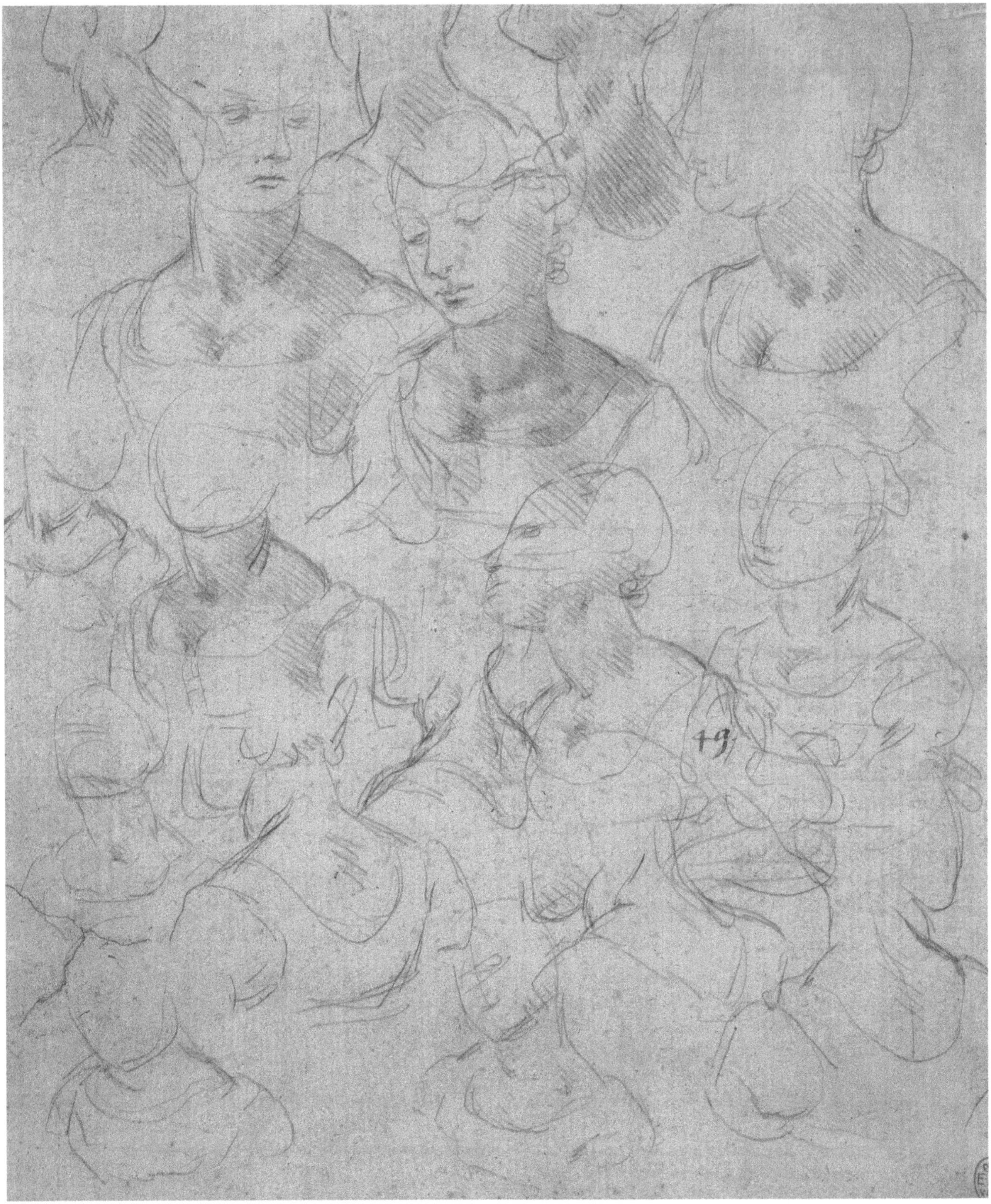

Fig. 3.3 A further example of pentimenti. RL 12513, Sketches of the bust of a woman. Leonardo da Vinci, c.1490. Metalpoint on pale buff prepared paper (Lent by Her Majesty The Queen. Royal Collection © Her Majesty Queen Elizabeth II)

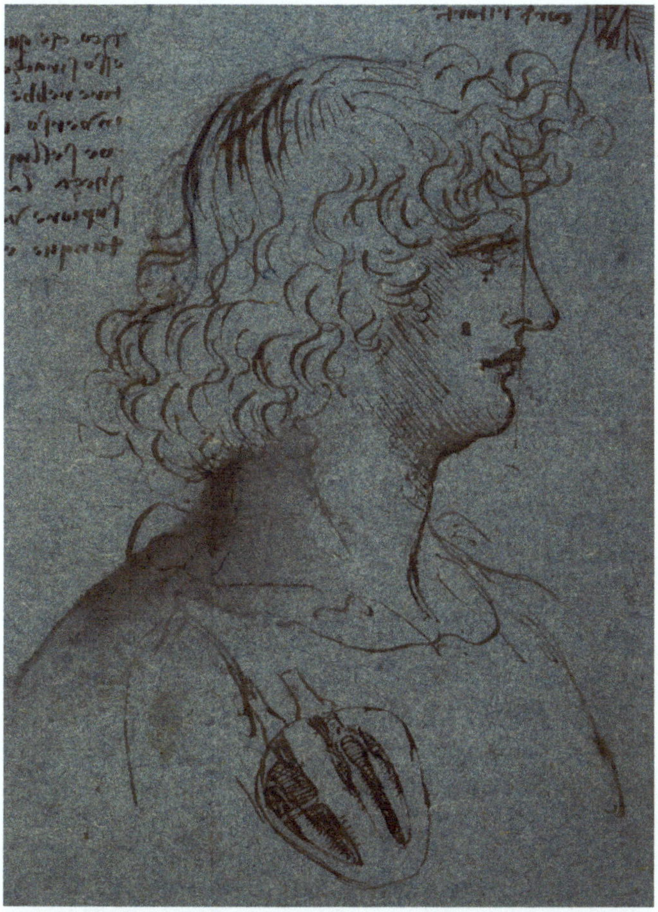

Fig. 3.4 This drawing of a young man enclosing the drawing of the heart may represent a likeness of Salai, Leonardo's young servant. Detail from RL 19093 recto. Studies of the heart, and the head of a youth in profile. Leonardo da Vinci, c. 1511–3. Pen and ink on blue paper (Lent by Her Majesty The Queen. Royal Collection © Her Majesty Queen Elizabeth II)

Other more important drawings illustrating thinking on paper are to be found in his working out of the mechanism of closure of the aortic valve. Here we can see analytical thought in process (See RL 19117 verso). These will be discussed at length in the next section.

Compilation and Summarized Memory

The well-known "Great Lady" (Fig. 2.6) is an example of a compilation drawing. It seems clear to me that this is a summary statement by Leonardo of his anatomical understanding of the interior of the body of a woman. Perhaps it was produced by studying separate drawings that he had made of individual structures and systems, or perhaps it represents his summarized memories in a single drawing. However it was produced, it is interesting in many ways.

This drawing contains several examples of style and substance and is worth dwelling on for a moment. Of significance is the fact that it demonstrates that at some stage of his development as an anatomist, Leonardo was of the common opinion (which is Galenic in influence) that the interior of all animals is similar and that man was no exception. The real fascination for anatomists of the day was the mystery of the uterus and the development of the foetus that was to be found therein.

In this drawing, a mixture of species and ages is represented. The vascular system and uterus are those of an ox. The arrangement of the vessels in the neck and in the abdomen is not human, but belongs to a quadruped. The relationships of the major abdominal organs (liver and spleen), however, are more humanoid. Also, the representational styles of each organ are different: The heart is in cross-section, but the thyroid gland and the liver and lungs are semi-translucent. Finally, the blood vessels swooping from the pelvic vessels (the iliac arteries) around and up to the umbilicus (in this case set off-centre) are those of a foetus.

Whether this drawing was produced from memory or compiled from his previous work we cannot know, but it does raise the interesting point of the integrative power of the brain when reassembling memorized facts in this way. Memory is often used as a test of learning and understanding. The use of spatial thinking in the arrangement of memories gives powerful reinforcement.

Schematic and Diagrammatic Drawing

Leonardo uses schematic diagrams frequently in this series of work. His very Galenic sketch of the heart and the other major thoracic and abdominal viscera (RL 19112 recto) is clearly a schematic drawing, perhaps from memory. A very similar representation is to be found in his earliest known anatomical illustration of the vascular tree superimposed upon a drawing of a nude man (RL 12597). Also in RL 19074 recto, there is a very faint drawing of a system of pulleys in the centre of the page. This is part of Leonardo's examination and explanation of the engineering mechanism involved in the opening and closure of the mitral and tricuspid valves. Here he has reduced his anatomical observations into a mechanical drawing that any secondary-school physics student would be familiar with.

An illustrative tool that Leonardo used to great effect in his anatomy (particularly the musculoskeletal drawings) is the exploded diagram. Today the exploded diagram is commonplace, but it was more unusual in his time. Certainly he and others had used it in architecture and mechanical drawing, but his use of it in anatomy appears to be completely novel. His use of this illusory technique, allowing a fuller examination of the relationship of one part to another, is an example of diagrammatic representation at its best (Fig. 3.5). Perhaps Leonardo learned the technique from a friend, Francesco di Giorgio Martini, who had learned it from his teacher Taccola.

A clue to the possible influence of contemporaries on Leonardo can be gained from the surviving work of Di Giorgio, particularly his *Trattato di Architettura*, which he probably gave to Leonardo in 1490 in Milan. This volume has survived, and in it can be seen Leonardo's notes on the sides of the pages. The book was written in 1480, whilst Francesco was at the court of Federico da Montefeltro in Urbino. Francesco is remembered for his stylized ideal city. His designs for fortress battlements and other features are very similar to those found in Leonardo's notebooks.

Representational Drawing

Leonardo's representational drawing is to be found in some of the anatomical drawings that he may have thought were ready for publication. These are to be found particularly in the musculoskeletal work (Fig. 3.6). Here there is a finish not seen in many of the others. They are aesthetically so pleasing that one can understand a reluctance on Leonardo's part to have their refinement reduced by conversion to woodcut, which was the common mode of transferring illustration in that early time of printing. Just as with every other area of his work, he became engrossed in perfecting the tools for execution of the project as much as in the project itself. An example is the design and building of the scaffolding for both the *Last Supper* in Milan and the *Battle of Anghiari* in Florence. The frustration caused by his painfully slow progress was commented upon at the time. The pride that Leonardo had in his prowess as a skilled draughtsman meant that many of his drawings were much more than mere illustrations. One can hear his salutary words echoing in our ears: "And though you have a love for such things, you will perhaps be impeded by your stomach…or perhaps the lack of draughtsmanship which appertains to such representation and even if you have the skill in drawing, it may not be accompanied by a knowledge of perspective or geometry.".[11] Perhaps Leonardo became so preoccupied with trying to find ways of transferring the detail of line in his drawings that it interfered with the possibility of their publication.

Four-Dimensional Drawing Implying Movement

Drawing in four dimensions is of course not possible without an artificial construct, such as the flipbooks that we see today mainly as amusements for children. However, if we read Leonardo's intention to present his anatomy in such a way as "to have the very person in front of the viewer" and we look at the sheets RL 19008 verso (Fig. 3.6) and RL 19005 verso, it is possible to imagine the arm there represented, turning in front of you. This movement through time represents a fourth dimension not remotely available in any other anatomical demonstration until the present time. There is something of this quality in the drawings of the vortex formation in the aortic valve sinuses and the opening and closing of the aortic/pulmonary valves. Hence his demand for many views of the same part of anatomy can be seen almost in a cinematographic sense.

It is clear to me that Leonardo derived amusement from some of his personal adventures on paper. A classic example is the pictogram, of which he completed a number (see RL 12692 recto). This child's game on paper has been used by generations and continues today. On a page of drawings of the aortic valve, he places two oblique strikes through the letter "Q" in one part of the notes as if mimicking the bileaflet aortic valve on another part of the page.

Leonardo's work contains a lot of lateral thinking. His architectural experience carries over into his anatomical work. As discussed earlier, the Vitruvian concept is a classic example. Again, an influence here was almost certainly Francesco di Giorgio Martini, as well as Alberti. Leonardo's use of cross sections and proportion theory also calls heavily upon his architectural practice and experience.

Leonardo's drawing of the mitral valve of an ox in RL 19080 recto, is reminiscent of the nave of a church. This very organic form can be seen reproduced in the nave of Gaudi's La Sagrada Familia in Barcelona. Gaudi's work is very organic, and his approach to the influence of nature on man is similar to that of Leonardo (Fig. 3.7).

Finally, there are a number of sketches amongst Leonardo's pages that are not so easy to work out. One amongst them is a small drawing that represents the interaction of the cords with the leaflets of the mitral valve and as such is another example of Leonardo's thinking on paper. This drawing is found on manuscript RL 19087 recto at the top of the page. With this drawing, Leonardo seems to be imagining the change in position of the chordae tendinae with the opening and closing of the mitral valve. It is likely that the two components of the drawing represent his perception of the position of the supporting cords of the valve in the two phases of the cardiac cycle, systole (the contracting phase) and diastole (the phase of relaxation).

Drawing with the Left Hand

Leonardo's use of his left hand to write and draw is well known. Much has been made of it, with some suggesting that it was a device to codify his ideas. Whilst he may have done this occasionally, it is much more likely that left-handedness

[11] RL 19070 verso.

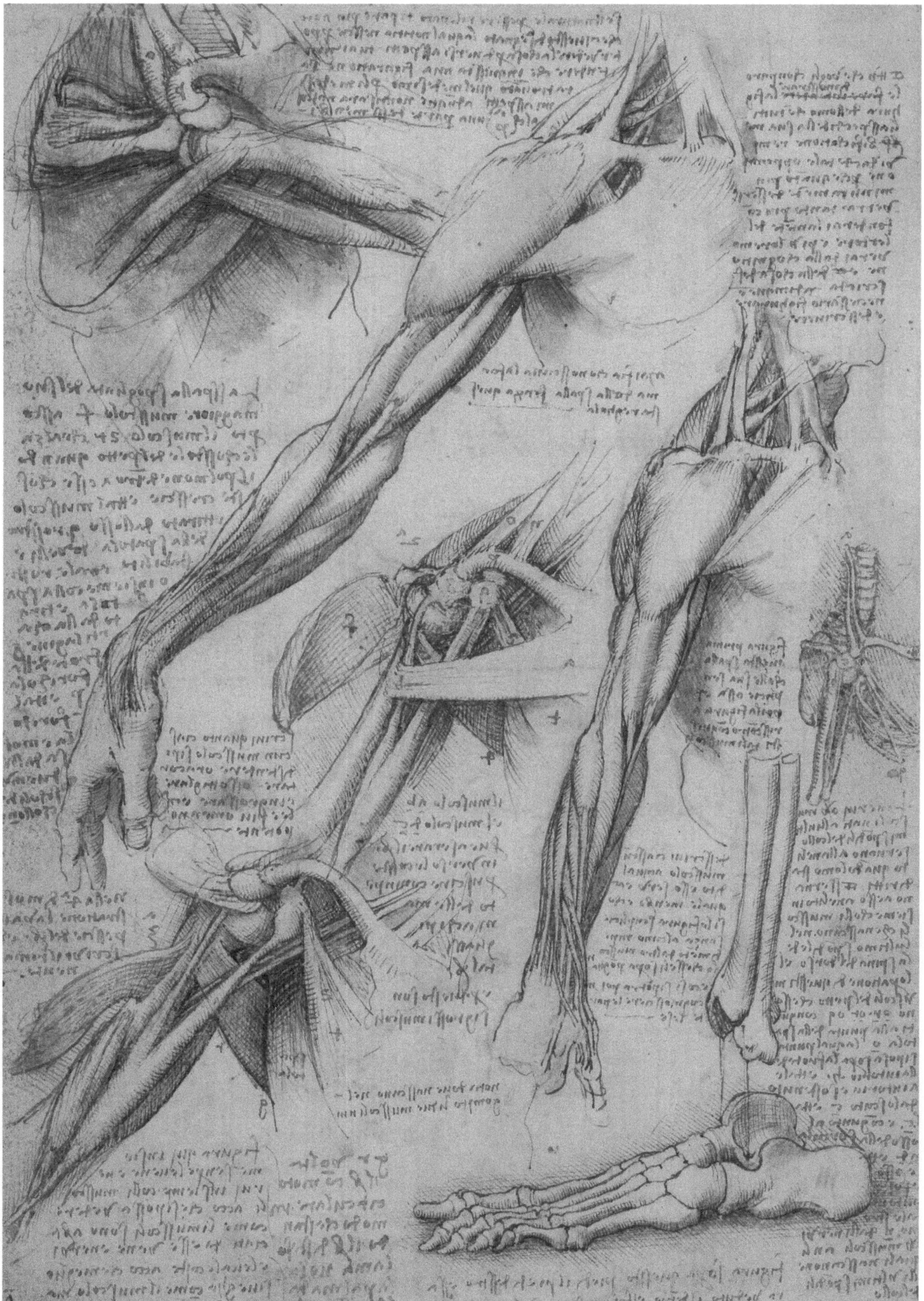

Fig. 3.5 An example of an exploded diagram used so effectively by Leonardo. RL 19013 verso, The muscles of the shoulder and arm, and the bones of the foot. Leonardo da Vinci, c.1510–1. Pen and ink with wash, over black chalk (Lent by Her Majesty The Queen. Royal Collection © Her Majesty Queen Elizabeth II)

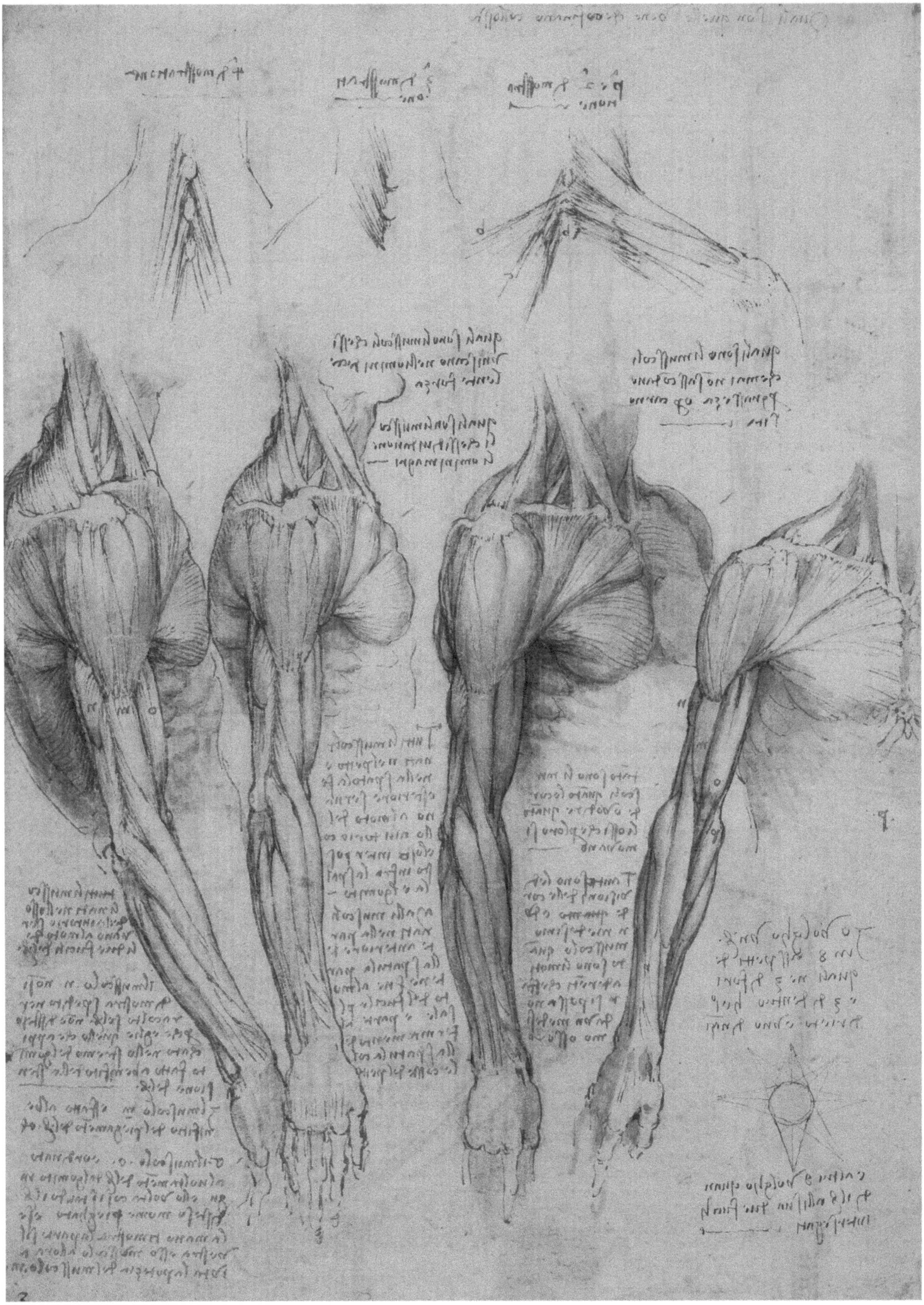

Fig. 3.6 Fine examples of (near) finished drawings, perhaps ready for publication. RL 19008 verso, The muscles of the shoulder, arm and neck. Leonardo da Vinci, c.1510–1. Pen and ink with wash over black chalk (Lent by Her Majesty The Queen. Royal Collection © Her Majesty Queen Elizabeth II)

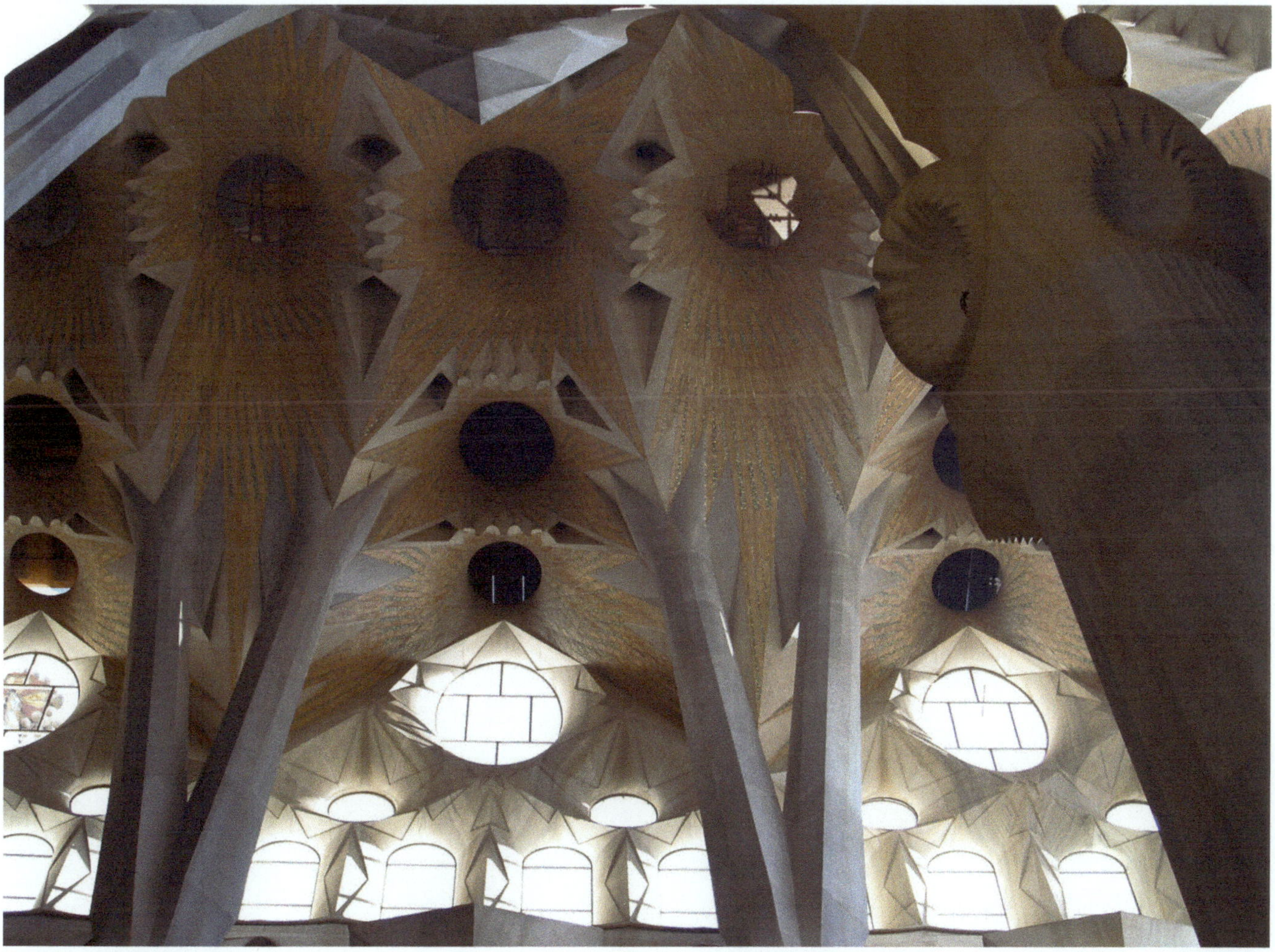

Fig. 3.7 The roof buttresses of La Sagrada Familia, Barcelona, designed by Antoni Gaudi. These wonderful organic shapes mirror the underside of the mitral valve

was his natural form. Left-handed people have to hold the pen in an awkward position to be able to draw it across the page from left to right, the natural thing for right-handed people. To write from right to left with the left hand reproduces this ease of movement and works with the nib or quill of the pen. The mirror-image component of his writing can be found naturally in about 10 % of left-handers and again is therefore not that unusual even today. The use of his left hand can be also found in the shading of his drawings, which pass from top right to bottom left. However, in works prepared for demonstration, Leonardo can be found to be writing and drawing with his right hand.

Conclusions

In critical viewing of Leonardo's anatomical drawings, it is important to keep in mind the considerations discussed in this chapter. Most are not finished drawings like those found in the *de Fabrica* of Vesalius or even the works of Mundinus. The extant pages of Leonardo's anatomical notes are a mixture of work in progress and developing ideas, with perhaps some completed work. Both animal and human anatomy is present, and most importantly, it is likely that a significant amount of the work has been lost. What we have left is simply a collection of pages from notebooks that have been disassembled. They represent work carried out over at least three decades and added to in different periods. They contain ad hoc notes as well as high-quality drawings, with occasional household scribbles. Not only do these pages tell us something of the breadth and depth of work that Leonardo had completed and the knowledge that he had derived from it, but they also shed a chink of light on his daily life. The humanity to be found in them is informative and sometimes amusing and moving.

The reader will likely find other examples of Leonardo's many uses of drawing, but it is my hope that this general discussion will open the eyes of those coming anew to these wonderful visual demonstrations.

Leonardo Da Vinci Drawings

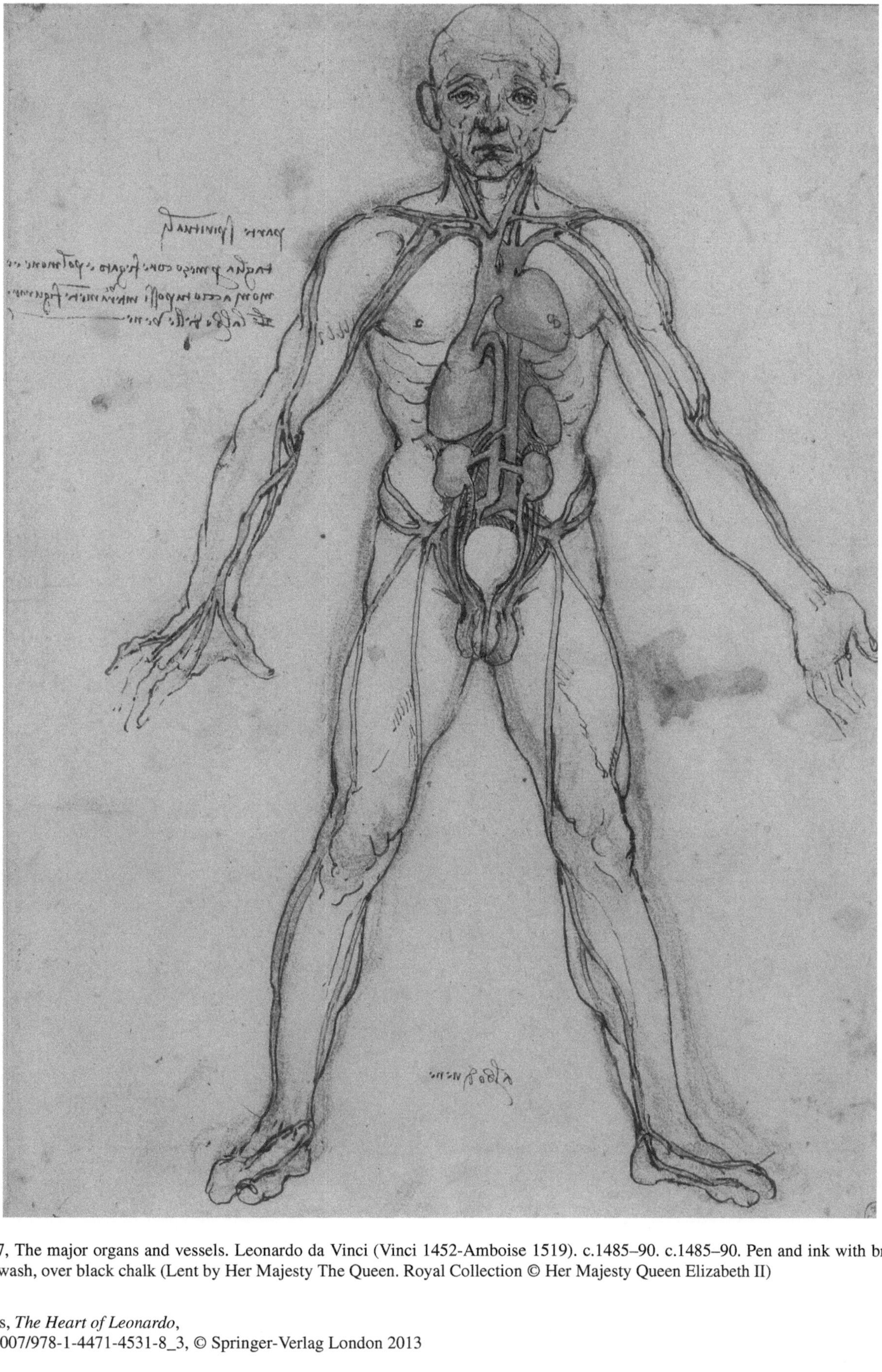

RL 12597, The major organs and vessels. Leonardo da Vinci (Vinci 1452-Amboise 1519). c.1485–90. c.1485–90. Pen and ink with brown and greenish wash, over black chalk (Lent by Her Majesty The Queen. Royal Collection © Her Majesty Queen Elizabeth II)

F.C. Wells, *The Heart of Leonardo*,
DOI 10.1007/978-1-4471-4531-8_3, © Springer-Verlag London 2013

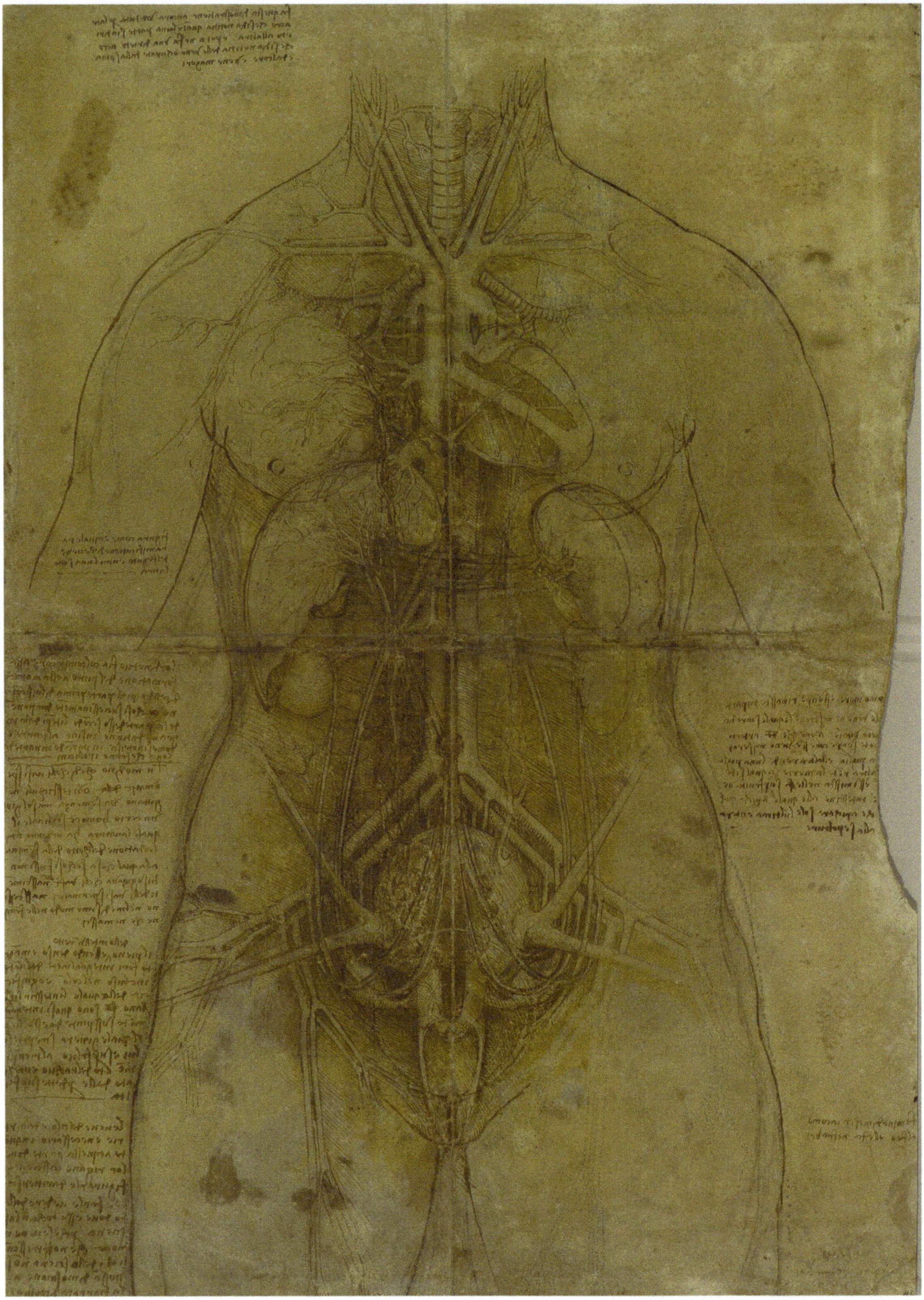

RL 12281 recto, The cardiovascular system and principal organs of a woman. Leonardo da Vinci (Vinci 1452-Amboise 1519). c.1509– 10. Black and red chalk, ink, yellow wash, finely pricked through (Lent by Her Majesty The Queen. Royal Collection © Her Majesty Queen Elizabeth II)

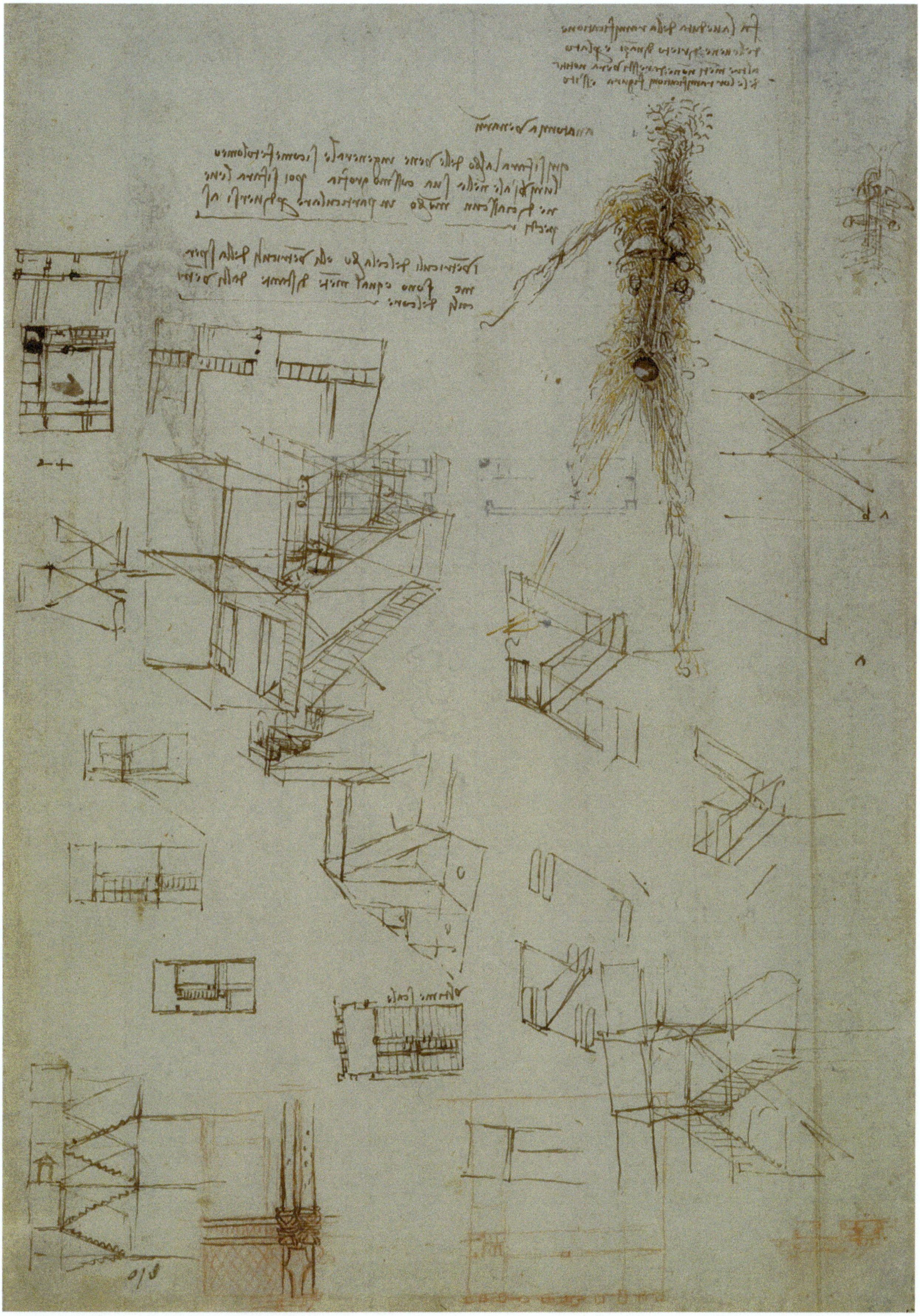

RL12592 recto, Studies of the vessels of the body, and of staircases. Leonardo da Vinci (Vinci 1452-Amboise 1519). c.1506–10. Pen and ink and red chalk (Lent by Her Majesty The Queen. Royal Collection © Her Majesty Queen Elizabeth II)

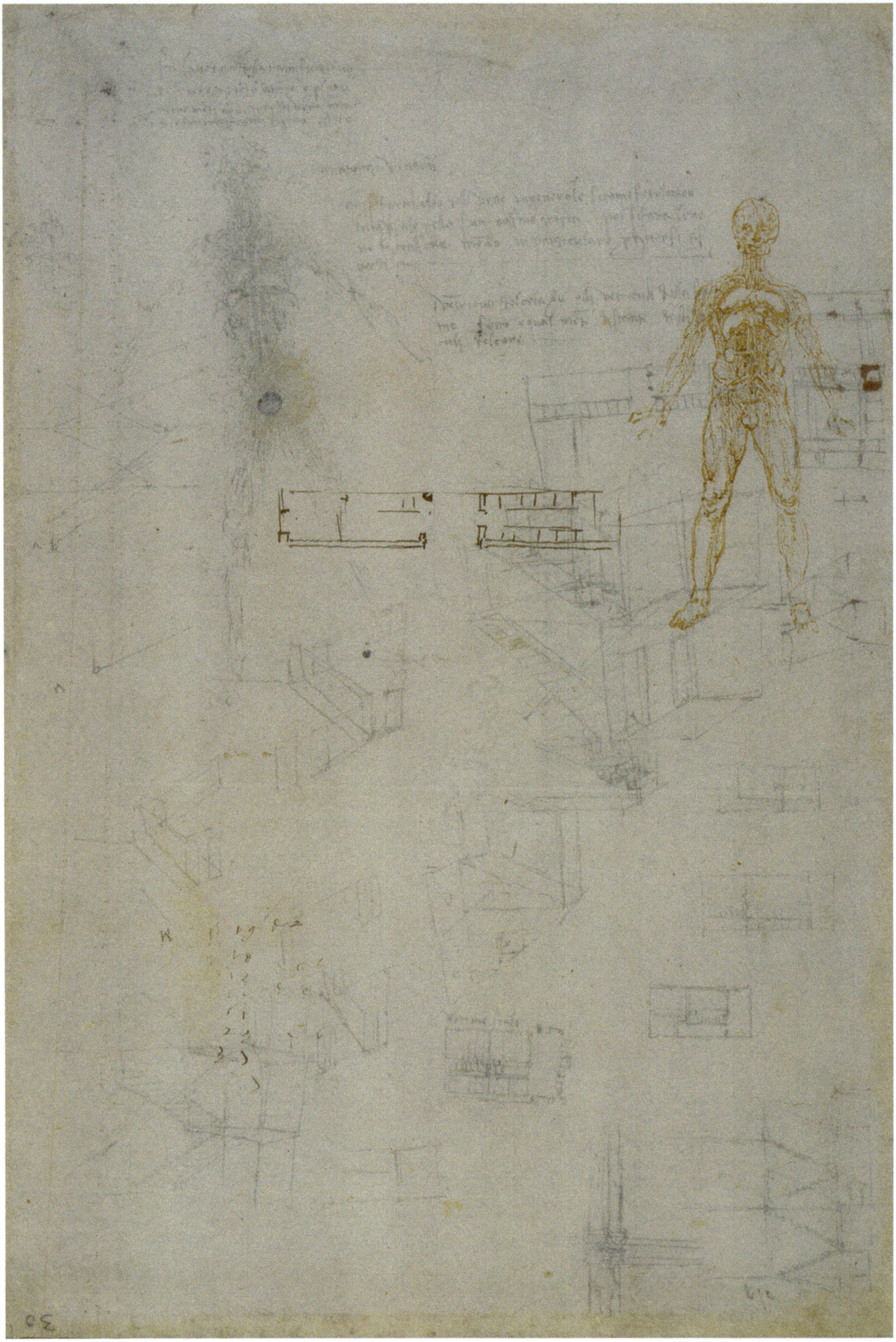

RL 12592 verso, The organs and vessels of the body. Leonardo da Vinci (Vinci 1452-Amboise 1519). c.1506–10. Pen and ink (Lent by Her Majesty The Queen. Royal Collection © Her Majesty Queen Elizabeth II)

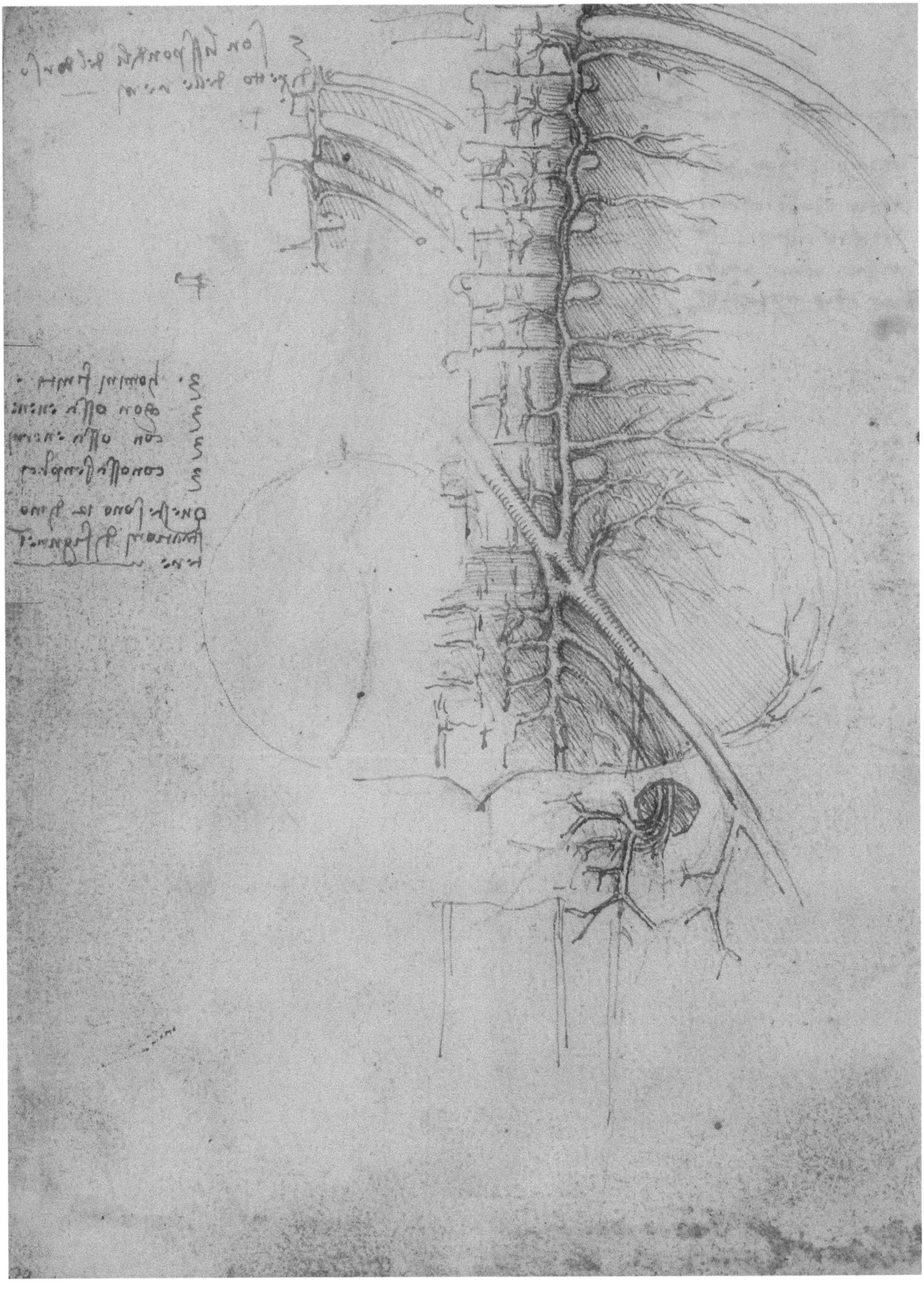

RL 19023 verso, The veins of the pelvic and lumbar region. Leonardo da Vinci (Vinci 1452-Amboise 1519). c.1508. Pen and ink over black chalk (Lent by Her Majesty The Queen. Royal Collection © Her Majesty Queen Elizabeth II)

19023v (B 6 v). Pen and brown ink (two shades) over traces of black chalk.	
[Fig. 1, 2]	[Fig. 1, 2]
[I] 5 son lisspondjli del dorso | dj retto delle renj –	[I] There are five vertebrae of the back directly in the loins.
[II] 3 · homjni finjtj	[II] three men finished
3 chon ossa euene	three with bones and veins
3 con ossa eneruj	three with bones and nerves
3 con ossa senplicj	three with bones alone
Queste sono 12 djmo|strationj dj figure ĩ|tere –	These are 12 demonstrations of the whole [human] figure.

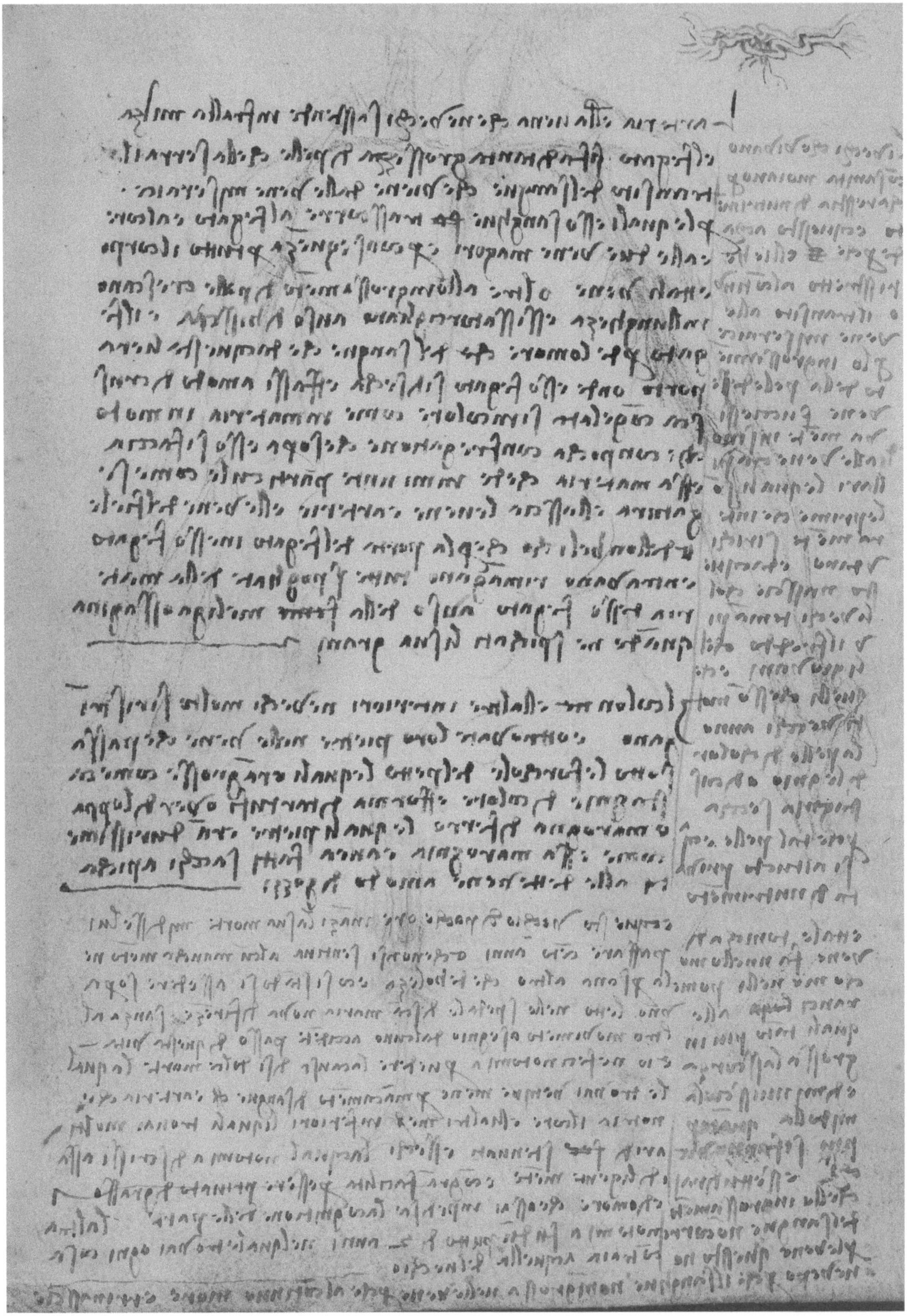

RL19027 verso, Notes on the death of a centenarian. Leonardo da Vinci (Vinci 1452-Amboise 1519). c.1508. Pen and ink over traces of black chalk. Lent by Her Majesty The Queen. Royal Collection © Her Majesty Queen Elizabeth II

<table>
<tr><td colspan="2">19027 v (B 10 v). Pen and brown ink over black chalk.</td></tr>
<tr><td>[Fig. 1, 2]</td><td>[Fig. 1, 2]</td></tr>
<tr><td>[I] Larteria ella uena che ne vechi sasstende infralla milza | el fegato · sifa dj tanta grossezza dj pelle chella serra il | transito delsangue che viene dalle vene mjseraice · | perle quali esso sanghue (*sa*) trasscorre al fegato e al core | e alle due vene magori e per conseguēza per tutto ilcorpo | e ttali vene oltre alloingrossamēto dj pelle crescano | inl-lungheza essissatorcigliano auso dj bisscia e il fe|gato perde lomore (*che*) del sangue che dacquesta li era | porto onde esso fegato sidjsecha effassi amodo dj crus|sca cōgelata si incolore come in materia in modo | che con pocha confregatione chesopra esso sifaccia | essa materia chede [sic: chade] in minute particule come se|gatura ellasscia leuene e arterie elle vene del fiele | e dellonbelicho cheperla porta del fegato inesso fegato | entravano rimāgano tutte spogliate della mate|ria desso fegato avso della (*seme*) meliga ossagina | quādo ne spichati li sua granj –</td><td>[I] The artery and vein which in the old extend between the spleen and the liver generate so thick a coat that it closes the passage of the blood which comes from the mesenteric veins through which blood passes to the liver and heart and to the two greater vessels [aorta; vena cava], and consequently through the whole body. And these veins, as well as thickening the coat, grow in length and become twisted like a snake, and the liver loses the humours of the blood which was carried there by the vein. Whence the liver is desiccated and becomes like congealed bran both in colour and substance so that when but a little friction is made on it this substance falls away in minute particles like sawdust leaving behind the veins and arteries. And the bile ducts and the vein from the umbilicus, which enter the liver through the porta hepatis, remain wholly deprived of liver substance like millet or broom-corn when their grains have been pulled off.</td></tr>
<tr><td>[II] Jlcolon (*ne*) ellaltre interiori ne vechi molto siristrī|gano e ottrovate loro pietre nelle vene che passa | sotto le forchole delpetto lequali erā grosse come ca|stagnje dj colore efforma djtartufi over di loppa | o marogna djferro le quali pietre erā durissime | come essa marognia e auea fattj sacchi apicha|ti alle dette vene amodo dj gozzj –</td><td>[II] The colon and other intestines in the old are very contracted. And I have found stones in the veins which pass beneath the forks of the chest [clavicles] which were as large as chestnuts, of the colour and shape of truffles or of dross or iron clinkers, which stones were very hard, like clinkers, and had formed bags attached to the said veins like goitres.</td></tr>
<tr><td>[III] ecquesto vechio dj poche ore ināzi lasua morte mj djsse lui | passare cēto anni e chenonsi sentiua alcū manchamēto ne|la persona altro che deboleza e cosi stādosi assedere sopra | vno letto nello spedale djscā [sic: di santa] maria nova djfirēze sanza al|ltro movimeto osegnjo dalcuno accidēte passo dj questa vita –
e io ne feci notomja per uedere lacausa djsi dolce morte la qual|le trovai venjre mene per mācamēto djsangue (*ch*) e arteria che | notria ilcore elli altri mēbri inferiori li quali trouai moltj | aridi (*sec*) stenuati essechi lacqual notomja djscrissi assa|i diligente mēte e cō grā facilita peressere priuato djgrasso | edjomore che assai inpedjsce lacognitione delle parte laltra | notomja fu dū putto dj 2 annj nelquale trovai ognj cosa | cōtra<r>ia acquella del uechio –</td><td>[III] And this old man, a few hours before his death, told me that he was over a 100 years old and that he felt nothing wrong with his body other than weakness. And thus, while sitting on a bed in the Hospital of Santa Maria Nuova in Florence, without any movement or other sign of any mishap, he passed out of this life.
And I made an anatomy of him in order to see the cause of so sweet a death. This I found to be a fainting away through lack of blood to the artery which nourished the heart, and other parts below it, which I found very dry, thin and withered. This anatomy I described very diligently and with great ease because of the absence of fat and humours which greatly hinder the recognition of the parts. The other anatomy was on a child of 2 years in which I found everything contrary to that of the old man.</td></tr>
<tr><td>[IV] livechi che vivano | cōsanjta moiano per | charesstia dj nutrimē|to e cquesto acha|de perche (*il*) ellie he | risstretto alcōtinv|o il transito alle | vene mjseraice | per lo ingrossamē|to della pelle desse | vene successi | vamēte insino | alle vene chapi|llari le quali sō | leprime che inte|ramēte sirichi|vdano e dacque|sto nasscie chel|li vechi temā pi|v ilfreddo chel|li giovanj e che | quellj chessō mol|ti vecchi anno | la pelle dj cholor | dj legnjo o di cas|tagnja seccha | perche tal pelle e cq^{a}|si [sic: quasi] altucto priva|ta dj nutrimēto</td><td>[IV] The old who live in good health die through lack of nourishment. And this occurs because the passage of the mesenteric veins is continually constricted by the thickening of the coats of these veins successively as far as the capillary veins, which are the first to be completely closed. And from this it comes about that the old fear the cold more than the young; and that those who are very old have skin the colour of wood or dried chestnut because this skin is almost completely deprived of nourishment.</td></tr>
<tr><td>ettale tonicha dj | vene fa nnellomo chome nelli pome|rancj (*leqa*) alle | quali tāto piu in | grossa lasscorza | e djmjnuissciela | mjdolla quātop|piu sifanno vec|chi essettu djraj | chello ingrossamēto | delsangue nō corre | perlevene quessto no|ne vero perche il sanghue non@grossa nelle vene perche al cōtinuo more errinasscie</td><td>And this coat on the vessels acts in man as it does in oranges, in which as the peel thickens so the pulp diminishes the older they become. And if you say that it is the thickened blood which does not flow through the vessels, this is not true, because blood does not thicken in the vessels since it continually dies and is renewed.</td></tr>
</table>

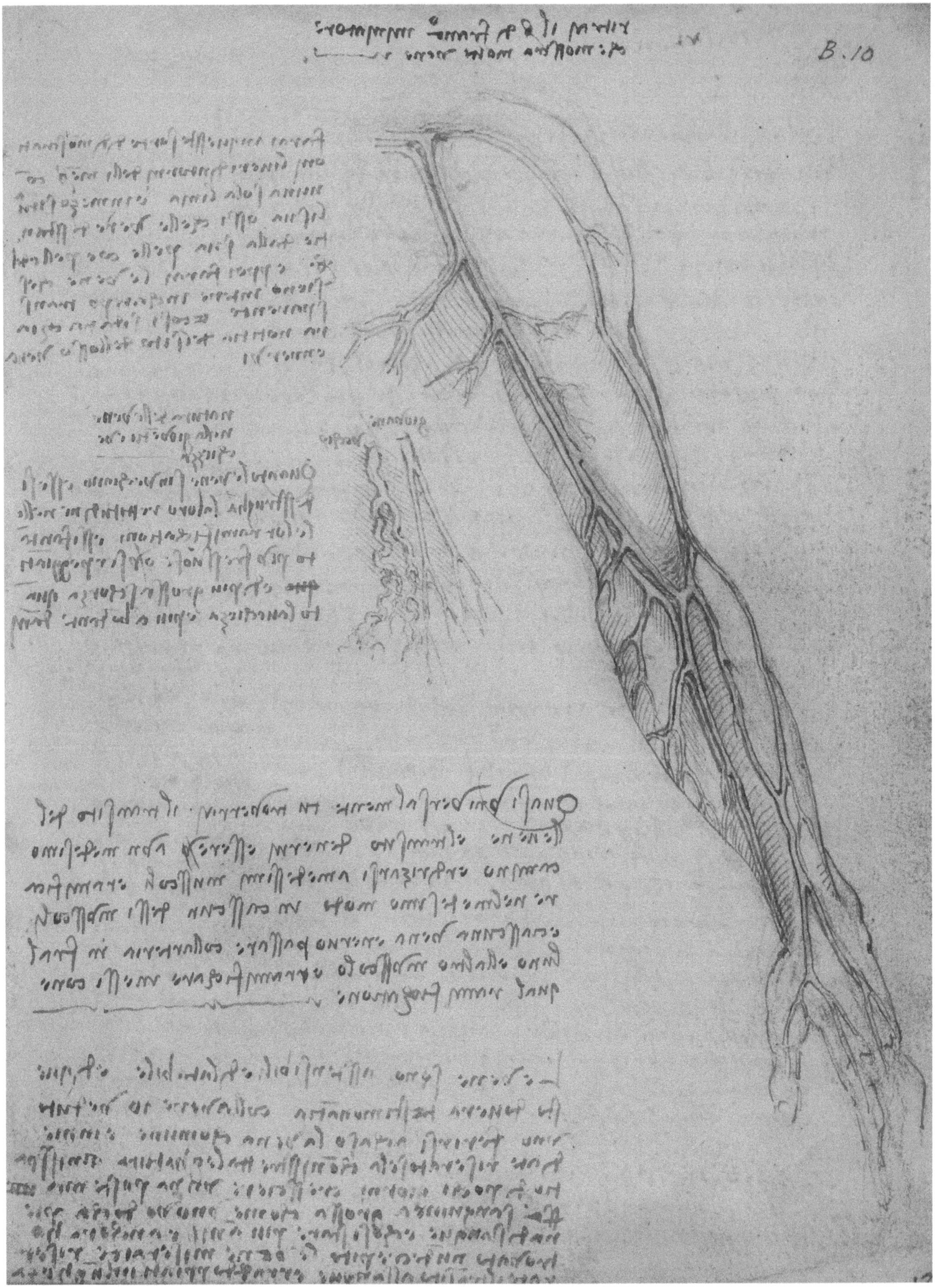

RL 19027 recto, The veins of the arm. Leonardo da Vinci (Vinci 1452-Amboise 1519). c.1508. Pen and ink over traces of black chalk (Lent by Her Majesty The Queen. Royal Collection © Her Majesty Queen Elizabeth II)

19027 r (B 10 r). Pen and brown ink (two shades) over traces of black chalk. 192×141 mm.	
[I] ri ritraj il braccio dj franc° mjnjatore | che mosstra molte vene –	[I] Portray the arm of Francesco the miniaturist which shows many veins
[Fig. 1]	[Fig. 1]
[II] Quasi vniversalmente tu troverraj · il transito del|lle uene el transito deneruj essere (*v*) avn medesimo | camjno er<a>djrizarsi a medessimj musscoli eramjfica|re nelmedesimo modo in casscun dessi mvsscolj | e ciasscuna vena eneruo passare collarteria in fral|luno ellaltro mvsscolo erramjfichare inessi cone|qual ramj fichatione –	[II] You will find almost universally that the passage of the veins and nerves is along the same path, and is directed to the same muscles, and ramifies in the same way in each of these muscles. And each vein and nerve passes with the artery between one and another muscle and ramifies within it with equal ramification.
[III] Le vene sono asstensibili e djlatabile e dj que|sto donera testimonãtia collavere io veduto | vno ferirsi achaso la vena chomune e inme|diate riseratosela chõnjsstretta leghatura einisspa|tio dj pochi giornj cressciere vn na poste ma (*ro|ssa*) sanguinea grossa chome vnovo docha pie|na dj sangue e chosi stare piu ãnj// e anchora ho | trovato nũ decrepito le vene mjseraice riser|rate iltrãsito alsangue erraddoppiati inlũgheza	[III] Veins are extensible and dilatable. And testimony of this is given by my having seen a man who by chance was wounded on the common vein [? median cubital vein] and it was immediately bound up with a tight bandage. In the space of a few days there grew up a [red] sanguineous abscess as large as a goose's egg, full of blood; and so it remained for several years. I also found in a decrepit old man the mesenteric veins obstructing the passage of blood, and doubled in length.
[Fig. 2] giovane	[Fig. 2] youth
[Fig. 3] vechio	[Fig. 3] old man
[IV] natura delle vene | nella giovẽtu e ve|chiezza – Quanto le vene sinvechiano esse si | desstrughã laloro rettitudjne nelle | le lor ramjfichationi essifan tã|to piv fressuose over serpeggiãti | (*qua*) e dj piu grossa schorza quã|to lauechieza e piu abõdante dãnj	[IV] The nature of the veins in youth and age As the veins grow old they destroy the straightness of their ramifications and they produce as much more tortuosity or windings, with thicker coats, as age is more abundant in years.
[V] faraj in quesste sorte djdjmõstrati|onj liueri djntornj delli mẽbri cõ|nuna sola linja e in mezo situ[a] | lisua ossi cholle vere djsstan|tie dalla sua pelle coe pelle del | · braccio · eppoi faraj le vene ches|sieno intere inchanpo trans|sparente e cosi sidara chia|ra notitia delsito dellosso vena | ennervi	[V] In this kind of demonstration you will make the true contours of the limbs with a single line; and in the middle place the bones at their true distances from the skin, that is, the skin of the arm. And you will do the veins which should be complete in a transparent field of vision; and in this way clear knowledge of the position of the bone, vein and nerves will be given.

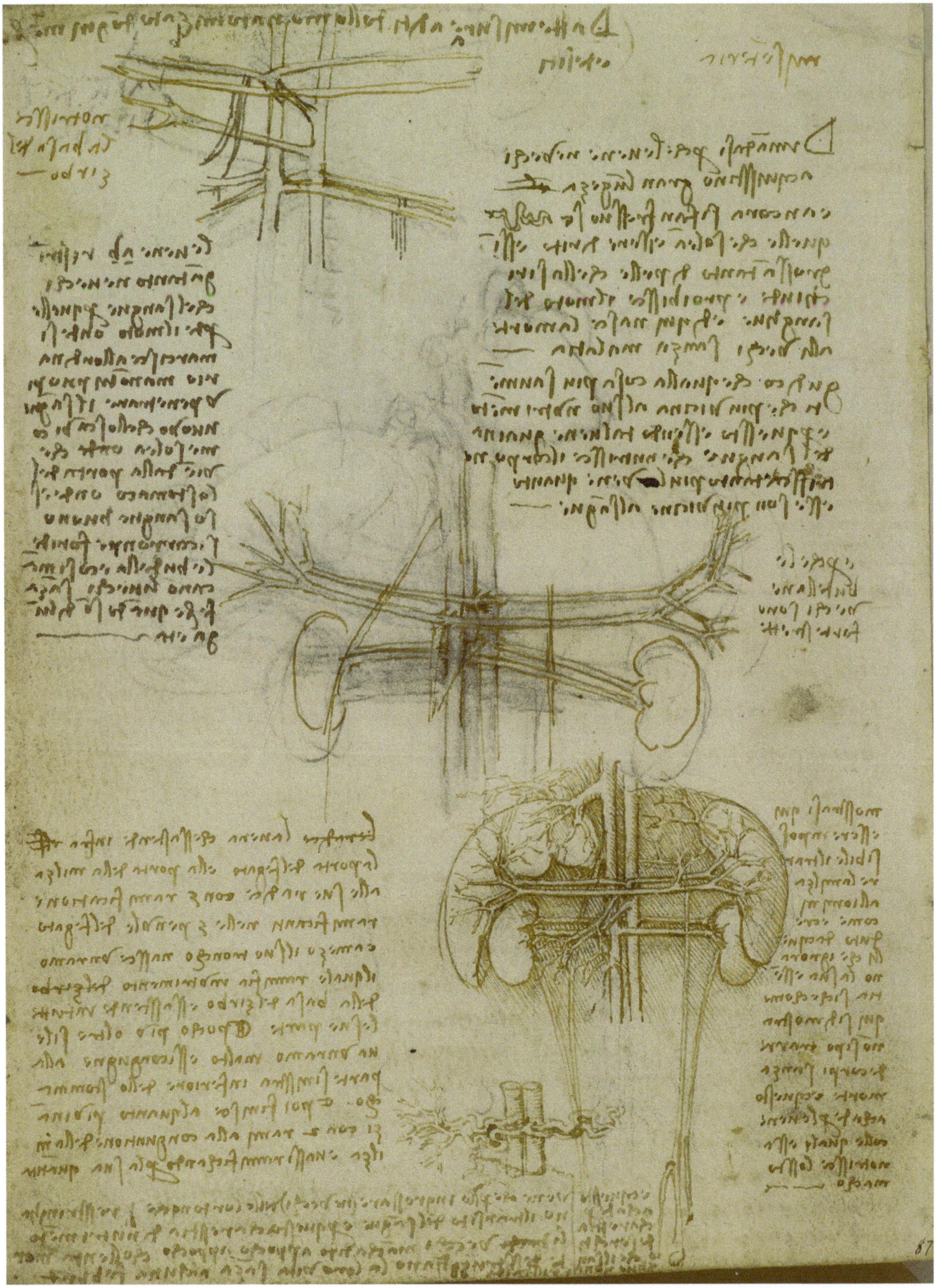

RL 19028 verso, The vessels of liver, spleen and kidneys. Leonardo da Vinci (Vinci 1452-Amboise 1519). c.1508. Pen and ink over black chalk (Lent by Her Majesty The Queen. Royal Collection © Her Majesty Queen Elizabeth II)

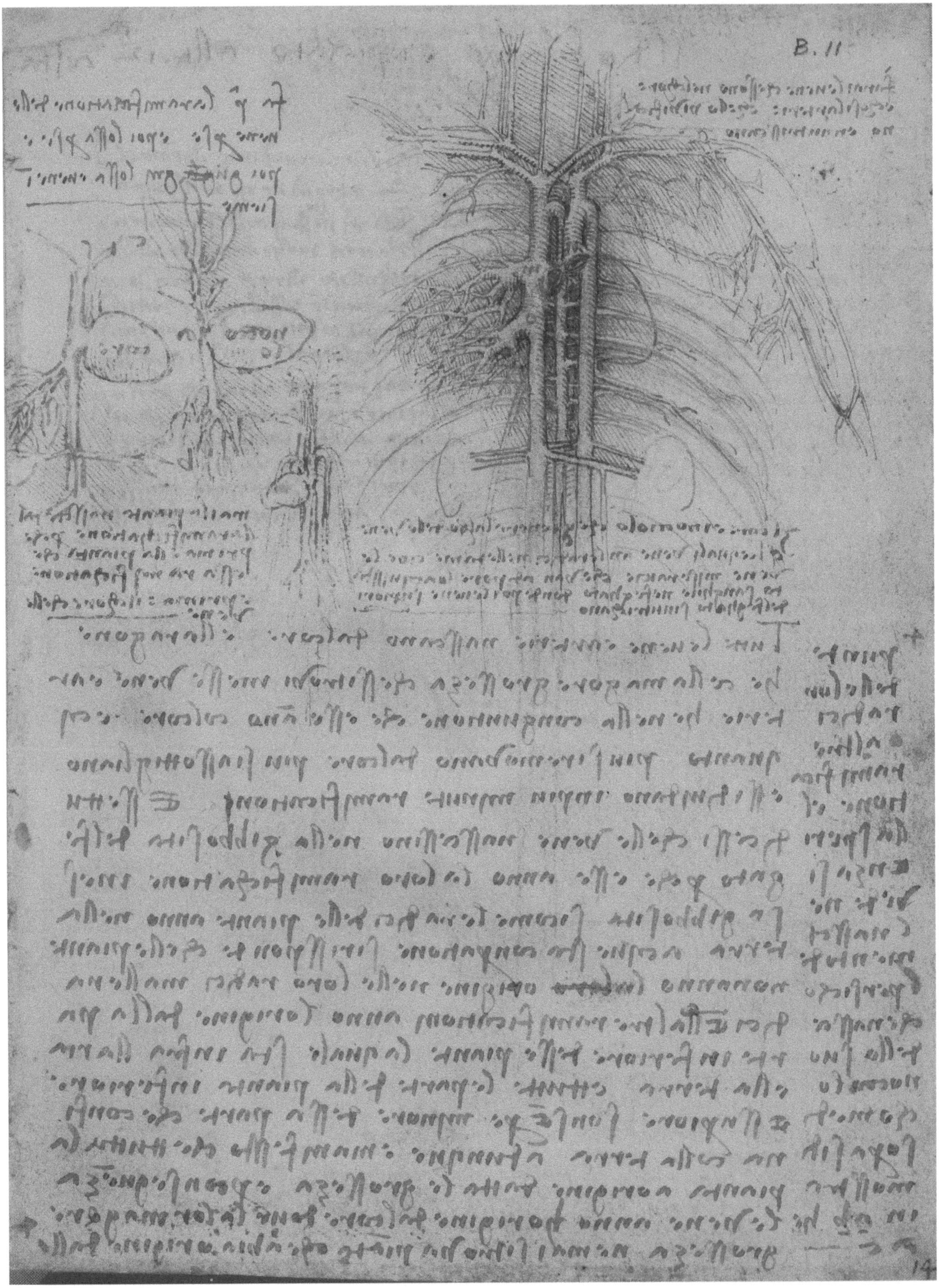

RL 19028 recto, The heart compared to a seed. Leonardo da Vinci (Vinci 1452-Amboise 1519). c.1508. Pen and ink over black chalk (Lent by Her Majesty The Queen. Royal Collection © Her Majesty Queen Elizabeth II)

<table>
<tr><td colspan="2">19028 r (B 11 r). Pen and brown ink (two shades) over traces of black chalk. 192 × 140 mm.</td></tr>
<tr><td>[I] Il djaf[ran]a e apichato alla 12ª costa</td><td>[I] The diaphragm is attached to the 12th rib.</td></tr>
<tr><td>[II] farai leuene chessono nelchore | echosi larterie chello vivifjca|no ennutrisscano</td><td>[II] Draw the veins which are in the heart, and also the arteries, which vivify and nourish it.</td></tr>
<tr><td>[Fig. 1] m n o</td><td>[Fig. 1] m n o</td></tr>
<tr><td>[Fig. 2] nocco|lo a b c</td><td>[Fig. 2] nut a b c</td></tr>
<tr><td>[III] Tutte le uene e arterie nasscano dalcore e llaragone | he c<h>ella magore grosseza chessitrovi inesse vene e ar|terie he nella conguntione che esse āno col core · e cq|quanto piu siremovano dal core piu si assottigliano | essj djujdano inpiu mjnute ramjficationj E sse ttu | djcessi chelle vene nasscessino nella gibbosita del fe|gato perche esse anno la loro ramjfichatione ines|sa gibbosita sicome le radjci delle piante anno nella | terra cque sta conparatione sirissponde chelle piante | nonanno la(loro) origine nelle loro radjci ma lle ra|dici ellaltre ramjficationj anno lorigine dalla pa|rte inferiore desse piante la quale sta infra llaria | ella terra ettutte le parte della pianta inferiore | e ssuperiore sonsē pre mjnore dessa parte che confi|na colla terra adunque e manjfessto che ttutta la | pianta a origine datta le grosseza e perconseguēza | le vene anno horigine dal core doue la lor magore | grosseza ne maj sitrova piāta che abia origine dalle 4| [^ –]4 punte | delle lor | radjci | o oltre | ramjfica|tione el|lla speri|enza si | vede ne|l nasscj|mento de|l persicho | che nassce | dello suo | noccolo | chomedj | sopra sidj|mosstra | in a b he | a c –[– ^]</td><td>[III] All the veins and arteries arise from the heart; and the reason is that the biggest veins and arteries are found at their conjunction with the heart. And the more they are removed from the heart the finer they become, and they divide into very small branches. And if you say that the veins arise in the gibbosity of the liver because they have their ramifications in this gibbosity, just as the roots of plants have in the earth, the reply to this comparison is that plants do not have their origin in their roots, but the roots and other ramifications take origin from that lower part of the plant which is situated between the air and the earth. And all the lower and upper parts of a plant are always less than this part which adjoins the earth. Therefore it is evident that the whole plant takes origin from such a size and in consequence the veins take their origin from the heart where they are biggest. Never does one find a plant which has its origin from the tips of its roots or other branches; and this is observed by experience in the sprouting of a peach which arises from its nut as is shown above at a b and a c.</td></tr>
<tr><td>[Fig. 3] core</td><td>[Fig. 3] heart</td></tr>
<tr><td>[Fig. 4]</td><td>[Fig. 4]</td></tr>
<tr><td>[IV] mai la pianta nasscie dal|llaramjfichatione perche | prima e lla pianta che | essa ramj fichatione | e prima e ilchore chelle | vene –</td><td>[IV] A plant never arises from its branches, for the plant exists before the branches and the heart exists before the veins.</td></tr>
<tr><td>[V] Jl core einocciolo che gienera lalbero delle vene | (e) Lequalj vene an leradici nelletame cioe le | vene mjseraicie che van adjpore loacqujssta|to sanghue nefeghato donde poi le uene superiori | del feghato sinutrichano –</td><td>[V] The heart is the nut which generates the tree of the veins; which veins have their roots in the manure, that is the meseraic [mesenteric] veins go on to deposit the blood they have acquired in the liver, whence then the upper veins of the liver [hepatic veins] are nourished.</td></tr>
<tr><td>[VI] fa pª laramjfichatione delle | uene perse e poi lossa perse e | poi gu(gli)gnj lossa e uene ī|sieme –</td><td>[VI] Draw the first ramification of the veins by itself, and then the bones by themselves, and then join the bones and the veins together.</td></tr>
</table>

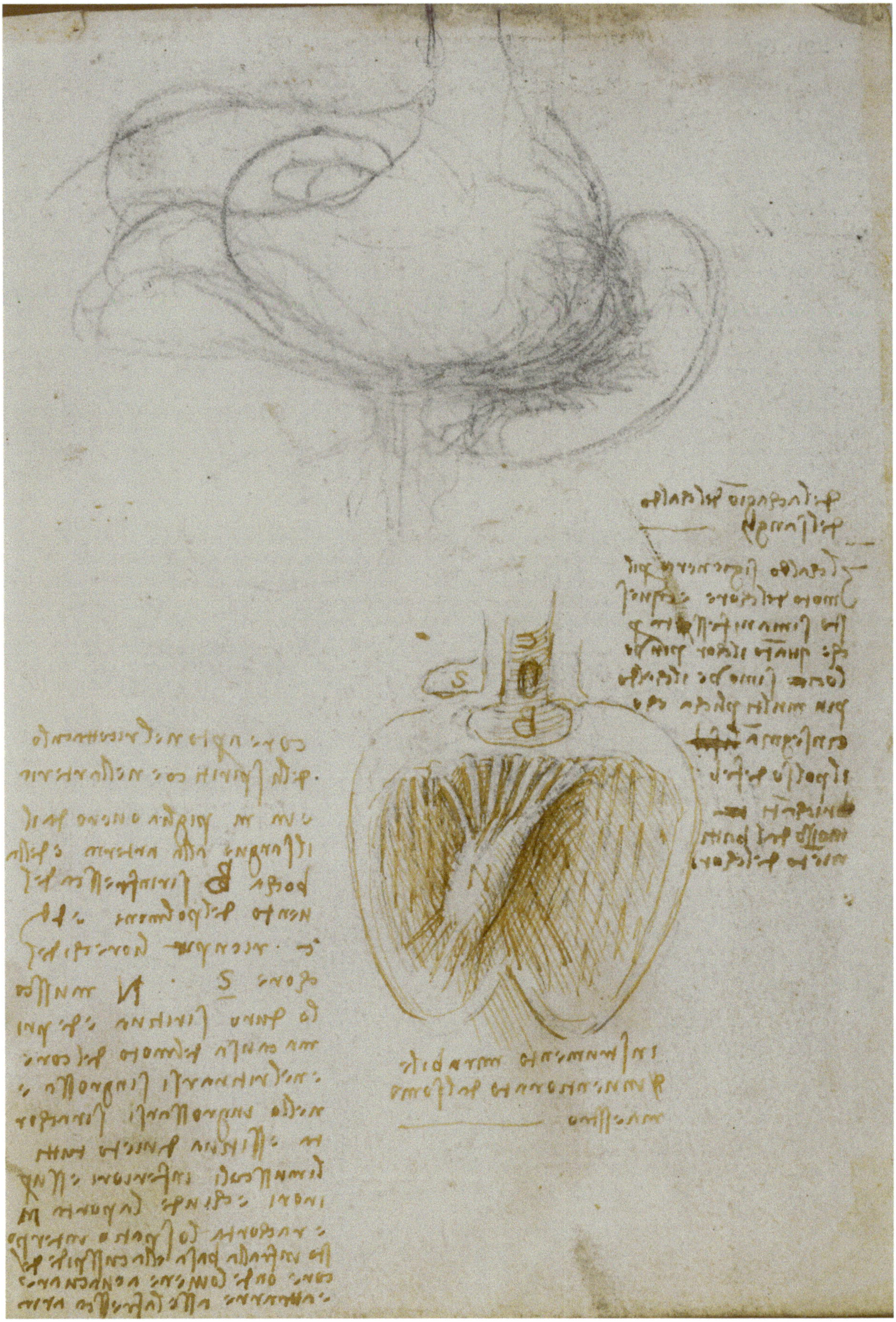

RL 19029 recto, The stomach, and the heart. Leonardo da Vinci (Vinci 1452-Amboise 1519). c.1508. Pen and ink with black chalk (Lent by Her Majesty The Queen. Royal Collection © Her Majesty Queen Elizabeth II)

<table>
<tr><th colspan="2">19029 r (B 12 r). Pen and brown ink (two shades) with black chalk. 191 × 138 mm.</th></tr>
<tr><td>[Fig. I]</td><td>[Fig. I]</td></tr>
<tr><td>[I] delachagiō del chaldo | del sangh<ue>–
Jl chaldo sigienera peril | moto del chore e cques|sto simanifessta per|che quāto ilchor piu ve|locie simove ilchaldo | piu multi plicha cho<me>| cinsegnjā (lifeb) | il polso defeb|brichāti (n) | mosso dal battj|mēto delchore | e</td><td>[I] On the cause of the heat of the blood
The heat is generated by the movement of the heart; and this is evident because the more quickly the heart moves the more the heat multiplies, as the pulse of febrile persons moved by the beating of the heart teaches us.</td></tr>
<tr><td>[Fig. 2] C S B M N</td><td>[Fig. 2] C S B M N</td></tr>
<tr><td>[II] instrumento mirabile | (d) inuentionato dalsomo | maestro –</td><td>[II] Admirable instrument, invented by the Supreme Master</td></tr>
<tr><td>[III] core aperto nel ricettaculo | dellj spiriti coe nellarteria | e in m piglia o uero da il | ilsangue alla arteria e della | bocha B sirinfressca del | uento del polmone eddj |· c · rienpie liorechi del | chore S · N mussco|lo duro siritira e de pri|ma causa delmoto del core | e nel ritirarsi singrossa e | nello ingrossarsi sirachor|ta essitira djrieto tutti | li musscoli inferiori essuper|irori e chiude la porta M | e rachorta lo spatio interpo|sto infralla basa ella cusspide del | core onde loujene a euacuare | e attrarre asse lafressca aria</td><td>[III] The heart, opened in the receptacle of the spirit, that is, the artery [left ventricle]. And at M it either takes the blood or gives it to the artery; and by the mouth B [mitral orifice] it is refreshed by the wind of the lungs; and through C it fills the auricles of the heart S.N, a hard muscle, contracts and is the primary cause of the movement of the heart, and in contracting it enlarges, and in enlarging it shortens and draws back all the muscles below and above it, and closes the gate M [mitral valve] and shortens the distance interposed between the base and the apex of the heart, whence it comes to empty itself, and to attract to itself fresh air.</td></tr>
</table>

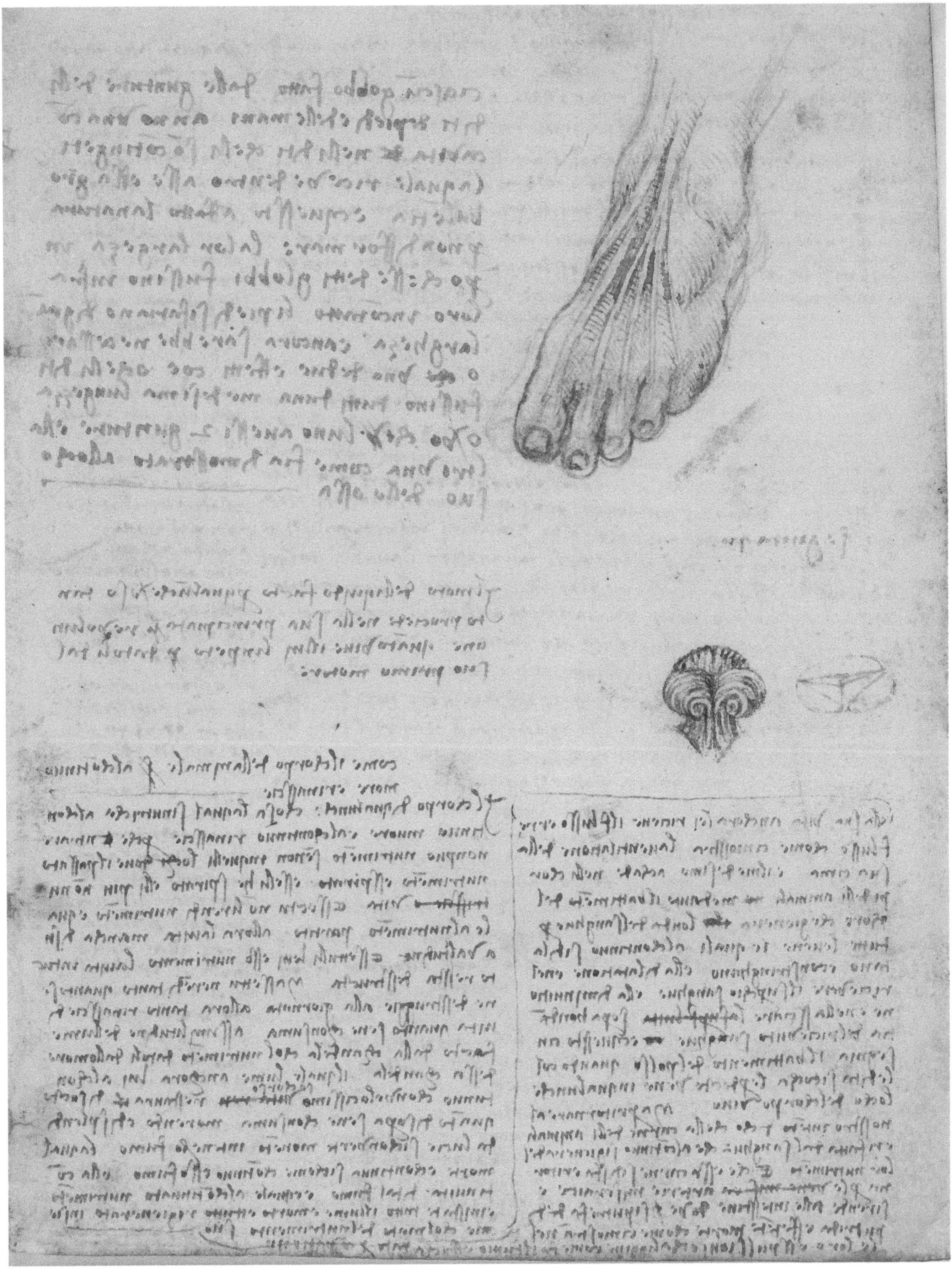

RL 19045 recto, A study of the tendons of the foot, etc. Leonardo da Vinci (Vinci 1452-Amboise 1519). c.1508. Pen and ink over traces of black chalk (Lent by Her Majesty The Queen. Royal Collection © Her Majesty Queen Elizabeth II)

19045 r (B 28 r). Pen and brown ink (three shades) over traces of black chalk. 192 × 139 mm.

[Fig. 1]	[Fig. 1]
[I] ciascū gobbo fatto dalle gunture dellj | djti depiedj e delle mani anno vna cō|cavita (*de*) nelli djti chellj sō cōtingēti | la quale riceve dentro asse essa gro|bulētia e cquessto affatto la natura | per non djffor mare la lor largeza in|pero chesse dettj globbi fussino infra | loro incōtatto li piedj sifariano dj grā | largheza · e ancora sarebbe necessari|o (*che*) vno de due effettj coe ochelli djti | fussino tuttj duna medesima lungezza | over che (*v*) luno auessi 2 gunture ella|ltro vna come fia djmosstrato allocho | suo delle ossa –	[I] Each protuberance made by the joints of the digits of the feet and hands has a hollow in the digits contiguous to it which receives into itself this rounded protuberance; and Nature has done this in order not to deform their width. For if the said protuberances were in contact with one another the feet would become of great width, and one of two further effects would necessarily occur; that is, either the digits would all be of the same length, or one would have two joints and the other one, as is demonstrated in its place on the bones.
[II] seguita qua	[II] It follows here …
[Fig. 2, 3]	[Fig. 2, 3]
[III] Jlmoto delliqujdo facto perqualūche verso tan|to prociede nella sua principiata (*li*) revoluti|one quāto viue illuj linpeto (*p*) datoli dal | suo primo motore	[III] The movement made by a liquid in any direction proceeds as far in its original rotation as the impetus given to it by its prime mover.
[IV] come ilchorpo dellanjmale (*s*) alchōtinuo | more erinasscie – Jl chorpo dj qualunche chosa laqual sinutrjcha alchon|tinuo muore e al chontinuo rinasscie perche entrare | nonpuo nutrimēto sēnon inquellj lochi doue ilpassato | nutrimēto esspirato · esselli he spirato ellj piu nō nu | (*trisscie e*) vita essectu nō li rendj nutrimēto equa|le al nutrimēto partito · allora laujta mancha djsu|a valitudjne essettullj leuj esso nutrimento laujta intuc|to ressta desstructa Massettu nerēdj tanto quanto se | ne desstruggie alla giornata allora tanto rinasscie dj | ujta quanto sene chonsuma assimjlitudjne dellume | facto dalla chandela chol nutrimēto datolj dallomore | dessa chandela il quale lume anchora luj alchon|tunuo chonvelocissimo (*vita ren*) ^sochorso^ restaura (*il*) dj socto | quāto djsopra sene chonsuma morendo e dj splendi|da lucie sichonverte morēdo intenebro<so> fumo laqual | morte e chontinua sichome chōtinuo esso fumo ella cō|tinuita · dj tal fumo e equale alchōtinuato nutrimēto | einjsta<n>te tutto illume e morto ettutto rigienerato insie|me chol moto delnutrimento suo [: – :] ella sua vita anchora lej ricieue ilflusso erre|flusso chome ci mosstra lauentilatione della | sua cime e ilmedesimo achade nelli chor|pi delli animali (*ne*) medjante il battimēto del | chore che gienera (*iln*) londa delsanghue per | tutte leuene le quali alchontinuo sidjla|tano econstringhano ella djlatatione enel | ricievere ilsuperchio sanghue ella djmjnuitio|ne e nellassciare (*lasuperflujta*) soprabondā|tia delricievuto sanghue (*co*) ecquessto cin|segnja il battimento del polso quando col|le djta sitocha le predecte vene in qualunche | locho del chorpo viuo Ma perritornare al | nosstro intēto djcho chella carne dellj anjmalj | e rifatta dal sanghue che alcōtinuo sigienera del | lor nutrimēto E che essa carne sidjsfa eritor|na perle (*vene mjs en*) arterie mjseraice e | sirende alle intesstine dove (*e*) siputrefa (*dj*) dj | putrida effetēte morte chome cimostrā nel|le loro esspulsionj e chaliggine come fa ilfumo effocha dato per cōperatione	[IV] How the body of an animal continually dies and is reborn The body of anything which is nourished continually dies and is continually reborn, for nourishment cannot enter except into those places where past nourishment has been exhausted, and if it has been exhausted it no longer (nourishes) or has life. Unless, therefore, you supply nourishment equal to the nourishment that has departed, life loses its vigour, and if you take away nourishment life is totally destroyed. But if you supply just as much nourishment as is destroyed daily, then as much life is reborn as is consumed – just as the light of a candle is made from the nourishment from the humour given it by the candle. And this light also is itself continually restored with swiftest succour from below by as much as is consumed above in dying; and in dying it is changed from bright light to dark smoke. And this death continues as long as the smoke continues; and its continuity is equal to the continuity of nourishment. And at any instant the whole light is dead and completely regenerated together with the movement of its nourishment. Its life, furthermore, receives from it flux and reflux as the to and fro movement of its tip shows us. And the same thing happens in the bodies of animals by means of the beating of the heart which generates a wave of blood through all the vessels which continually dilate and contract. And dilatation occurs on the reception of superabundant blood and diminution occurs on the departure of the superabundance of the blood received. This the beating of the pulse teaches us when we touch the aforesaid vessels with our fingers in any part of the living body. But to return to our intention. I say that the flesh of animals is being continually re-made from the blood which is continually generated from their nourishment. And that this flesh is unmade, and returns through the mesenteric arteries and is supplied to the intestines where it becomes putrified into putrid and foetid death, as their expulsions and steams show us, like the smoke and fire given for comparison.

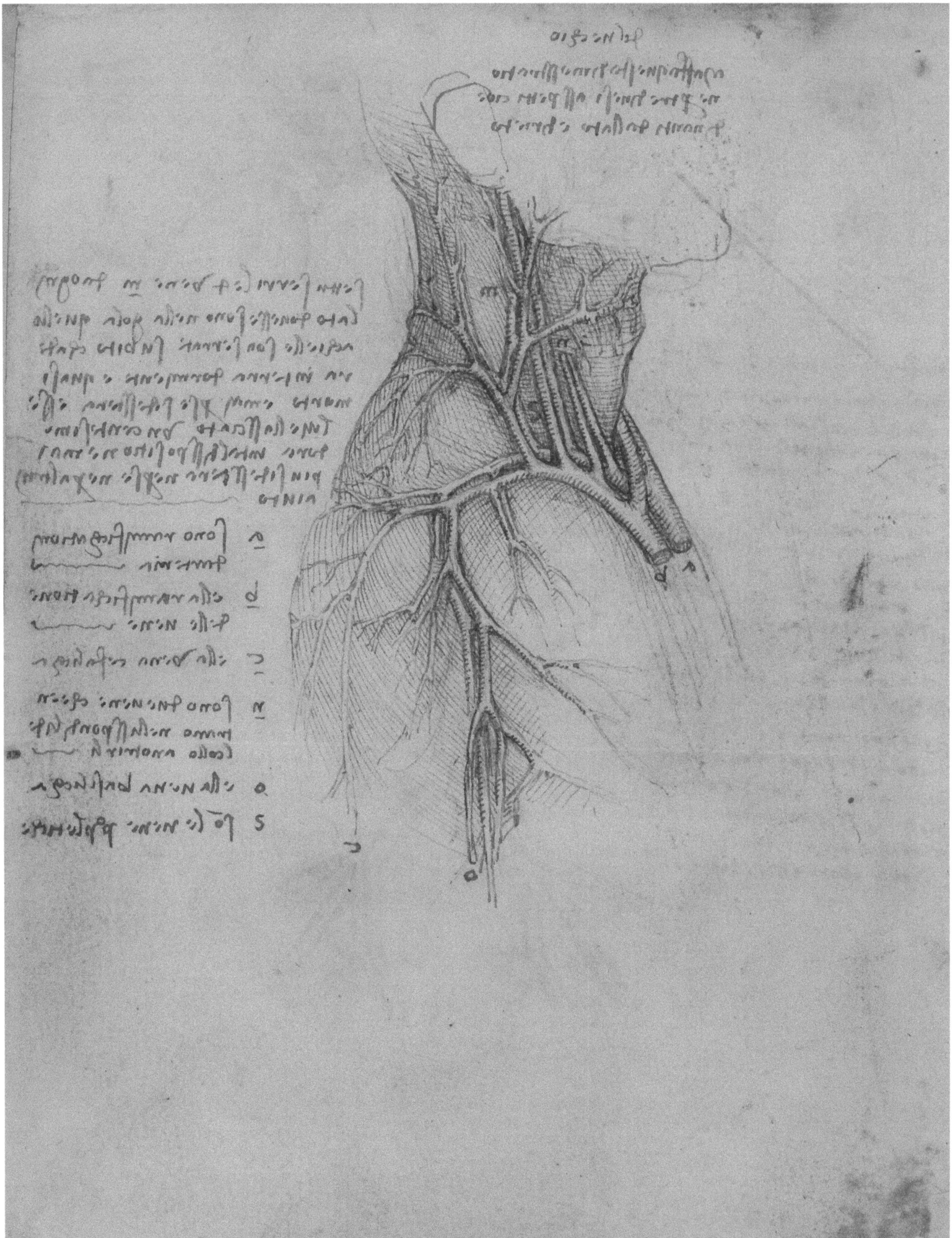

RL 19049 verso, The vessels of the neck and shoulder. Leonardo da Vinci (Vinci 1452-Amboise 1519). c.1508. Pen and ink over black chalk (Lent by Her Majesty The Queen. Royal Collection © Her Majesty Queen Elizabeth II)

19049v (B 32 v). Pen and brown ink (two shades) over traces of black chalk.

[I] del uechio Maffa questa djmosstratio|ne per tre djue<r>si asspetti cioe | djnanti dallato e djrieto	[I] Of the old man But make this demonstration from three different aspects, that is from the front, side, and back.
[Fig.] a b c m n o s	[Fig.] *a b c m n o s*
[II] settu serri le 4 vene m da ognj | lato douesse sono nella gola quello | achielle son serrate subito chade|ra in terra dormjente e quasi | morto e maj perse sidesstera esse | luj e llassciato vn centesimo | dora intal djssposítione maj | piu sidesstera neperse neperaltruj | aiuto –	[II] If you close the four vessels *m* on each side where they are in the throat, he who has them closed will fall to the ground immediately in sleep [coma] as if dead and he will never wake up on his own. And if he is left for one-hundredth of an hour in such a condition he will never wake again, neither on his own nor with the help of others.
[III] a sono ramjfichationj | darteria – b ella ramjfichatione | delle uene – c ella vena cefalicha n sono due uene che en|trano nellisspondjli de|l collo anotrirli – o ella uena basilicha s sō le uene p^o^pletiche	[III] *a* are the ramifications of the artery *b* is the ramification of the veins *c* is the cephalic vein *n* are the two vessels which enter the cervical vertebrae to nourish them *o* is the basilic vein *s* are the apoplectic vessels

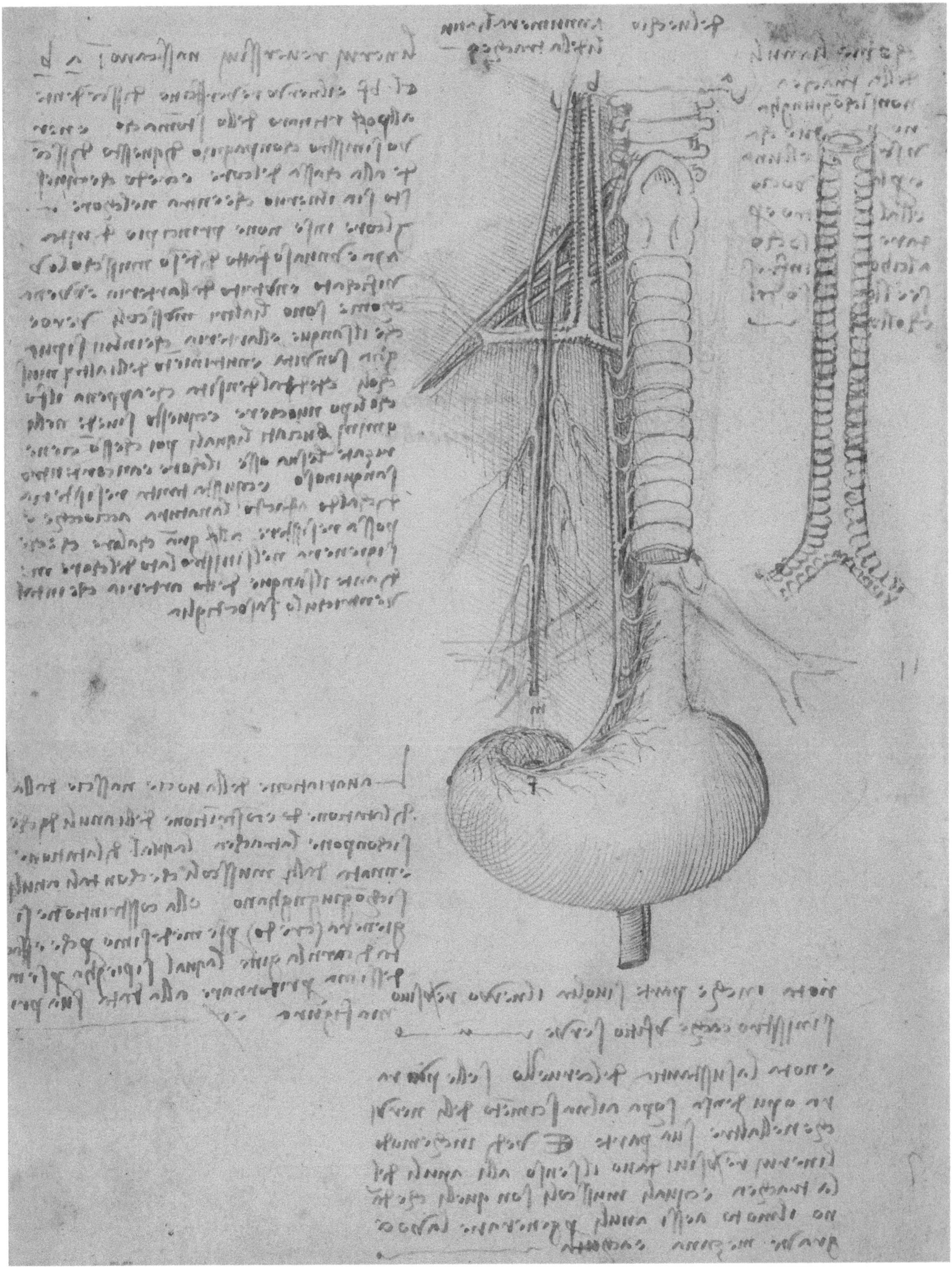

RL 19050 verso, The distribution of the right vagus and right phrenic nerves. Leonardo da Vinci (Vinci 1452-Amboise 1519). c.1508. Pen and ink over black chalk (Lent by Her Majesty The Queen. Royal Collection © Her Majesty Queen Elizabeth II)

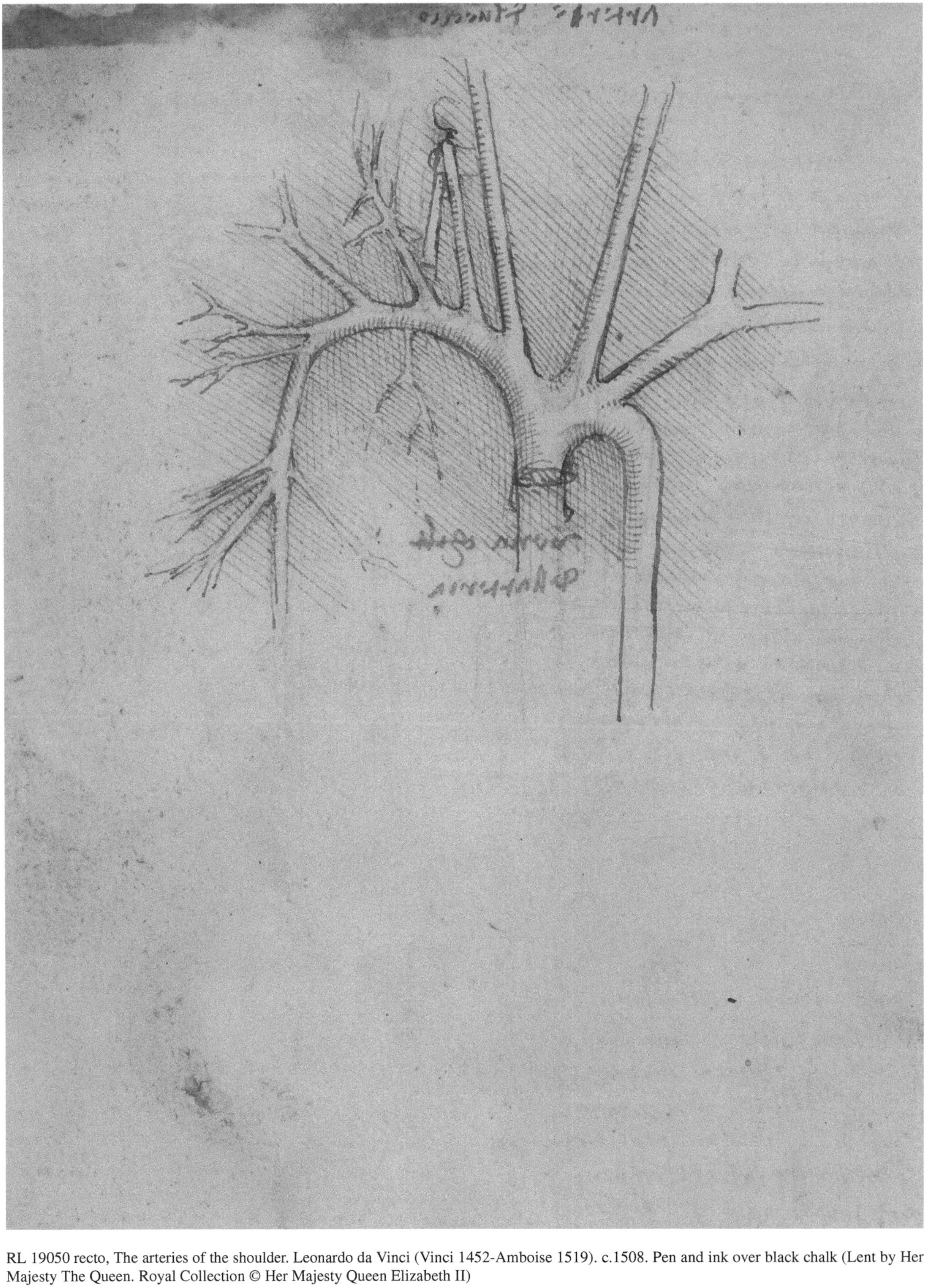

RL 19050 recto, The arteries of the shoulder. Leonardo da Vinci (Vinci 1452-Amboise 1519). c.1508. Pen and ink over black chalk (Lent by Her Majesty The Queen. Royal Collection © Her Majesty Queen Elizabeth II)

19050r (B 33 r). Pen and brown ink (two shades) over traces of black chalk. 193 × 133 mm.	
[I] arterie deluechio	[I] Artery of the old man
[Fig.]	[Fig.]
[II] vena (*chili*) dellarteria	[II] Vessel of the artery

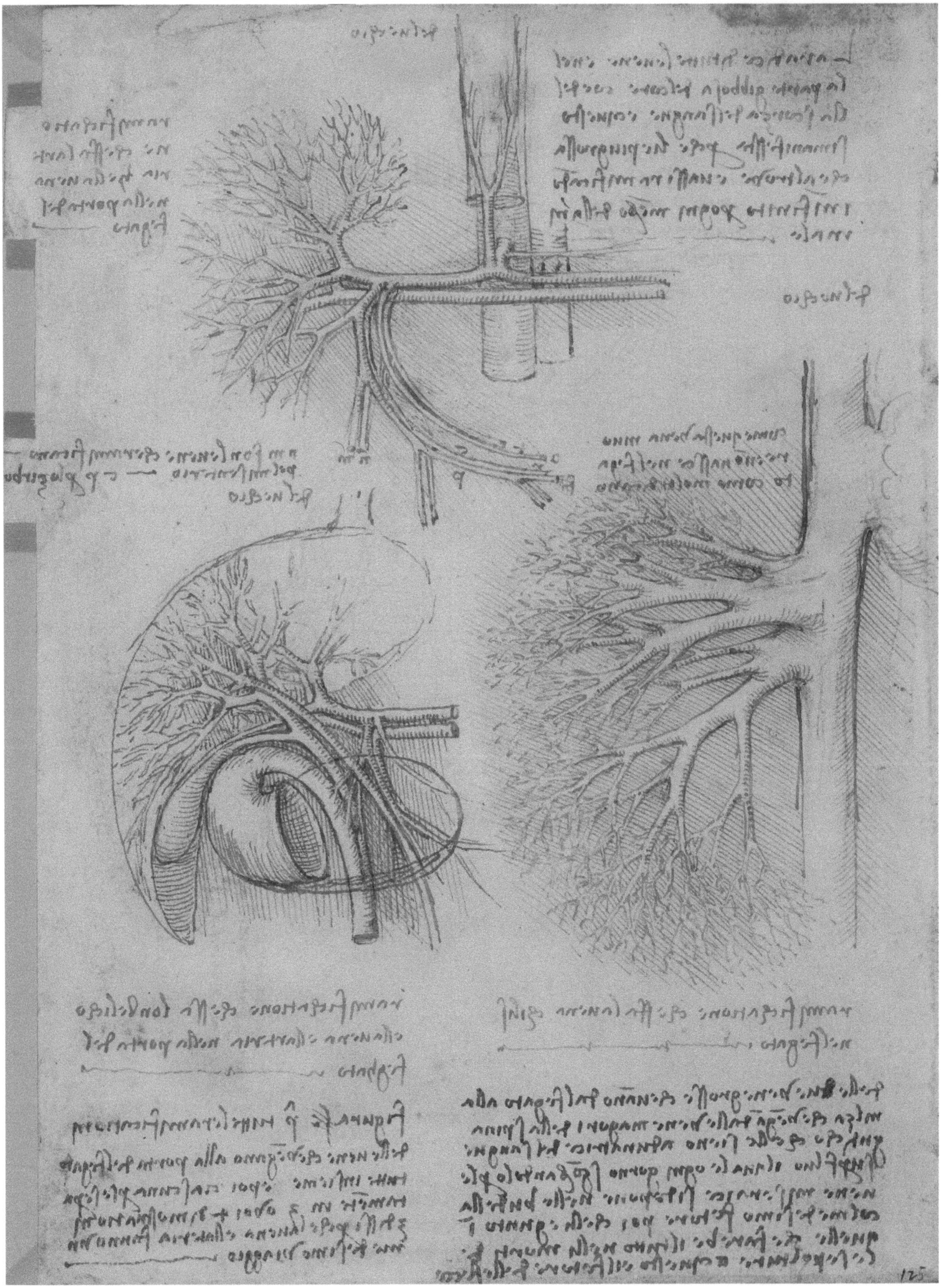

RL 19051 recto, The vessels and nerves of the neck. Leonardo da Vinci (Vinci 1452-Amboise 1519). c.1508. Pen and ink. Lent by Her Majesty The Queen. Royal Collection © Her Majesty Queen Elizabeth II

19051r (B 34 r). Pen and brown ink over traces of black chalk. 191 × 135 mm.	
[I] deluechio	[I] Of the old man
[Fig. I]	[Fig. I]
[II] deluechio	[II] Of the old man
[Fig. 2] a b	[Fig. 2] *a b*
[III] neruo djscēdēte alla cassa delcore | in mezo allarteria euena –	[III] Nerve descending to the capsule of the heart between the artery and the vein.
[IV] a ella vena b ellarteria	[IV] *a* is the vein *b* is the artery
[V] nota selle piu grossa larteria | chella vena olla vena chellarte|ria eilsimjne [sic: simile] fa ne fancullj | govanj e vechi e massci effe|mjne eanj mali dj terra eda|ria e dacqua –	[V] Note whether the artery is bigger than the vein, or the vein that the artery; and do the same in children, young and old, both male and female, and animals of the earth, air and water.

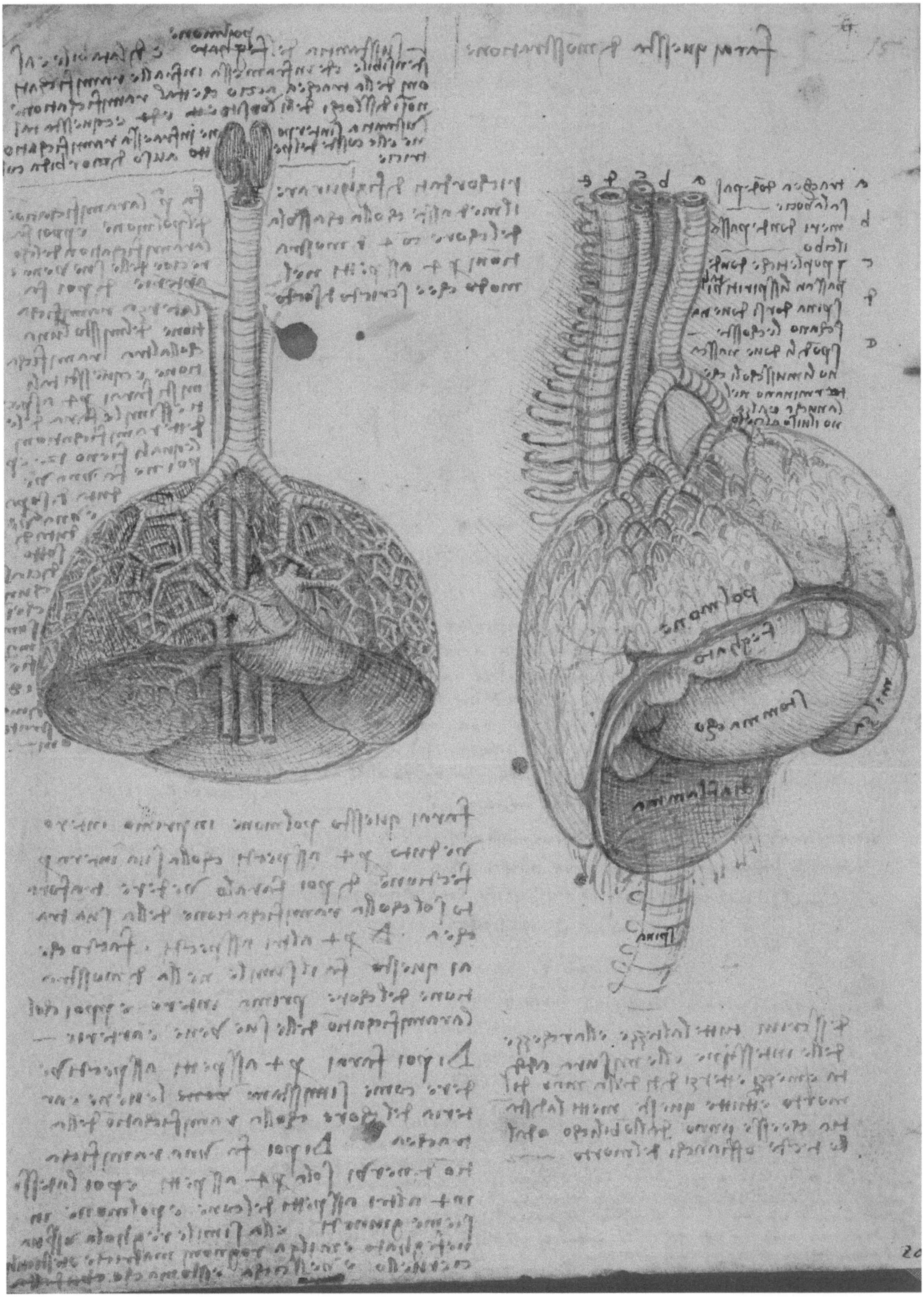

RL 19054 verso, The lungs. Leonardo da Vinci (Vinci 1452-Amboise 1519). c.1508. Pen and ink over traces of black chalk (Lent by Her Majesty The Queen. Royal Collection © Her Majesty Queen Elizabeth II)

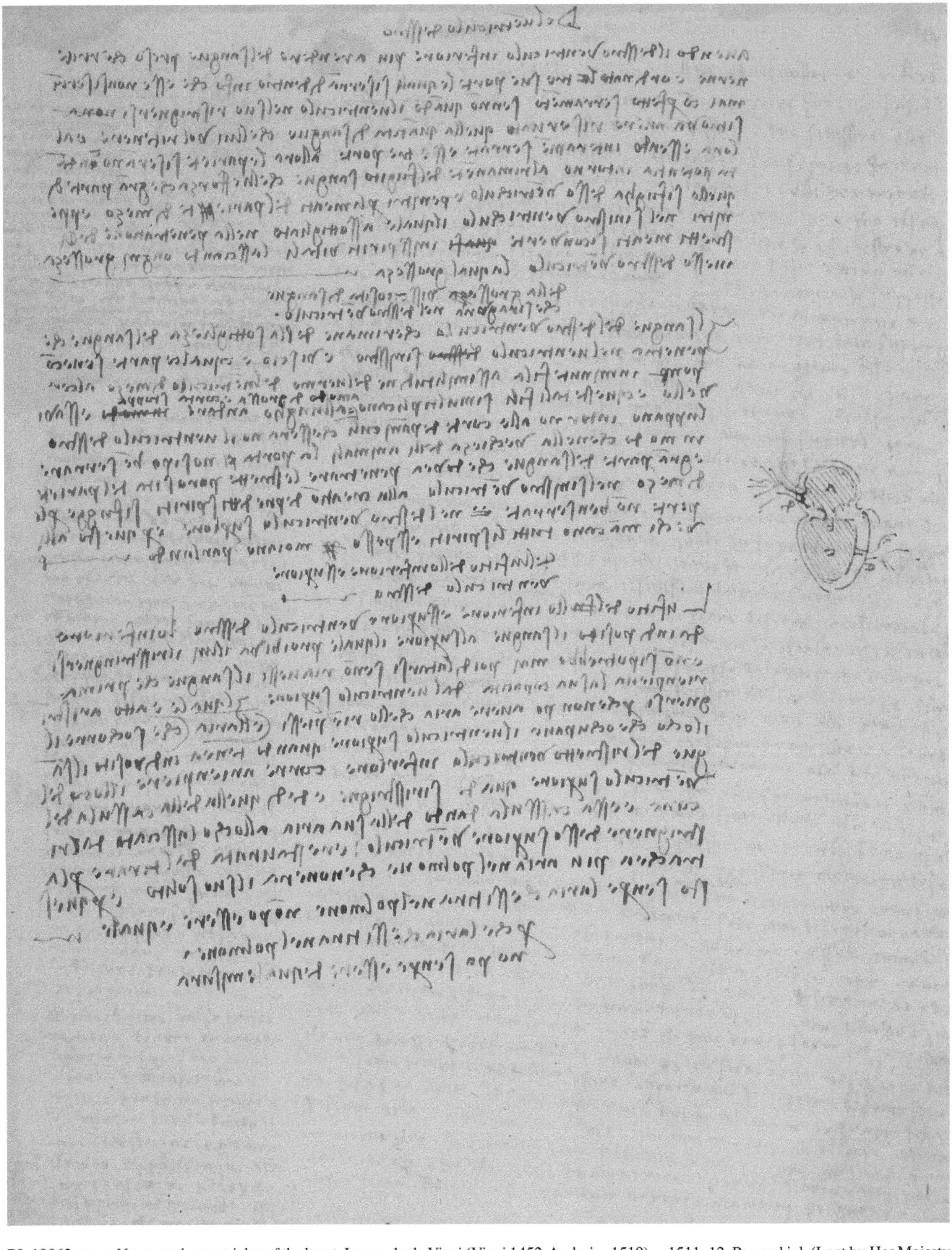

RL 19062 verso, Notes on the ventricles of the heart. Leonardo da Vinci (Vinci 1452-Amboise 1519). c.1511–12. Pen and ink (Lent by Her Majesty The Queen. Royal Collection © Her Majesty Queen Elizabeth II)

19062 v (C I 3 v). Pen and brown ink

[Fig.] a b c d e

[I] Del uētriculo desstro
auendo il desstro ventriculo inferiore piu a rendere del sangue preso che rrite|nerne e ordjnato le tre sue porte le qualj si serrā dj dentro infi [sic: infino] che esse non siserrī | mai cō perfetto serramēto sennō quādo iluentriculo nel suo ristrignersi (*non a*) | sitrova auere riseruato quella quātita dj sangue che llui vol ritenere e al|lora essendo interamē [sic: interamente] serrate esse tre porte allora le pariete siserāno ^c^on tā|ta potentia intorno alrimanēte del fugito sangue chelli e fforza che grā parte dj | quello sifugha desso vētrichulo e penjtri per li meati del parie(*de*)te dj mezo eppe|njtri nel sinistro ventrichulo ilquale assottigliato nella penetratione dellj | stretti meati siconverte (*quasi*) in jsspiriti vitali lassciando ongnj grosseza | inesso desstro vētriculo laqual grosseza –

[Fig.] a b c d e

[I] On the right ventricle
The right ventricle, having to give back more of the blood taken in than it retains, has its three doors [tricuspid valve-cusps] which are closed from within so arranged that they never shut with perfect closure until and unless the ventricle in its contraction is found to have retained that quantity of blood which it needs to retain. Then these three gates [cusps] being completely closed, the walls [of the right ventricle] close with such power round the remainder of the escaped blood with such a force as to make a great part of it issue from this ventricle and penetrate through the meati [pores] of the middle wall [interventricular septum] and so penetrate into the left ventricle. This [blood] subtilised by penetrating through the narrow meati is converted into vital spirits leaving all gross material in the right ventricle, which coarse material …

[II] della grosse(*za*) vissciiosita dj sangue | ce siragva nel desstro vētriculo ·
Jl sangue del destro ventriculo cherimane della sottiglieza del sangue che | penetra nel uentriculo (*desstro*) sinjsstro e viscio | e cqualc<h>e parte se ne cō|pone immjnute fila assimjlitudine del uermo deluētriculo dj mezzo alcer|vello e cqueste tali fili simultiplicano ^a modo di grossa e corta stoppa^ e allungho andare (*in modo*) e ssavi|luppano intorno alle corde de pa$jculi chesserano il uentriculo desstro | in modo che nella vechieza delli anjmalj la porta (*e*) nosipo bē serrare | e grā parte del sangue che dovea penetrare le strette porosita del pariete | di mezo nelsinjstro vētriculo alla creatiō depredetti spiriti sifugge per le | porte nō ben serrate (*e*) nel destro ventriculo superiore e per questo allj | vechi mācano tutti li spiritj esspesso (*p*) moiano parlando –

[II] On the gross viscosity of the blood which collects in the right ventricle
The blood in the right ventricle which remains behind after the subtilised blood penetrates into the left ventricle is viscous and a certain part of it is composed into minute fibres [fibrin threads] like the worm of the middle ventricle of the brain [the choroid plexus in the third ventricle]. And these fibres multiply like thick short tow and lengthen and wrap themselves around the cords of the membranes [chordae tendineae of the valve-cusps] which close the right ventricle [tricuspid valve] in such a way that with the ageing of the animal the orifice [of the valve] cannot close well and a large part of the blood which ought to penetrate the narrow porosities of the middle wall into the left ventricle for the creation of the above-mentioned spirits, escapes through the imperfectly closed valve orifice into the right upper ventricle [right atrium]. For this reason all the spirits are insufficient in the aged and they often die whilst speaking.

[III] dellufitio dello inferiore e ssuperiore | ventriculo desstro –
Lufitio del(*su*)llo inferiore essuperiore ventriculo desstro loinferiore | da in djposito il sangue al superiore ilquale proibiva illuj ilrisstringnersi e nō si potrebbe maj poi djlatarsi senō riauessi ilsangue che prima | rienpiena la sua capacita dal uentriculo superiore Il quale e atto a ristrī|gnersi perche non po auere aria chello riēpiessi ∂ e llaria © che sochorre il | ilocho che ochupaua il uentriculo superiore quando tenea indjposito il sā|gue del ristretto ventriculo inferiore corre a reienpiere illoco del | vētriculo superiore quādo sirisstrigne ed e dj quella della cassula del | core e essa casssula dando della sua aria allocho lassiato dal ri|strignere desso superiore vētriculo: e restaurata dal tirare per la | trachea piu aria nel polmone che non era ilsuo solito e per ques|sto senpre laria chessitira nelpolmone nō po essere equale –

[III] On the function of the lower and upper right ventricle
The function of the lower and upper right ventricle. The lower one gives a deposit of the blood which prevented it contracting to the upper one, and afterwards it would never be able to dilate unless it got back from the upper ventricle the blood which first filled its capacity. This [upper ventricle] is ready to contract because it cannot get air to fill it. Air which takes the place occupied by the upper ventricle when it holds the blood deposited in it by the contraction of the lower ventricle, flows in to replace the upper ventricle when it contracts, and it comes from that in the capsule of the heart [pericardium]; and this capsule having given away its air to the place left by the contraction of the upper ventricle is restored by drawing more air into the lung through the trachea than usual; therefore the air drawn into the lung cannot always be equal.

[IV] perche laria chessi tira nel polmone | no po senpre essere dequale mjsura

[IV] Why the air which is drawn into thelung cannot always be of equal measure

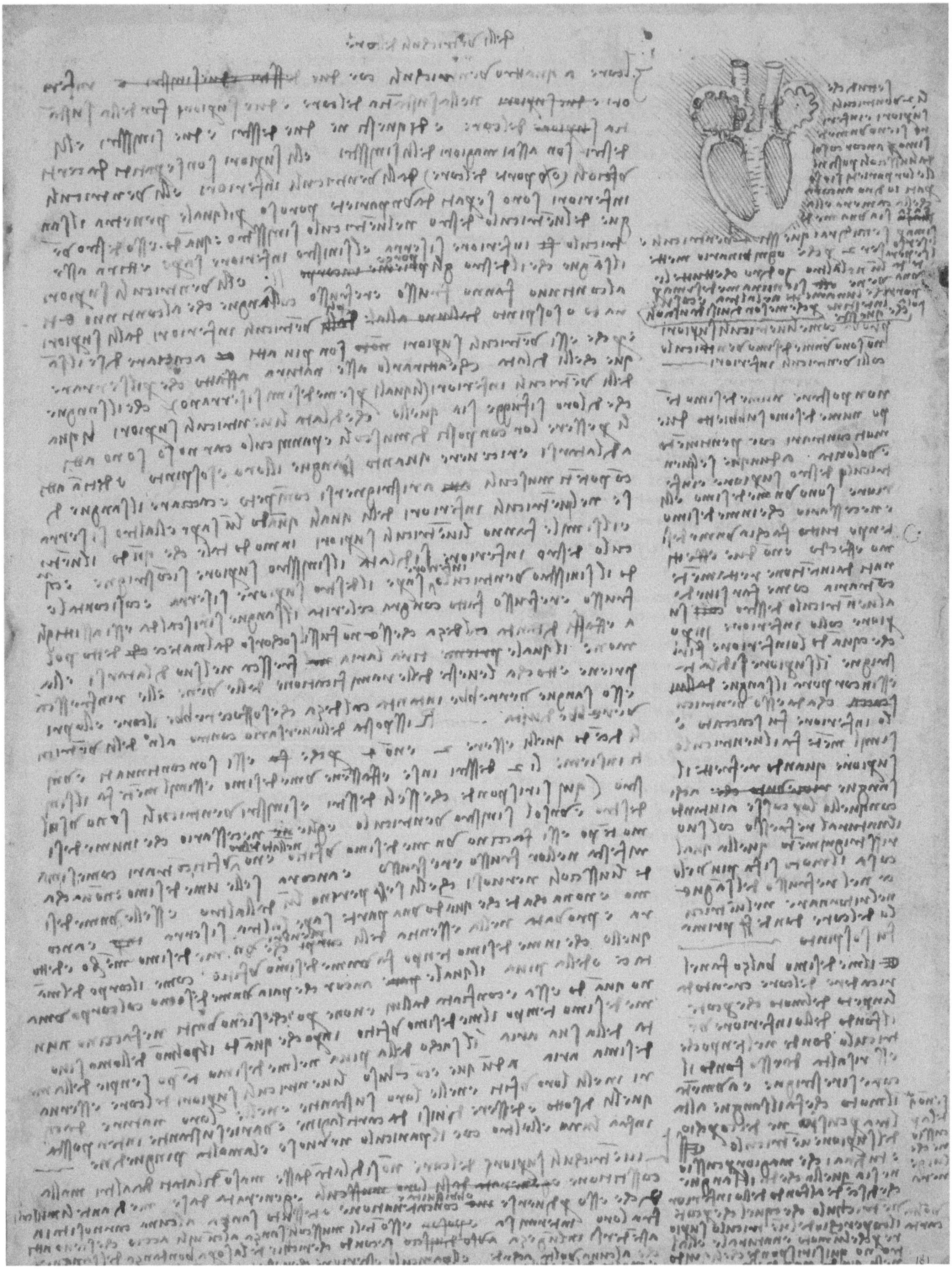

RL 19062 recto, The atria and ventricles of the heart. Leonardo da Vinci (Vinci 1452-Amboise 1519). c.1511–12. Pen and ink (Lent by Her Majesty The Queen. Royal Collection © Her Majesty Queen Elizabeth II)

<table>
<tr><th colspan="2">19062 r (C 13 r). Pen and brown ink 288×215 mm.</th></tr>
<tr><td>[Fig.]</td><td>[Fig.]</td></tr>
<tr><td>[I] delli vētrichuli del core</td><td>[I] On the ventricles of the heart</td></tr>
<tr><td>Il core a quattro ventrichuli coe due (desstri e due sinjstri e) inferi|ori e (due superiori) nellasustātia del core e due superiorj for della susā|tia (superiore) del core e dj questi ne due desstrj e due sinjsstrj ellj | destri son assai magiori delli sinjsstri ellj superiori son separati da certi | vscioli (over porte del core) dalli ventriculi inferiori elli ventriculi | inferiori sono separatj da vn pariete poroso per il quale penetra ilsan|gue del uētriculo destro neluētriculo sinjsstro e quādo esso destro vē|triculo (si) inferiore si serra elsinjstro inferiore sapre ettira asse | ilsāgue che il destro gli (prieme in corpo) ^porge^ //.</td><td>The heart has four ventricles, that is two lower in the substance of the heart and two upper [atria] outside the substance of the heart, and of these two are on the right and two on the left. The upper ones [atria] are separated by certain little doors (or gateways of the heart) from the lower ventricles. And the lower ventricles are separated by a porous wall through which the blood of the right ventricle penetrates into the left ventricle; and when the right lower ventricle shuts, the left lower one opens and draws into itself the blood which the right offers it.</td></tr>
<tr><td>elli ventriculi superiori | alcontinuo fanno frusso e refrusso colsangue che alcontinuo e ti|rato o sospinto (dal uno allaltr dallj) ^per lj^ vētriculi inferiori dallj superiori | e perche essi vētriculi superiori (nōte) son piu atti (ac) acaciare dj se il sā|que [sic: sangue] chelli djlata che attirarlo asse natura affatto che per il serrare | delli vētriculi inferiori (li qualj per se medesimo si serrano) che il sangue | che dj loro sifugge sia quello che djlata li uentriculi superiori li qua|li per essere lor conposti dj muscoli e pannjculo carnoso sono attj | a djlatarsi ericeuere quanto sangue illoro e sospinto od etiā attj | cō potēti musculi (atti) a ristrignersi con@peto e cacciare il sangue dj | se neljuētriculi inferiori delli quali quādo lū sapre ellaltro siserra | e il simjle fanno liuētriculi superiori in modo tale che qūdo [sic: quando] il uētri|culo destro inferiore si djlata ilsinjsstro superiore sicōstrigne e cquan|do il sinisstro ventriculo ^inferiore^ sapre ildestro superiore siserra e cosi con tale | frusso e refrusso fatto con gra celerita il sangue siriscalda essi assittigli|a effassi dj tanta caldeza chesse nō fussil sochorso dal matece (che) detto pol|mone il quale (prieme) tira laria (nel) fressca nel suo djlatarsi ella | priene [sic: prieme] e ttocha leueste delle ramjficatione delle vene elle rinfressca | esso sangue verrebbe intanta caldeza che soffocerebbe ilcore ello pri|verebbe dj ujta –</td><td>The upper ventricles [atria] continually make a flux and reflux of blood which is continually pulled or pushed through the lower ventricles from the upper. And since the upper ventricles [atria] are more suited for driving out of themselves the blood which dilates them than pulling it into themselves, Nature has so made it that by the closure of the lower ventricles (which close on their own) the blood which escapes from them is that which dilates the upper ventricles. These through being composed of muscles and fleshy membranes are suitable for dilation and for receiving as much blood as is pushed into them: they are also suited by their powerful muscles for contracting with impetus and driving out of themselves the blood into the lower ventricles, of which when one opens the other closes. And the upper ventricles [atria] do the same thing in such a way that when the right lower ventricle dilates the left upper ventricle contracts, and when the left lower ventricle opens, the right upper one closes. And so by flux and reflux made with great rapidity the blood is heated and subtilised and is made so hot that but for the help of the bellow called lungs, which draw in fresh air by dilating and pressing it into contact with the coats of the ramifications of the vessels refreshing them, the blood would become so hot that it would suffocate the heart and deprive it of life.</td></tr>
<tr><td>Rissposta dellauersario contro al numero delli vētricu|li dicēdo quelli essere 2 e nō 4 perche (so) essi son continuati e vnj|ti insieme li 2 desstri inse effassene £ medesimo e ssimjlmēte fa il sinj|stro ∂ quj si risponde che sse li desstri e sinjstri ventriculi sono u sol | destro vn sol sinjstro ventriculo egli e (ue) necessario che in u medesi|mo tēpo essi faccino vn medesimo vfitio e nō ^nellato destro^ vfiti cōtrari come si ma|njfesta nellor frusso e refrusso e ancora selle u medesimo e nō uacha|de li usscioli neruosi chelli si(se)perino lū dallaltro esselle vnmedesi|mo e non achade che quādo vna parte sapre laltra si serra (inper) e anco|ra e provata nella essentia delli (corpi) ^menbri^ che vn medesimo mēbro e detto | quello che in medesimo tenpo fa vnmedes-imo vfitio come il corpo del mā|tace o della piua ilquale (pare) ancor che paia vn medesimo colcorpo vma|no quādo essa e confiata dalluj e none pero che sieno vnjti ne faccino in un | medesimo tempo il medesimo vfitio inperoche quādo il polmō dellomo siuo|ta della sua aria il sacho della piua nel medesimo tēpo senpie della me|desima aria adūque e cōcluso li uentriculi superiori del core esser ua|ri inellj loro vfiti e nelle loro sustantie e nello loro natura da cq|quelli dj sotto edessere djuisi da cartilagine e varie sustantie interpossta | infra luno e llaltro coe il paniculo neruoso e lamolta pinguedine –</td><td>Reply to the adversary against the number of ventricles saying that they are two not four because they are continuous and united together, the two right by themselves making the same one, and the left doing likewise. Here one replies that if the right and left ventricles are one single right and one single left ventricle only, it is necessary that during one and the same time they should perform one and the same function, and not opposite functions on the right side as is manifested by their flux and reflux. Furthermore, if it is one and the same, there is no need for the membranous doors [valves] which separate one from the other; and if it is one and the same [ventricle] there is no need for one part to open when the other closes. This is further proved by the essential nature of a part of the body in that one and the same part is called that which at the same time performs one and the same function. Just as the body of a bellows or the bagpipes which may appear to be one and the same with the human body when it is inflated by it, but is not united with it, nor does it perform at the same time the same function; for when the man's lungs are emptied of air the bag of the bagpipes at the same time is filled with the same air. Therefore it is concluded that the upper ventricles of the heart [atria] are different in their functions, in their substance, and in their nature from those below, and that they are separated by gristle and various substances interposed between one and the other, that is sinewy membrane and much fat.</td></tr>
</table>

<table>
<tr><th colspan="2">19062 r (C 13 r). Pen and brown ink 288×215 mm.</th></tr>
<tr><td>Li uētrichuli superiorj del core nō sidjlatā dasse ma sō dilatati da altri malla | cosst<r>itione (e generata dalli loro mussculi) e generata da se medjante li muscolj | dj che esso per djuerse (inc) ^obbliquita e^ concatenatione o tessuto sanza alcuna carnosita in|fra lora intermjsa (e poson) essō tali musscoli sanza alcū ujli accioche sieno attj | a stēdersi in lūgeza (a vso dj ujscio) secondo che richiede la soprabondanza del sangue | che alcuna volta achade ell panjculo esteriore che ueste tali musscoli e pelliculoso carnoso | e molto djlatabile –</td><td>The upper ventricles of the heart [atria] do not dilate by themselves, but they are dilated by the other ventricles. However, their contraction is generated by themselves by means of their muscles. These through their different obliquities, interlacing or interweaving, without any fleshiness intervening between them, are muscles without threads so that they are suitable for stretching lengthwise in accordance with the demands of the superabundance of blood, which sometimes occurs. And the exterior membrane that coats these muscles is membranous, fleshy and very dilatable.</td></tr>
<tr><td>[II] se ttu djrai quessti 4 ventriculi e|ser 2 perche ognj binario mette | lū nelaltro jo djro che ttutte le | vene (elle) sien una medesima per|che lunamette nelaltra e cosi le | intestine perche me son djuisi da usciolj | che quesste –</td><td>[II] If you say that these four ventricles are two because each pair leads from one into the other, I will say that all the veins are one and the same because one leads into the other, and likewise the intestines because they are divided up by little doors likewise.</td></tr>
<tr><td>[III] se ttu djche | li 2 ventriculj | superiori e inferi<o>|re sieno vn mede|simo (per) ancor co sol | tali ussciolj posti ne|lle lor parietj siē se|parati io djro ancora | chella camera ella | (piaza) sala sia vna mede|sima per | eser so|l sepera|te da |vna | porta|pol^a^</td><td>[III] If you say that the two upper and lower ventricles are one and the same, although they are separated only by one such little door placed in their walls, I will further say that the chamber and the hall are one and the same because they are separated only by a small doorway.</td></tr>
<tr><td>[IV] prova come li uētriculj superiori | nō sono vnmedesimo ventriculo | colli ventriculi inferiori –
non po stare<i>numedesimo tē|po<i>nu medesimo subbietto due | moti contrari coe pentimēto | e volonta · adunque se lj uen|triculj destro superiore e infe|riore sono vn medesimo ellj | e necessario che inmedesimo | tenpo tutto fachia vn medesi|mo effecto e nō due effettj | nati da intētione rettamēte | cōtraria come far si uede | al ue$triculo desstro (col s) su|periore collo inferiore inpero|che cquādo lo inferiore si ri|strigne il superiore si djlata | a ssincorpora il sangue (da lluj | scacca) che da esso ventricu|lo inferiore fu scaccato e | simjlmēte fa iluentriculo | superiore quando refrette il | sangue (ricievuto che) a chi | conquello lopercosse aiutando | ilnatural refresso col suo | risstrignjmēto quella qual|cosa il moto si fa piu velo|ce nel refrusso del sāgue | nelritornare nel uētricu|lo del core donde (f) prima | fu sospinto –</td><td>[IV] Prove how the upper ventricles are not one and the same ventricle with the lower ventricle
At one and the same time in one and the same cause two contrary movements cannot exist, that is repentance and desire. Therefore if the right upper and lower ventricles are one and the same it is necessary that at the same time the whole should produce one and the same effect and not two effects arising from directly contrary tendencies as one sees produced by the right upper ventricle with the lower, since when the lower one contracts the upper one dilates and incorporates the blood which has been driven out of the lower ventricle. And the upper ventricle does likewise when it reflects the blood back to that ventricle from which it was percussed, helping the natural reflection with its contraction. This movement of the reflection of the blood is made somewhat more swiftly in its return into the ventricle of the heart from which it was first pushed out.</td></tr>
<tr><td>E ilmedesimo balzo fa nel | ricadere delcore creato da|linpeto delmoto che percote | il fondo dello inferiore vē|triculo donde nel tenpo che | ess<o>risalta daesso fondo il | core sirestrigne e avmēta | il moto che fa ilsangue alla|ltra percusjone delcoperchio | delsuperiore uētriculo E ss|e tu djrai che magior percussio|ne sia quella che da il sangue | che djscēde alfondo dello inferior | vētrichulo che cquel che percote | ilcoperchio deluētriculo superio|re perche lū moto e naturale ella|ltro no qui sirisponde che ljqujdo | nelliqujdo non pesa [:–:] se non quan|te la per|cussio|ne che | lui ge|nera</td><td>And the same rebound is made in the falling back of the heart created by the impetus of the movement which percusses the bottom of the lower ventricle from which, during the time that it leaps back from this base the lower ventricle is greater than that which percussed the cover of the upper ventricle because the one movement is natural and the other is not, here one replies that liquid is weightless in liquid [:–:] except by the amount which percussion generates.</td></tr>
<tr><td>[V] volta | carta</td><td>[V] Turn the page.</td></tr>
</table>

[illegible]
Majesty The Queen. Royal Collection © Her Majesty Queen Elizabeth II

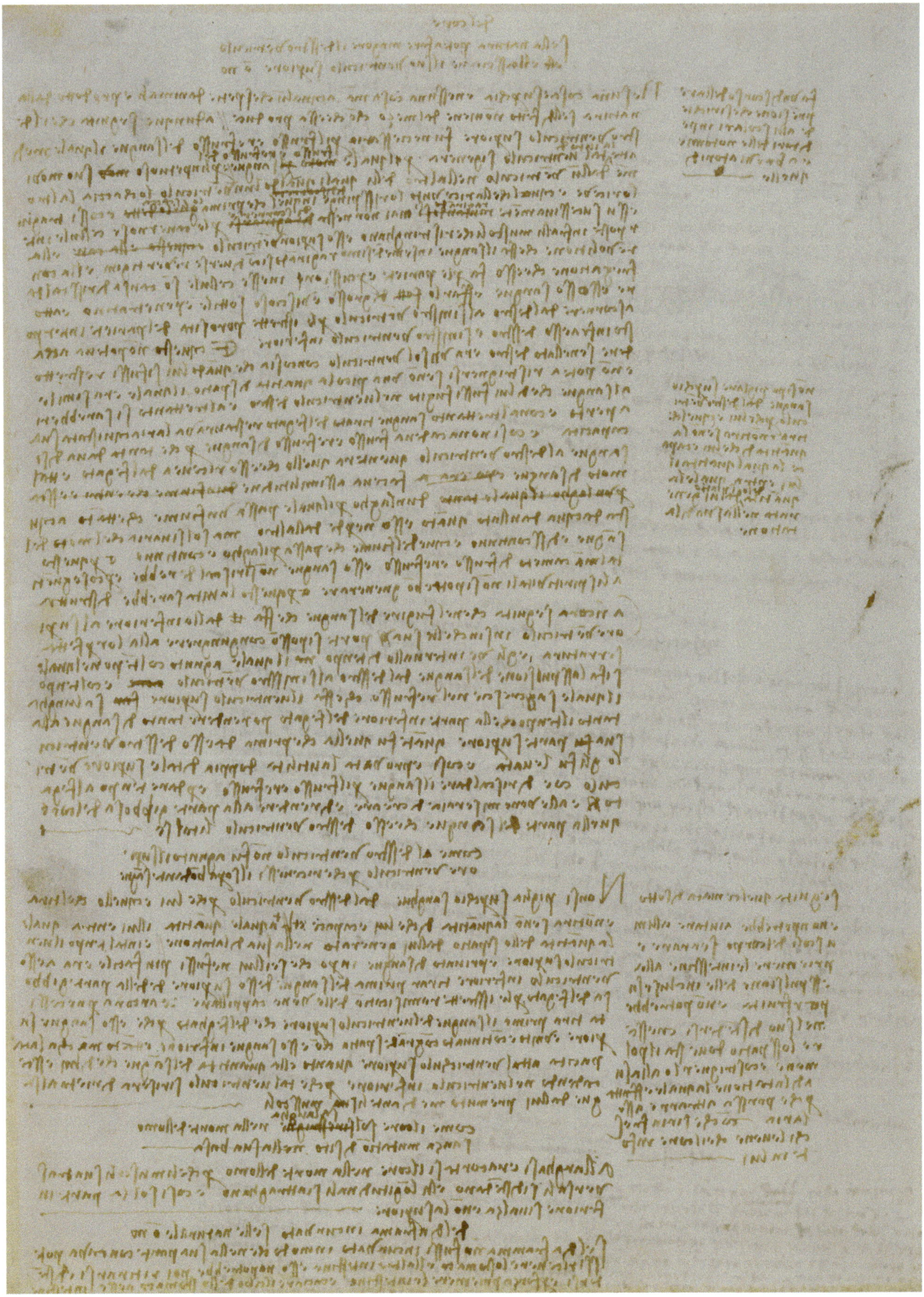

RL 19063 verso, Notes on the action of the heart, etc. Leonardo da Vinci (Vinci 1452-Amboise 1519). c.1511–13. Pen and ink (Lent by Her Majesty The Queen. Royal Collection © Her Majesty Queen Elizabeth II)

19063 v (C I 4 v). Pen and brown ink 29×212 mm.

[I] del core Sella natura potea fare magore il desstro vētriculo | (*e ll*) ellassciare il suo ventriculo superiore ō no Nesuna cosa e superchia e nessuna cosa mā [sic: manca] acqualūche spetie danimali e prodotto dalla | natura sell difetto nō ujene dal mezo chō che essa produce // adunque seguita che il de|stro ventriculo superiore fu neciessario per il frusso e refrusso del sangue ilquale medj|atte [sic: medjante] ^laiuto di^ tal uentriculo sigenera · per ilquale (*moto il*) ^frusso e refrusso del^ sangue e per linpetuoso (*mov*) movi|mē [sic: movimento] dallū vētriculo nellaltro delli quali quādo lun vētriculo lo chaccia laltro | lo riceve e cquel chella ricevuto lo rjsspinge ^(*di djscorrere*)^ inquel che prima (*glielo dette*) ^lo scacca^ e cossi tra giu | e ssu sucessiuamēte (*rantandosi*) ^ragirādosi^ mai non resta (*dj ragirarsi*) ^di scorrere^ per la cauernose cellule inte|rposte infralli muscoli che ristringhano esso superior vētriculo (*e cquesto e lla cau*) ella | revolitione che ffa il sangue in se medesimo ragirādosi cō djuerse revertigini ella con|fregatione che esso fa per le pariete e percussionj in esse cellule sō causa dj risscalda|re essosso sangue effarlo (*sott*) di grosso e viscioso sottile e penetratiuo e atto | a scorrere dal destro alsinjstro vētriculo per le istrette porosita del pariete interpo|sto infra esso desstro e sinjstro ventriculo inferiore E cquesto nō poteua acca|dere se nellato destro era vn sol ventriculo concosia che quādo lui sifussi restretto | e nō potea ristrignersi senō vna picola quatita di spatio ilquale era simile | al sangue che di lui fussi fugito neluentriculo destro e altrettanto si sarebbe ri|aperto e conaltrettanto sangue tirato del fegato restaurava lariacquistata sua | capacita e cosi non acadeua frusso e refrusso dj sangue perche tanta daua dj se | sangue al destro ventriculo quentera [sic: quant era] quello che esso riceuea dal fegato ettal | moto di sangue (*era era a*) faceua assimjlitudine (*dun fiume che entra e essca | per vn lagho il quale tanto*) dun lagho per ilquale passa vnfiume chettāto acqu|sta [sic: acquista] dacqua da ullato quāto esso neperde dallaltro ma sol siuaria chel moto del | sāgue e djsscontinuo e cque del fiume che passa per ilagho e continuo e per questo | tal mācameto dj frusso e refrusso esso sangue nō si riscalderebbe e per coseguēti|a [sic: consequentia] li spiriti vitali nō sipotrebō gnenerare e per questo la ujta sarebbe distrutta.	[I] On the heart Whether Nature could make the right ventricle larger and leave out its upper ventricle or not Nothing is superfluous and nothing is lacking in any species of animals or product of Nature, unless the defect comes from the means by which she produces it. Therefore it follows that the upper right ventricle was necessary for the flux and reflux of the blood which is generated through the help of this ventricle. Through this flux and reflux of the blood, and through the impetus of its movement from one ventricle to the other, when one ventricle drives it out the other receives it, and that which has received it repels it back into that which first drove it out; and so between revolving up and down successively it never ceases to flow through the cavernous recesses interposed between the muscles [pectinate] which contract the upper ventricles. And the revolvings made by the blood whirling round in itself in different eddies, and the friction that it makes with the walls, and the percussions in the recesses, are the cause of the heating of the blood, and of making the thick viscous blood fine and penetrating and suitable for flowing from the right to the left ventricle through the narrow pores of the wall interposed between the right and left lower ventricles. And this could not happen if there were only one single ventricle on the right side, for when it had narrowed itself down it would not be able to contract except by a small quantity of space equal to that of the blood which had escaped from it into the right [sic: left] ventricle; and it would open again by as much [space] as that of the blood drawn from the liver to restore its re-acquired capacity. And thus flux and reflux of blood would not occur because as much blood is given to the right ventricle as it has received from the liver. And such movement of the blood would behave like a lake through which flows a river which acquires as much water from one side as it loses from the other. However, the only difference is that the movement of the blood is discontinuous and that of the river flowing through the lake continuous, and from this lack of flux and reflux the blood would not be heated and consequently the vital spirits could not be generated and for this reason life would be destroyed.
∂ ancora seguita che nel fugire del sangue che ffa (*il*) dallo inferiore al superi|ore vētriculo insin chelle sua (*v*) porte sipossō congungnere alla lor perfetta | serratura egli ve interuallo dj tenpo (*ne*) ilquale agunto col tēpo nelquale | sifa lasspulsione delsangue dal desstro al sinjsstro vētriculo (*acre*) e col tenpo | ilquale sa (*g*)crescie nel refrusso cheffa iluentriculo superiore (*san*) salungha | tanto iltenpo chella parte inferiore del fegato po rendere tanto dj sangue alla | sua (*su*) parte superiore quāto fu quella che prima daesso desstro ventricu|lo gli fu leuata e cosi e provata la utiljta doppia di tale superiore vētri|culo coe dj riscaldare ilsangue per il frusso erefrusso e per dare tenpo alfega|to (*di*) e alle vene mjserajce di creare e dj rendere all parte gibbosa del core | quella parte delsangue che esso desstro ventriculo li tolse –	Furthermore it follows that in the escape of blood from the lower to the upper ventricle which goes on until its valves can join together to make a perfect closure, there is an interval of time. When this interval is added to the time in which the expulsion of the blood from right to left ventricle takes place, and to the time which is increased in its reflux by the superior ventricle, this prolongs the time so much that the lower part of the liver can give to its upper part as much blood as that which was first taken from it by the right ventricle. And so the double utility of the right upper ventricle is proved, that is to heat the blood by flux and reflux, and to give time for the liver and mesenteric [portal] veins to create and to give back to the gibbous part of the heart [atrium] that part of the blood which the right ventricle took from it.

19063 v (C I 4 v). Pen and brown ink 29×212 mm.

[II] come al desstro ventriculo nō fu agunto il superi|ore ventriculo perche riceussi il soprabōdante sāgue
Non si piglia superchio sanghue dal desstro ventriculo perche lui ecquello chel tira | e nō tira senō laquātita dj che luj e capace et(*L*)^t^a [sic: ttale] quale quātita illui entra quale | la quātita dello spatio dalluj generato nella sua djlatatione // e in tal tenpo iluen|triculo superiore e priuato dj sangue inperoche seilluj ne fussi piu facile era a esso | ventriculo inferiore tirar prima delsangue desso superiore de [sic: che] della parte gibbo|sa del fegato per le istrette ramjficatiō delle vene cappillare : e ancora per necessi|ta tira prima ilsangue del uentriculo superiore che delfeghato perche esso sangue su|periore e vnjto e cōtinuato cō grāde spatio chō esso sangue inferiore ettāto mācha la ca|pacita attal ventrichulo superiore quanto ella quantita delsāgue che dj luj essce | cadendo neluentriculo inferiore perche tal uentriculo sirisera djrieto al sā|gue da lluj premuto medjante lisua musscoli –

[II] How the [upper] ventricle was not added to the right [lower] ventricle in order to receive the superabundant blood
Superfluous blood is not taken up by the right ventricle because it is that which draws it in, and it does not draw it in unless it has the capacity for its quantity; and that quantity enters it which is the quantity of space [volume] generated by it in its dilation [diastole]. And in that time the upper ventricle [atrium] is deprived of blood because any blood that might be in it would be drawn more easily first into the lower ventricle from the upper than from the gibbous part of the liver through the narrow ramifications of its capillary veins; and furthermore it necessarily first draws blood from the upper ventricle rather than from the liver because this upper blood [in the atrium] is united and continuous with the lower blood [in the ventricle] over a large area. And the capacity of the upper ventricle [atrium] diminishes by the same amount as the quantity of blood which issues from it and falls down into the lower ventricle because this [upper] ventricle contracts behind the blood which is pushed out by means of its muscles.

[III] come il core sol (*si restrigne*) ^salargha^nella morte dellomo | sanza mutatiō dj sito nella sua basa –
Allarghasi e racortasi il core nella morte dellomo perche li muscolj sua tras|versalj sidjstēdano elli lōgitudjnali siattraghano e cosi sol la parte in|feriore sinalza e nō la superiore –

[III] How the heart widens only at the death of a man without changing its position at its base
The heart widens and shortens at the death of man because its transverse muscles extend and its longitudinal muscles are drawn up and thus only the lower part is raised and not the upper.

[IV] deldjaframa incurvato selle naturale ō no sel djaframma nō fussi incurvato inmodo che nella sua parte concava pote|ssi riceuere lo stomaco e llaltre intestine esso nopotrebbe poi ritirarsi e djstē|dersi e per forza priemere le intestine e caciare ilcibo dello stomaco aesse intestine [:–:] seguita quel c<h>e māca dj sotto | e non potrebbe aiutare alli m|uscoli delcorpo serrare e | priemere leintesstine alla | esspulsione delle incluso su|perfruita e nō potrebbe | nelsuo djstēdersi cressce|re losspatio doue sta il pol|mone e costrignerlo alla su|a djlatatione laquale effata | perche porssa [sic: possa] attrarre asse | laria cō che si rinfres|chi le uene che il core infō|de in lui –

[IV] On the curved diaphragm, whether it is natural or not
If the diaphragm were not curved in such a way that it could receive the stomach and other intestines into its concavity it would not then be able to retract and expand, and with force press the intestines and drive the food from the stomach into the intestines [:–:] Here follows what is lacking, below. And it would not be able to help the muscles of the body to squeeze and press on the intestines for the expulsion of their enclosed superfluities, and it would not by being stretched be able to increase the space where the lung is situated and compel its dilatation, which is made in order that is should draw air into itself, with which the veins which the heart distributes to it are refreshed.

[V] fa vn djcorso della re|prēsione che sirichie|de alli scolari inpe|djtori delle notomje | e abreviatori dj | quelle –

[V] Make a discourse on the censure required by students who hinder anatomy, and abbreviators thereof.

[VI] nō sipo piglare superchio | sangue dal destro vētri|culo perche lui e cquel che | tira e nō la | quātita dj che lui e capa|ce la qual quātita il|lui entra quale la | quātita ^de lo spatio^ da lui gene|rata nellasua djla|tatione

[VI] Superfluous blood cannot be taken up by the right ventricle because this is the one which draws it in, and it draws in only the quantity for which it has the capacity which quantity enters it; and this is the quantity of space produced by it in its dilatation.

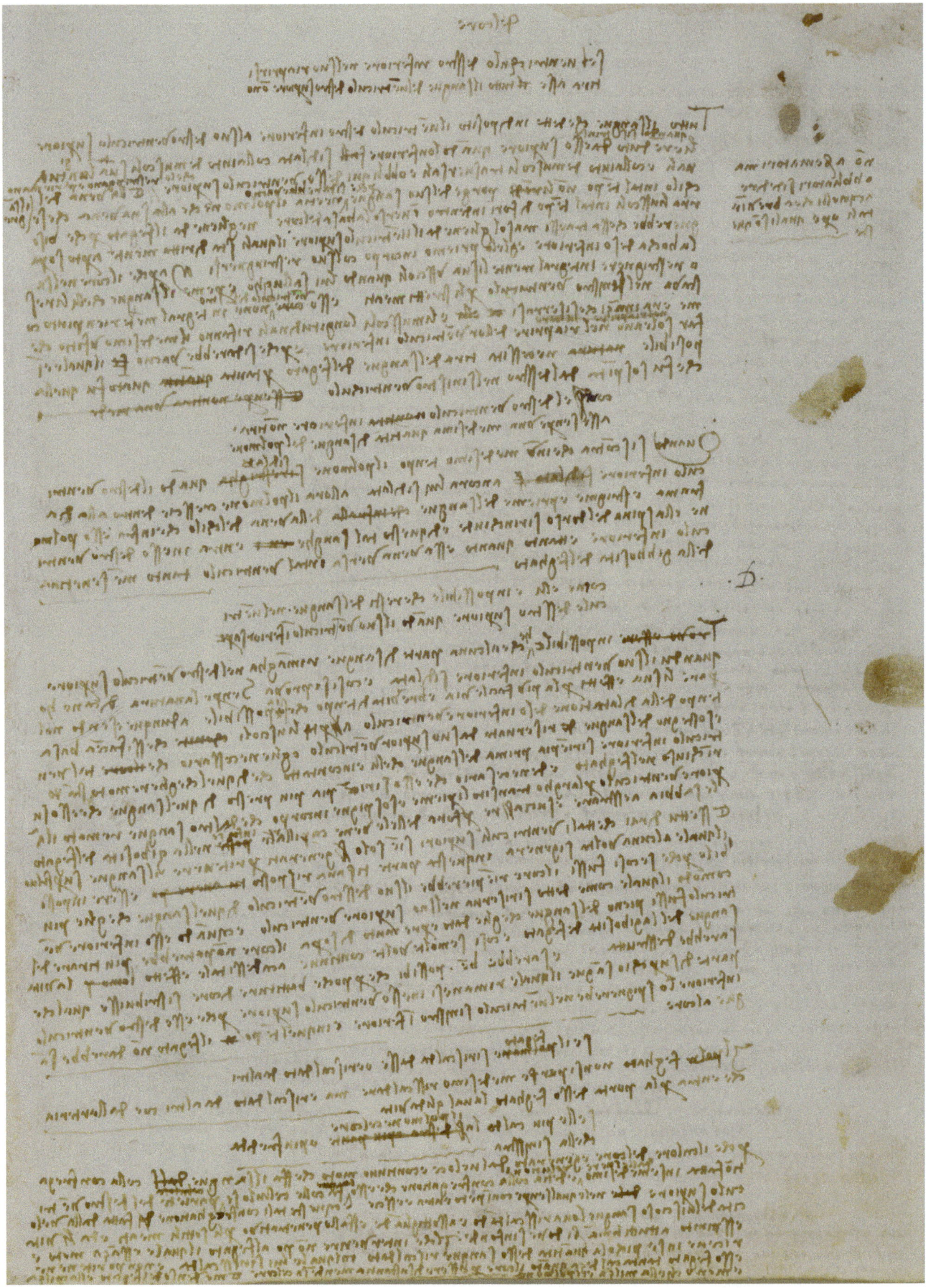

RL 19063 recto, Notes on the heart and motion of the blood. Leonardo da Vinci (Vinci 1452-Amboise 1519). c.1511–13. Pen and ink (Lent by Her Majesty The Queen. Royal Collection © Her Majesty Queen Elizabeth II)

19063 r (C I 4 r). Pen and brown ink 29×212 mm.

[I] del core
sel uentrichulo desstro inferiore nelsuo riaprirsi | tira asse (*il*) tutto il sangue deluētriculo destro superiore ō no
Tutto ilsangue che dette in djposito il uētriculo destro inferiore alsuo destro ventriculo superiore | ^quando si strinse^ li e rēduto da esso superiore quādo lonferiore (*sall*) sidjlata collaiuto demuscoli sua (*tr*) lon^gi^tudj|nali e collaiuto demuscoli trasuersali e obbliqui desso ventriculo superiore ^che lo restringano e priemano | il sā|gue^ //
E lla vena del | chilo intal tēpo nō li (*rēde*) porge delsuo sangue ^perche si darebbe vacuo^ neetiā il polmone che alla sua vena che se|rra liusscoli intal tēpo dj fori in dentro ĩuerso la basa del core ne gliene da il fegato perche biso|gnerebbe chessa traessi ma sol gliene da il uētriculo superiore ilquali sta djritta mente aperto sopra | la bocha deso inferiore e glelo prieme incorpo colsuo restringnersi Ma perche il core nelsu|o restrignere integralmente lisua vsscioli quando lui sallugho e preme il sangue chellj lj res|stava nel sinjstro ventriculo per li stretti meati esso (*core*) ^vētriculo des tro^ none integralmēte rienpiuto co|me era ināzi che si serrasi (*e elli*) e li musscolj longitudjnali rifanno il medesimo vfitio che | far soleano nel riaprire ^(*nel riaprire il loro*) ^ dellor vētriculo inferiore e perche si darebbe vacuo (*fe*) ilquale e @|posibile (*natura*) necessita tira delsangue del fegato per tanta (*quātita*) quāto fu quella | che fu sospĩta dal desstro nelsinistro ventrichulo (*E ssenpre non tira vna mede*) –

[I] On the heart
Whether the right lower ventricle in its reopening draws in all the blood of the upper right ventrical [atrium] or not
All the blood which the right lower ventricle gave in deposit to its right upper ventricle when it contracted is returned by the upper one when the lower one dilates with the aid of its longitudinal muscles. And this is assisted by the transverse and oblique muscles of the upper ventricle which contract it and press out the blood.
And the vena cava during this time does not put any of its blood into it [the lower ventricle] because it would give rise to a vacuum; neither does the lung which has its own vessel [pulmonary artery] the little doors of which [cusps of the pulmonary valves] during that time close from without inwards towards the base of the heart; nor does the liver give any [blood] to it because it would have to be drawn from it; indeed only the upper ventricle gives it any [blood]. This opens directly above the mouth of the lower [ventricle] and presses it [the blood] into its body by its contraction. But because of the heart's complete contraction of its little doors [valves] when it lengthens and presses the blood remaining in it into the left ventricle through the narrow pores, the right ventricle is not completely refilled as it was before it closed; and the longitudinal muscles again perform the same function as usual in the reopening of their lower ventricle. And since a vacuum would be created, which is impossible, of necessity as much blood is drawn from the liver as that which was pressed out of the right into the left ventricle.

[II] (*comes*) el destro ventriculo (*nontira*) inferiore nō tira · | asse senpre vna medesima quātita di sangue del polmone
Quando siscōtra che in£ medesimo tenpo il polmone (*si restringha*) ^si dilata^ quādo il destro ventri|culo inferiore (*si dilata s*) ancora luj sidilata allora il polmone cresscie dentro alla dja|frama e strignie e prieme delsangue (*che infralla*) della vena del chilo che infra esso polmo|ne ella spina deldorso sirinchiude e dj questo tal sanghe (*en i*) entra in esso destro ventri|culo inferiore ettanto quanto essa vena versa intal ventriculo tanto mē senetira | della gibbosita del feghato –

[II] The right inferior ventricle does not always draw into itself one and the same quantity of blood from the lung
When it happens that at one and the same time the lung is dilated when the right lower ventricle also dilates then the lung is increased inside the diaphragm and it squeezes and presses blood out of the vena cava which is enclosed between the lung and the dorsal spine. Some of this blood enters the right lower ventricle [via the upper] and just as much blood as this vein pours into that ventricle so much the less is drawn from the gibbosity of the liver.

[III] come elli e inpossibile che resti delsangue nel uētri|culo desstro superiore quādo ilsuo vētriculo ĩferior sapre
(*Trovo essere*) inpossible ^he^ che alcuna parte dj sangue rimāgha nel destro vētriculo superiore | quandu ilsuo ventriculo inferiore si djlata e cosi si prova Senpre la natura djcano ho|perare lisua effetti per piv facile via e brevita dj tenpo che s(*o*)^ia^ possibile adunque esendo nel | tenpo della djlatatione delo inferiore ventriculo a(*b*)pertj li uscioli (*che rite*) chessi facēa basa | e sostegno delsangue (*de*) riseruato dasuo superior vētriculo egli e necessario che (*il core*) tal ven|triculo inferiore si riēpia prima delsangue chelli e incontatto che dj quel che gli e remoto stādo | rĩchiuso nel feghato ed e necesario che esso si riēpia piu presto dj quel sangue che esso su|periore ventriculo per largho transito li prieme e sospigne incorpo che daltro sangue remoto ilqua|le sabbia a esstrare e sucia(*s*)re per fora dellele vene capilla^r^(*l*) e (*poste*) ^infuse^ nella gibosita delfegato | Essettu djrai chettali ventriculi superiori siē solo (*s*) generati per ritenere il sangue superfluo | ilquale alcuna volta sigenera in questa parte tisara risposto (*tu auere po*) essere inpossi|bile perche se cosi fussi il core riēpierebbe ilsuo desstro vētriculo dj quel sangue che glie piu | comodo ilquale come detto siriserua nelsuo superiore ventriculo ec quādo esso inferiore vē|triculo fussi pieno del sangue che gli e dato e premuto dj sopra il core nō potrebbe piu tirare del | sangue della gibosita de fegato e cosi se molte volte continue acadessi tale effetto (*lomo p*) la vita | sarebbe desstrutta e sarebbe bē · possibi [sic: possibile] che per poche battiture dj core si stribuisse qualche | parte dj superchio sāgue ilquale rimanesi inesso ventriculo superiore perche esso destro ventriculo | inferiore lo spignerebe nel uētriculo sinjstro ĩferiore e inquel tēpo (*n*) il fegato nō darebbe sā|gue al core –

[III] How it is impossible for some blood to remain in the right upper ventricle when the right lower ventricle is opened
It is impossible for any part of the blood to remain in the right upper ventricle when its lower ventricle dilates, and this is proved thus: Nature, they say, always works her effects in the easiest way and in the shortest time possible. Therefore, since in the time of dilating the lower ventricle the little doors [valve-cusps] which provide the supporting base to the blood kept in the upper ventricle are opened, it is necessary that this lower ventricle should be filled first with the blood which is in contact with it rather than that which is enclosed within the liver and is distant from it. And it is necessary that it should be filled more rapidly with that blood which the upper ventricle presses and impels into its body through a wide passage, than with other distant blood which would have to be extracted and sucked out of the capillary veins intermingled in the gibbosity of the liver. And if you say that these upper ventricles are generated solely for retaining superfluous blood which is sometimes generated in this part, you will get the reply that this is impossible, because if it were so the heart would fill its right ventricle with that blood which is most convenient to it which, as has been said, is kept in its upper ventricle. And when the lower ventricle is full of blood, given to it and pressed into it from above, the heart cannot draw more blood from the gibbosity of the liver. Thus if this event were to happen many times continuously life would be destroyed. It would be quite possible that through a few beats of the heart some part of the superfluous blood which is kept in the upper ventricle would be distributed, because the lower right ventricle would press it into the lower left ventricle, and during that time the liver would not give any blood to the heart.

19063 r (C I 4 r). Pen and brown ink 29×212 mm.

[IV] se il (*polmone*) ^fegato^ siriscalda dasse o eriscaldato da altri Il (*polm*) feghato non si per se medesimo risscaldare ma e riscaldato da altri coe dallarteria | che entra per lo porta desso feghato lanal [sic: laqual] gli da vita –	[IV] Whether the liver heats itself spontaneously or is heated by others The liver cannot heat itself but is heated by others, that is by the artery which enters by the gateway of the liver [hepatic artery] which gives it life.
[V] sell piu caldo la (*s destra o piu parte*) ^il polmone c<h>el core^ o piu freda | chella sinjsstra perche il calore del core e generato dal ueloce e continuo moto che ffa ilsangue (*dal s*) colla confrega|tiō fatta in se medesimo ^dalle sue re volutioni^ ed etia colla confregatione che esso ^(*sangue*)^ fa colle cellulose ^(*celulosa*)^ pariete del destro vētri|culo superiore (*dele*) nelequal senpre con@peto entra e esscie e cqueste tali confreghatione (*dj*) fatta dalla velo|cita del uiscioso sangue lo ua risscaldādo e assottigliādo effallo penetrativo per li sottilj meati e da (*lj*) vita | esspirito a ttutti li mēbri dove sjnfonde : Il che intervenre nō po al fegato il quale essanza moto e | riceue in se pichola quātita desso sangue riscaldato colquale lui sirisscalda e non po ritenere | esso fegato tanta caldeza quāto il core per essere di sustanti mendēsa c<h>elcore e mē denso del fegato ella milza | e meno chella milza e il polmone –	[V] Whether the lung is hotter or colder than the left heart Because the heat of the heart is generated by the swift and continuous movement which the blood makes by its friction within itself through its eddyings, also by the friction which it makes with the pitted walls of the right upper ventricle into which it continually enters and escapes with impetus, these frictions made by the velocity of the viscous blood heat it, subtilise it, and make it penetrate through the fine meati [pores] giving life and spirit to all the organs into which is it infused. This cannot occur in the liver which is without movement and receives into itself a small quantity of the heated blood with which it is heated. And the liver cannot retain so much heat as the heart because it is of a less dense substance than the heart, and the spleen is less dense than the liver; and the lung is less dense than the spleen.
[VI] nō abreuiatori ma | obbliatori si de djre | a cquelli che abreviā | tali opere quali sō que|ste	[VI] Those who abbreviate such works as these should be called obliterators rather than abbreviators.

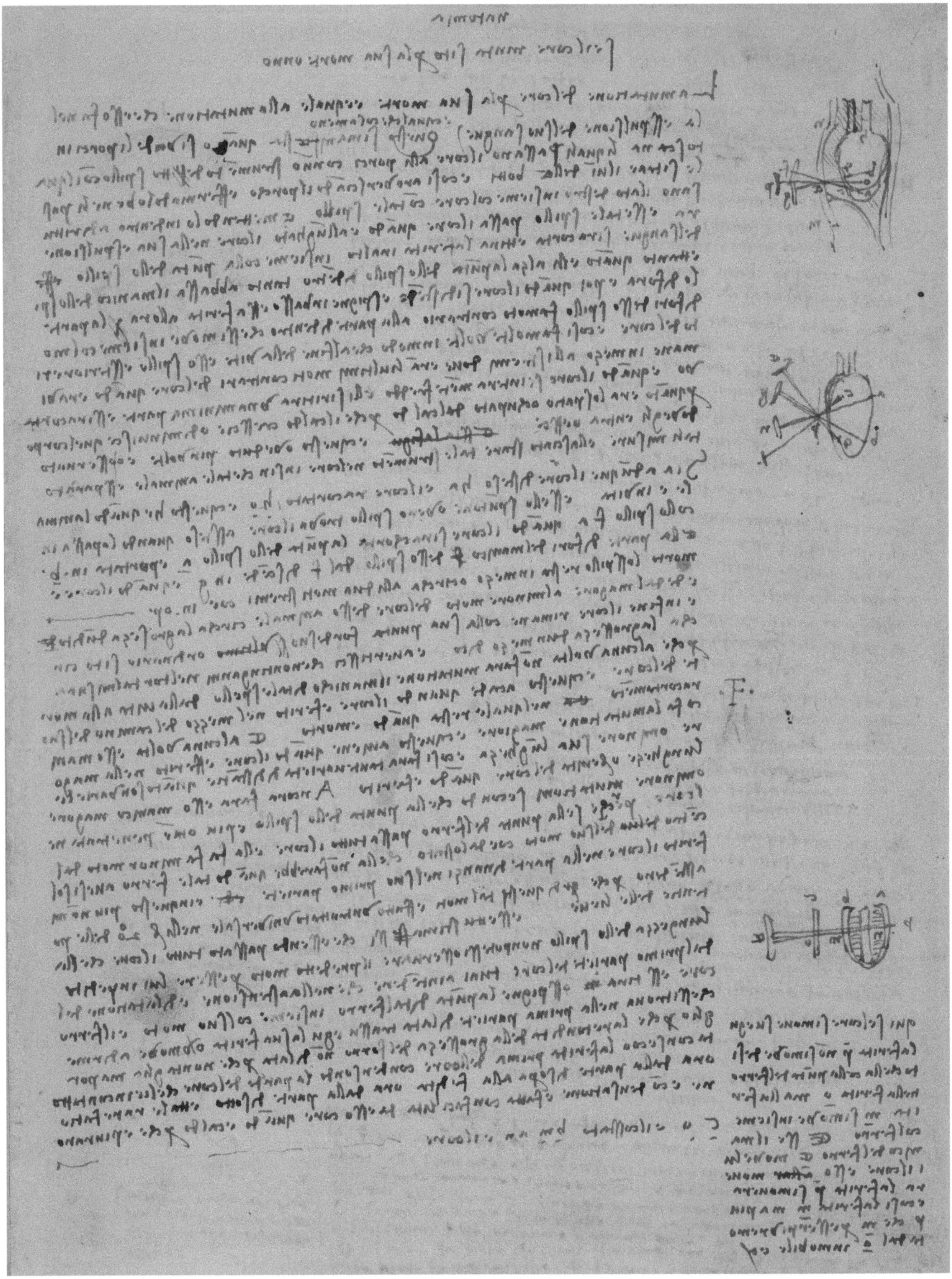

RL 19065 recto, The movement of the heart. Leonardo da Vinci (Vinci 1452-Amboise 1519). c.1511–12. Pen and ink (Lent by Her Majesty The Queen. Royal Collection © Her Majesty Queen Elizabeth II)

19065r (C I 6 r). Pen and brown ink. 29×211 mm.

[I] natomja
seil core muta sito per la sua morte o nno
La mutatione del core per la sua morte e equale alla mutatione che esso fa nel|la esspulsione delsuo sangue ^e cqualche cosa meno^ Questo simanjfesta quãdo si vede li porci in | toscana liqualj Passano ilcore alli porci conũo strumẽto de(*s*)tto spillo cõ ilqua|le sitrae iluī delle botti e cosi aroverscãdo il porcho effermadolo bene li pas|sano ilato [sic: illato] destro insieme col core cõtale spillo e mettendolo indentro adjrittu|ra esse tale spillo passa ilcore quãdo e allũghato ilcore nella sua espulsione | delsangue siracorta ettira laferita inalto insieme colla pũta dello spillo (*esse*) | e ttanto quãto elli alza la pũta dello spillo di dẽtro tanto abbassa ilmanico dello spi|llo djfora e poj quãdo ilcore sidjstẽde e spigne inbasso essa ferita allora la parte | djfori desso spillo famoto contrario alla parte djdentro che ssi move insieme col mo|to del core e cosi fa molte volte inmodo che alfine della vite [sic: vita] esso spillo essteriore ri|mane in mezo allj stremj doue erã liultimj moti contrari del core quãdo era vi|vo e quãdo ilcore si<a>interamẽte freddo elli siritira vna mjnima parte e ssiracorta | per quãto era lospatio ochupato dalcaldo perche ilcaldo cresscie o dimjnuisce quelcorpo | dovegli entra o essce (*Essia la figu*) e cquesto o veduto piu volte e o osseruato | tali mjsure e llasciato stare tale strumẽto nel core insin che tale anjmale essparato

[I] Anatomy
Whether the heart changes its position at death or not
The change of the heart through its death is equal to the change that it makes in the expulsion of its blood, and somewhat less. This is shown when they kill pigs in Tuscany where they pierce the hearts of the pigs with an instrument called a 'spillo' with which wine is drawn from casks. And so turning the pig on its back and fastening it down firmly they pierce its right side together with its heart with this 'spillo', pushing it straight inside. And if this 'spillo' pierces the heart when it is elongated, in its expulsion of blood the heart shortens and draws the wound upwards together with the point of the 'spillo'; and as much as it raised the point of the 'spillo' inside so much it lowers the handle of the 'spillo' outside. And then when the heart is distended and pushes this wound downwards then the part of this 'spillo' outside makes a movement contrary to that of the part inside which moves together with the movement of the heart. And this it does many times in such a way that at the end of life the external part of the 'spillo' remains in the middle between the two extremes at which were the last contrary movements of the heart when it was alive. And when the heart becomes quite cold it retracts by a very small fraction and is shortened by as much space as was occupied by its heat, because heat increases a body which it enters and diminishes it when it leaves. And this I have seen many times; and I have observed such measurements, and I have let such an instrument remain in the heart until the animal was cut up.

[Fig. I] a b c d e f g h m n o

[Fig. I] *a b c d e f g h m n o*

[II] Sia adũque il core djsteso h a e ilcore racortato h o e cquesto he quãdo lanima|le e in vita essello spũtone overo spillo trova ilcore assteso quando lo passa in | collo spillo f a quãdo il core sirachorta la pũta dello spillo a e portata in·b·| e lla parte djfori del manjco (*f*) desso spillo dal f djscẽde in g e quãdo ilcore e | morto losspillo resta in mezo o circha alli dua moti stremi coe in·o p·–

[II] Therefore let the distended hear be *h a* and the shortened heart be *h o*, and this is when the animal is alive. If the pointed tool or 'spillo' finds the heart extended when it is pierced by the 'spillo' *f a*, when the heart shortens the point of the 'spillo' *a* is carried to *b* and the outside part of the handle of the 'spillo' descends from *f* to *g*. And when the heart is dead the 'spillo' remains in the middle, approximately, of the two extreme movements, that is at *o p*.

[Fig. 2] a b c d e g n o

[Fig. 2] *a b c d e g n o*

[III] ede dal magore alminore moto delcore desso anjmale circha la groseza dũ djto (*de*) | e infine ilcore rimane colla sua punta for de suo (*s vltimo*) ordjnario sito cir|cha la grosseza dun mezo djto e auertissci che nontingannj neltor talmjsura | perche alcuna volta nõ fara mutatione ilmanicho dj tale spello dalla ujta alla mor|te delcore e cquesto acade quando il core e ferito nel mezzo del camino del suo | racortamẽto (*ea*) nelquale resta quãdo e morto E alcuna volta esso manj|co fa lamutatione magiore e cquesto aujene quãdo il core efferito nella mago|re o mjnore sua lũgheza e cosi fara tanteuarieta dj djstãtie quãto son variete | lungheze o breujta del core quãdo e ferito Ancora fara esso manjco magore | o mjnore mutationj secondo chella punta dello spillo e piu o mẽ penetrata ne|l core pe^r^che se lla punta delferro passa tutto il core ella fa fa minor moto dal | cẽtro (*del uo*) delsuo moto coe dalostato chella nõ farebbe quãdo tale ferro auesi sol | ferito il core nella parte djnanzi nel suo primo pariete (*e dj*) e inquesto piu nõ mj | asstẽdero perche (*gr*) dj questj tal moti effatto vntrattato vniversale nellibro 20° delle po|tentie delle lieue e ssettu stima(*st*)ssj che essendo passato tutto il core che lla | lungezza dello spillo non potessi osseruare il predetto moto per essere lui inpedjto | dal primo pariete del core tuai a intẽdere che nella astensione e djlatatione del | core ess<o>tira (*n*) osspigne la pũta dj tal ferro insieme colsuo moto e il ferro | chessitroua nella prima pariete djlata tra ssu e gu lasua ferita overmove a djr me|glio perche la retondita della grosseza del ferro nõ djlata perche nontaglia ma por|ta consecoo laferita prima dellcore condensondo [sic: condensando] la parte del core che liein contatto | ora dalla parte djsopra alla fedjta ora dalla parte djsotto ettale rarefatio|ne e cõdensatione e fatta con facilita da esso core quãdo e caldo perche e piuraro

[III] And from the greatest to the smallest movement of the heart of the animal there is about the thickness of a finger, and finally the heart remains with its apex out of its usual position by about a half a finger's breadth. Take heed that you are not mistaken in your measurements, because sometimes the handle of the 'spillo' will not make any change from the lift to the death of the heart. And this happens when the heart is wounded in the middle of the path of its shortening, where it remains when dead. And sometimes the handle makes a greater change, and this happens when the heart is wounded at its greatest or smallest length. And so it will move to many different distances as there are different lengths or shortnesses of the heart when it is wounded. Furthermore, the handle will make greater or lesser changes according to whether the point of the 'spillo' had penetrated more or less into the heart, because if the iron point passes through the whole heart [Fig. 2 and 3] it makes a smaller movement from the centre of the motion, that is, from its fulcrum than it would do if the iron had wounded the heart only in the front part of its anterior wall; and on this I will not expatiate further because a complete treatise on such movements has been made in the 20th book, On the Powers of Levers. If you should estimate that when the whole heart has been pierced the length of the 'spillo' could not indicate the above-mentioned movement because of being impeded by the front wall of the heart, you must understand that on extension and dilatation of the heart it pulls or pushes the point of the iron together with its own movement, and the iron which is in the front wall [of the heart] enlarges its wound upwards and downwards, or to express it better, moves it, for the roundness of the thickness of the iron does not enlarge it because it does not cut but carries with it the front wound of the heart, condensing that part of the heart in contact with it, now on the upper part of the wound, now on the lower part; and such rarefaction and condensation is easily made by the heart when it is hot because it is more rarefied.

19065r (C I 6 r). Pen and brown ink. 29×211 mm.	
[Fig. 3] a b c e m n o p	[Fig. 3] *a b c e m n o p*
[IV] c o e il cosstato b m a n e il core –	[IV] *c o* is the rib-cage, *b m a n* is the heart
[V] qui sel core si moue su e gu | la ferita # nō simove desi|to chella cōlla pūta delferro | nella ferita o ma lla fer|ita m si move insieme | col ferro E sse il ma|njco del ferro e move lu|i il core esso (*allar*) moue|ra la ferita # simouera | e cosi la ferita m ma piu | p che m per esser piv remo|ta dal o immobile &c	[V] Here if the heart moves up and down the wound *p* does not move away from the place where the point of the iron is in the wound *o* [in the chest wall]. But the wound *m* moves together with the iron. And if the handles of the iron, *e*, moves itself the heart will move the wound; *p* will be moved, and also the wound *m*, but *p* more than *m* because it is more distant from *o* the immovable point, etc.

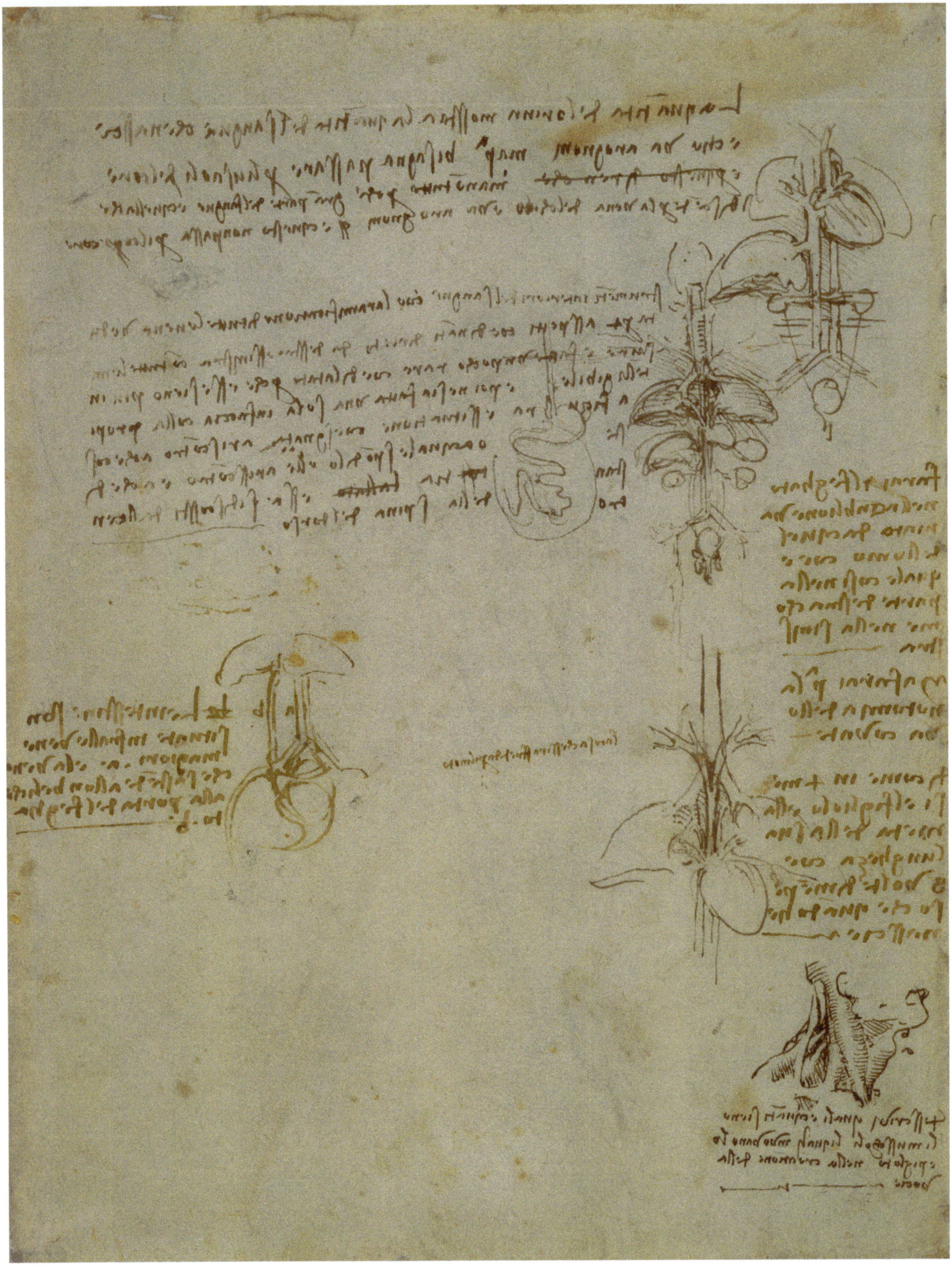

RL 19069 recto, Studies of the internal organs and great vessels. Leonardo da Vinci (Vinci 1452-Amboise 1519). c.1511–13. Pen and ink (Lent by Her Majesty The Queen. Royal Collection © Her Majesty Queen Elizabeth II)

19069r (C I 10 r). Pen and brown ink (three shades). 27×201 mm.	
[Fig. 1]	[Fig. 1]
[I] La quãtita delorina mosstra la quãtita delsangue che nassce | e che va arognonj ma prima bisagna [sic: bisogna] passare per liuscioli delcore | e (*per questo djren che*) manõ tutto perche grã parte delsangue ecquellache | djscẽde per la vena delchilo e va arognonj (*o*) e cquesto nonpassa per ilco (*q*) core	[I] The quantity of urine shows the quantity of blood which is produced and goes to the kidney; but first it must pass through the valves of the heart. But not all, for a great part of the blood descends by the vena cava and goes to the kidneys, and does not pass through the heart.
[Fig. 2]	[Fig. 2]
[II] strumẽti interiori delsangue coe laramjficatione djtutte leuene vedu|ta per 4 asspetti coe djnãti dirieto da desstra e ssinjsstra cõ tutte le mj|sure e fatte vnpocho rare coe djlatati perche esse sieno piu in|telligibile e poi nesia fatta vna sola infaccia colla propi|a figu ra essituatione coesegnata ariscõtro ache cos|ste o acquale spõdjlo elle arisscõtro e a che dj|stan (*to*)tia (*lallato*) essa sidoscossti dallcen|tro della spina del dorso	[II] Internal instruments of the blood, that is, the ramifications of all the vessels seen from four aspects, that is, from the front, behind, from the right and from the left, with all the measurements, and a little opened up, that is spread out, because they will be more intelligible. And then make a single one from the front with its own natural shape and position, that is, shown opposite those ribs or those vertebrae which it is opposite to, and by what distance it is separated from the spine of the back.
[Fig. 3, 4]	[Fig. 3, 4]
[Fig. 5] a b	[Fig. 5] *a b*
[III] (*be*) Leintesstine son | situtate infrallevene | magiori · a · ella vena | che sastẽde allonbelicho |<d>alla porta del fegha|to · b · –	[III] The intestines are situated between the major vessels *a* and the vein *b* which extends to the umbilicus from the gateway of the liver [porta hepatis].
[IV] faraj ilfeghato | nellaenblione va|riato dacquel | dellomo coe e|quale cosi nella | parte destra cho|me nella sinjs|tra –	[IV] You will make the liver in the embryo different from that in a man, that is, equal in its right and left parts.
[V] Ma farai prima la | notomja dello|va covate –	[V] But first make the anatomy of hatched eggs.
[VI] dj come in 4 me|si elfigliolo ella | meta della sua | lungheza coe | 8 volte dj mẽ pe|so che quãdo he | nasscie –	[VI] Tell how in 4 months the infant is half its length; that is eight times less weight than when it is born.
[VII] la cosa chessi raffredda per il moto –	[VII] That thing which is cooled by movement…
[Fig. 6] a b c	[Fig. 6] *a b c*
[VIII] desscrivi quali ecquãti sieno | li musscholi liqualj mvovano lo | epigloto nella creatione della | vocie –	[VIII] Describe which and how many are the muscles which move the larynx [*a*] in the creation of the voice.

RL 19071 verso, A sketch of the heart and great vessels. Leonardo da Vinci (Vinci 1452-Amboise 1519). c.1511–13. Black chalk on blue paper (Lent by Her Majesty The Queen. Royal Collection © Her Majesty Queen Elizabeth II)

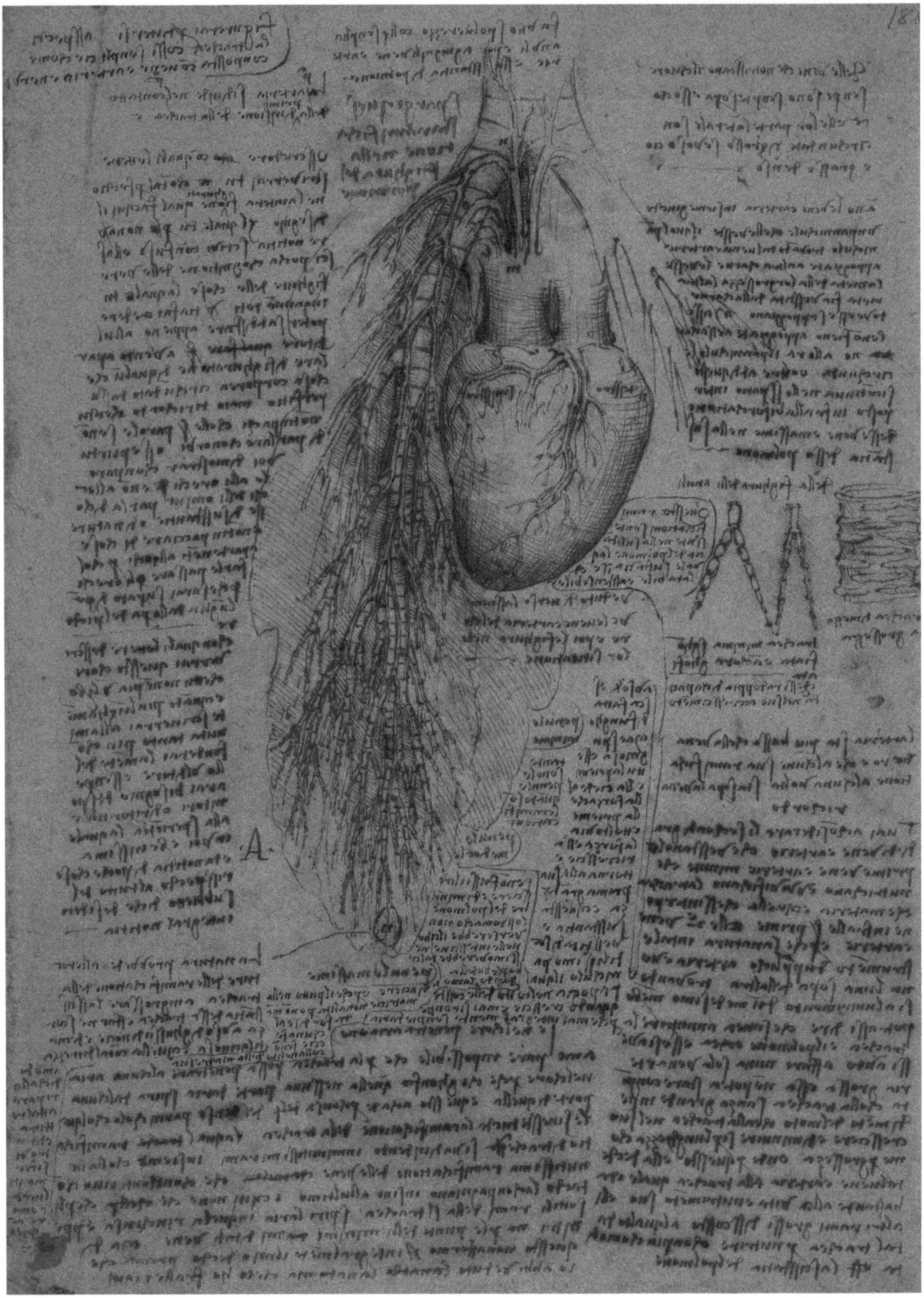

RL 19071 recto, The heart, bronchi and bronchial vessels. Leonardo da Vinci (Vinci 1452-Amboise 1519). c.1511–13. Pen and ink on blue paper (Lent by Her Majesty The Queen. Royal Collection © Her Majesty Queen Elizabeth II)

19071 r (C II I r). Pen and brown ink, on blue paper. 288×203 mm.

[Fig. 1] m n desstro | sinisstro

[Fig. 1] *m n* right; left

[I] delle vene che notrisscano ilchuore
senpre sono scoperte sopra esso cho|re elle lor parte laterale son | circhundate dj grasso sevoso cio|e grasso denso –

[I] On the veins which nourish the heart
They are always exposed on the heart; and their lateral parts are surrounded by tallowy fat, that is, dense fat.

[II] āno le vene e arteria insieme giu<n>cte | vnpannichule chelle vesste ilqual pā|njchulo trovādo taluena e arteria | appoggiate a altra charne levesste | lameta della lor grossezza laltra | meta fia vesstita dalla charne | dove esse sapoggiano Masse|le nō sieno appoggiate a essachar|(*cha*)ne allora ilpannjchulo le | circhunda o oltre addj questo | sicōtinua nellosspatio inter|posto infralla biforchatione | desse vene e massime nella sus|tātia desso polmone –

[II] The veins and the artery joined together have a membrane which clothes them. This membrane, finding this vein and artery apposed to the other flesh, clothes half their thickness; the other half being clothed by the flesh to which they are applied. But if they are not applied to this flesh, then this membrane surrounds them, or is continued farther and is continued into space interposed between the bifurcation of the vessels and maximally in the substance of the lung.

[III] larteria sta piu bassa chella vena | vero e che alchune sua ramjficha|tione alchune volta sta sopra la vena
richordo
Tu aj achōsiderare lisechondj gra|dj · di vene e arterie che vesstano le | prime vene e arterie mjnute che | nutrichano evvivifichano latrachea | che material ecqualla chessinterpo|ne infralle (*s*) prime elle 2^e vene | e arterie e perche lanatura intale | strumēto dupplicho arteria e ve|na luna sopra dellaltra trovando|si alnutrimento dū medesimo mēbro | potrassi djre che sauea annutrire la | trachea e il polmone (*e che a*) e sse siave|ssi avto affare nuna sola venarte|ria grossa essa nō potea stare congū|ta cholla trachea sanza grande inpe|djmēto delmoto chealla trachea nel suo | cressciere edjmjnuire si per lunghezza cho|me per grosseza onde per quessto ella decte | taluene e arteria alla trachea quale era | bastante alla vita e nutrimēto suo ellj | altri ranj grossi djsscossto alqualto [sic: alquanto] da | tal trachea per nutrire chonpiu chomodj|ta (*ess*) lasusstātia delpolmone

[III] The artery lies below the vein; it is true that some of its branches sometimes lie above the vein.
Memorandum
You have to consider the second order of veins and arteries which cover the first [i.e. bronchial veins and arteries] minute veins and arteries which nourish and vivify the trachea [the whole bronchial tree] the substance of which is interposed between the first and second [order] of veins and arteries; and why Nature duplicated artery and vein in such an instrument, one above the other, for the nourishment of one and the same organ. You could say that the trachea and the whole lung had to be nourished and if this had to be done by a single large venous artery this could not be joined to the trachea without great impediment of the movement which the trachea has in its increase and decrease in length as well as breadth. Wherefore this reason she [Nature] gave such veins and arteries to the trachea sufficient for its life and nourishment [bronchial veins and arteries] and the other big branches she separated somewhat from the trachea in order to nourish the substance of the lung more conveniently.

[IV] Osscrictore (*che*) cō quali lettere | scriverraj tu (*n*) chō tal perfectio|ne la intera fi^ghurati^one qual facquj il | djsegnjo Jl quale tu per (*la*) nō naver|re notitia scrivi confuso e llas|sci pocha chognjtione delle vere | fighure delle chose laquale tu | ingannādoti (*v*) ti a fai credere | poter saddjsfare appieno allul|djtore (*qual fare f*) avendo apar|lare dj fighuratione dj qualūche | chosa corporea circhūdato dassu|perfitio ma io tirichordo chetctu | nō tinpachi cholle (*s*) parole se nō | di parlare chon orbi osse purtu | voi djmostrar chonparo|le alli orechi (*di*) e nō alloc|chi delli omjni parla dj cho|se dj susstantie o dj nature | e nō tinpacciare dj chose | apartenēti allj ochi (*per*) chol | farle passare per li orechi | perche saraj superato djgrā | lūgho dallopera del picto|re –
chon quali lettere desscri|verrai quessto chore | chettu non ēpia £ libro | e quāto piu lūghamē|te scriuerrai alla mj|nuta tanto piu chō|fonderai lamēte del|llo vldjtore essenpre | arai bisognjo dj spo|njtori o djritornare | alla speriētia la quale | in voi e breujssima | e danotitia dj poche chose | risspecto altutto del | subbietto diche desideri | integral notitia –

[IV] O writer, with what words will you write with such perfection of the complete configuration which this drawing here composes. This you describe confusedly having no knowledge; and you leave little conception of the true shapes of things which you, deceiving yourself, make yourself believe capable of fully satisfying your auditors when you have to speak of the configuration of any corporeal thing surrounded by surfaces. But I remind you not to involve yourself in words unless you are speaking to the blind. If however you want to demonstrate with words to the ears and not to the eyes of men, speak of the essence or nature of things, and do not busy yourself with things belonging to the eyes by making them enter through the ears, for you will be far surpassed by the work of the painter.
With what words will you describe this heart so as not to fill a book? And the longer you write on the detail the more you will confuse the mind of the auditor, and you will always be in need of commentators, or of returning to experience, which with you is very brief and gives knowledge of few things relating to the whole of the subject of which you desire complete knowledge.

[V] Se nelchore penetra aria o no
ame pare inpossibile che per la trachea possa penetrare alchuna aria | nelchore perche chighonfia quella nessuna parte daria spira dalchuna | parte di quella e quessto achade per chausa (*delj*) del denso pan-njcholo chol qua|le siuesste tucta laramjfichatione della trachea la qual (*trache*) ramjficha|tiō djtrache^a^ (*cha*) si va djujdendo in mjnutissimj ramj insieme cholla mi|nutissima ramjfichatione delle vene (*che in chon*) che chon chontinuo chō|tacto lachonpagnjano insino allultimo e cquj none che (*chell per*) che per lj | sottili ramj della (*c*) trachea spiri laria in quella rinchiusa e ppe|njtri (*ne*) perle punte delli mjnjmi ranj dj talj vene Ma dj | quessto non affermo (*i*) integralmēte ilmjo decto prima che | io abbi veduto (*lanache*) lanatomja cheio ho fralle manj

[V] Whether air penetrates into the heart or not
To me it seems impossible that any air can penetrate into the heart through the trachea because if one inflates it, no part of the air escapes from any part of it. And this occurs because of the dense membrane with which the entire ramification of the trachea is clothed. This ramification of the trachea as it goes on divides into the most minute branches together with the most minute ramification of the vessels which accompany them in continuous contact right to the end. It is not here that the enclosed air is breathed out through the fine branches of the trachea and penetrates through the pores of the smallest branches of such vessels. But concerning this I shall not wholly affirm my first statement until I have seen the anatomy which I have in hand.

<table>
<tr><td colspan="2">19071 r (C II I r). Pen and brown ink, on blue paper. 288 × 203 mm.</td></tr>
<tr><td>[VI] Quesste ramj|fichationj son te|ssute nella susstā|tia del polmone la q|vale sustātia ī se e dj|latabile easstensibile [:–:] a vso dj es|sca fatta | dj fungho | Ma espu|gnjosa esse | ttu la priemj | ella ciede al|lla forza che | lla prieme | e ttolto via | la forza essa | ricresscie e | ttorna alla sua | prima grādez|za e cquessta | susstantia e | vesstita dj soc|tilissimo pa|njchulo ilqual | sapogia nelrotto delle cosste | quando cresscie e mai si ronpe | perche mai integralmente senpie daria</td><td>[VI] These ramifications are woven into the substance of the lung, which substance is in itself dilatable and extensible [:–:] like tinder made from fungus. But it is spongy and if you press it yields to the force that presses it; and if the force is taken away it increases again and returns to its previous size. And this substance is clothed by a very thin membrane [pleura] which is applied to the space between the ribs. When it [the lung] enlarges it is never ruptured because it is never completely filled with air.</td></tr>
<tr><td>[Fig. 2, 3]</td><td>[Fig. 2, 3]</td></tr>
<tr><td>[VII] trachea mjnjma sghō|fiata e āchora ghōfi|ata – | chessi radoppia dj chapaci|ta nelsuo accresscimēto</td><td>[VII] Smallest trachea uninflated and inflated; which is doubled in capacity in it its expansion.</td></tr>
<tr><td>[Fig. 4]</td><td>[Fig. 4]</td></tr>
<tr><td>[VIII] della fighura delli anulj</td><td>[VII] On the shape of the rings [of the trachea]</td></tr>
<tr><td>[IX] trachea dj mezza|na grossezza</td><td>[IX] Trachea of medium size</td></tr>
<tr><td>[X] fa vno spoluerezzo collj senplici | anvli e poi agiugnjli vene e arte|rie e ssusstantia dj polmone ·</td><td>[X] Make a pounce of the simple rings and then add the veins and arteries and substance of the lung.</td></tr>
<tr><td>[XI] spargi ques|sta ramjficha|tione nella | fighura del | pormone</td><td>[XI] Spread out this ramification in the drawing of the lung.</td></tr>
<tr><td>[XII] figurerai per diuersi asspectj | la trachea cossi senplicie chome | conpossta cō uene e arteria e nervj</td><td>[XII] You will draw from different aspects the trachea, both simple and compound, with veins, artery and nerves.</td></tr>
<tr><td>[XIII] la ^p^a^ arteria sidjujde nel ^(*p^o*)^ contatto | della ^prima^ djujsione della trachea e […]</td><td>[XIII] The first artery divides at its point of contact with the first division of the trachea and …</td></tr>
<tr><td>[XIV] Lanatura provvede alleroc|ture delle ramjfichatione della | trachea a ingrossare lassus|stātia dessa trachea e ffarne scor|za a vsodj ghussci dj nocie ed e car|tilacinosa e cquella cōtal durezza | a modo | dj challo | ripara | a ttal ro|ttura | e di den|tro vi | si fer|ma po|luere | e omo|re ac|queo</td><td>[XIV] Nature prevents ruptures of the ramifications of the trachea by thickening the substance of this trachea and making thereof a crust like a nutshell; and it is cartilaginous, and this, with hardness like a callus, repairs the rupture, and dust and watery humour remain inside.</td></tr>
<tr><td>[XV] veduto dj uerso la ssciena
ve le uene e arteria del cho|re e poi le fighura nelle | lor situatione –</td><td>[XV] Seen from the back.
Observe the veins and the artery of the heart and then draw them in their situations.</td></tr>
<tr><td>[XVI] penule | mjnjme</td><td>[XVI] Smallest lobes</td></tr>
<tr><td>[XVII] penule | medjocle</td><td>[XVII Medium lobes</td></tr>
<tr><td>[XVIII penule massime</td><td>[XVIII] Largest lobes</td></tr>
<tr><td>[XIX] tante | sono le | penule | quāte sō | le ramjfi|cationj</td><td>[XIX] There are as many lobes as ramifications</td></tr>
<tr><td>[XX] senō fussi il cre|sciere e djmjnu|ire delpolmone | lo sstomacho non | verserebbe ilcibo | nelle intesstine ne | ssimoverebbe talci|bo perle budella | stando lomo a djaciere e perche il putto nella | matricie nonalita pero nō mā|da for djse al|chuna fe|ccie se nō | collanelito della madre sua</td><td>[XX] But for the increase and diminution of the lung the stomach would not pour food into the intestine, nor would such food be moved through the bowel when a man is lying down. And since the child in the womb does not breathe, therefore it does not expel any faeces except through the breathing of the mother.</td></tr>
</table>

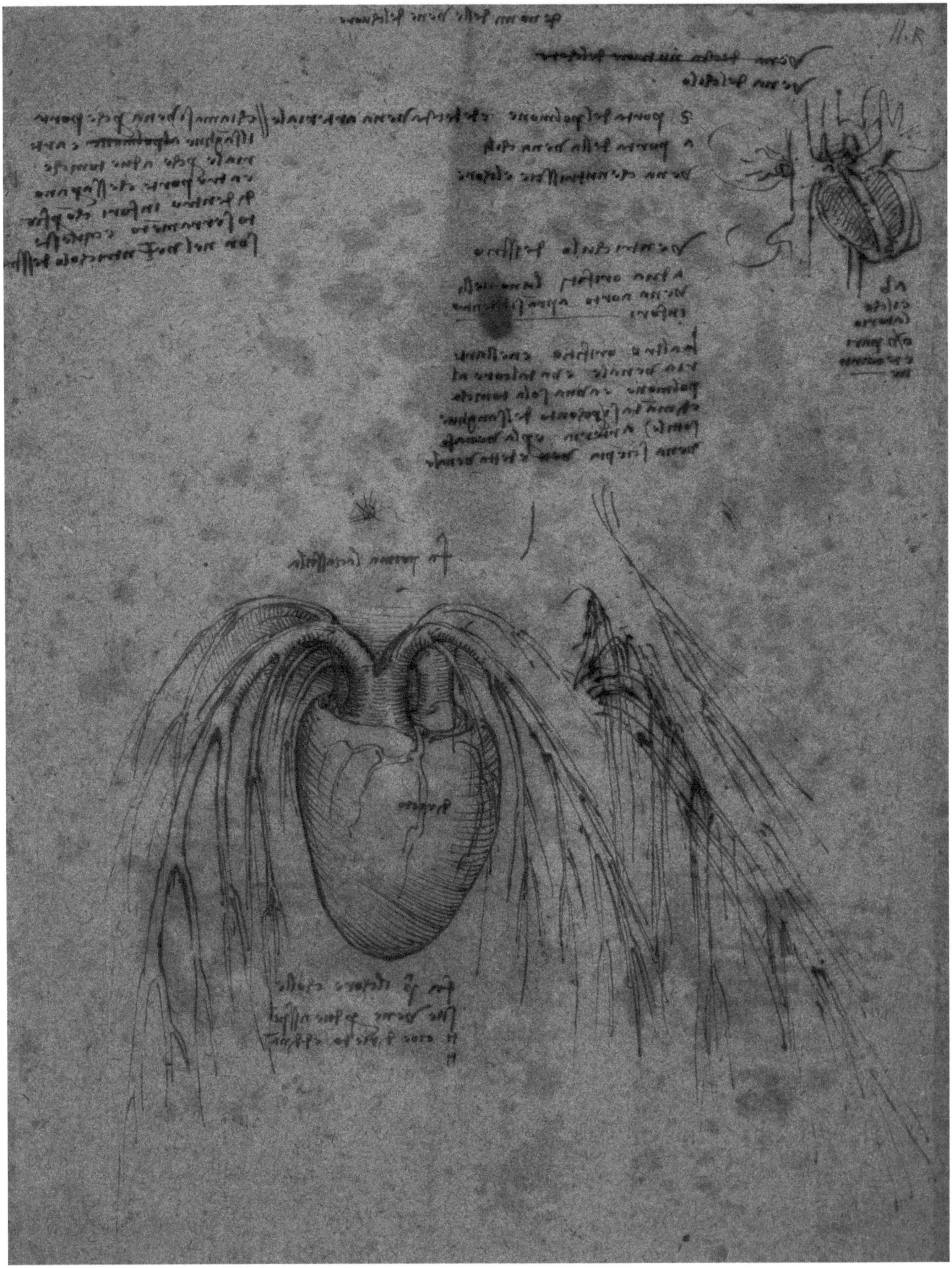

RL 19072 verso, Studies of the heart and pulmonary vessels of an ox. Leonardo da Vinci (Vinci 1452-Amboise 1519). c. 1511–13. Pen and ink on blue paper (Lent by Her Majesty The Queen. Royal Collection © Her Majesty Queen Elizabeth II)

	19072 v (C II v). Pen and brown ink.
[I] de nomi delle vene del chuore	[I On the names of the vessels of the heart
[Fig. 1] a b s	[Fig. 1] *a b s*
[II] (*vena decta nuticie del chore*) Vena del chilo s porta del polmone ed e decta vena arteriale //chiamasi vena perche porta | il sãghue (*alpolmone*) e arte|riale perche a due tonjche | e atre porte che ssaprano | dj dentro in fori chõ perfec|to serramẽto e cqueste | son nelve( )ntricholo desstro a porta della vena chilj vena che nutrisscie el chore	[II] (*Vein called the nourisher of the heart*) Vein of the chyle [vena cava] *s*, the gate of the lung, and it is called arterial vein. It is called vein because it carries blood (*to the lungs*) and arterial because it has two tunics; and it has three gates [valve-cusps] which open from within outwards with perfect closure, and these are in the right ventricle. *a*, gate of the vena chilis [vena cava] Vein which nourishes the heart.
[III] a b | e il cho|latoria | over pari|ete comu|ne -	[III] *a b* is the sieve or the common wall [septum].
[IV ventrichulo desstro | a dua orifitj luno nelli | vena aorta aprãsi dj dentro | in fori -	[IV] The right ventricle has two [valvular] orifices, one into the 'vena aorta'; they open from within outwards.
[V] laltro orifitio enelarte|ria venale e va dalcore al | polmone e a vna sola tonicha | e djmãdasi (perchonto delsanghue | sottile) arteria e per la (*vena sc*) | vena sciẽpia (*ven*) e detta venale	[V] The other orifice is in the venous artery which goes from the heart to the lung and has only one tunic and it is called artery (because of the subtilised blood) and vein because of the enlarged vessel [atrium].
[VI] fa prima lachassula	[VI] First make the capsule [pericardium].
[Fig. 2] djrieto	[Fig. 2] behind
[Fig. 3]	[Fig. 3]
[VII] fa p il chore cholle | sue vene per due asspe|ti cioe djrieto eddjnã|ti	[VII] First make the heart with its vessels from two aspects, that is from behind and in front.

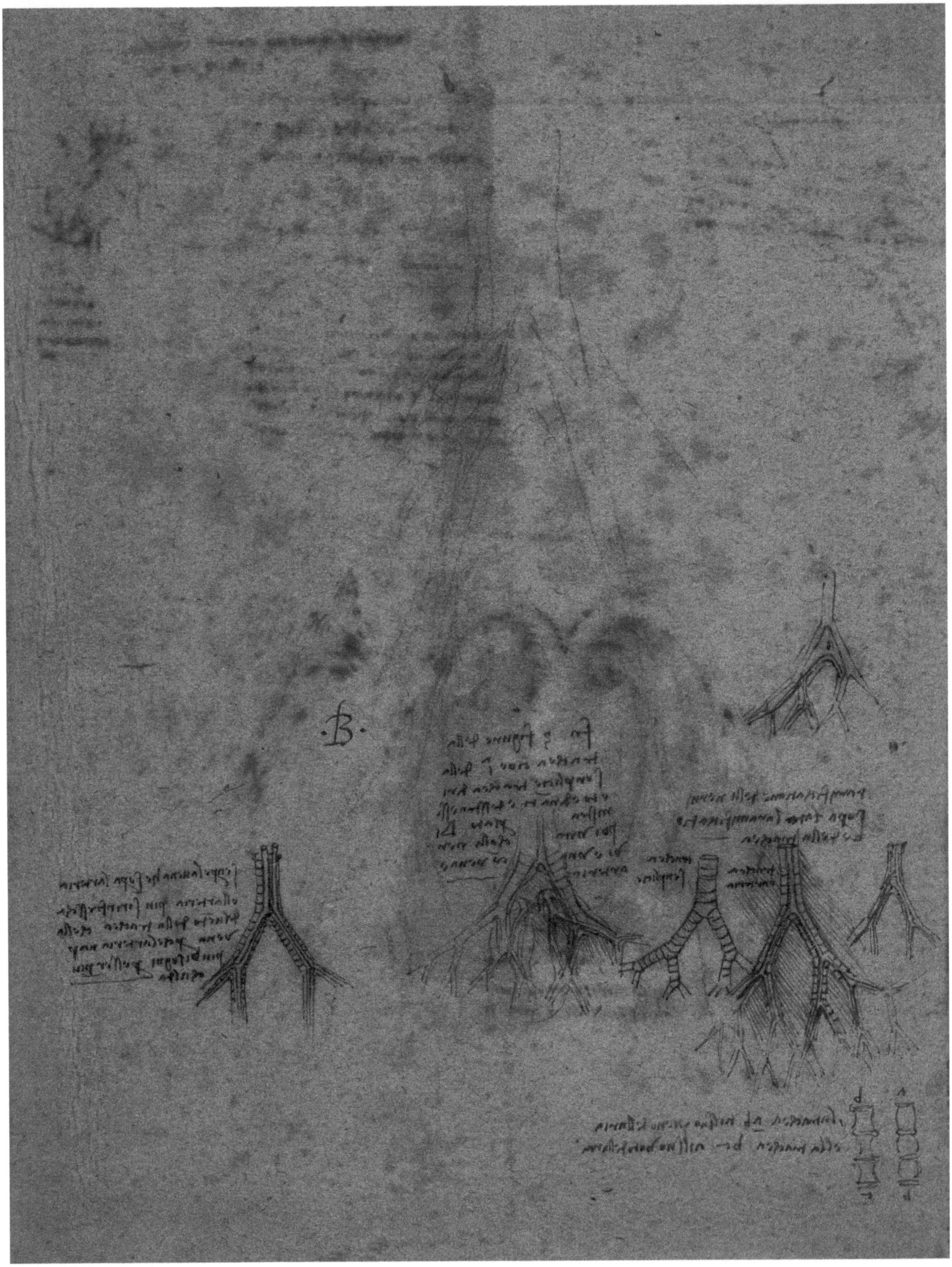

RL 19072 recto, Studies of the trachea, and associated vessels. Leonardo da Vinci (Vinci 1452-Amboise 1519). c. 1511–13. Pen and ink and black chalk on blue paper (Lent by Her Majesty The Queen. Royal Collection © Her Majesty Queen Elizabeth II)

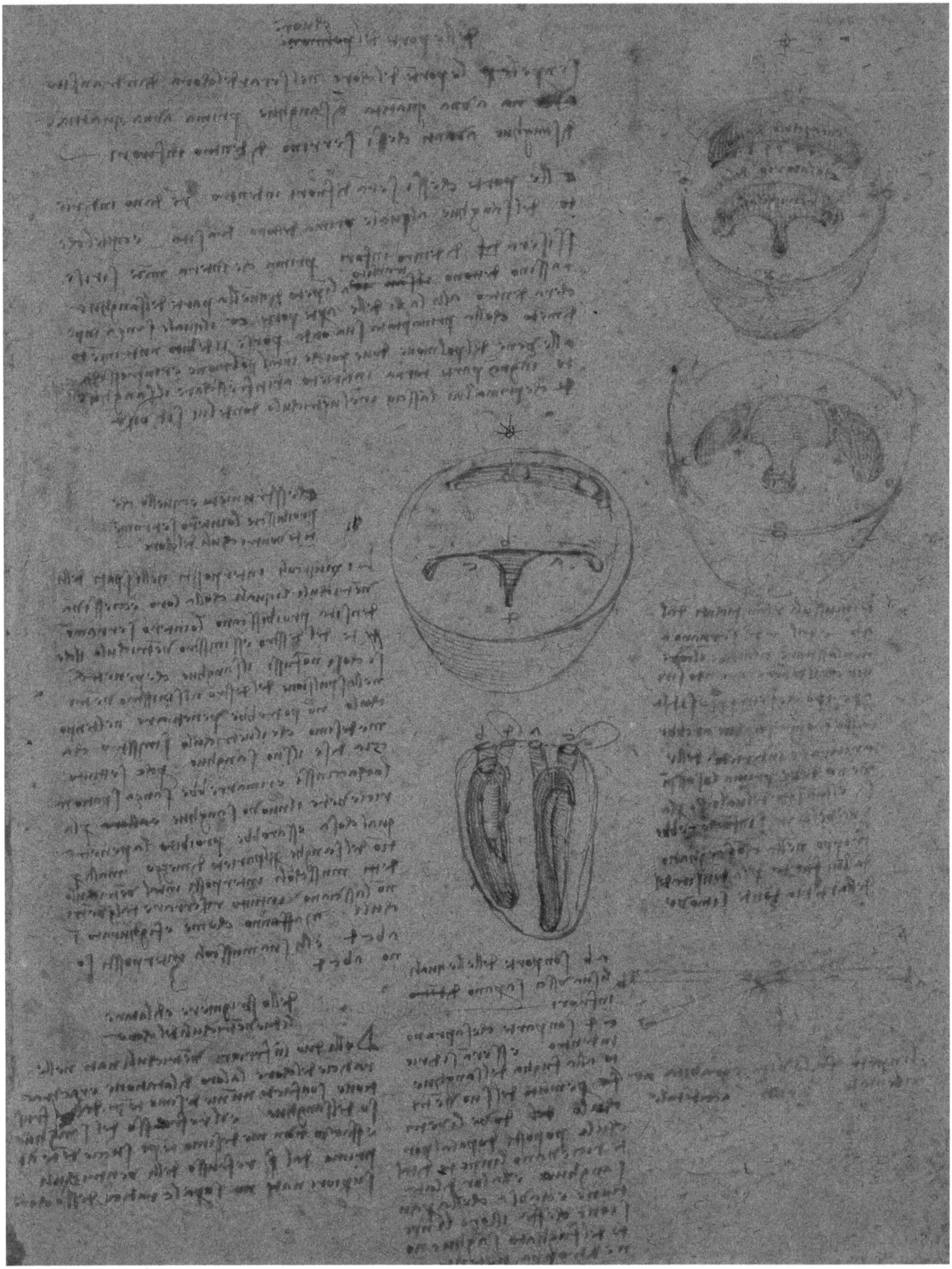

RL 19073 recto and RL 19074 recto, The ventricles, papillary muscles and tricuspid valve. Leonardo da Vinci (Vinci 1452-Amboise 1519). c. 1511–13. Pen and ink on blue paper (Lent by Her Majesty The Queen. Royal Collection © Her Majesty Queen Elizabeth II)

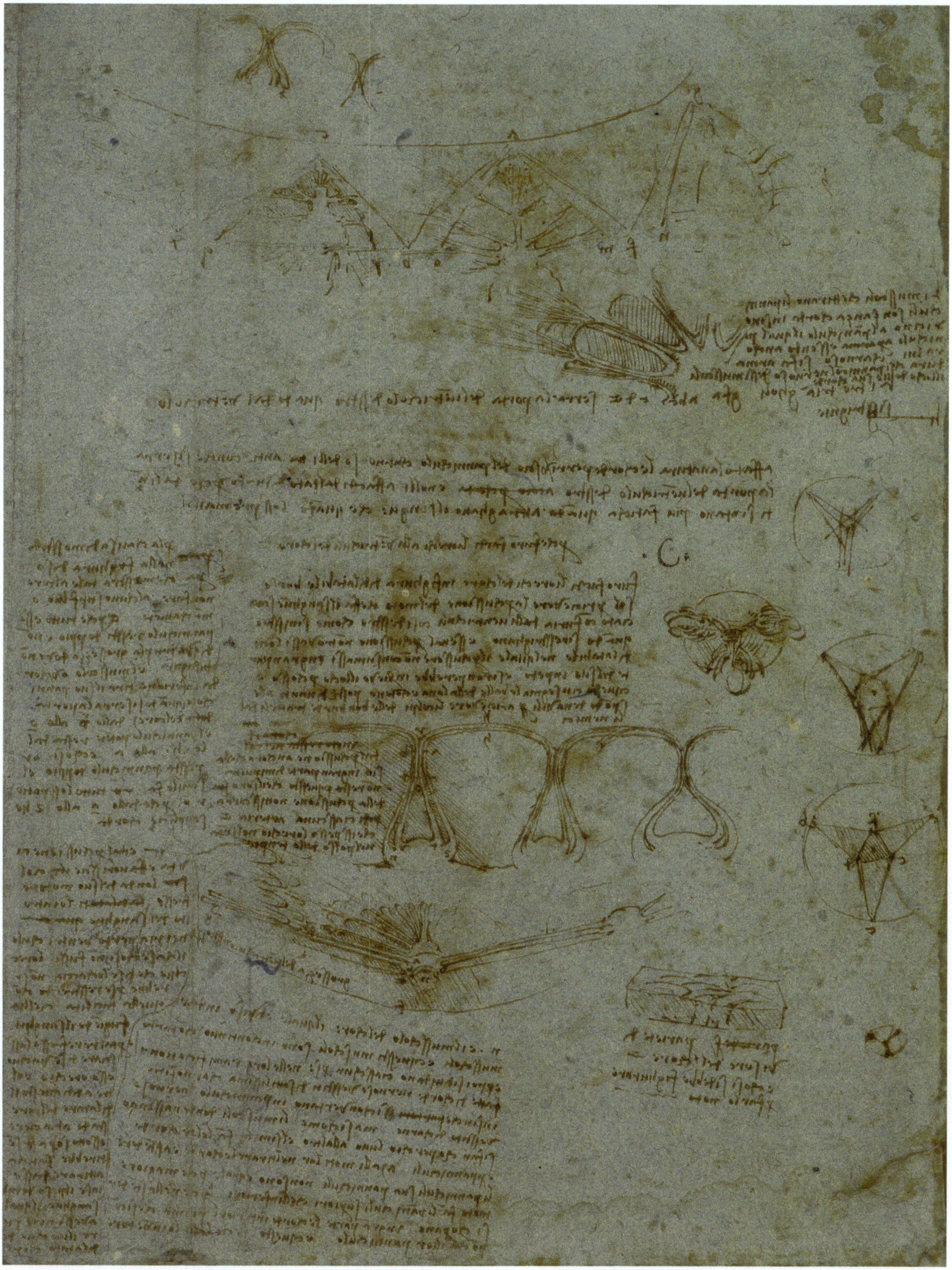

RL 19073 recto and RL 19074 recto (continued)

19073–74 r (C II 4 v and C II 3 r). Pen and brown ink, on blue paper. 288×413 mm

<table>
<tr><td>[Fig. 1, 2]</td><td>[Fig. 1, 2]</td></tr>
<tr><td>[Fig. 3] a b c d e f g h m n o p q</td><td>[Fig. 3] a b c d e f g h m n o p q</td></tr>
<tr><td>[I] Li tre triāgholi g f a a b c c d e serrā la porta del uētricholo desstro quādo tal vētricholo | sisstrignie</td><td>[I] The three triangles g f a, a b c, c d e shut the gate of the right ventricle [tricuspid orifice] when that ventricle contracts.</td></tr>
<tr><td>[Fig. 4]</td><td>[Fig. 4]</td></tr>
<tr><td>[II] Li musscoli chettirano li pannj|chuli son sanza chorda insino | vicino alpānjchulo ilqual pa|nichulo (a per arma) essendo ancho|ra lui charnoso sifa arma|dura chelpannjcōl neruoso dessi musscoli | illocho delle sua chorde</td><td>[II] The [papillary] muscles which pull the membranes [valve-cusps] are without tendons until close to the membrane [valve-cusp], which membrane being itself still fleshy makes for itself a protective armature which consists of the sinewy membrane of these muscles in place of their chordae tendineae [see Fig. 10].</td></tr>
<tr><td>[Fig. 5, 6, 7]</td><td>[Fig. 5, 6, 7]</td></tr>
<tr><td>[Fig. 8] a b c d e f</td><td>[Fig. 8] a b c d e f</td></tr>
<tr><td>[Fig. 9]</td><td>[Fig. 9]</td></tr>
<tr><td>[III] affato lanatura le chorde per riverscio del pannjchulo charno so delli tre anti con che siserra | la porta del uētrichulo desstro (a ccio perche ta) e nolli affacti dallato djricto perche tali ā|ti sentano piu faticha quādo attraghano ilsangue che quādo losspremano</td><td>[III] Nature has made the chordae tendineae on the back side of the fleshy membrane of the three gates [cusps] with which the gateway of the right ventricle is shut; and she has not made them on the front because these cusps feel more strain when they draw the blood in than when they push it out.</td></tr>
<tr><td>[IV] perche furō facti liorechi alli vētrichuli del chore
furō facti li orechi del chore in fighura dj djlatabile borse | sol per ricievere la perchussione del moto cheffa ilsanghue scac|ciato cō furia dalli uentrichuli cosi desstro chome sinjsstro | quādo siresstri<n>ghano essettal perchussione nō trovassi locho | djlatabile nelquale il perchussore nō consumassi in gran par|te delsuo inpeto esironperebbe in brieve illocho perchosso e | cquesto cinsegnia le balle della lana e chotone poste djnanti alle | spōde de navili per aricievere licolpi delle bonbarde tratteli dal|li nemicj</td><td>[IV] Why the auricles were made on the ventricles of the heart
The auricles of the heart were made in the shape of dilatable purses solely in order to receive the percussion of the movement which the blood makes driven violently out of both the right and left ventricles when they contract. And if such percussion did not find a dilatable region in which the impetus of the percussor was in large part consumed, the place percussed would in a short time be broken. And this we are taught by the bales of wool or cotton placed in front of the sides of ships to receive the blows from the guns fired at them by the enemy.</td></tr>
<tr><td>[V] ma (an|chora il | moto resssta nel col) | tal perchussione anchora chella | sia in gran parte djmjnuita | nō resta per quessto cheil sono (sp) | della perchussione non jsscorra | per (che) ciasscuna arteria e | che isspesso lorechio nol sēta | nel polso delle tenpie [:–:] e ttal perchussione ai|vta effauorisscie (il r) chol | (sua) londa delsuo moto re|fresso (la djlatati) lointro|ito del sanghue (qua ch) | nel riaperto ventrichulo | ilche se chosi nō fussi lorec|chio che djse lochaccia nō sa|rebbe per se resstrecto chō | quella facilita chelli a | fugiēdo ilsanghue | e per tal refresso lass|sciare di se vachuo | esso orechio e ol|tre addj questo il | djlatare delcore | stādo a diacere | ossotto sopra se sē|tirebbe faticha | attirare jn sse | inse il peso di tal | sangue ilquale | avessi a riēpie|re iluachuo del | djlatato chore</td><td>[V] But although this percussion may be greatly diminished, it does not result from this that the sound of the percussion does not run through every artery and that the ear does not often hear it in the pulse of the temples. [:–:] And such percussion assists, and with the wave of its reflected movement favours the entrance of the blood into the reopened ventricle. If this were not so the auricle which drives it away from itself would not on its own contract with such facility as it does on the escaping blood, by such reflection leaving the auricle empty. Besides this, dilatation of the heart if it were lying flat or upside down would be found laborious in drawing into itself the weight of that blood which would be necessary to refill the vacuum of the dilated heart.</td></tr>
<tr><td>[Fig. 10] a b c d h h n o r s t v</td><td>[Fig. 10] a b c d h h n o r s t v</td></tr>
<tr><td>[Fig. 11] a b c d e f g h n grossezza del uētrichulo desstro</td><td>[Fig. 11] a b c d e f g h n thickness of the right ventricle</td></tr>
<tr><td>[VI] n · e il musscholo del chore ilquale e diviso in due | musscholi e cquessti muscholi sono in chontinuo chōtacto | e ppoi sidjujdano ciasschun per se nelle loro ramijfichatione | fatte dj chorde neruose vesstiti dj sottilissima charnosita | insin che (partori) ssichonvertano in pannichulo neruoso | vesstito dj charne ma sichome limuscoli donde nasscano | sifan choperchio luno allaltro elsimjle fā lelor chorde | eppannjchuli Malli moti lor neltirare le chorde e asstēdere | li pannjchuli sua pannjchuli non sono equale perche magiore | moto fā li pānjchuli superiori chellinferiori perche nellastēder|si choprano in grā parte le chorde inferiori prima che si cō|po (cha) illor pannjchulo e cquesto fecie (lo lo) lo invētore [:–:] (perche) per la chausa djmosstra | nella fighura dj so|pra che mosstra tale altore | non fare alchuno superfluo o | māchamēte E perche tutto esso | pannjchulo ressti doppio e nō | dj qvadruple grossezza dove nō | bisognia elmusscolo over chor|da inferiore tira ilsuo panni|cholo (quādo siserra la porta | (della) delcore) dallo (p) ^r^ allo ō | el pannjchulo potēte resta dal|lo · h · allo a e chosi a r | ressta pannichulo doppio el | simile fa (r o) tutto lo sspatio | r o perche dallo <u>a</u> allo s he | senplicie chorda</td><td>[VI] n is the muscle of the heart which is divided into two muscles, and these muscles are in continuous contact [n is the papillary muscle in contact with a or e f which is the thickness of the right ventricular wall]. Then they are divided each into their own ramifications of sinewy cords clothed by the finest flesh [endocardium] until they are converted into sinewy membrane clothed by flesh. But just as the muscles from which they arise form a covering for one another, so do their cords and membranes [chordae tendineae and valve-cusps]. But the movements in drawing on the cords and stretching the membranes are not equal because the upper membranes make a greater movement than the lower because in being stretched they cover a large part of the lower cords before these have been formed into their membrane. And this the Inventor did [:–:] by means of the cause demonstrated in the above illustration [Fig. 11] which shows that the Author makes nothing superfluous or deficient. And in order that the complete membrane should be double and not quadruple in thickness where it is not needed, the muscle or inferior cord pulls on its membrane (when the gate of the heart is shut) from r to o and the powerful membrane from h to a remains [Fig. 10]. Thus a r remains a double membrane and likewise does the whole space r o because from a to s is one single cord.</td></tr>
<tr><td>[Fig. 12]</td><td>[Fig. 12]</td></tr>
<tr><td>[VII] (pariede f) pariete dj|visore del chore– | e chosi sidebbe fighurare per farlo noto–</td><td>[VII] The dividing wall of the heart – and thus it must be drawn in order to make it understood.</td></tr>
</table>

[Fig. 1] a c d n | uētrichulo desstro | cholatorio del chore | vētrichulo sinistro	[Fig. 1] *a c d n* right ventricle; sieve of the heart; left ventricle
[I] delle porte del (*polmone*) ^chuore^ senpre (*le p*) leporte del chore nel serrar delchore dan transito | (*a vn na*) a vna quātita | dj sanghue prima avna quātita | dj sanghue avanti chessi serrino dj infuori – E lle porte chessi serā dj fuori indentro rēdano indjrie|to del sanghue alquale prima dettono trāsito e cquel che | ssiserā (*dj f*) dj dentro infori prima che intera mēte si rise|rassino dettono (*il san in*) ^transito^ al īpeto dj quella parte delsanghue | chera dentro alli labri delle aperte porte c o ilquale sanza inpe|djmēto cholla principiata sua onda porse il debito nutrimēto | alle vene del polmone doue poiche intal polmone e rinfresscha|to ingrā parte torna indjrieto a rinfresschare il sanghue | (*de*) che prima lui lasscio inel ūetrichulo donde lui si divise	[I] On the valves of the heart In the shutting of the heart the valve-cusps of the heart [mitral and tricuspid] always give passage first to a quantity of blood before they shut from within outwards. And the valve-cusps which shut from without inwards [aortic and pulmonary] return back some of the blood to which they previously gave passage. Those which close from within outwards before they were shut completely gave way to the impetus of that part of the blood which was within the lips of the opened valve-cusps *c o*. This, without any impediment at the beginning of its wave, delivered due nourishment to the vessels of the lung, where after being refreshed in the lung, a large part of it returns to refresh the blood which it previously left in the ventricle from which it was separated.
[Fig. 2] a b c d	[Fig. 2] *a b c d*
[Fig. 3] a b c d	[Fig. 3] *a b c d*
[II] che sstrumēto cquello che | probisscie lointero serramē|to de ventrichuli de chore – Li musscolj interpossti nelli spati delli | vētrichuli liquali cholla loro ecciessiva | densita proibisscano lointero serramē|(*de*)to del desstro essinjsstro vētrichulo ilche | se chosi nō fussi il sanghue che penetra | nella spulsione del destro alsinjsstro vētri|chulo nō potrebbe penetrare neltenpo | medesimo che il uētrichulo sinjsstro cha|ccia djse ilsuo sanghue perche settucto | lo chacciassi e rimarrebbe sanza spatio da | ricievere il nuovo sanghue (*e allora*) per la | qual chosa essarebbe proibito la penetra|tiō del sanghe per ilpariete dj mezzo malli pre|detti musscholi interposti intal vētrichulo | nō lassciano in tutto riserrare talj vētri|chulj Ma ffanno chome e fighurato ī | a b c d ellj sua musscoli interpossti so|no a b c d	[II] Which is the instrument that prevents the complete closure of the ventricles of the heart The muscles interposed within the ventricular spaces are those which by their excessive density prevent complete closure of the right and left ventricles. If this were not so, the blood which in its expulsion from the right into the left ventricle penetrates [the septum] would not be able to penetrate it at the same time as the left ventricle was driving its blood out of itself because if it drove all of it out it would be left with no space for receiving the new blood; whereby the penetration of blood through the middle wall [septum] would be prevented. But the aforesaid muscles interposed in the ventricles do not let these ventricles close completely. But they do what is drawn at *a b c d*; and the interposed muscles are *a b c d* [Fig. 2 and 3].
[III] dello strignjere e djlatare | li due vētrichuli del chore – Delli due inferiori vētrichulj nati nelle | radjcie delchore laloro djlatatione eractrac|tione sonfacte nū medesimo tēpo dal frus|so delsanghue e il refrusso del sanghue | effacto nvn medesimo tēpo succiedēte al | primo dal (*f*) refrusso delli ventrichuli | superiori natj (*no*) sopra le radjce desso chore	[III] On the contraction and dilatation of the two ventricles of the heart Of the two lower ventricles arising from the roots of the heart, their dilatation and contraction are made at one and the same time by the flux of the blood. And the reflux of the blood is made at one and the same time, following the first by reflux from the upper ventricles arising on top of the roots of the heart.
[Fig. 4] a b c d	[Fig. 4] *a b c d*
[IV] a b son porte delle lle quali | li sua vssci saprano (*dj dentro*) | infuori – c d son porte chesaprano | in dentro esserāsi djrie|to alla fugha del sanghue | (*se*) premuto del suo vētri|chulo (*del*) dove li vētri|chuli poposti dopo tal por|te ricievano linpeto dj tal | sanghue e lla lor djlata|tione e chavsa chella perchu|sione che ffa illoro linpe|to del fughato sāghue no|ne di troppa potentia –	[IV] *a b* are the gates the doors of which open outwards [aortic and pulmonary valves] *c d* are gates opening inwards [mitral and tricuspid valves]. And they shut behind the escape of blood pressed out of its [upper] ventricle [atrium] where the [lower] ventricles, placed beyond such gates, receive the impetus of that blood. And their dilatation is the reason that the percussion which the impetus makes on them through the escaped blood is not too powerful.
[V] Li muss<c>oli ramjfichatj dal | a b e dal c d seruano a | non lassciare dilatare ilcore | piu che il dovere quando siri|apre īpero che se ttroppo sidjla|tassi troppo sāghue arebbe | a ritirare indjrieto delle | vene dove prima lososspī|se el trāsito del<q>uale (*si*) per la | sua velocita sinfocherebbe | troppo nella chōfreghatio<ne>| da llui facta per la densita del | dellarteria donde si move	[V] The muscles ramifying from *a b* and from *c d* serve to prevent the heart dilate more than it should when it reopens. For if it dilated too much it would have to draw too much blood back from the vessels into which it first pressed it. And the passage of this [blood] through its velocity and too much friction, with the density of the artery through which it moves, would make it catch fire.
[Fig. 5] a b b c d e	[Fig. 5] *a b b c d e*
[VI] linpeto fa levita e gravita ac|cidentale Levita accidetale	[VI] Impetus makes lightness and accidental weight. Accidental lightness

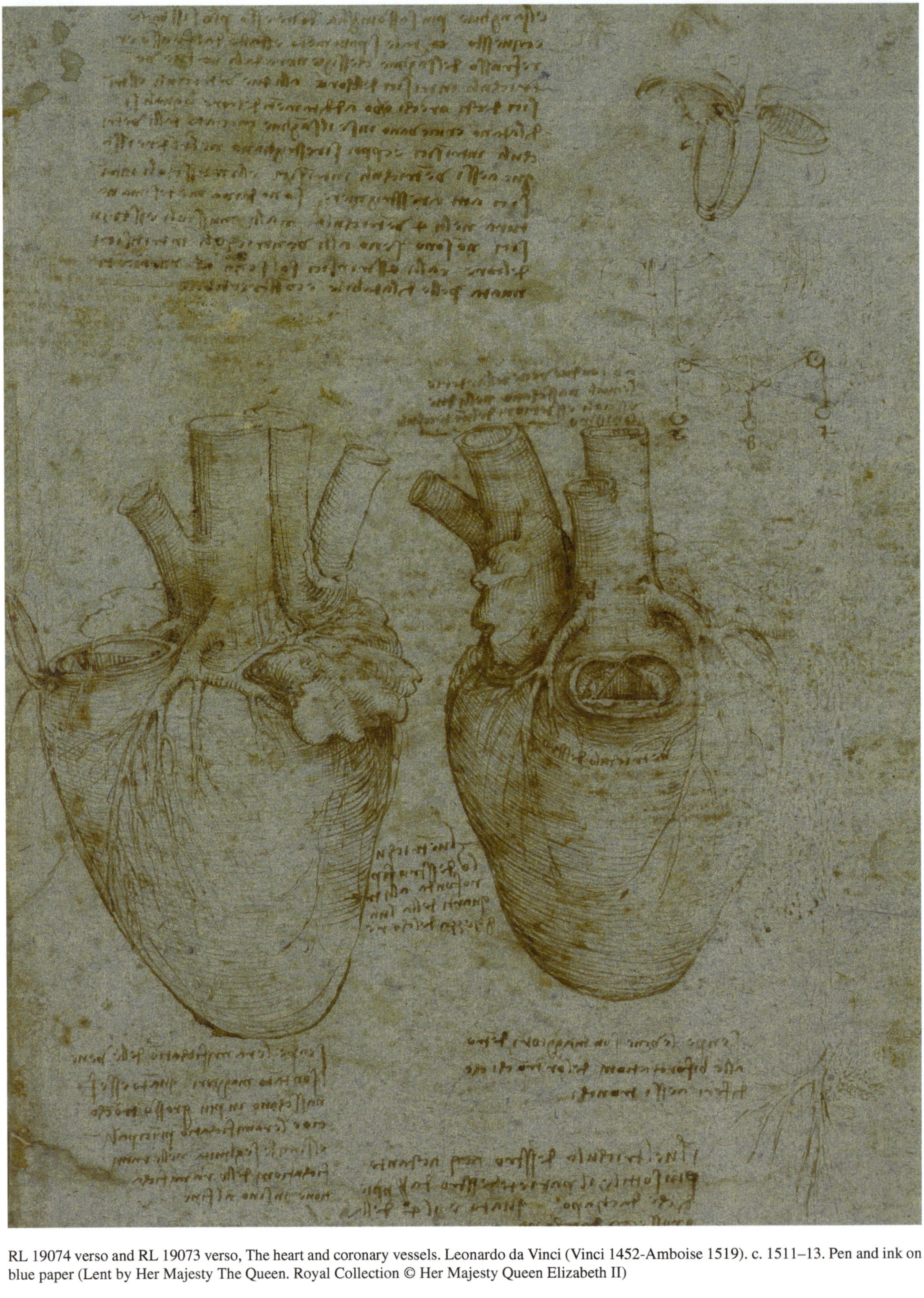

RL 19074 verso and RL 19073 verso, The heart and coronary vessels. Leonardo da Vinci (Vinci 1452-Amboise 1519). c. 1511–13. Pen and ink on blue paper (Lent by Her Majesty The Queen. Royal Collection © Her Majesty Queen Elizabeth II)

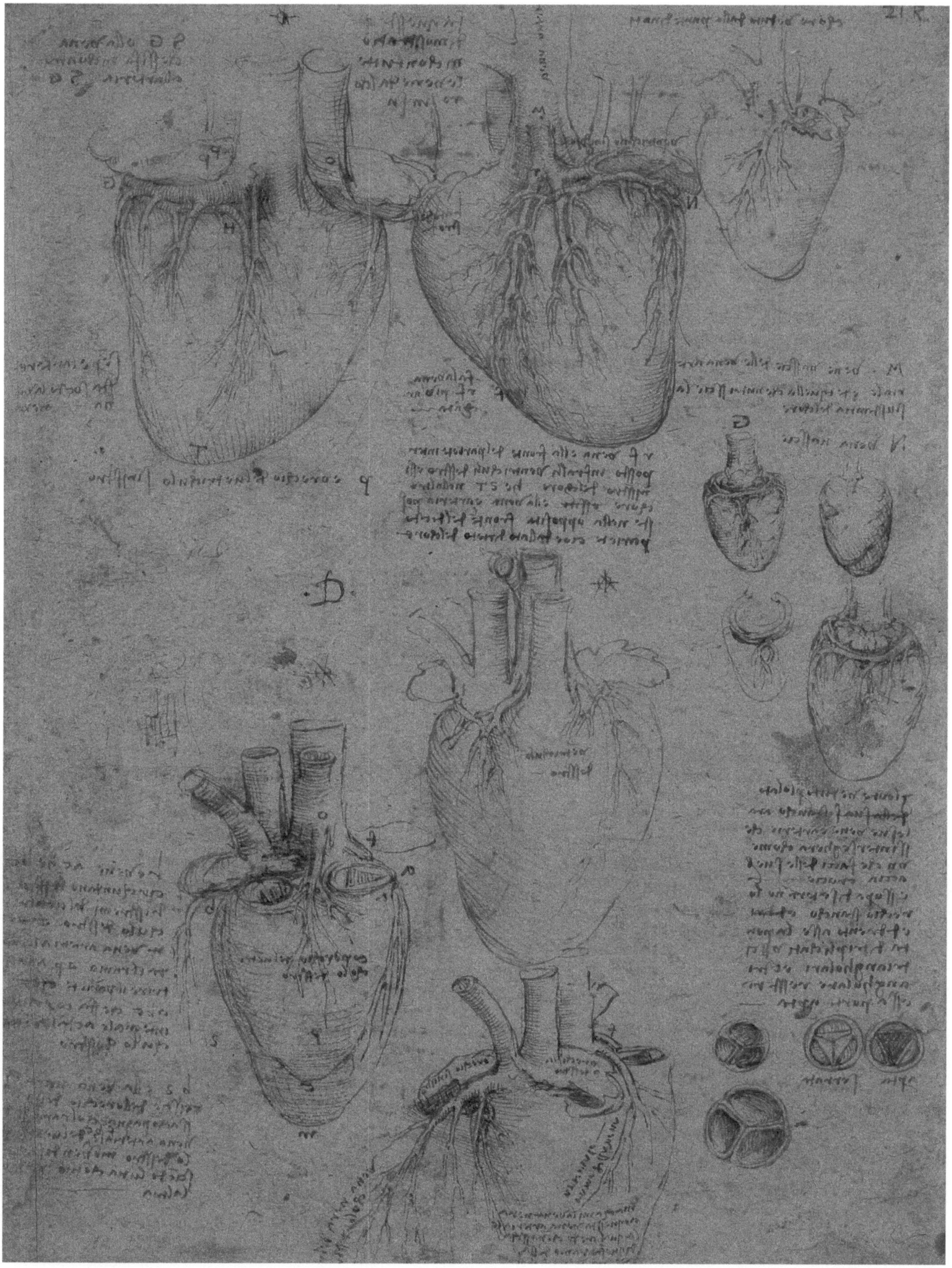

RL 19074 verso and RL 19073 verso (continued)

19073 – 74 v (C II 3 v and C II 4 r). Pen and brown ink.

[I] elsanghue piu sassottiglia dove esso piu sissbacte | ecquessto (*co*) tale spattimẽto effatto dal frusso er|refrusso delsãghue chessigienera dalli (*ve*) due vẽ|trichuli intrisici dellchore alli due vẽtrichuli esstrĩ|sici decti orechi overo addjtamẽti del core liquali si | djlatano e ricievano inse ilsãghue cacciato dalli vẽntri|chuli intrĩsici ecppoi sirestrĩghano nelrẽdere ilsã|gue a essi vẽtrichuli intrĩsici elli musscholj intrĩ|sici atti a resstrigniere sono duna medesima na|tura nellj 4 vẽtrichuli mallj musscholi estrin|sici nõ sono se nõ alli ventricholj intrinsicij | delcore e alli esstrinsici sol sono (*al*) vnacõti|nuata pelle djlatabile e cõsstrectibile -	[I] The blood is more subtilised where it is more beaten, and this percussion is made by the flux and reflux of the blood generated from the two intrinsic ventricles of the heart of the two extrinsic ventricles called auricles or additions of the heart which are dilated and receive into themselves blood driven from the intrinsic ventricles; and then they contract, returning the blood to those intrinsic ventricles. And the intrinsic muscles suitable for contracting are of one and the same nature in the four ventricles. But the extrinsic muscles exist only on the intrinsic ventricles of the heart. On the extrinsic ventricles there is only a continuous dilatable and contractible skin.
[Fig. 1, 2]	[Fig. 1, 2]
[Fig. 3] a b vẽtrichulo desstro	[Fig. 3] *a b* right ventricle
[II] a b son due vene dellarterie | le quali nasschano nellj due | vsscioli essteriori del uẽtrichulo | sjnisstro -	[II] *a b* are two arterial vessels [coronary arteries] which arise in two openings outside the left ventricle.
[III] Jluẽtrichu|lo desstro sip|rofonda allj tre | quartj della lun|ghezza del chore	[III] The right ventricle extends to a depth of three-quarters of the length of the heart.
[Fig. 4]	[Fig. 4]
[IV] il ueltrichulo desstro (*a cq*) actanto | piusottile il pariete desstro da (*l*) ppie|dj che da chapo quãto e il 4° della | grossezza dacchapo -	[IV] The right ventricle has its right wall much thinner at its base than its apex – It is as much as a quarter of the thickness at the apex.
[Fig. 5]	[Fig. 5]
[V] senpre leramjfichatiõ delle vene | son tãto maggiori quãto esse (*s*) | nasschano in piu grosso trõcho | cioe leramjfichatiõ prĩcipalj - | elsimjle seghuita nelle ramj|fichationj delle ramjficha|tione insino al fine	[V] The ramifications of the vessels are always that much larger as is the larger trunk from which they arise, that is the main vessels; it follows likewise in the ramifications of the ramifications right to the end.
[VI] senpre le vene son maggiori dẽtro | alle biforchationj delor trõchi che | dj fori a essi tronchi	[VI] Vessels are always larger within the bifurcations of their trunks than they are beyond these trunks.
[Fig. 6] f g s	[Fig. 6] *f g s*

[Fig. 1]	[Fig. 1]
[I] chore veduto dalla parte djnātj	[I] Heart seen from the front side.
[Fig. 2] M N f r ventrichulo sinistro vena arteriale lato des|stro	[Fig. 2] *M N f r* left ventricle; arterial vessel; right side
[II] fa la vena | r f piv ar|chata -	[II] Make the vessel *r f* more arched.
[III] M · vene nasscie delle vena arte|riale ed e cquella che nutrisscie la | susstantia del chore – N vena nasscie	[III] *M* vessel [left coronary artery] arises from the arterial vessel, and it is that which nourishes the substans of the heart. *N* vessel arises.
[Fig. 3] o p p G H S T	[Fig. 3] *o p p G H S T*
[IV] S G e lla vena | chessifa inchontro | allarteria S G	[IV] *S G* [coronary sinus] is the vein which goes together with the artery *S G*
[V] sēpre larteria | sta socto la ve|na – nera	[V] The artery [left coronary, *H*] is always beneath the black vein [great cardiac vein].
[VI] p e orechio del uētrichulo sinjsstro	[VI] *p* is the auricle of the left ventricle
[VII] r f vena ella fronte del pariete inter|posto infralli ventrichulj desstro e ssi|njsstro del chore he S T nellaltro | chore (*essitu*) ella vena e arteria pos|ste nella opposita fronte del decto | pariete cioe dallato dirieto del chore	[VII] *r f* [Fig. 2] vessel is at the front of the wall interposed between the right and left ventricles of the heart [interventricular septum]. And *S T* in the other heart [Fig. 3] are the vein and artery placed on the opposite side of the said wall, that is, on the back side of the heart [the posterior interventricular branch of the right coronary artery, and the middle cardiac vein joining the coronary sinus at *S*]
[VIII] fa quesste 3 | djmosstratio|nj chon tutte | le vene dalcho|re in su	[VIII] Make these three demonstrations with all the vessels of the heart on it
[Fig. 4, 5, 6, 7]	[Fig. 4, 5, 6, 7]
[IX] Jl core venduto per lo lato | (*della sua s*) stancho ara | lesue vene e arterie che |ssintersegherā chome | vn che facci delle suebr|accia crocie – | e ssopra di se terano lo|rechio stancho (*eddjri*) | e ddrento asse la por|ta dj triplichatj vsscj | triangholari ectri|angholare resstera | essa porta aperta -	[IX] The heart viewed from the left side will have its veins and arteries intersecting like someone crossing his arms. And they will have above them the left auricle and inside them the gateway with triplicated triangular doors [aortic valve-cusps] and the opened gateway will remain triangular [see Fig. 8 and 9].
[Fig. 8] aperta	[Fig. 8] opened
[Fig. 9] serrata	[Fig. 9] shut
[Fig. 10] vēntrichulo | desstro -	[Fig. 10] right ventricle
[Fig. 11] a b c f m o p s coperchio del uētri|cholo desstro	[Fig. 11 *a b c f m o p s* covering of the right ventricle
[X] Le vene a c he b c | circhundano (*li sstre*) | li sstremj del uentri|chulo desstro (*e lla v*) e | (*n*) vena arteriale mā|da ilramo o p a nnu|trire il pariete (*che*) | a b c che ffa coperchio | integrale actal vētri|chulo desstro b s ella vena nera che | esscie dellorechio desstro e | ssachōpagna chol ramo della | vena arteriale f b c del uentrichu|lo desstro movendosi e cre|sciēdo luna chōtro ddal|laltra –	[X] The vessels *a c* [anterior interventricular branch of the left coronary artery] and *b c* [posterior interventricular branch of the right coronary artery] surround the edges of the right ventricle. And the arterial vessel [aorta] sends the branch *o p* [right coronary artery] to nourish the wall *a b c* which makes the compete covering of this right ventricle. *b* is the back vein which issues from the right auricle [middle cardiac vein] and is accompanied b a branch of the arterial vessel *f b c* [posterior interventricular branch of right coronary. See also Fig. 10] of the right ventricle proceeding and increasing close to one another.
[Fig. 12] f orechi|o desstro orechio sinjstro ramo della vena | arteriale – vena nera del uēt<ri>|cholo desstro māchamj la vena nera | a cquessta vena arteriale | laqual credo che nasscha | dj questo ramo della | vena nera sinistra	[Fig. 12] *f* right auricle; left auricle; branch of the arterial vessel; black vein of the right ventricle I lack the black vein to this arterial vessel which I believe arises by this branch of the left black vein.
[Fig. 13, 14, 15, 16]	[Fig. 13, 14, 15, 16]

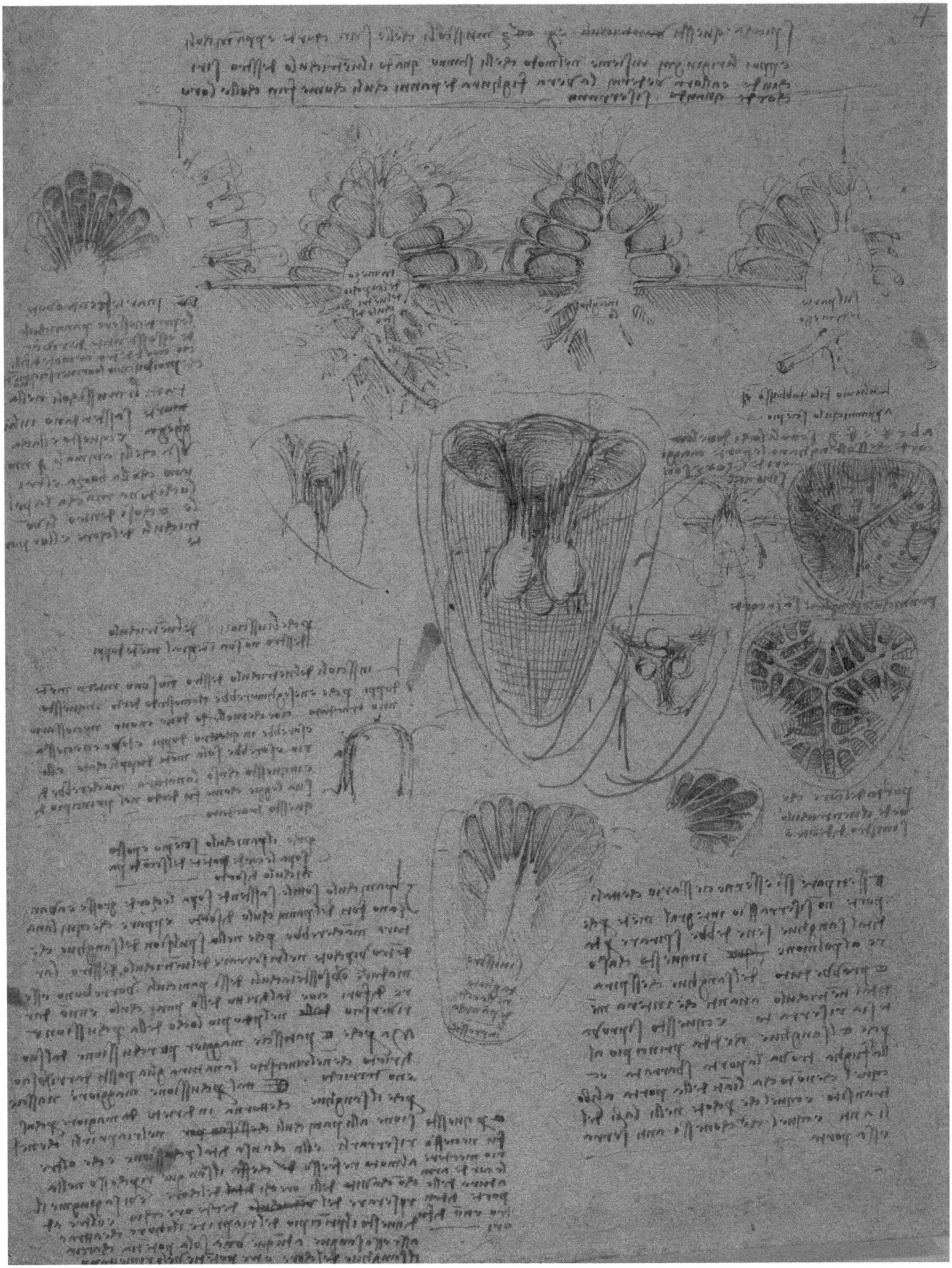

RL 19078 verso, The right ventricle and tricuspid valve. Leonardo da Vinci (Vinci 1452-Amboise 1519). c.1512–13. Pen and ink on blue paper (Lent by Her Majesty The Queen. Royal Collection © Her Majesty Queen Elizabeth II)

19078 v (C II 8 v). Pen and brown ink.	
[I] giegrofia del core spiccha quessti (*ventrichuli e g cō*) 3 musscoli cholle sua chorde eppānjcholj | eppoi li rigiugnj insieme nelmodo chelli stanno quādo iluētrichulo desstro siri|chiude e allora vederaj la vera fighura de pannichulj chome fan cholle loro | chorde quando siserrano –	[I] Geography of the heart Cut out these three muscles with their cords and membranes [chordae tendineae and valve-cusps] and then join them together in the way in which they are placed when the right ventricle is closed. Then you will see the true shape of the [valve] membranes and what they do with their chordae [tendineae] when they are shut.
[Fig. 1] sul parie|te di mezzo in āgho|lo in mezo | del coperchio | del vētri|chulo des|stro	[Fig. 1] On the middle wall [septum]; in the angle; in the middle of the covering of the right ventricle.
[II] Lultimo filo dabbasso (*oj*) | appannjchulo sciēpio	[II] The last thread [chorda] below has a broad membrane [moderator band or septo-marginal trabecula].
[Fig. 2]	[Fig. 2]
[III (*i vn*) jnante sien trovate | le porte essere pannichula|te e ssosstenute da river|cio cioe di dētro armate dj fili | che proibiscano loariuersciamēto | suo	[III] In front [atrial side] the gates [valves] are found to be membranous and they are held from behind, that is, protected from within, by threads [chordae tendineae] which prevent their reversal.
[IV] tutti li musscholi nella | morte sasstendano in lū|gheza e cquesto ella cha|vsa chellj anjmali (*j*) mo|iano cholla bocha eltri | lochi dove māscha la pel|le e chosi dentro li vē|trichuli delchore ellor por|te	[IV] All muscles extend in length at death. This is the reason that animals die with the mouth open, and other places where there is a gap in the skin. And such is the inside of the ventricles of the heart and their orifices [valve-rings].
[Fig. 3] a b c d e f g	[Fig. 3] *a b c d e f g*
[V] a b c d e f g sono li lochi doue lor | chorde che ssostenghano le porte maggi|ori del chore son | situate	[V] *a b c d e f g* are the places where the chordae which hold the large gates [valves] of the heart are situated.
[VI] pannjchulo senplicie sāza corde	[VI] The simple membrane, without chordae
[Fig. 4]	[Fig. 4]
[VII] porta del core che | vede il uentrichulo | sinjstro dj dentrō	[VII] Gate [valve] of the heart which is seen from within from the left [right] ventricle.
[Fig. 5, 6, 7, 8]	[Fig. 5, 6, 7, 8]
[Fig. 9] a b c	[Fig. 9] *a b c*
[VIII] perche li usscioli del uētrichulo | desstro nō son ītegralmēte doppi Li usscioli del uētrichulo desstro nō sono interamēte | doppi perche e ne seghuiterebbe chome stato decto in quessto | mio tractato cioe che in ellocho doue e non e neciessario | e sarebbe in quatro doppi e dove a neciessa|rio e sarebbe sola mēte dupplichato (*e lli*) | e in quessto chaso lanatura mācherebbe dj | sua leggie chome fu decto nel principio dj | quessto tractato –	[VIII] Why the little doors [valve-cusps] of the right ventricle are not entirely double The valve-cusps of the right ventricle are not entirely double because it would follow, as has been said in this treatise of mine [166r (A), Note III], that at the place where it is not necessary it would be doubled to four [layers]. And where it is necessary it would only be doubled. And in this case Nature would fail in her law, as was said at the beginning of this treatise.
[Fig. 10] sinisstro | tagliato | nellarcho | depanjchu| li – | in proffilo	[Fig. 10] Left: cut in the curve of the valve-cusps – in profile
[Fig. 11]	[Fig. 11]
[IX] perchē il panichulo sciēpio e posto | sopra le corde potēte del secōdo pa|nichulo djsocto – Jl panjchulo sottile sasstende sopra le chorde grosse e avan|zano fori del pannjchulō dj socto eppare che cquj lana|tura mācherebbe perche nella spulsion del sanghue che | dētro vi perchote nelriserrare del uētrichulo desstro lar|madure over sosstētachuli dessi panichuli vorrebono esse|re dj fori cioe dal djritto desso pani chulo ennō da r|riverscio (*dalle*) nel propio locho della perchussione | Ma perche e patisscie maggior perchussione dalsuo | djricto chēnelriuerscio la natura gli a possti darriverscio | e nō darricto · E ttal perchussione maggiore nasscie | perche il sanghue che ttorna in dirieto da magiore perchus|sione alli panjchuli (*chessi fan por*) nel riaprirli chenel | riserrarli ella chausa dj tal perchussione e che oltre | almoto refresso (*de*) che ffa il sangue riperchosso nella | chōchavita delli orechi (*dj tal*) delchore e vi sagiugnie il | rjserare del (*vētrichulo*) decto orecchio e oltre a d|djquesto ilprīcipio del riaprire ilchore cheattrae | asse esso sangue adūque vna sola potētia chaccia | il sanghue del chore e tre potētie ve lo rimettano [:–:] e per questo | fu neciessa|rio mectere | le corde arm|adure delle | porte dj den|tro e nō dj fu|ori –	[IX] Why the expanded membrane is placed on the powerful chordae of the second, lower membrane The fine membrane extends on the thick chordae which spread out on the lower membrane. And it appears that Nature would fail here, because on the expulsion of the blood which percusses there at the closing of the right ventricle, the protection and support of these membranes should be on the outside, that is on the front of the membrane and not at the actual place of percussion at the back [ventricular surface]. But because it [the valve] suffers greater percussion from its front than its back, Nature has placed them on the back not on the front. And the greater percussion arises because the blood which returns back gives a greater percussion to the membranes in their reopening than in their closing. And the cause of this percussion is that besides the reflected movement which the blood makes repercussed in the concavity of the auricles of the heart, there is added the contraction of the said auricle: and besides this, the beginning of the reopening of the heart which attracts the blood into itself. Therefore one single power drives the blood out of the heart and three powers send it back there. [:–:] For this reason it was necessary to put chordae protecting the valves on the inside and not on the outside.
[X] e sse ti paressi esser neciessario chettali | porte nō siserrassi<n>o integralmēte perche | djtal sanghue se ne debbe spirare per da|re alpolmone (*Il que*) in questo chaso | e provveduto delsanghue chesspira | dj tal uētrichulo auanti che interamē|te sia riserrato e cquessto siprova | perche el sanghue che dda principio al|lla fugha trova la porta sbarrata e c|cquel che nō tocha li āti della porta a libero | transito e cquel che perchote nelli labri del|li anti e cquel che chonessi anti serra | essa porta –	[X] And if it should seem necessary to you that these gates [valves] should not be completely closed because some of such blood ought to be breathed out and given to the lung, in this case blood is provided which is blown out of that ventricle before it is completely closed. And this is proved since the blood which begins the escape finds the gateway unbarred, and that which does not touch the cusps of the valve has free passage, and that which percusses on the lips of the cusps is that which by means of these cusps closes the gateway [valve].

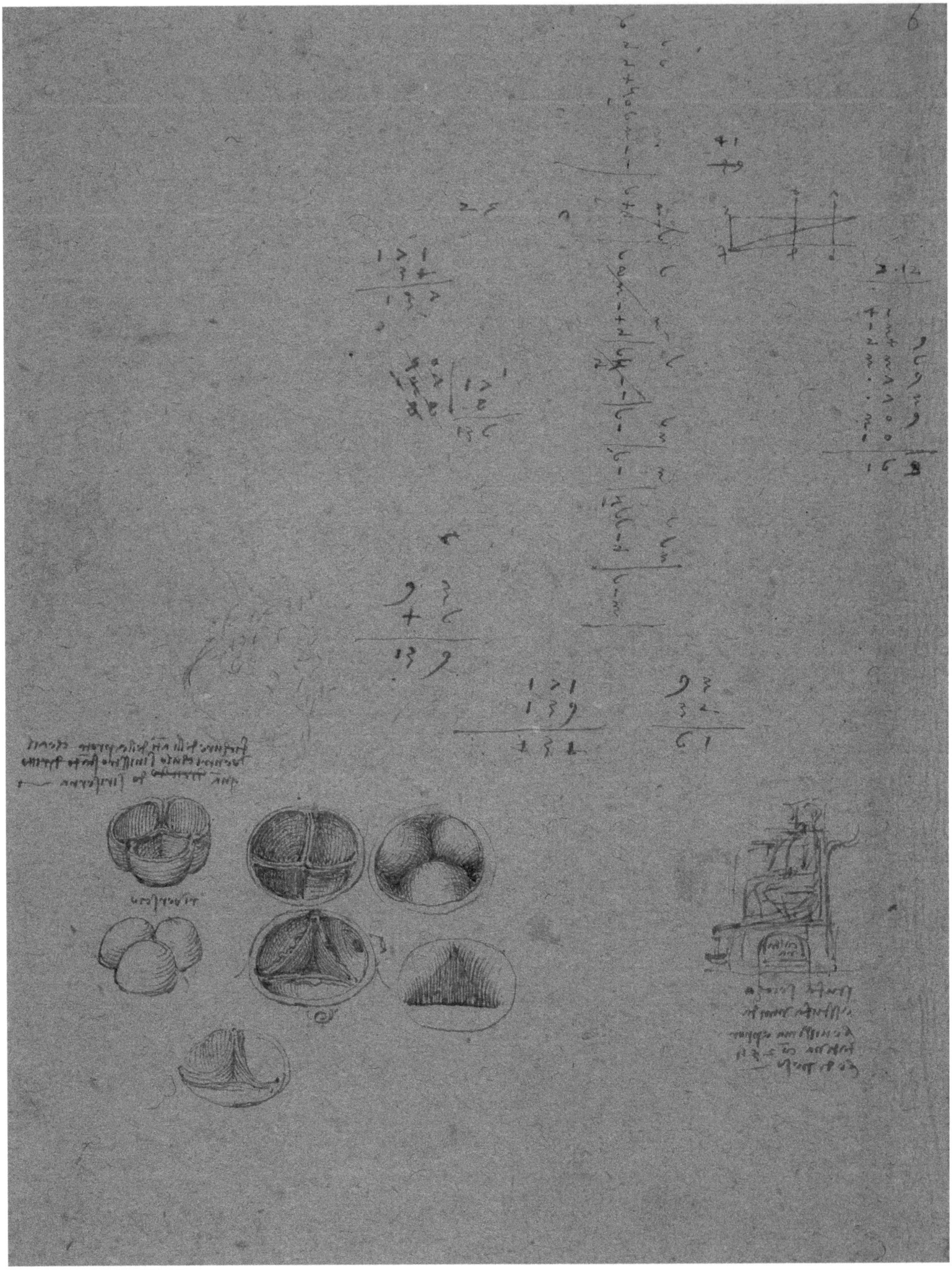

RL 19079 verso, Studies of the valves of the heart, and sketch of a suditary, and calculations. Leonardo da Vinci (Vinci 1452-Amboise 1519). c. 1511–13. Pen and ink on blue paper (Lent by Her Majesty The Queen. Royal Collection © Her Majesty Queen Elizabeth II)

19079 v (C II 9 v). Pen and brown ink.	
[Fig. 1]	[Fig. 1]
[I] fichure delli āti della prota [sic: porta] che ail | ventrichulo sinisstro (*stādo djritto*) | quā(*diricto*)do siriserra –	[I] Shapes of the entry doors [cusps] of the gateway which the left ventricle has when it is shut.
[II] riverscio	[II] Back
[Fig. 2–7]	[Fig. 2–7)
[Fig. 8] calda|ra	[Fig. 8] Cauldron
[III] stufa seccha | esstufa vmjda | breuissima eppor|tativa cō 25 lj|bre dj peso –	[III] A dry sudatory, a moist sudatory, very short and portable weighing twenty-five pounds.
[Fig. 9] a b c d e f	[Fig. 9] *a b c d e f*
[Fig. 10]	[Fig. 10]

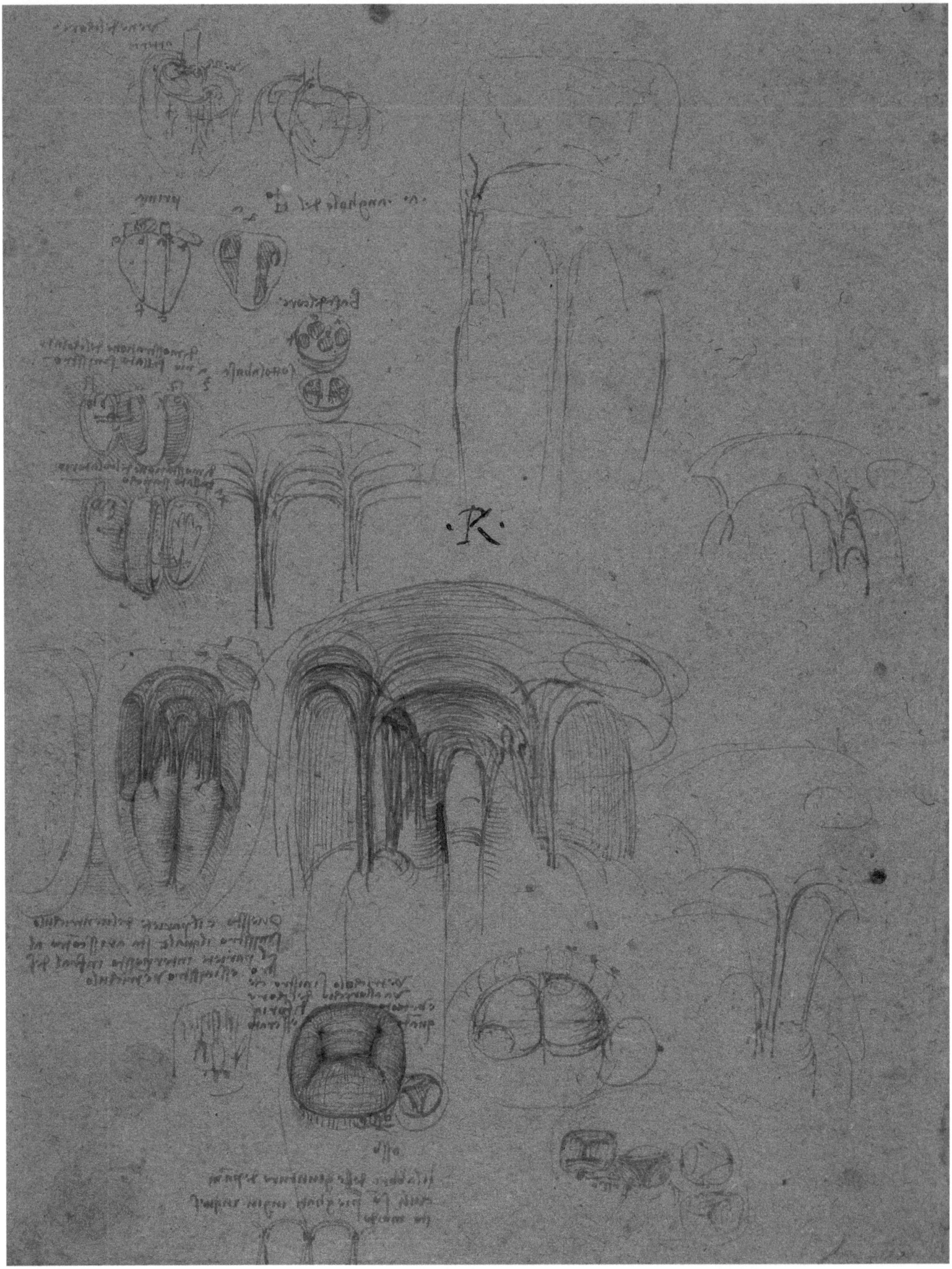

RL 19080 recto, The left ventricle and mitral valve. Leonardo da Vinci (Vinci 1452-Amboise 1519). c.1512–13. Pen and ink on blue paper (Lent by Her Majesty The Queen. Royal Collection © Her Majesty Queen Elizabeth II)

19080r (C II 10 r). Pen and brown ink, on blue paper. 284×209 mm.	
[I] vene del chore	[I] Vessels of the heart
[Fig. 1]	[Fig. 1]
[Fig. 2] vena | arteria	[Fig. 2] vein; artery
[II] prima	[II] First
[Fig. 3] a b c d e f ×	[Fig. 3] *a b c d e f* ×
[III] a · angholo el del [to]	[III] *a*, angle of the rectangle [marked with × in Fig. 3]
[IV] 2ª	[IV] Second
[Fig. 4]	[Fig. 4]
[V] basa del core	[V Base of the heart
[Fig. 5] a b c d	[Fig. 5] *a b c d*
[VI] sottolabasa	[VI] Below the base
[Fig. 6]	[Fig. 6]
[VII] [^–] 3ª [–^] dimosstratione delcholato|rio dallato sinjsstro [sic: destro]	[VII] Third: demonstration o the sieve from the left [sic: right] side
[Fig. 7]	[Fig. 7]
[VIII] [^–] 4ª [–^] dimosstratione delcolatorio | dallato stancho–	[VIII] Fourth: demonstration of the sieve from the left side
[Fig. 8]	[Fig. 8]
[Fig. 9, 10, 11]	[Fig. 9, 10, 11]
[Fig. 12]	[Fig. 12]
[IX] Quessto e il pariete del uentrichulo | sinjsstro il quale sta a risscōtro al | al pariete interpossto infral des|stro essinjsstro vētrichulo	[IX] This is the wall of the left ventricle which is placed opposite the wall interposed between the right and left ventricle [septum].
[Fig. 13, 14, 15]	[Fig. 13, 14, 15]
[Fig. 16 osso | osso	[Fig. 16] bone; bone
[X] vētricholo sinistro che | va allorecchio delchore | e veduto dj fori q|quādo esserrato	[X] The left ventricle which goes to the auricle of the heart is seen from outside when it is closed.
[Fig. 17]	[Fig. 17]
[Fig. 18] osso	[Fig. 18] bone
[Fig. 19]	[Fig. 19]
[XI] li labbri delle giunture de pānj|chulj sō pieghati in giu in ques|sto modo	[XI] The lips of the junctions of the membranes [of the valve-cusps] are bent down in this manner.

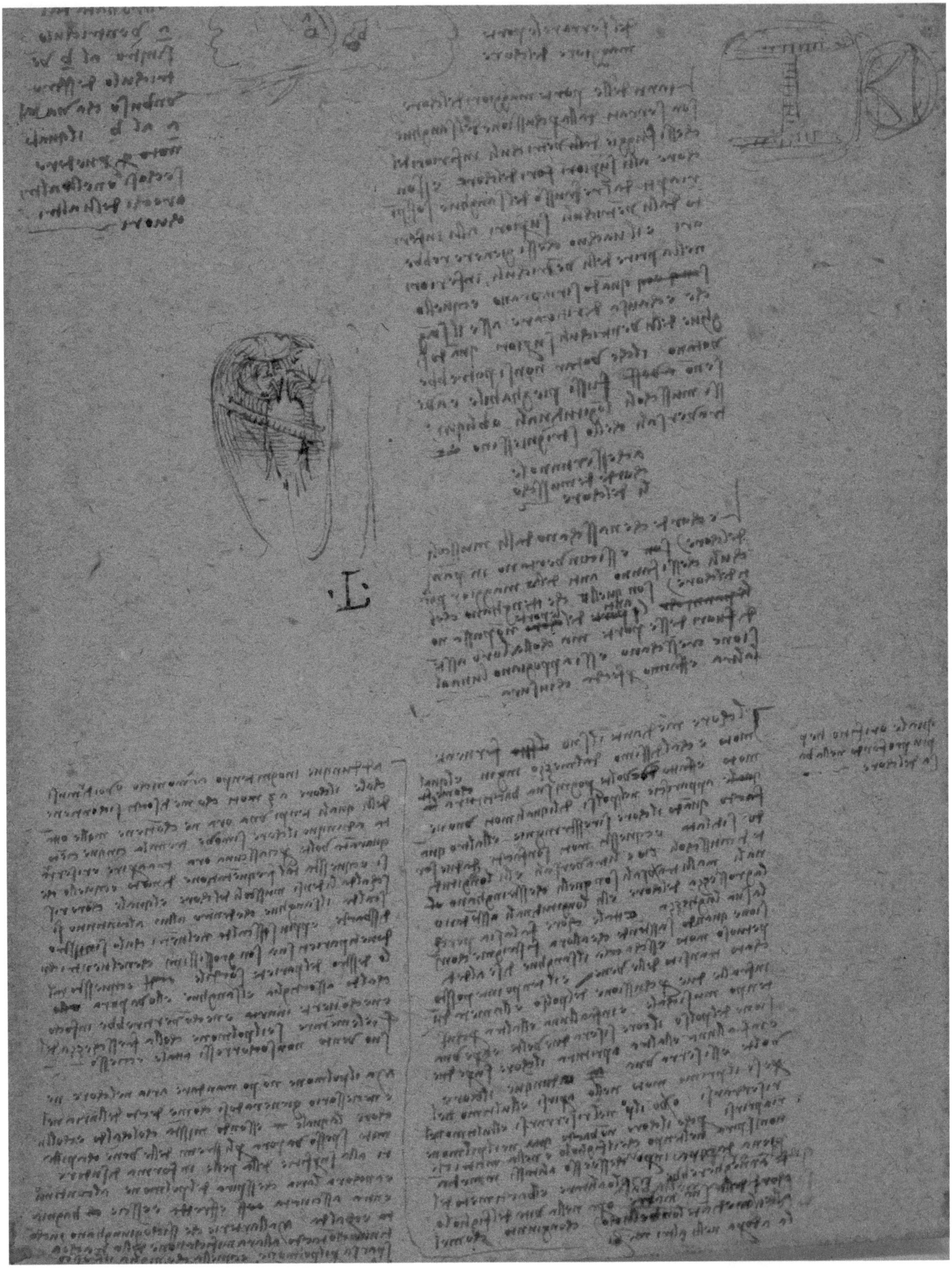

RL 19081 Studies of the motion of the heart. Leonardo da Vinci (Vinci 1452-Amboise 1519). c. 1511–13. Pen and ink and red chalk on blue paper (Lent by Her Majesty The Queen. Royal Collection © Her Majesty Queen Elizabeth II)

19081r (C II 11 r). Pen and brown ink and red chalk, on blue paper. 280×205 mm.

[Fig. 1, 2]	[Fig. 1, 2]
[I] del serrareleporte | maggiore delchore Lianti delle porte maggiori delchore | son serrati dalla perchussione del sangue | chessi fuggie dalli vētrichuli inferiori del | chore alli superiori fori delchore e sson | riaperti dal refrusso del sanghue sospī|to dalli vētrichuli superiori alli inferi|ori e il uachuo chessi gienererebbe | nellaprire delli vētrichuli inferiorj | (*son q e cq*) quādo siriaprano ecquello | che e chausa di ritirare asse il sāg|ghue delli ventrichuli superiori quādo si | votano ilche votar non si potrebbe | se nō (*avess*) fussi pieghabile e ave|ssi musscholi lōgitudjnali obliqui e | traversali chello strīgnjessino &c –	[I] On the closing of the greater gates [valvular orifice] of the heart The cusps of the greater valves of the heart [mitral and tricuspid] are closed by the percussion of the blood which escapes from the lower ventricles of the heart to the upper ventricle outside the heart [atria]. They are reopened by the reflux of blood pushed from the upper ventricles into the lower. And the vacuum which would be generated by the opening of the lower ventricles when they reopen is the cause of pulling back into themselves the blood of the upper ventricles when they are emptied. These could not empty themselves if they were not pliable and had longitudinal, oblique and transverse muscles which contracted them, etc.
[Fig. 3]	[Fig. 3]
[II] a che sseruano le | chorde de musscho|li del chore – Le chorde che nasschano dalli musscolj | del chore) (*son*) e ssiconvertano in pānj|chuli chessi fanno anti delle maggior por|ti del chore) son quelle che ttenghano chel | (*le pannjchu*)lj (*porte*) ^anti^ del (*chore*) ^le porte^ nō passano | di fuori desse porte ma cholla lora asstē|sione cresschano essi appogiano luna al|laltra e ffanno perfecta chiusura –	[II] What function do the chordae tendineae of the muscles of the heart serve? The chordae which arise from the muscles of the heart and are converted into the membranes which are made into the cusps of the great valves of the heart, are those which hold the cusps of the valve so that they do not pass outside these gates [valve orifices]; but with their extension they increase and are applied to each other, making perfect closure.
[III] Il chore medjante il suo (*il suo*) feruente | moto e chaldjssimo dalmezzo ingiu elqual | moto effato dvevolte per ognj sua bactitura (*la*) ^chome (*si*)^ | (*quale*) appariscie neljposi dj liquali moto vno ne | facto quādo ilchore siresstringnie ellaltro quā|do sidjlata e cquessti moti sonfacti dadue sor|te di musscholi cioe litraversali elli longitudj|nali malli traversali son quelli che sstringhano (*el*) | lagrossezza delchore elli longitudjnali asstēdano | la sua lūghezza Ettale chore fa la sia [sic: la sua] perch<u>|sione quando sasstende che allora si strignie chon @|petuoso moto esschaccia ilsanghue dj se aldedj|chato transito delle vene e il tenpo inte<r>possto | infralle due perchussione delpolso ella meta dū | tenpo musichale e infralluna ellaltra perchus|sione del polso il core sisera due volte e apre vna | e infralluna ellaltra apritura ilchore sapre due | volte essiserra vna (*M*) adunque ilchore | prese ilprimo moto nello aprirsi ellultimo nel | riserrarsi overo il p° nelriserrarsi ellultimo nel | riaprirsi perche ilchore nō bacte (*qua*) neilpolmone | non jspira neltenpo che il figliolo e nella matricie | piena dacqua inpero che sse esso alitassi inmedjā|te annegherebbe Ma lo alitare el bactimēto del | chore della sua madre opera nella vita del figliolo | allei (medjāte lonbellicho) chongiunto chomel|la adopera nelli altri mēbri [:–:] addunque inognj tenpo armonicho o voi di^r^ musi|chale ilchore a 3 moti chome dj socti [sic: sotto] sichontiene | delli quali tenpi vna ora ne chōtiene mille ottā|ta adunque il chore simove tre mila cinque ciēto | quarāta volte per ciasscuna ora traaprire e riserra^r^|si e cquessta tal frequētatione di moto ecquello che | schalda li densi musscoli del chore el quale chore ris|scalda ilsanghue chedentro allui alcontinuo si | djssbacte eppiu sjsscalda nel uētrichulo sinjsstro | doue li parieti sua son grossissimj cheneluentrichu|lo desstro del pariete soctile (*e cq*) e cquesto tal | chaldo assoctiglia elsanghue ello vapora (*ello*) | e ne chōuerte inaria e ne chōvertirebbe infocho | (*s*) elemētare seil polmone cholla fresschezza del | suo vento non sochorressi attale eccesso – Ma ilpolmone nō po mandare aria nelchore ne | e neciessario gienerādosi chome decto dellaria nel | chore la quale (*m*) essendo mjssta cholchaldo e chollu|mjto spesso vapora per li stremi delle vene chapilla|ri alla superfitie della pelle in forma dj sudore | e anchora laria chesspira delpolmone al continu^o^ | entra assciucta (*e ess*) e ffredda e esscie (*ch*) bagnia|ta e chalda Mallarterie chessi chōgiunghano in chō|tinuo chōtacto allaramifichatione della trachea | sparsa per ilpolmone e cquella che piglia il fresco | dellaria che ētra ī tal polmone	[III] The heart owing to its heating movement is hottest from the middle downwards [in the lower ventricles]; and this movement is made twice for each of its beats as appears in the pulses. Of these movements one is made when the heart contracts and the other when it dilates. And these movements are made by two sorts of muscles, that is transverse and longitudinal. But the transverse are those which narrow the size of the heart and the longitudinal ones extend it in length. And the heart makes its percussion when it is extended, for then it squeezes by movement with impetus, and drives the blood from itself into the assigned passage of the vessels. And the time interposed between two percussions of the pulse is half one musical tempo. And between one and the next percussion of the pulse the heart is closed twice and opens once, and between one and the next opening the heart opens twice and closes once. Therefore the heart took its first movement in opening and its last in closing again, or its first in closing and its last in reopening. Why the heart does not beat nor the lung breathe during the time when the infant is in the uterus. This is full of water, therefore if it breathed it would immediately be drowned. But the breathing and beating of the heart of its mother operates the life of the infant joined to her (through the umbilicus) as it operates in other organs. [:–:] Therefore in every harmonic or, as you might say, musical tempo, the heart has three movements which are contained below. Of these tempi, 1 hour contains one thousand and eighty. Therefore the heart moves three thousand five hundred and forty times during each hour in opening and shutting. And this frequency of movement is that which heats the dense muscles of the heart, and the heart heats the blood which is continually beaten up inside it. And it is heated more in the left ventricle where its walls are very thick than in the thin-walled right ventricle. And this heat subtilises the blood and vaporises it and converts some of it into air [water vapour] and it would convert some of it into the element of fire if the lung with the freshness of its wind did not relieve such an excess. But the lung cannot send air into the heart, nor is it necessary because, as said, some air [vapour] is generated in the heart, which being mixed with heat and dense moisture, evaporates through the ends of the capillary vessels at the surface of the skin in the form of sweat. Furthermore the air which is breathed out of the lung continually enters dry and cold, goes out moist and hot. But the arteries which are joined in continuous contact with the ramifications of the trachea spread throughout the lung are those which take up the freshness of the air which enters the lung.
[Fig. 4] a b	[Fig. 4] *a b*
[IV] o ttrovata dal | a ventrichulo | sinj<s>tro al b vē|trichulo desstro | vn buso chava dal | a al b ilquale | noto qui per uedere | se chosi e nelli altri | orechi delli altri | chuori –	[IV] I have found from *a* left ventricle to *b* right ventricle a deep hole from *a* to *b* which I note here in order to see whether this is so in the other auricles of other hearts.
[V] quale orifitio he p|piu profondo nella ba|sa delchore –	[V] Which orifice is deepest at the base of the heart?
[VI] quale orifitio piu | si profonda nella basa | del chuore –	[VI] Which orifice goes deepest in the base of the heart?

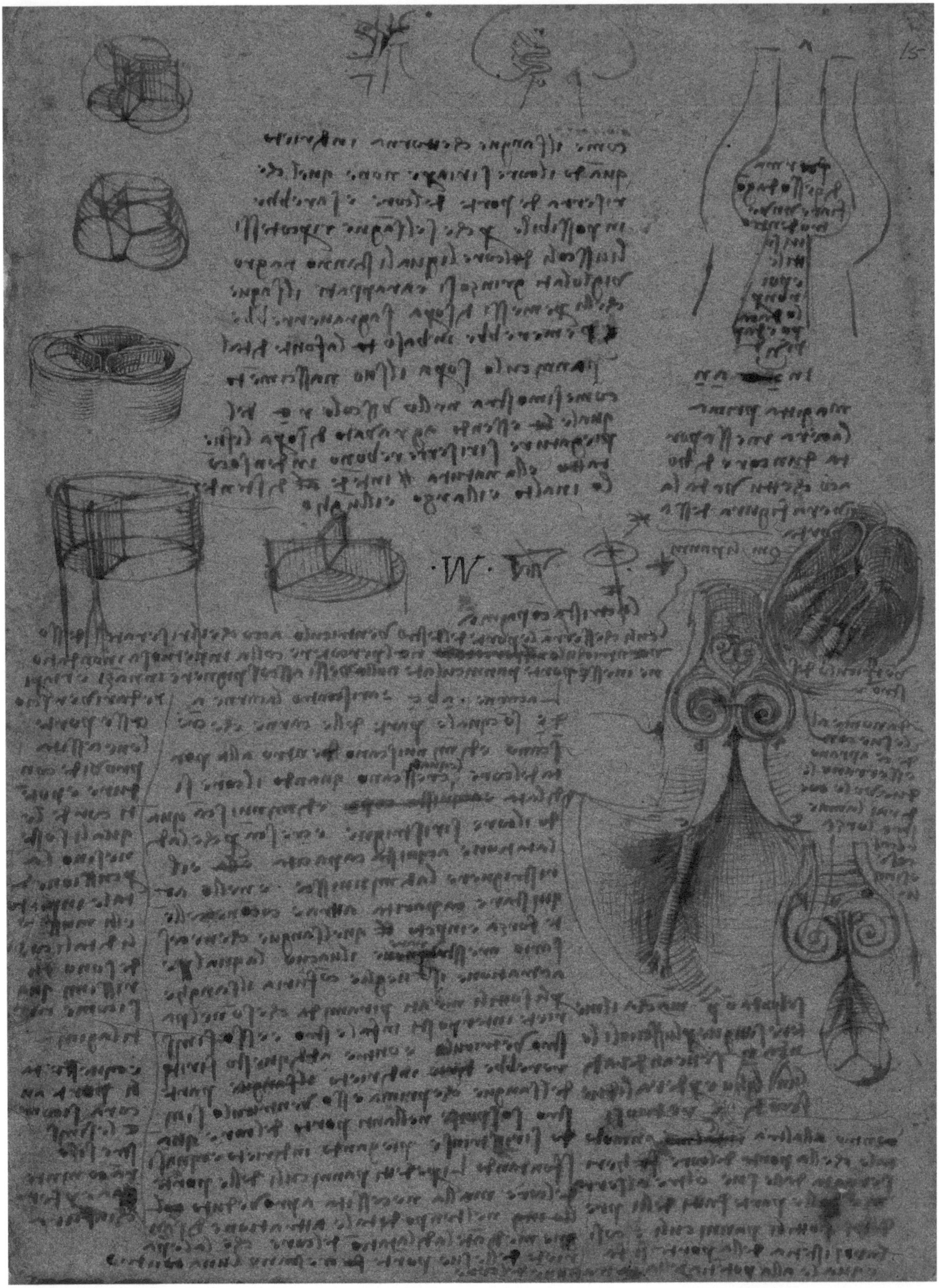

RL 19082 recto, The aortic valve. Leonardo da Vinci (Vinci 1452-Amboise 1519). c.1512–13. Pen and ink on blue paper (Lent by Her Majesty The Queen. Royal Collection © Her Majesty Queen Elizabeth II)

19082 r (C II 12r). Pen and brown ink, on blue paper. 283×204 mm.

[Fig. 1] a n

[Fig. 1] *a n*

[I] forma | dj gesso dagõ|fiare vnve|tro dentro|uj so|ttile | e poi | ronper|lo da ca|po e da p|piedj | in (n *a*) a n

Ma gitta prima | lacera inessa por|ta dun core dj bo | a co chettu veda la | vera figura dessa | porta –

[I] A mould of gypsum for blowing up inside with thin glass; and then break it from head to foot at *a n*.

But first pour wax into the gate of a bull's heart so that you may see the true shape of this gate [valve orifice].

[Fig. 2] r o

[Fig. 2] *r o*

[II] come ilsangue chettorna indjrieto | quãdo il core siriapre none quel che | riserra de [sic: le] porte delcore e sarebbe | inpossibile perche selsãgue ripercotessi | liusscoli delcore liquali stanno ragro|viglolati grinzosi earappati ilsãgue | chelli premessi dj sopra sagrauerebbe | e premerebbe in baso (*ta*) lafonte dj tal | pannjculo sopra ilsuo nasscimẽto | come simostra nello vsscolo r o del | quale (*le*) essendo agravato dj sopra lesue | piegature siriserrerebono in denso cõ|tatto ella natura (*a*) intẽde (*al*) djstender|lo inalto e illargo e illũgho –

[II] How the blood which turns back when the heart reopens [diastole] is not that which closes up the gates [valves] of the heart. This would be impossible, because if the blood percussed back on to the valve-cusps of the heart which stay curled, wrinkled and snatched up, the blood which pressed on them from above would weigh on them and press the surface of each membrane down on to its origin, as is shown by the valve-cusp *r o* which, being weighed down from above its folds, would shut up in close contact. And Nature intends to stretch it out in height, breadth and length.

[Fig. 3–8]

Fig. [3–8]

[Fig. 9]

[Fig. 9]

[III] la ctrista cõpagnja

[III] Bad company

[Fig. 10] vẽtriculo des|stro

[Fig. 10] Right ventricle

[IV] da nome al|le sue cor|de ce aprano | esserrano le | due vele coe | diraj lamae|stra lorza | el cal|cese | e simj|li –

[IV] Give names to the chordae, which open and shut these two sails, that is, call the main one the bowline, topmast, and the like.

[V] Qui li panni[:–:]culj cheserrã le porte deldestro ventriculo acco cheil riserare desso | ventrichulo (*nel percotere*) nelpercotere colla inpetuosa inondatio|ne inesse porte | panniculate nollavessi assospig-nere innãzi e riapri|re dariverscio | esse porte | lenecessita | provide con | dure e potẽ|ti corde le | qualj soste|nesino la | percussione di | tale inpeto | elli mussco|lj dj tali cor|de sono du|rissimj qua|si come car|tilaginj –

e cqueste ta|lj porte an|cora sicom|e le sinjs|stre sise|rã conjnte|ra e perfetta | chiusura

[V] Here are the membranes which close the gates [valves] of the right ventricle [tricuspid]. In order that the re-shutting of this ventricle by the percussion of the wave of impetus on these membranous valves should not force them onwards and reopen the valves from behind, necessity provides them with hard and powerful chordae which can sustain the percussion of such impetus; and the muscles of these chordae are very hard almost like cartilage.

And these valves also, just like the left ones, shut with complete and perfect closure.

[Fig. 11] a b c d e

[Fig. 11] *a b c d e*

[VI] La carne·a b c·e ariscontro la carne a | d e sõ cquale parte delle carne che cre|scano e djmjnuiscano dentro alla por|ta del core ^lecquali^ cresscano quando ilcore si |djlata (*e acqujsta capa*) e djmjnuiscã quã|do il core siristrigne e crescã perche ladj|latatione acquista capacita (*e lla*) e il | ristrignere la djmjnuissce·e nello ac|qujstare capacita attrae co con eccellẽ|te forza einpeto (*il*) quelsangue che neces|sario a resstav(*gnare*)^rare^ iluacuo la qual(*r*)e | actratione is ueglie cõ furia ilsanghe | per li sottili meati piramjda<li>che sõ nelpa|riete interposti inf<r>a l<d>estro e esso sinjs|stro vẽtriculo e ontre [sic: oltre] addjquesto siritj|rerebbe (*djrie*) indjrieto (*il sangue*) parte | delsangue che prima esso ventriculo sinj|stro sospinse nellantiporto delcore quã|do sirisstrinse piegando in djrieto e cquasi | sfondando Lipredettj pannjculi delle porte | delcore malla necessita a proveduto (*col|llo ing*) nel tenpo dj tale attratione dj san|gue medjãte la djlatiatiõ delcore che la le pa|riete delle due porte (*si*) crescano luna contro [:–:] control allaltra (*in tal mo*) in modo | tale chella porta delcore (*fu*) heri | serrata dale sue oltre alserra|me delle porte fattj delli pre|detti sottili pannjculj e cosi | laresistẽtia della porta si fa | equale alla potẽtia della atratione del core

[VI] The flesh *a b c* and the opposite flesh *a d e* are equal fleshy parts which increase and diminish inside the gateway [valvular orifice] of the heart. They increase when the heart dilates and diminish when the heart contracts. And they increase because dilatation acquires capacity and contraction diminishes it; and in acquiring capacity it [the left ventricle] attracts with extraordinary force and impetus that blood which is necessary to make amends for the vacuum. And this attraction violently plucks out the blood through the fine pyramidal pores in the wall interposed between the right and left ventricle. And further part of the blood which the left ventricle previously pressed out of the cusps of the [aortic] valve when it contracted would be pulled back, bending back and almost uprooting the aforesaid membranes of the valve of the heart. But necessity, at the time of this attraction of the blood through dilatation of the heart, has provided that the walls of its gateway [valve-ring] increase, one opposite the other [*a b c* and *a d e*] in such a way that the gateway [valve orifice] of the heart is closed by these walls as well as the shutting of the valve orifice by the aforesaid fine membranes. And thus the resistance of the valve is made equal to the power of attraction of the heart.

[Fig. 12] n o p

[Fig. 12] *n o p*

[VII] selonda o p mãcha ilme|tere sangue per lussciioli lo|nda n se ne andera dj | lunghano e perdera lidue | secõdj retrosi

[VII] If the wave *o p* fails to send the wave of blood, *n*, through the cusps as it goes along, it will lose the two secondary vortices.

[Fig. 13]

[Fig. 13]

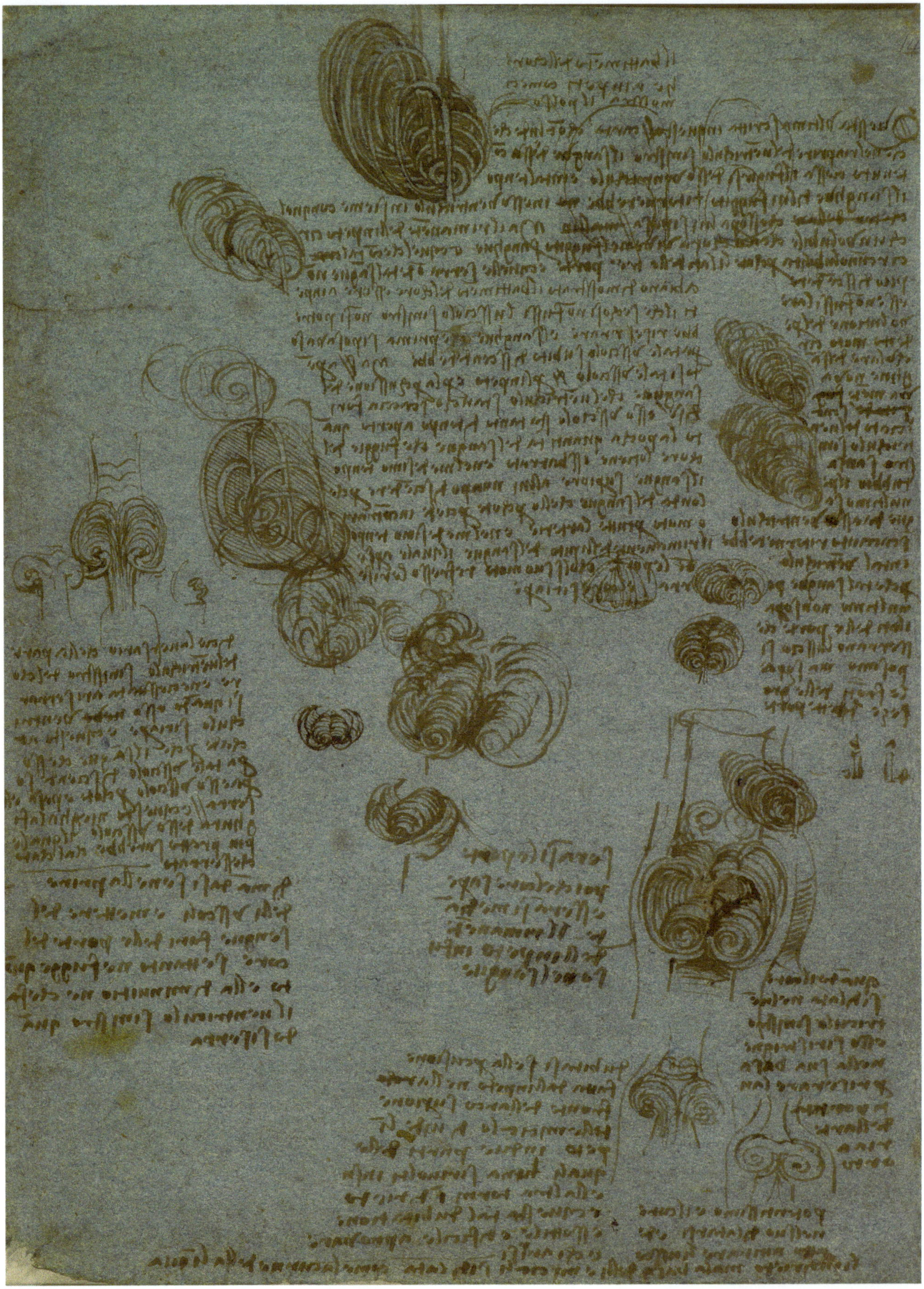

RL 19083 verso, Blood flow through the aortic valve. Leonardo da Vinci (Vinci 1452-Amboise 1519). c.1512–13. Pen and ink on blue paper (Lent by Her Majesty The Queen. Royal Collection © Her Majesty Queen Elizabeth II)

19083 v (C II 13v). Pen and brown ink.

[I] il battimēto dellchore | he a inpetj come ci | mosstra il polso

[I] The beating of the heart is by impetuses as the pulse shows us.

[Fig. 1–7]

[Fig. 1–7]

[II] Abiano djmosstrato il battimēto delchore essere a inpe|ti ilche se chosi nō fussi lussciolo sinjstro nō si potre|bbe rise(*s*)rrare elsanghue che prima siposava so|pra tale vssciolo subito djsscenderebbe Ma (*s*) aprē|dosi tale vssciolo (*a*) perlinpeto e per la perchussione del | sanghue chel uētrichulo stancho scaccia forj | dise esso vssciolo sta tanto dj tenpo aperto quā|to la pocha quantjta del sangue che fuggie del | chore lotiene issbarrato enel medesimo tenpo | ilsanghue superiore alluj non po djsciēdere perche | londa del sangue chello perchote prociede in cōtrari|o moto per tutte larterie e nel medesimo tenpo | ilrimanente delli<n>pito delsangue ilquale aperse | (*le*) leporte cholssuo moto refresso le rise|rra el core siriapre

[II] We have demonstrated the beating of the heart to be by impetuses. If this were not so the left [aortic] valve would not be able to shut and the blood which was at first located above the valve would immediately descend. But the valve being opened by the impetus and percussion of the blood which the left ventricle drives out of itself, the valve stays open just as long as a small quantity of blood which issues from the heart keeps it free from obstruction. And during this very same time the blood above it cannot descend because the wave of blood which percusses it proceeds in an opposite movement throughout all the arteries. And during this very same time the remaining impetus of the blood which opened the valves, with its reflected movement closes them again, and the heart [ventricle] opens again.

[Fig. 8, 9 10]

[Fig. 8, 9 10]

[III] djcie lauersario chella parte | deluētrichulo sinistro del cho|re e neciessitata a riserrar|si quādo esso (*ue ve*) ventri|chulo siriapre e cquesto ac|chade perche il sāgue chesso|pra tale vssciolo djsciende so|pra esso vssciolo perchote e pesa ello | serra //

ecquesto niegha la fi|ghura desso vssciolo ilquale | piu pre<s>tto sarebbe chalchato | chesserrato –

djmādasi se nellaprire | dellj vsscolj e mettere del | sengue fori delle porte del | core settanto ne fugge quā|to ella djmjnuitione che fa | il uentriculo sinjstro quā|do siserra

[III] The adversary says that part of the left ventricle of the heart is necessarily shut when this ventricle reopens; and this happens because the blood which is above this valve descends on to the valve, percusses, weighs on it and shuts it.

And this is denied by the drawing of the valve; it would be more readily trampled down than shut [see Fig. 8].

One asks whether, with the output of blood through the valves of the heart on opening, the amount that escapes equals the diminution of the left ventricle when it closes.

[Fig. 11–18]

[Fig. 11–18]

[Fig. 19]

[Fig. 19]

[IV] serāsi le porte | poichelcore sapre | esserāsi mediā|te il rimanēte | dellinpeto infu|so nelsangue

[IV] The gates [of the aortic valve] shut, then the heart opens; and they are shut by means of the remainder of the impetus infused into the blood.

[Fig. 20]

[Fig. 20]

[V] dubitasi sella percusione | fatta dallinpeto nella(*rcho*) | fronte dellarco superiore | dellemjciclo djujde līl|peto in due parti delle | quali luna siriuolti insu | ellaltra tornj ī djrieto | e cquesta tal dubitatione | essottile e djficile a provare | e chiarirsi –

[V] It is uncertain whether the percussion made by the impetus on the upper arch of the hemicycle [internal annular ring between aortic sinuses and trunk of aorta] divides this impetus into two parts, one of which revolves upwards and the other turns backwards. Such doubts are subtle and difficult to test and clarify.

[Fig. 21]

[Fig. 21]

[VI] quādo ilcore | si djlata nel uē|triculo sinjstro | esso siristrigne | nella sua basa | per riserare lan|tiportita | deellarte|ria a|orto

[VI] When the heart's left ventricle dilates it narrows at its base in order to close the entrance gate [valve-ring] of the aortic artery.

[VII] potentissimo e il core | nelsuo djlatarsi ede | atto attirare li ussco|li indjrieto mala basa delli emjciclj sidjlata come la carne della līgua

[VII The heart is very powerful in its dilatation, and it is apt to draw backwards the valve-cusps; but the base of the hemicycles dilates like the flesh of the tongue.

[VIII] Quessta vltima scritta in questa carta chōclude che | che nelriaprire del uētrichulo sinjstro il sangve desso cō|tenuto ciessa ilfugirsi desso ventrichulo e intal tenpo | ilsanghve djlui fuggito ritornerebbe (*ne*) inesso ventrichulo insieme conquel | (*che tre della re*) chessopra luj siposa // (*ma llin*) Ma ilrimanēte dellinpeto cir|chunvolubile che anchora vive nelfuggito sanghue e cquel che cō tal (*cir*) | circūuolubiljta perchote li lati delle tre porte e cquelle serra ōde tal sāgue nō | puo djsscēdere | esse nō fussi lare|volutione del pre|detto moto cir|cholare delsā|ghue nova|vamēte (*fug|gito de*) scac|ciato dal uen|trichulo sinj|stro sanza | dubbio ilpe|nultimo sa|<n>gue da esso ventrichulo | scacciato ritornerebbe | in tal vētrichulo | perche tal sangue pe|nultimo no sopra | ilati delle porte che | sserrano lusscio si | posano ma sopra | le frōti delle gro|seze djdette porte

[VIII] This, the last [note] written on this page, concludes that on the reopening of the left ventricle the blood contained in it ceases to issue from this ventricle, and at that time the blood which has issued from it would return into this ventricle together with that which is placed above it. But the remaining circumvolving impetus which still survives in the ejected blood is that with which such circumvolution percusses the sides of the three valves and closes them so that this blood cannot descend. And if there were not the revolving of the aforesaid circular movement of the blood newly driven out of the left ventricle, without doubt the penultimate blood driven out of this ventricle would return into the ventricle because such penultimate blood is not situated on the sides of the valves that close the valve orifice, but on the ends of the widths of the said valves.

[Fig. 22] a b c d

[Fig. 22] *a b c d*

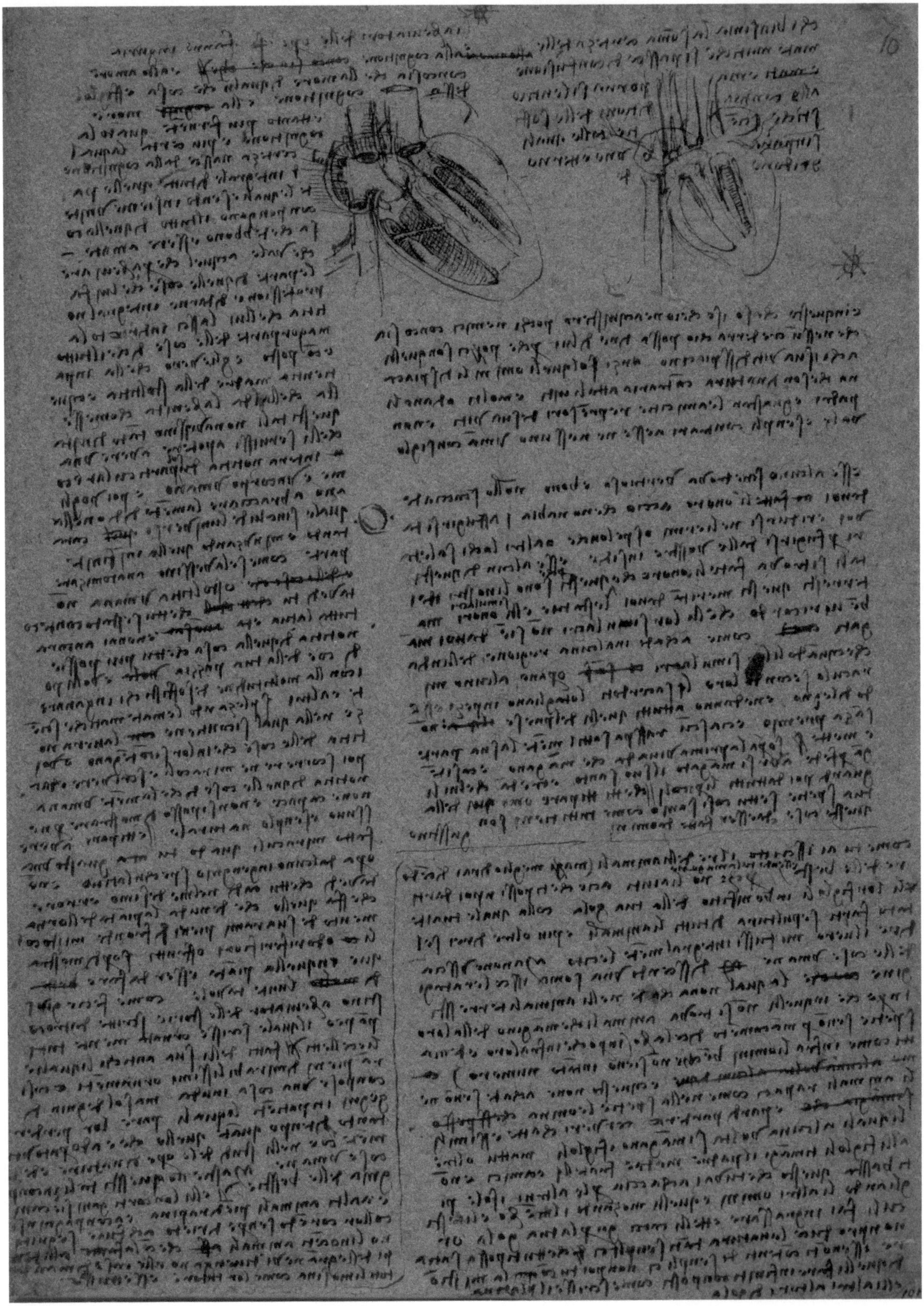

RL 19084 Studies of the heart, and notes on the vices of men. Leonardo da Vinci (Vinci 1452-Amboise 1519). c. 1511–13. Pen and ink on blue paper (Lent by Her Majesty The Queen. Royal Collection © Her Majesty Queen Elizabeth II)

19084r (C II 14 r). Pen and brown ink, on blue paper. 283 × 204 mm.

[Fig. 1] d	[Fig. 1] *d*
[Fig. 2]	[Fig. 2]
[I] chi biasima la sōma certeza delle | matematiche si passce dj confusione | (*e mati*) e mai porra silentio | alle contradjtioni delle soffi|stiche sciētie colle qualj | sinpara vno etterno | gridone	[I] He who blames the supreme certainty of mathematics feeds on confusion and will never be able to silence the contradictions of the sophisticated sciences which impart perpetual clamour.
[II] e inquesto chaso iso cheio ne acqujstero pochi nemjci conco sia | che nessū crederra possa djre djlui perche po(*s*)ci son quellj | a chi isua viti djsspiacino anzi sol queli omjni li djspiaca|no che son dj natura cōtraria attali ujtj e molti odjano li | padri e guastan leamjcitie reprēsori desua viti e non | vale esenpli contrari a esse ne nessuno vmā consiglo	[II] And in this case I know that I shall acquire few enemies for no one will believe that I can be speaking of him because there are few who are displeased by their own vices. On the contrary, they displease only those men who are of a nature contrary to such vices. Many hate their fathers and spoil friendships which are reproaches of their vices; and examples to the contrary are of no value to them, nor any human advice.
[III] esse alcuno sene trova virtuoso e bono nōllo scacciate | dauoi (*a c*) fatteli onore accio che nonabia affugirsi da | voj e ridursi nel ermj o spelonche oaltri lochi soleta|ri per fugirsi dalle vosstre insidje esse alcun dj questj | talj sitrova fateli onore che ^perche^ questi sono linostri idei | terresti questi meritā danoj le statue ^simulacri^ e lli onori ma | bē ui ricordo chelli lor simulacri nō siē dauoi mā|gati (*coe i*) come achade inalcuna regione dellindja | checquādo li lor simulacri (*cr sō sa*) operano alcuno mj|raculo secondo loro lj sacerdoti lo tagliano inpezi esse<n>|do dj legno e nedanno a tuttj quelli delpaese (*il qa*) e nō | sāza premjo e ciascū rasspa sottilmēte la sua parte | e mette (*s*) sopra la prima viuada che māgano e cosi tē|gā per fede aversi māgato ilsuo santo e credā che lui li | guardi poi dattutti li pericolj // chetti tti pare omo quj della | tua spetie settu cosi saujo come tutti tienj son | queste cose daesser fatte da ominj	[III] And if anyone among you is found to be virtuous and good do not drive him away from you but do him honour so that he will not have to flee from you and take refuge in hermitages or caves or other solitary places in order to escape from your perfidy; and if any such be found do him honour, for these are our earthly gods. These men deserve from us statues, images and honours. But I must remind you that their images are not to be eaten by you, as happens in a certain region of India, so that when their images work some miracle, according to them, their priests cut them up into pieces, being of wood, and give part of it to all those of the country– not without payment and each grates his portion very fine, and puts it on the first food that he eats. And thus according to their faith they maintain that they have eaten their saint, and they believe that he will then protect them from all dangers. What do you think, O Man, of your species? Are you as wise as you consider yourself to be? Are these things that should be done by men?
[IV] li abreuiatori delle opere (*s*) fanno inguria | (*allo amore e*) alla cognjtione (*con co sia che che v*) e allo amore | concosia che llamore dj qualūche cosa e ffiglol | dessa cognjtione e lla (*cognjt*) more | ettanto piu feruēte quāto la | cognjtione e piu certa la qual | certeza nassce dalla cognjtione | dintegrale djtutte quelle pa<r>|te lequali esendo insieme vnjte | conpongano iltutto dj quella co|sa che debbono essere amate– | che vale acquel che per abreujare | leparte dj quelle cose che luj fa | professione dj darne integral no|titia che lluj lassci in dirieto la | majorparte delle cose dj che il tutto | e cōposto egli e vero chella inpa|tientia madre della stoltitia e cque|lla chellalda la breujta chomesse | questi tali nonavessino tāto dj ujta | chelli seruissi apotere avere vna | (*n*) intera notitia dū ^sol^ particulare co|me e vncorpo vmano e poi vogli|ano abracciare lamēte dj djo nella | quale sinclude lunjverso (*quel*) cara|tando e mjnvzando quella in@finite | parte come se lavessino anatomizate | (*e delle cose che*) o stoltitia vmana nō | tavedj tu (*chett chell*) chettu sestato conteco | tutta la tua eta (*e nō fai*) e nonai ancora | notitia dj quella cosa chettu piu possie|dj coe della tua pazzia (*vole*) e volli po|i con lla moltitudjne desoffistichi inganare | te e altri splezando le matematiche sciē|ze nella qual sicontiene (*con*) lauera no|titia delle cose che in lor sicōtēgano o voj | poi scorrere ne miracoli e scrivere e dar | notitia dj quelle cose di che la mēte vmana | non e capace e non siposso<n>djmostrare per ne|ssuno esenplo naturale //	[IV] The abbreviators of works do injury to knowledge and love, for the love of anything is the offspring of knowledge, and love is as much more fervent as knowledge is more certain. And this certainty is born from the integrated knowledge of all those parts which, being united together, compose the whole of that thing which is to be loved. Of what value is he who in order to abbreviate the parts of those things of which he professes to give integrated knowledge leaves out the greater part of the matters of which the whole is composed? It is true that impatience, the mother of folly, is she who praises brevity, as though some people had not enough life to use to enable them to acquire a complete knowledge of a single particular subject, such as the human body. And then they want to embrace the mind of God in which is enclosed the universe, carefully weighing and minutely dissecting it into infinite parts as though they were making an anatomy of it. O human stupidity! Do you not see that you have spent your whole life with yourself and yet have no knowledge of that which you possess above all, that is your own folly? And then with the multitude of sophists you want to deceive yourself and others, despising the mathematical sciences in which is contained true knowledge of the things considered by them. Or you then survey miracles and write about them and give information about these subjects which the human mind is incapable of understanding and which cannot be demonstrated by any example in Nature.

19084r (C II 14 r). Pen and brown ink, on blue paper. 283 × 204 mm.

e ti pare avere | fatto mjraculi quãdo tu (*ai*) a guasto vna |opera dalcuno ingengnjo spechulatiuo e nõ | tavedj chettu cadj nelmedesimo errore | che ffa quello che denuda la piãta dellorna|mento de sua ramj pieni dj fronde misto col|li (*co*) odoriferi fiorj o ffrutti sopra djmostra | que inquella piãta esser da fare (*belle*) | di (*molto*) lnude tavole come fecie gives|stino abreuiator delle storie scritte da troco | põpeo ilquale scrisse ornatamente tutj | li ecelle<n>tti (*v*) fatti delli sua antichi liquali e|rã pienj dj mjrabilissimi ornamẽti e cosi | conpose vna cosa inuda ma sol degnia dj | gegnj inpatiẽti liquali pare lor perder | tanto dj tenpo quãte quello che e adoperato vtil|mẽte coe nelli studj de le operere dj nature e delle | cose vmane Ma stieno quessti tali inconpa|gnja delle bestie (*v*) ellj lor cortiganj siẽ canj | e aalti [sic: altri] anjmali piẽ di rapina e aconpagnjnsi | collor corẽdo senpre djrieto a ch<I>fuge seguita|no linocẽti anjmali (*ass*) che cõ la fame alli ten|pi delle grã nevi tiuengano alle case djmanda<n>|toti limosina come lor tutore essettu sse [:–:] come tu ai isscritto il re delli anjamli (ma (*gi*) meglio djrai dicẽdo | re delle besti<e>^essendo tu la magore^) perche nõ liaiuti acie che tipossin poi dartj | (*de*) li lor figloli in benjfitio della tua gola colla quale tu ai te<n>|tato farti sepulture dj tutti li anjmalj e piu oltre djrei sel | djre il uero mi fussi integralmẽte lecito Ma none vssciã | delle cose vmane (*M*) djsscendo vna soma issceleratagi|gine (*coe che*) la qual non achade nelli animali terressti | inpero che in quellj nõ si trova anjamli che mãgino della loro | spetie se nõ per mãcamẽto dj celebro (inperoche infraloro e de ma|tti come infra liominj bẽche nõ sieno intãto numero) (*co|me alcuna volta alcunj dj que*) e cquesto none achade senõ ne|li anjmali rapaci come nella spetie leonjna (*chesspesso | si mãgia che*) a pardj panterie cerveri chatte e ssimjlj | li quail alcuna volta simãgano I figlolj mattu oltre | allj figlolj ti mãgi il padre madre fratellj e amjci e nõ | ti bassta questo che tuvai a chaccia per le altrui isole pi|gliando li altri omjnj e quelli mozando il mẽbro e li testi|culi fai ingrassare ettelli cacci gu per la tua gola or | nonproduce lanatura tãti senplici (*d*) che ttu tipossa satia|re essenõ ti cõttnti de senplicj non poi tu cõ (*au*) la mjstiõ | di quelli fare infinjti conpostj come scrisse il platina | ellialtri altori dj gola

And it seems to you that you have performed miracles when you have spoiled the work of some contemplative genius; and you do not see that you are falling into the same error as one who strips a tree of its adornment of branches full of leaves mixed with fragrant flowers or fruit, in order to demonstrate that the tree is suitable for making bare boards; just as did Justinus, the abbreviator of the histories written by Trogus Pompeius, who wrote eloquently of all the extraordinary deeds of his ancestors, which were full of very wonderful ornate descriptions; and by so doing he [Justinus] composed a bare work indeed, fit only for such impatient minds as think they are wasting time when they use it usefully, that is, in the study of the works of Nature and human affairs. But let such as these remain in the company of beasts, and their courtiers be dogs and other animals full of rapacity, and let them keep company with them. Always running after what flees from them, they pursue innocent animals which because of hunger during the season of heavy snow, come to your houses to beg for alms from you as from a guardian. And if you are, [:–:] as you have written, king of animals (but it would be better for you to call yourself king of beasts, since you are the greatest of them) why do you not help them so that they may later give you their young for the benefit of your gullet, with which you have tried to make yourself the sepulchre of all animals. And I would say still more if speaking the whole truth were allowed me. But do not let us quit these human matters without speaking of a supreme wickedness which does not exist among the animals of the earth, since among them one does not find animals which eat their own species, except through a lack of brains (since there are mad ones amongst them, as among men, though not so numerous) and this does not happen except among the rapacious animals such as the leonine species, leopards, panthers, lynxes, cats and the like, which sometimes eat their young. But you, besides eating your children, eat father, mother, brothers and friends; and even this is not enough, so that you go hunting through foreign islands to capture other men, and then after cutting off the penis and testicles, you fatten them up and cram them down your gullet. Now does not Nature produce enough simple things to be able to satisfy you? And if you are not content with simple things can you not with a mixture of these make infinite compounds, like those described by Platina and other authors on gluttony?

[V] gusstino

[V] Justinus

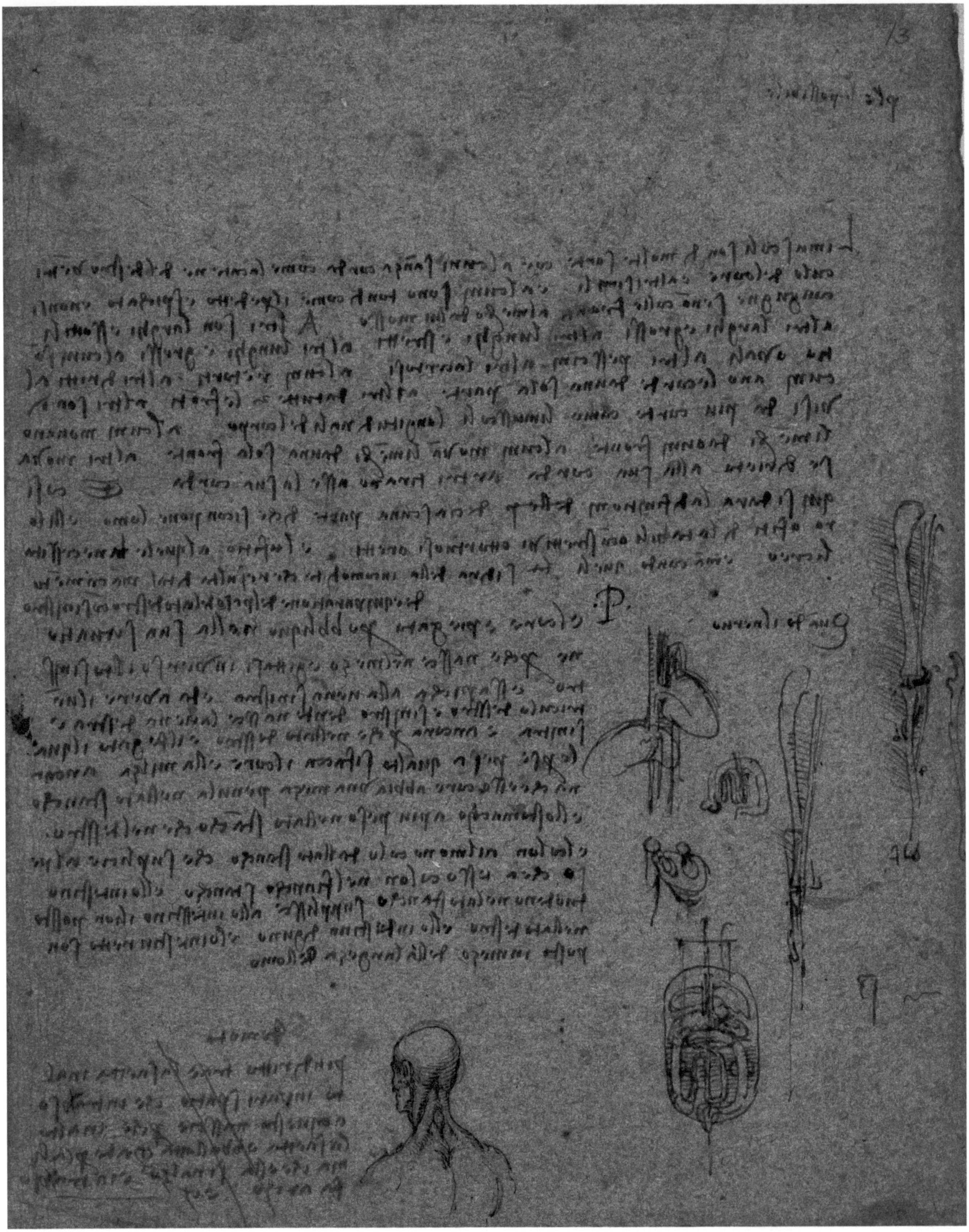

RL 19085 Notes on the muscles of the heart, and assymetry in the body. Leonardo da Vinci (Vinci 1452-Amboise 1519). c. 1511–13. Pen and ink with some red chalk on blue paper (Lent by Her Majesty The Queen. Royal Collection © Her Majesty Queen Elizabeth II)

19085r (C II 15 r). Pen and brown ink (two shades) over some red chalk, on blue paper. 269×210 mm.

[I] p l e i<n>possibile	[I] *p l e* impossible
[II] Limuscoli son dj molte sorte coe alcunj sa$za corda come lacatene del destro vētriĐĐĐ|culo delcore e altri simjli e alcunj sono tondj come ilpredetto e spichato enonsi | congugne senō colle freni almēbro da lui mosso Altri son larghi e ssottilj | altri larghi e grossi altri lunghi e stretti altri lunghi e grossi alcunj sō|no ovalj altri pesscinj altri lacertosi alcunj retorti altri djritti al|cunj ano lecorde dauna sola parte a(*r*)ltri datutte 2 le frōti altri son dj|visi da piu corde come limusscoli longitudjnali delcorpo alcunj mouano | li mēbri da onnj fronte alcunj movā li mēbri dauna sola fronte altri mova|se djrieto alla sua corda artri [sic: altri] tirano asse la sua corda E cosi | quj si dara la djfinjtionj (*delle p*) dj ciascuna parte dj che siconpone lomo e lli lo|ro ofiti djlatabili ocōstrettivi o ttortuosi oretti e lufitio alquale lanecessita | licreo e mācondo [sic: mancando] queli (*la*) sidjra della incomodjta che resulta dj tal mācamēto	[II] Muscles are of many sorts, that is, some are without tendons like the bands of the right ventricle of the heart and others like them [moderator band]. And some are round like that mentioned above, and detached and not joined except by bridles to the part moved by them [papillary muscles]. Others are broad and thin; others broad and thick; others long and narrow; others long and thick; some are oval; others pisciform [fish-like]; others lizard-like [lacerti]; some twisted; others straight. Some have their tendons on one side only, others on the whole of both surfaces; others are divided by several tendons like the longitudinal muscles of the body [e.g. rectus abdominis]. Some move the limbs from both sides, others move the limbs from one side only; others are moved backwards towards their tendons; others draw their tendons towards them. In this way there will be given here the definition of every part of which man is composed, and their functions in dilating or narrowing, bending or straightening; and the function for which necessity created them. And if these functions fail we shall speak of the trouble that results from such a deficiency.
[Fig. 1, 2, 3, 4]	[Fig. 1, 2, 3, 4]
[III] de equjparatione delpeso delato destro cō sinjstro elcore e piegato per obbliquo nella sua situation|ne perche nassce nelmezo e gittasi inverso il l<a>to sinjs|tro e ssapicha alla uena sinistra e da avere il uētriculo desstro e sinjstro donde nassce la uena destra e | sinj<s>tra e ancora perche nellato desstro e ilfegato ilqua|le per se pesa qualto [*sic*: quanto] sifacca il core ella mjlza ancor|ra che esso core abbia vna meza penula nellato stancho | ello stomacho a piu peso nellato stācho che nel desstro | elcolon ailmonocolo dallato stancho che suplicie alpe|so che a esso colon nel fiancho stancho elloíntestino | duodeno nelato stancho supplissce allo intesstino ilion possto | nellato destro ello intestine djguno elointestin retto son | posto inmezo della largeza dellomo	[III] On the equilibration of weight on the right and left sides The heart is inclined obliquely in its situation, because it originates in the middle and is thrown towards the left side. It is attached to the left vessel [aorta] by having the right and left ventricles from which the right and left great vessels arise [superior vena cava and aorta; see Fig. 1]. Furthermore, on the right side is the liver which by itself weighs as much as do the heart and spleen. Further the heart has one half lobe [ventricle] on the left side [Fig. 3]. The stomach has more weight on the left side than on the right, and the colon has the caecum on the left [sic: right] side which makes up for the weight of the colon in the left flank [Fig. 2]. The duodenal intestine on the left side compensates for the intestinal ileum placed on the right side. And the intestine, jejunum and the rectum are placed in the middle of the width of a man [Fig. 4].
[IV] [*–] de moto Piu djritto trae lasaetta in al|to inparj spatio che intraverso | e cquesto nasscie perche inalto | la saetta o bballota chade per lalj|nja che ella sinalzo ein traverso | fa archo &c–[–*]	[IV] [*–] On movement The arrow goes more straight upwards than transversely in an equal distance; and this happens because the arrow, or small ball, follows upwards along the line which elevates it and in the transverse it makes a curve, etc. [–*]
[V] Quādo il neruo	[V] When the nerve
[Fig. 5]	[Fig. 5]
[Fig. 6] a b c d e f	[Fig. 6] *a b c d e f*
[Fig. 7, 8]	[Fig. 7, 8]

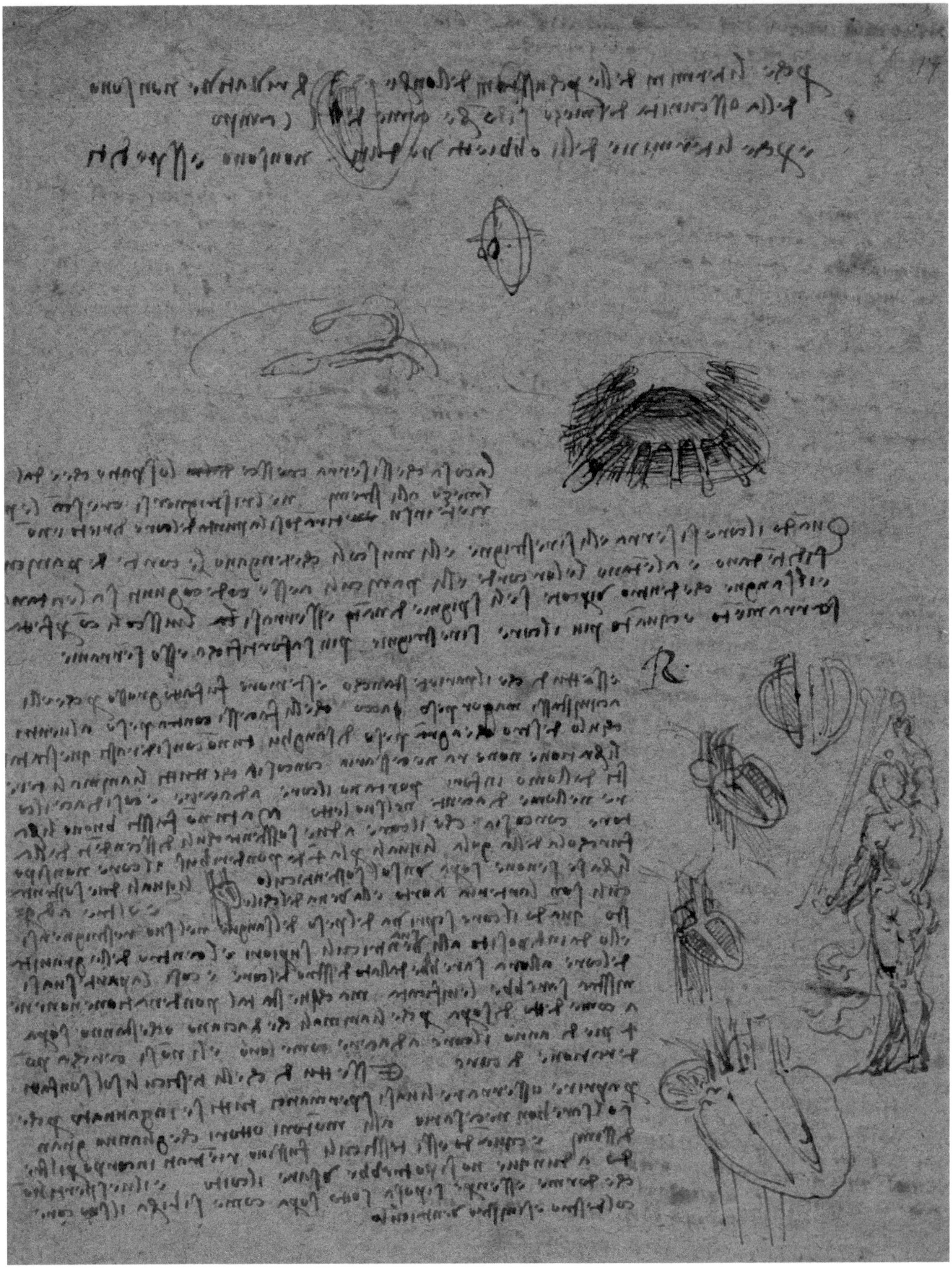

RL 19087 recto, Studies of the heart, notes on light and shade, and a standing figure. Leonardo da Vinci (Vinci 1452-Amboise 1519). c. 1511–13. Pen and ink on blue paper (Lent by Her Majesty The Queen. Royal Collection © Her Majesty Queen Elizabeth II)

19087 v (C II 17 v). Pen and brown ink.

[Fig. 1, 2]

[Fig. 1, 2]

[I] il core a 4 ve<n>triculi (*du*) coe dua superiori detti orechi | del core e dua inferiori acquelj detti desstro e sinjstro | ventriculo

[I] The heart has four ventricles, that is two upper called auricles of the heart, and two below these called the right and left ventricles.

[II] fu fatto piu grave ildes|stro che il sinjstro ventri|culo acco che il core ste|ssi per obbliquo · esse lsē|gue sileua del destro vē|triculo ello alleuja esso sangue ne va ī|verso la parte sinjstra colcentro della sua grauita | quādo e nellj ventriculi superiori

[II] The right ventricle was made heavier than the left ventricle in order that the heart might be placed obliquely; and if blood is taken away from the right ventricle and lightens it, some of the blood goes towards the left side with its centre of gravity when it is in the upper ventricles.

[III] djco che lla magor capacita dellj | ventriculi (*e capace dj ri*) ^e attra e atta a rri^ceuere | (*vna*) ^inse^ quātita dj sangue la qual si | fugie dj lor (*nella minor lor ca|pacita coe quādo si risstringano*) ^quādo sirestringhano^ | il qual (*fugimeto s^caccare^*) e riceuuto ^in djposito^ | delli ventriculi superiori dettj o|rechi del core e inquelli siriser|va insino che e restituito alla | ^sucedēte^ djlatation del core ma cō tanta | mē quātita quāto fu quella | che ssatrebuirono li nutritori | della ujta ilqual mācamēto | e ricōpensato dal fegato gene|ratore dj tal sangue –

[III] I say that the greatest receptive capacity of the ventricles is the quantity of blood which escapes from them when they contract. This is received in deposit by the upper ventricles called the auricles of the heart, and it is kept in these until it is returned on the succeeding dilatation of the heart, but with so much less quantity as was contributed to the nourishers of life [i.e. the veins]. This deficiency is compensated by the liver, generator of such blood.

[IV] Se il core nelrestrignersi djmjnuissce lasua | capacita e da (*il suo*) sangue (*a*) ^da llui scaccato^ in djposito alli | ventriculi (*p djpositari del sangue da lluj | scaccato*) essi ventriculi liele restjtuis|scano con(*ta*)tanta mjnor quātita quanto fu | quella che ssatrebui lanecessita delnutrimē|to della vita la quale e rēduta dal fegato | tesoriere

[IV] If the heart in contracting diminishes its capacity and gives blood driven out of it in deposit to the [upper] ventricles [i.e. atria] these ventricles return it to the heart as much reduced in quantity as was necessary for the contribution to the nourishment of life. This is restored by the liver, the treasurer.

[V] si elcore nello strignersi da ilsuo sangue | in djposito allj sua orechi (*e*) liquali liene | restituissca ^Ma^ tanto meno che quel che [^–] fu loro | dato [–^] (*ricevano*) quāte lla parte che lortolto | dale uene superiori e auendo il core a e|ssere rifatto della soma intera che solea | riēpiere ilsuo vacuo e necessario chel | vētriculo sinjsstro inpremuti dal des|stro e che il uētriculo desstro siarifatto | dal fegato dj quel che gli fu tolto dalla(*rte*) | vena arteriale e daluētriculo sinjstro | adunque iluētriculo desstro a duppia perdj|ta a cquella deluētriculo sinjsstro –

[V] If the heart in its contraction gives its blood in deposit to its auricles, they return some of it, but so much less than was given to them as the portion which was taken by the superior veins. And the heart having to be resupplied with the complete quantity which formerly filled its vacuum, it is necessary that the left ventricle be replenished from the right, and that the right ventricle be resupplied from the liver for that which was taken away by the arterial vein [? pulmonary artery] and by the left ventricle. Therefore the right ventricle has lost twice that lost by the left ventricle.

[VI] Il cor cacia fori ilsangue nel suo risstrignersi e cquāto piu siristrignje i pannjculi | chesserā dj dentro in fori piu perfettamēte siserrano adunque ilsangue integral|mēte non siuota ma cquel che rimane e premuto dal uētriculo desstro nel sinjsstro | perche non po penetrare essi vsscioli onde e necessario chellj musscoli ^de panjcolj^ intaltenpo sidjlughino

[VI] The heart drives out the blood on contracting, and the more it contracts the more perfectly do the membranes [valve-cusps], which close from within outwards, shut. Therefore the blood is not entirely emptied out, but what remains is pressed from the right ventricle into the left. As it cannot pass through these valve-cusps it is necessary that the muscles of the valve-membranes [the papillary muscles] should at this time be elongated.

[Fig. 3, 4]

[Fig. 3, 4]

[VII] perche son fattj li ussciolj principalj deluētriculo desstro di si poco pānj|culo e dj tanto traforamēto dj corde // fv ordjnato talcosa dalla natura | aco che com@cando ariserrarsi ildesstro vētriculo non siserrassi subito | lusscita del sangue della sua grā capacita perche dj quel sangue sauea da|re alpolmone e nō se ne sarebbe dato se llusscolo auesssi proibito lusscita | (ma ssisero esso vētriculo quādo ilpolmone ebbe ricevuto lasua quā|tita dj sangue e che ildesstro vētriculo potea premere per li meati de | pariete dj mezo nel uentriculo desstro e in questo tenpo lorechio des|stro sifece djpositario della sopra bondanza delsangue che avanzo | alpolmone ilqual subito rēde allo aprimēto desso desstro ventri|culo ristorādosi per il sangue · chelli porge il (*polmone*) ^fegato^ –

[VII] Why are the principal valve-cusps of the right ventricle [tricuspid] made of so little membrane and so much interlacing of chordae [tendineae]? This was arranged by Nature so that when the right ventricle begins to contract, the exit of its large volume of blood should not be shut off suddenly since some of the blood has to be given to the lung and it would not be given if the valve-cusps prevented its exit. However, this ventricle is closed when the lung has received its quantity of blood. And the right ventricle is able to push it through the meati of the middle wall of the right ventricle [interventricular septum]. And during this time the right auricle provides the depository of the excess blood which goes out into the lung, which immediately returns it to the opening of this right ventricle replenished by the blood which the liver delivers to it.

[VIII] quāto sangue li po porgere il (*po*) fegato per apritura del core | rēdeglene quāto esso esso neconsuma coe vna mjnjma parte perche in u|nora achade circha dumjla aprimēti dj core va gran peso

[VIII] How much blood is the liver able to deliver owing to the opening [diastole] of the heart? It restores to it as much as it consumes, that is a very small part. Since in 1 hour about two thousand openings of the heart occur, it is a great weight.

[IX] (*14*) 25
12
300
7 once per ora

[IX] 25
12
300
7 ounces per hour

[X] figura il cor per 4 asspecti

[X] Draw the heart from four aspects.

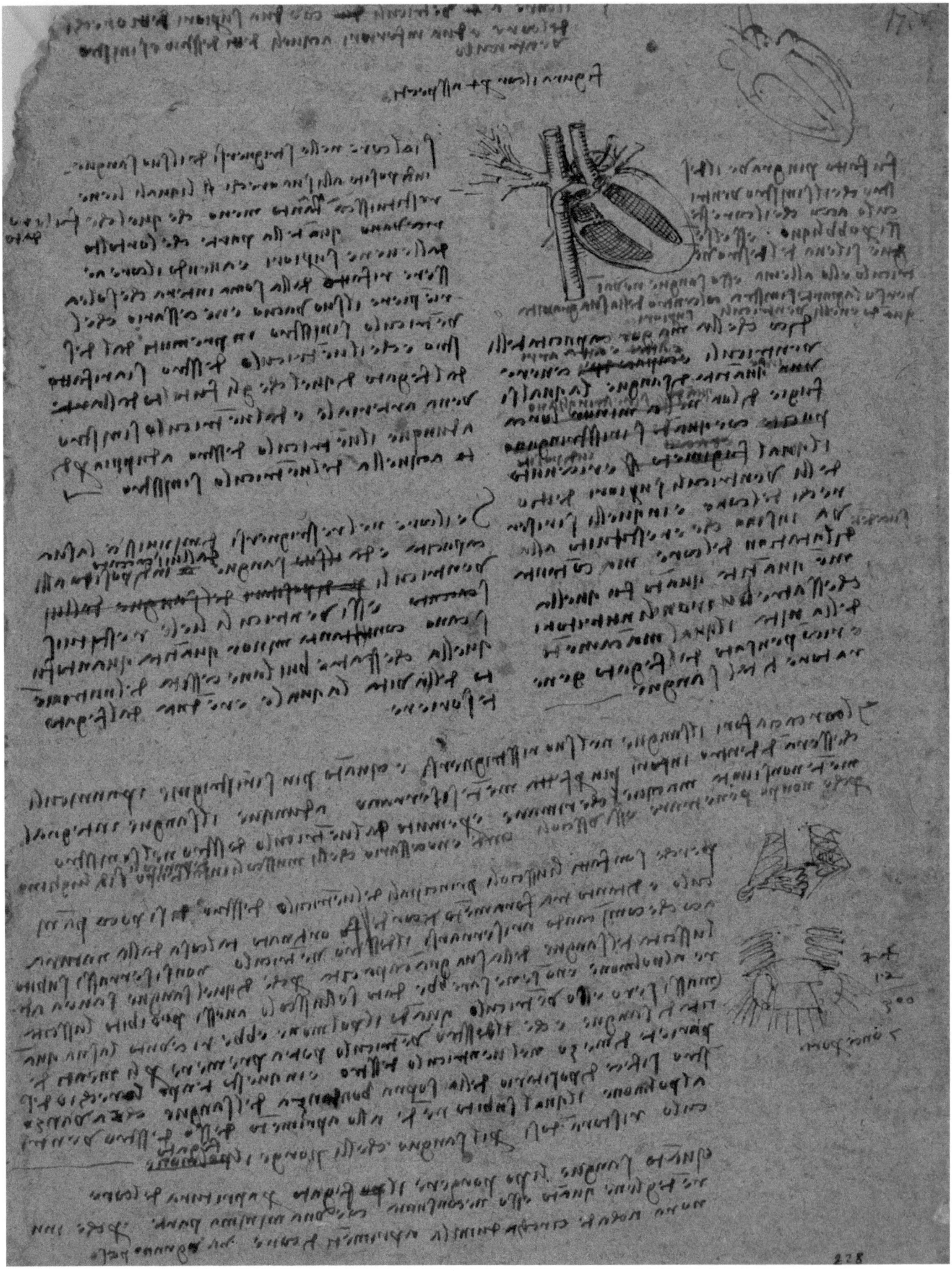

RL 19087 verso, Studies of the motion of the heart. Leonardo da Vinci (Vinci 1452-Amboise 1519). c. 1511–13. Pen and ink on blue paper (Lent by Her Majesty The Queen. Royal Collection © Her Majesty Queen Elizabeth II)

19087 r (C II 17 r). Pen and brown ink, on blue paper. 285×212 mm.

<table>
<tr><td>[I] perche li termjnj delle perchussionj dellonbre djrivative non sono | della osscurita delmezo si dōbre come del campo | e perche li termine delli obbietti vedutj nonsono esspedjtj</td><td>[I] Why the edges of the percussions of derivative shadows are not of the darkness of the centre of both the shadows and the background. This is because the outlines of objects seen are not distinct</td></tr>
<tr><td>[Fig. 1, 2, 3, 4, 5]</td><td>[Fig. 1, 2, 3, 4, 5]</td></tr>
<tr><td>[II] la cosa che ssi serra cressce (dal m) lospatio che e dal|l mezzo alli stremj nel ristrignersi crescā le p[a]|riete insu (e re) tirādosi la punta delcore djrieto e nō […]</td><td>[II] An object which is squeezed increases the distance which exists between its centre and its ends. In contracting, the walls increase upwards drawing behind them the point of the heart and not …</td></tr>
<tr><td>[III] Quādo il core si serra ellj si restrigne ellj muscoli chetengano le corde de panjcu[li] | sidj<s>tēdano e alētano lelor corde ellj panjculj a esse co<r>de cōgunti salentano | e il sangue che dentro vi percote se li spignie dinātj e sserrasi (la) liusscolj cō perfetta | serramēto e cquāto piu ilcore sirestrignie piu safortificha esso serrame</td><td>[III] When the heart closes it narrows, and the muscles which hold the chordae tendineae of the valve-cusps stretch and relax their chordae, and the valve-cusps joined to these chordae relax. And the blood that percusses within, pushes them in front of itself and closes the cusps with perfect closure. And the more the heart contracts, the more is the lock-gate strengthened.</td></tr>
<tr><td>[Fig. 6, 7, 8]</td><td>[Fig. 6, 7, 8]</td></tr>
<tr><td>[IV] e sse ttu dj che ilpariete stancho esteriore fu fatto grosso perche elli | acqujstassi magor peso (p) acco chellj facessi contrapeso aluentri|chulo destro che a grā peso dj sanghu<e>tu nō considerasti questa tal | liberatione non era necessaria concosia che ttutti li anjmali tere|sti dellomo in fori portano ilcore a diacere e cosi djace ilco|re nellome [sic: nellomo] djacente nelsuo letto Ma tu nō fussti buono libra|tore concosia che ilcore a due sosstentachuli disscendēti dalla | furchola della gola li quali perla 4ª de ponderibus il core non si po | libra<r>se se non e sopra vn sol sostentaculo [Fig. 9] liquali due sostenta|culi son larteria aorto ella vena delchilo e oltre a dj que|sto quādo il core sipriva del peso delsangue nel suo restrignersi | ello da in djposito alli ^sua^ ventriculi superiori el centro della graujta | delcore allora sarebbe dallato desstro delcore e cosi la parte sua si|njsstra sarebbe leujficata ma cquesta tal ponderatione non e ue|<r>a come detto djsopra perche li anjmali che djaciano o che stanno sopra | 4 piedj anno ilcore a djacere come loro eli nō si cercha pō|deratione di core E sse ttu dj che lli testiculi sol sonfatti | per aprire o sserrare liuasi spermatici tutti se ingannato perche | sol sarebon necesarie alli mōtoni o ttori che glianno gran|djssimj ecquādo essi tessticuli fussino riētrati in corpo per il fre|do adunque nō si potrebbe vsare ilcoito e iluespertjliō | che dorme essenpre siposa sotto sopra come si Libra ilsuo core | col destro e sinjstro ventriculo</td><td>[IV] And if you say that the left exterior wall was made thick in order that it should acquire greater weight to form a counter-weight to the right ventricle which has a great weight of blood, you have not considered that such a balancing was unnecessary inasmuch all terrestrial animals except man carry their hearts lying horizontally. And likewise the heart of man lies horizontally when he lies in bed. But you would not be a good balancer because the heart has two suspensoria descending from the pit of the throat, and according to the fourth [proposition] of On Weights, the heart cannot be balanced unless it has above it one suspensorium only [Fig. 9]. These two suspensoria are the aortic artery and the vena cava. Furthermore when the heart is deprived of the weight of blood on its contraction and gives it in deposit to the upper ventricles, the centre of gravity of the heart would then be on the right side of the heart and so the left side would be made lighter. But such weighting is not true, because as stated above, animals which lie down or stand on four feet have their hearts lying down [horizontally], and no such weighting of the heart is to be looked for. And if you say that the testicles are made only to open or close the spermatic ducts, you are mistaken, because they would only be necessary to rams or bulls which have very big [testicles]. And when these testicles had re-entered the body on account of the cold then coitus could not be performed. And the bat which sleeps and always places itself upside-down, how does it balance the right and left ventricles of its heart?</td></tr>
<tr><td>[Fig. 10]</td><td>[Fig. 10]</td></tr>
</table>

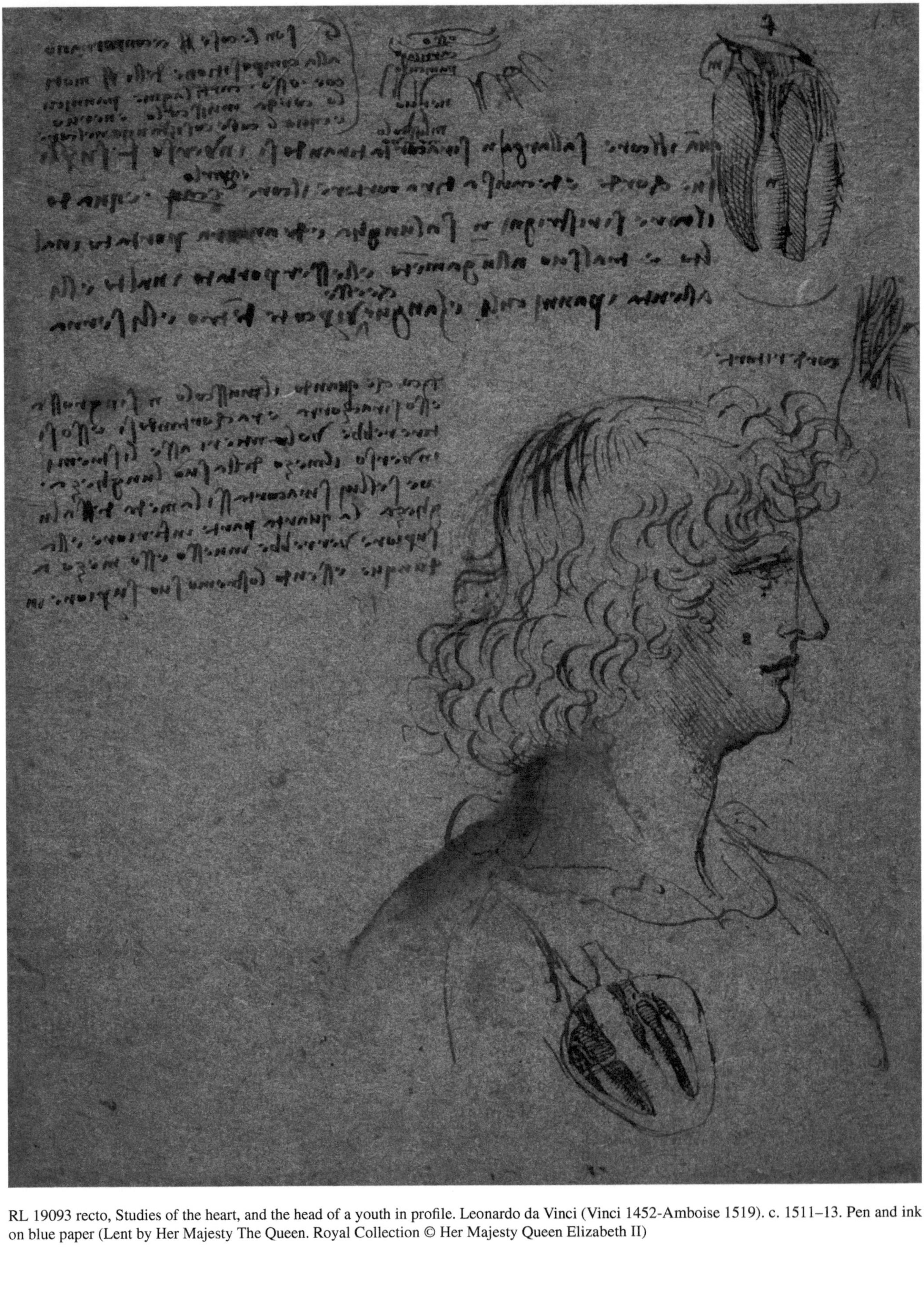

RL 19093 recto, Studies of the heart, and the head of a youth in profile. Leonardo da Vinci (Vinci 1452-Amboise 1519). c. 1511–13. Pen and ink on blue paper (Lent by Her Majesty The Queen. Royal Collection © Her Majesty Queen Elizabeth II)

<table>
<tr><th colspan="2">19093 r (C II 23 r). Pen and brown ink, on blue paper. 260×200 mm.</th></tr>
<tr><td>[Fig. 1] f m n</td><td>[Fig. 1] f m n</td></tr>
<tr><td>[I] quā [sic: quando] il core sallarga n siracorta tirandosi inverso f su perli | sue chorde ed e causa djracortate ilcore ^e aprirlo^ (e caq) e quādo | ilcore siristrignj <u>n</u> salungha ed e ancora portato inal|lto e tral suo allūgamēto ellesser portato inalto ellj | allenta i pannjculi e sangue ^che esse^ vi percote dētro ellj serra</td><td>[I] When the heart broadens n [papillary muscle] is shortened, drawing itself up towards f through its chordae, and this is the cause of shortening and opening the heart. And when the heart narrows [contracts] n elongates and is carried farther upwards, and through its elongation and being carried upwards it relaxes the valve-cusps, and the blood which percusses them from within shuts them.</td></tr>
<tr><td>[II] djco che quando ilmusscolo n singrossa | esso sirachorta e rachortandosi esso si | tirerebbe volontieri asse li stremj | inverso ilmezo della sua lungheza·|<c>oe selluj siracortassi la meta dessa lū|gheza la quarta parte inferiore e lla | superiore verrebbe inue<r>sso esso mezo a|dunque essendo lostremo suo superiore in […]</td><td>[II] I say that when the muscle n enlarges, it shortens; and shortening, it would spontaneously draw its ends towards the middle of its length, that is, if it shortened itself by half its length, the lower quarter and the uppe r quarter would come towards its middle, therefore its upper end being in …</td></tr>
<tr><td>[Fig. 2]</td><td>[Fig. 2]</td></tr>
<tr><td>[III] corde ritorte</td><td>[III] plaited chordae [tendineae]</td></tr>
<tr><td>[Fig. 3] osso | cartilago | paniculo | neruo | musscolo | nervo</td><td>[Fig. 3] bone; cartilage; membrane; tendon; muscle; nerve</td></tr>
<tr><td>[IV] ∂ 6 son le cose (di) ce concorrano | alla compositione delli (os) moti | coe·osso·cartilagine pannjcu|lo corda musscolo e neruo | e cque 6 cose cosi stanno nelcore ©</td><td>[IV] The things that go together in composing movements are six; that is bone, cartilage, membrane, tendon, muscle and nerve; and these six things are in the heart.</td></tr>
<tr><td>[Fig. 4, 5]</td><td>[Fig. 4, 5]</td></tr>
</table>

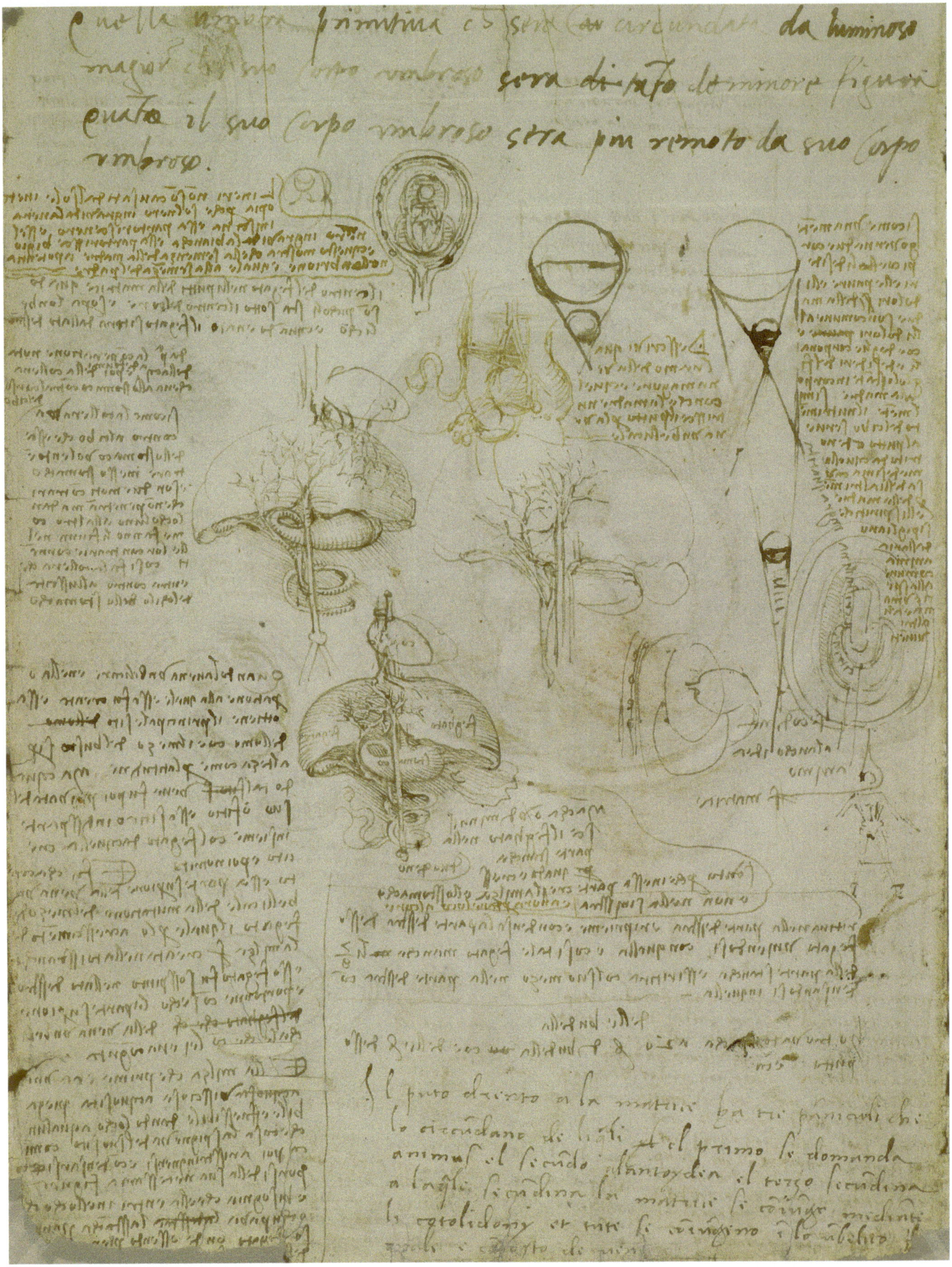

RL 19102 verso, Notes on reproduction, with sketches of a foetus in utero, etc. Leonardo da Vinci (Vinci 1452-Amboise 1519). c.1511. Pen and ink, with some offset red chalk (Lent by Her Majesty The Queen. Royal Collection © Her Majesty Queen Elizabeth II)

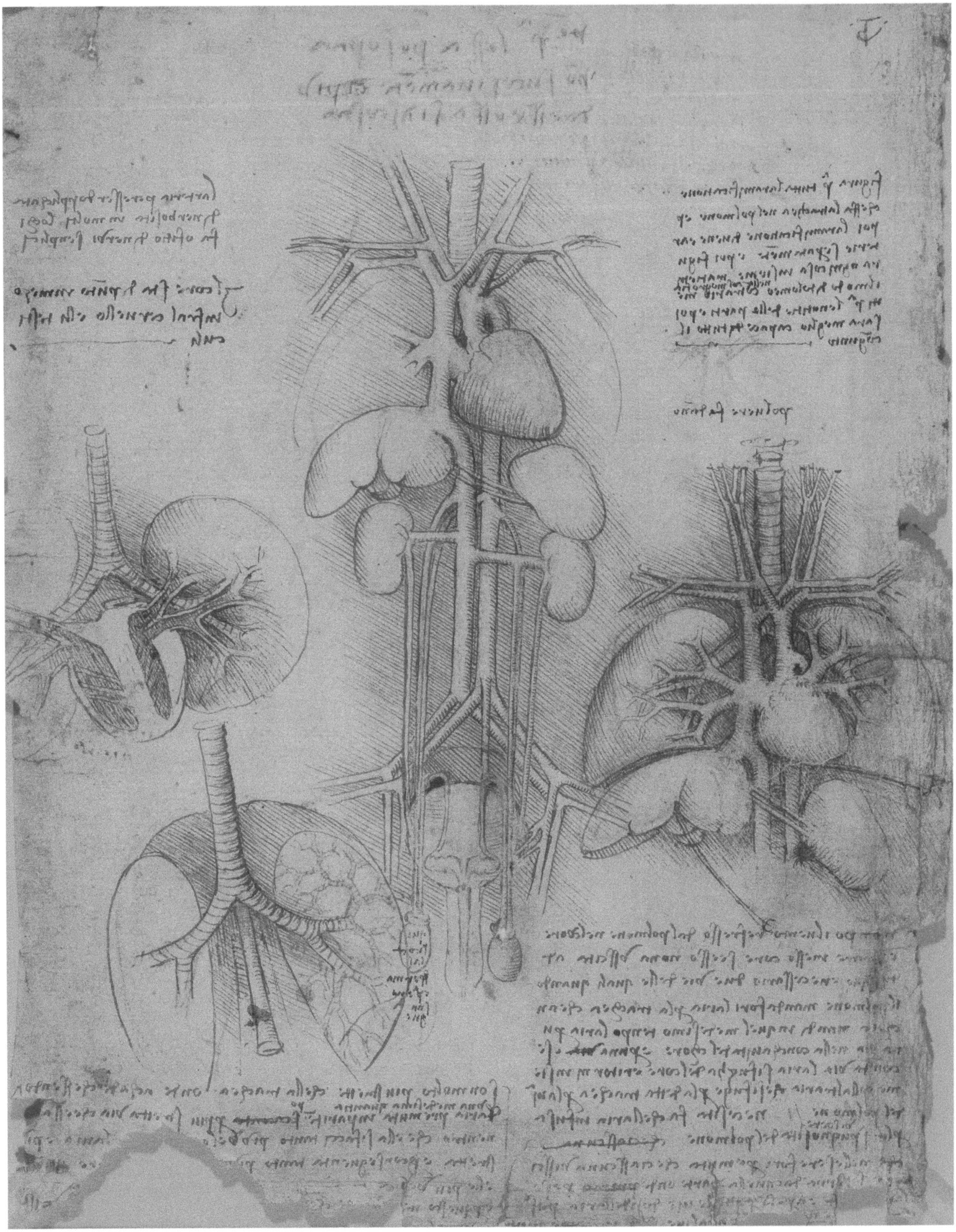

RL 19104 verso, The heart, lungs and other organs. Leonardo da Vinci (Vinci 1452-Amboise 1519). c.1508. Pen and ink over black chalk (Lent by Her Majesty The Queen. Royal Collection © Her Majesty Queen Elizabeth II)

<table>
<tr><td colspan="2">19104 v (C III I0v). Pen and brown ink (two shades) and black chalk. 283×219 mm.</td></tr>
<tr><td>[I] pō pª lossa po sopra | pō sucesiuamēte ch<e>piv | da esse ossa si djcosta</td><td>[I] First put in the bones, then gradually put on them structures which are more remote from them.</td></tr>
<tr><td>[II] figura pª tutta laramjficatione | che ffa la trachea nel polmone ep|poj la ramjficatione djuene e ar|terie separatamēte e poi figu|ra ognj cosa insieme ma tienj | il modo dj tolomeo ^nella cosmogrofia^ cōtrario me|ttj pª lenotitie delle parti e poj | sara meglio capace djtutto il | cōgunto –</td><td>[II] First draw the whole ramification which the trachea makes in the lungs and then the ramification of the veins and arteries separately. Then draw each of these structures together. Follow the method of Ptolemy in his Cosmography but in the opposite order, first putting information about the parts, and then you will get a better capacity for a knowledge of the whole joined together.</td></tr>
<tr><td>[III] larteria per esser dopplichata | djnerbosita in moltj lochi | fa ofitio dj nervj senplicj</td><td>[III] The artery being double-coated with sinew functions in many places as a simple tendon.</td></tr>
<tr><td>[IV] Jl core sta dj pūto inmezo | infral ceruello Slli testj|culi –</td><td>[IV] The heart is placed exactly in the middle between the brain and the testicles.</td></tr>
<tr><td>[Fig. 1]</td><td>[Fig. 1]</td></tr>
<tr><td>[V] qui | sicoce | las|sperma | e pª era | san|gue</td><td>[V] Here the sperm is concocted; and it was first blood.</td></tr>
<tr><td>[Fig. 2] m n</td><td>[Fig. 2] m n</td></tr>
<tr><td>[VI] poluere fa dāno</td><td>[VI] Dust does damage.</td></tr>
<tr><td>[VII] non po iluento refresso dal polmone nelcvore | entrare inesso core seesso non a vsscita ad|dūque enecessario due vie delle quali quando | il polmone mandafori laria per la trachea che an|chora mandj in quel medesimo tenpo laria per u|na via nella conchaujta del chore e per una (via) se|conda via laria sifugha del core e ritornj insie|me collaltraria che sifuge per la detta trachea per la ujª | del polmone //
necesita fa chellaria infusa | per le spugnosita ^viscere^ del polmone (che ciasscuna | che) nellesere fori premuto che ciasscuna vissci|cola si serra dacquella parte onde (primo) prese | lar[ia …] te e apresi per quelle uie dondellaria pos|[sa…]e ri[…]cquali ue[…]e insieme gunte [:–:] son molto piu strette chella trachea onde achade chessendo | ^vna medesima quanti<t>a^ daria premuta in pari tē^po^ (scaccata) per piu stretta via cheessa[…] | n entra che ella sifaccj tanto piv velo[ce quanto] la uia e pi[v] | stretta e per conseguenta tanto piv[…]re quā[to] | elle piu veloce | e per questo ne[…]s[…]de che[…]ess | rip[…]</td><td>[VII] The wind reflected from the lung to the heart cannot enter the heart unless there is an outlet. Therefore two passages are necessary. Of these, one which when the lung sends out air through the trachea at the same time sends air into the concavity of the heart; and a second passage through which air issues from the heart and returns together with the other air which escapes through the said trachea, the passage of the lungs.
Necessity causes the air infused though the visceral sponginess of the lung to be pressed out of it so that each vesicle is closed in that part which took in the air … and opened by those passages whence the air can … and joined together they are much narrower than the trachea, whence it happens that when one and the same quantity of air is pressed in an equal time through a passage narrower than that by which it enters, it makes so much the more velocity as the passage is more narrow and consequently so much the more … the more is its velocity. And for this reason…</td></tr>
<tr><td>[Fig. 3, 4]</td><td>[Fig. 3, 4]</td></tr>
</table>

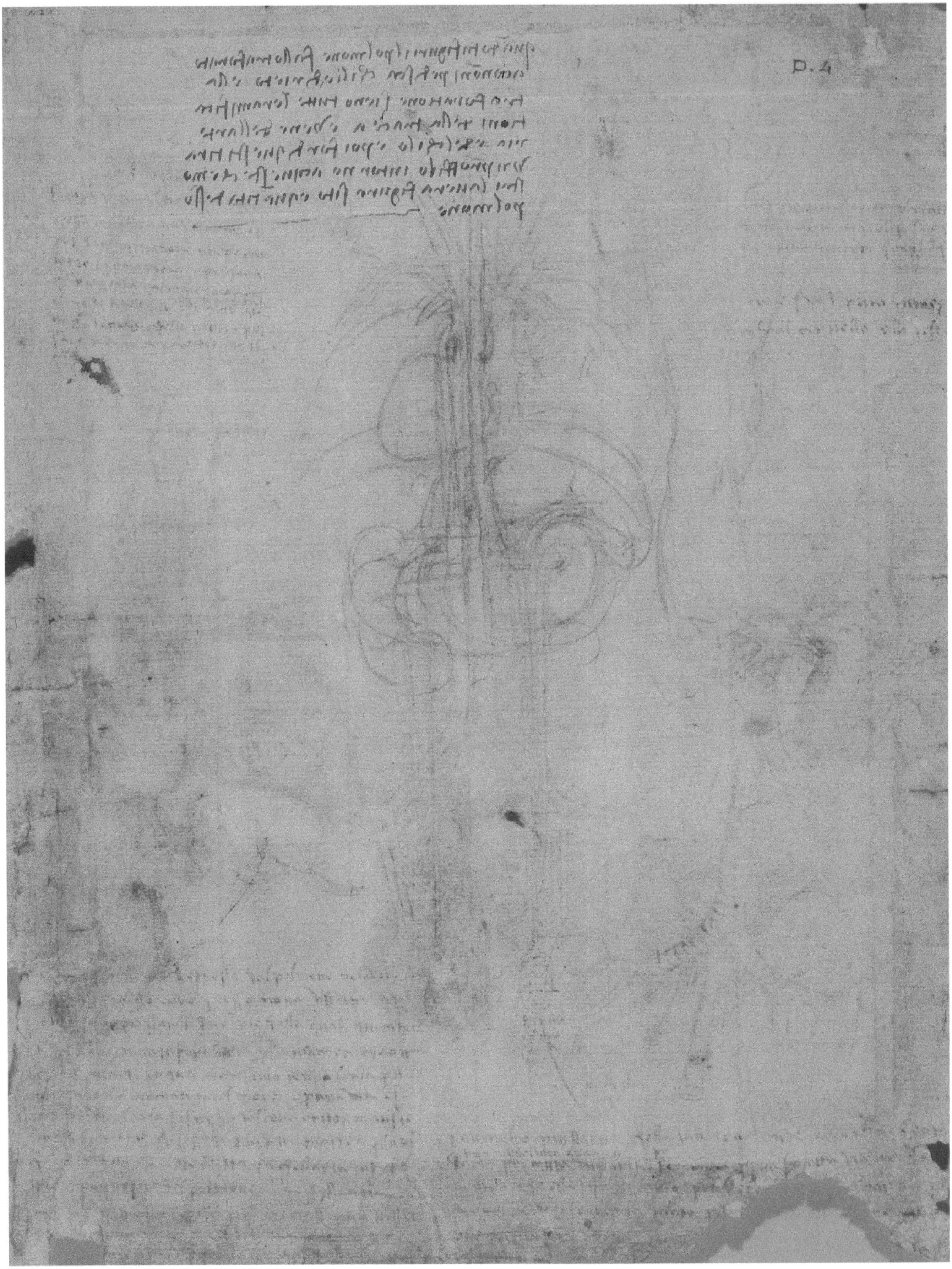

RL 19104 recto, Studies of the heart, lungs, liver, spleen, etc. Leonardo da Vinci (Vinci 1452-Amboise 1519). c.1508. Black chalk and pen and ink (Lent by Her Majesty The Queen. Royal Collection © Her Majesty Queen Elizabeth II)

19104 r (C III I0r). Pen and brown ink and black chalk.	
[I] quādo tu figuri il polmone fallo traforato | a cio nōn@pedjsca chiliedjrieto e lla | traforatione sieno tute leramjfica|tioni della tracea e vene dellarte|ria e delchilo e poi for dj questi tira | vn proffilo intorno acqueste che mo|stri lauera figura sito e quātita desso | polmone –	[I] When you draw the lungs make a penetrating view of them so that they do not impede what is behind them, and let the penetrating view consist of all the ramifications of the trachea and of the arterial vessels and of the vena cava. And then outside these draw a profile around them which shows the true shape, position and size of the lungs.
[Fig. 1, 2]	[Fig. 1, 2]

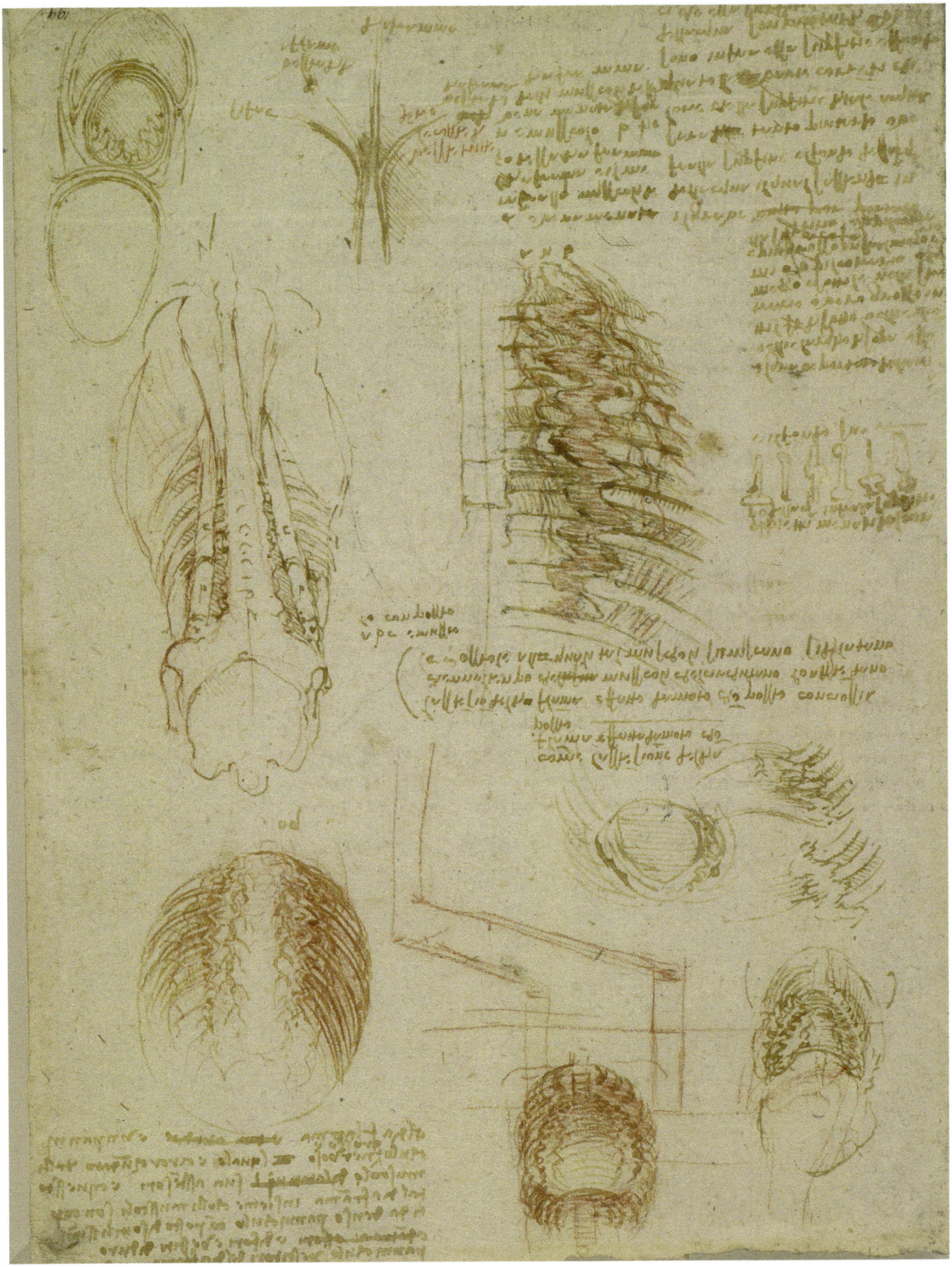

RL 19108 verso, The thoracic and abdominal cavities of an ox. Leonardo da Vinci (Vinci 1452-Amboise 1519). c.1512–13. Pen and ink with red chalk (Lent by Her Majesty The Queen. Royal Collection © Her Majesty Queen Elizabeth II)

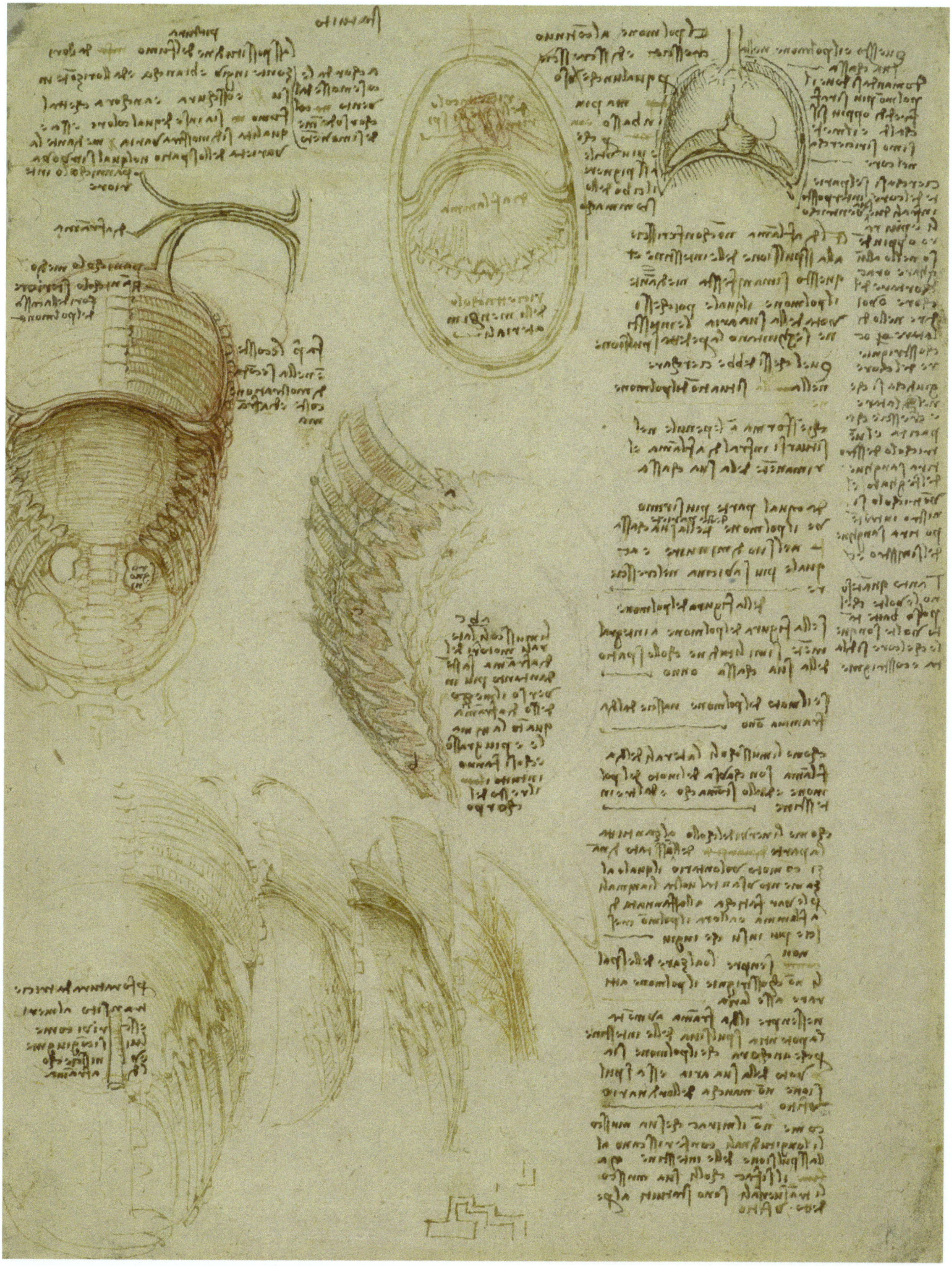

RL 19109 recto, The thoracic and abdominal cavities of an ox. Leonardo da Vinci (Vinci 1452-Amboise 1519). c.1512–13. Pen and ink with red chalk (Lent by Her Majesty The Queen. Royal Collection © Her Majesty Queen Elizabeth II)

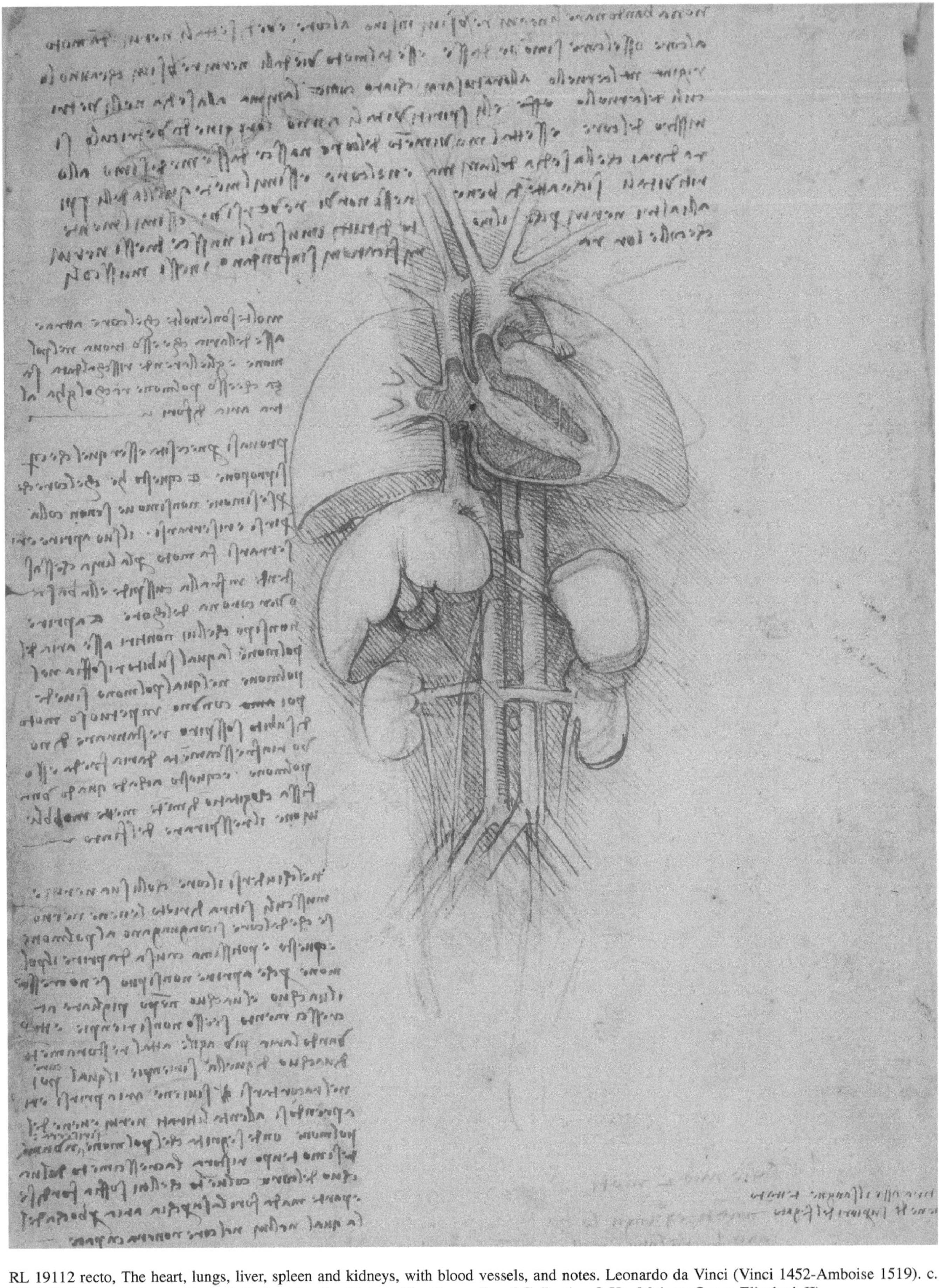

RL 19112 recto, The heart, lungs, liver, spleen and kidneys, with blood vessels, and notes. Leonardo da Vinci (Vinci 1452-Amboise 1519). c. 1508–10. Pen and ink over black chalk (Lent by Her Majesty The Queen. Royal Collection © Her Majesty Queen Elizabeth II)

19112 r (C IV 7r). Pen and brown ink (two shades) and black chalk. 275 × 190 mm.

[Fig. 1, 2]	[Fig. 1, 2]
[I] molte son le uolte chelcore attrae | a sse dellaria che esso troua nel pol|mone e gliella rende risschaldata sā|za chesso polmone racholgha al|tra aria djforj –	[I] Many are the times that the heart attracts to itself the air which it finds in the lungs and returns this to it, heated, without this lung collecting any other air from outside.
[II] prouasi per necesita esser quel che cqui | sipropone e cquesto he chel core che | per se simoue nonsi moue senon colla|prirsi e riserrarsi · ilsuo aprire eri|serarsi fa moto per la linja chessas|stende infralla cusspide ella basa | over corona delchore E aprire | non sipo chelluj nontiri asse aria del | polmone laqual subito risoffia nel | polmone nelqual polmone siuede | poi (*a mo*) convno inpetuoso moto | djsubito sosspire restaurare dj no|vo rinfresscamēta daria freda esso | polmone e cquesto achade quādo vna | fissa chogitatiō dj mēte mette in obble|ujone ilresspirare del fiato –	[II] It is proved by necessity that what is here proposed must occur. This is that the heart which moves on its own does not move except in opening and closing again. Its opening and closing make a movement along the line extending between the apex and the base or crown of the heart. It cannot open without drawing into itself air from the lung which it blows out again into the lung immediately afterwards. Then, with the movement and impetus from a sudden respiration, the refreshment of cold air is restored into the lung; and this happens when a fixed thought in the mind makes one forget to breathe in the air.
[III] nelchiudersi ilcore cholli sua neruj e | mussculi sitira djrieto le uene neruo|se che dalcore si congungano al polmone | e cquesto e potissima causa daprire il pol|mone perche aprire nonsipuo se nō cressce | iluachuo el uachuo nō po pigliare ac|cresscimento seeso nonsi rienpie e ttro|vando laria piv agile attal restoramēto | djuachuo djquella sirienpie ilqual ^core^ poj | nelracortarsi (*ch*) siuiene a riaprirsi e ri|aprendosi allenta li tirati nerui e uene del | polmone onde seguita chel polmone ^siriserra e ^ n vn me|desimo tenpo ristora lacresscimēto delua|chuo delcore coluēto che llui soffia fordj se | e parte ma<n>da fori lasuperchia aria per bocha del|la qual ne lluj nel core nonera capace –	[III] In closing, the heart with its tendons and muscles draws behind it the most sinewy vessels which from the heart unite with the lung; and this is a most powerful cause of opening the lung because it cannot open unless the vacuum increases, and a vacuum cannot get larger unless it is refilled. And the air being more nimble in returning into such a vacuum, it is refilled with that. The heart after contracting comes to reopen itself, and in reopening it relaxes the tension of the sinews and vessels of the lung, whence it follows that the lung closes again and at the same time returns to the enlarged vacuum of the heart the wind which it blows out of itself. Part of the superfluous air it sends out of the mouth; that for which neither it nor the heart has room.
[IV] e tira asse ilsangue tettato | [dalle] vene (*de*) superiori del fegato –	[IV] And it draws into itself blood sucked from the upper veins of the liver.
[V] non abandonare li nerui reversiuj insino alcore e vedj se ttalj neruj dā moto | al core ossel core si move dasse esse tal moto viē dalli neruj reversiuj cheanno lo|rigine nel ceruello allora tusaraj chiaro come lanima alla sedja nellj vētri|culi del ceruello (*e sse*) e llj spiritj vitalj anno lorjgine da vētriculo si|njsstro delcore e sse ttal movimēto delcore nassce dasse medesimo allo|ra djrai chella sedja dellanjma e nelcore e ssimjlmēnte quella delli spi|riti vitali siche attēdj bene a essi nervi reversivj e ssimjlmente | allialtri neruj perche il mo to dj tuttj i muscoli nassce daessi neruj | che colle lor ra mjficationj sinfondano inessi musscolj	[V] Do not fail to follow up the reversive nerves [vagi] as far as the heart; and see whether these [nerves] give movement to the heart or whether the heart moves by itself. And if such movement comes from the reversive nerves which have their origin in the brain, then you will make it clear how the soul has its seat in the ventricles of the brain and the vital spirits have their origin in the left ventricle of the heart. And if this movement of the heart arises from the heart itself then you will say that the seat of the soul is in the heart and likewise to the other nerves because the motion of all muscles arises from these nerves which with their ramifications are infused into the muscles.
[VI] lalie ano 2 moti | 1 ināti e 1° in gu lo | ingu he per loro ssenpli|ce calare lo ināci | [...]	[VI] The wings have two motions, one forwards and one downwards. The downward one for their simple lowering, the forward one ...

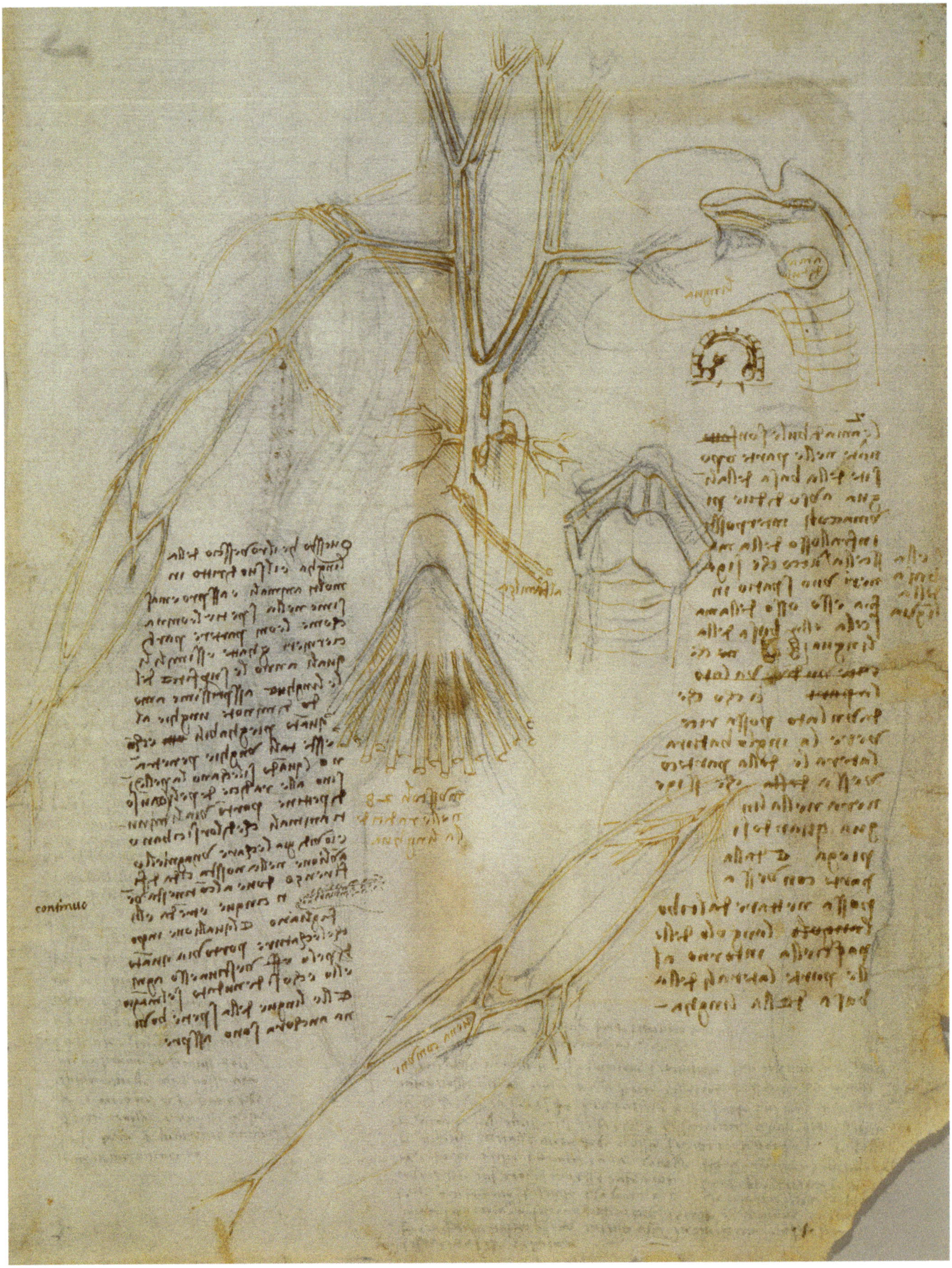

RL 19114 verso, The main blood vessels of the thorax and superficial veins of arm, etc. Leonardo da Vinci (Vinci 1452-Amboise 1519). c. 1508–10. Pen and ink over black chalk (Lent by Her Majesty The Queen. Royal Collection © Her Majesty Queen Elizabeth II

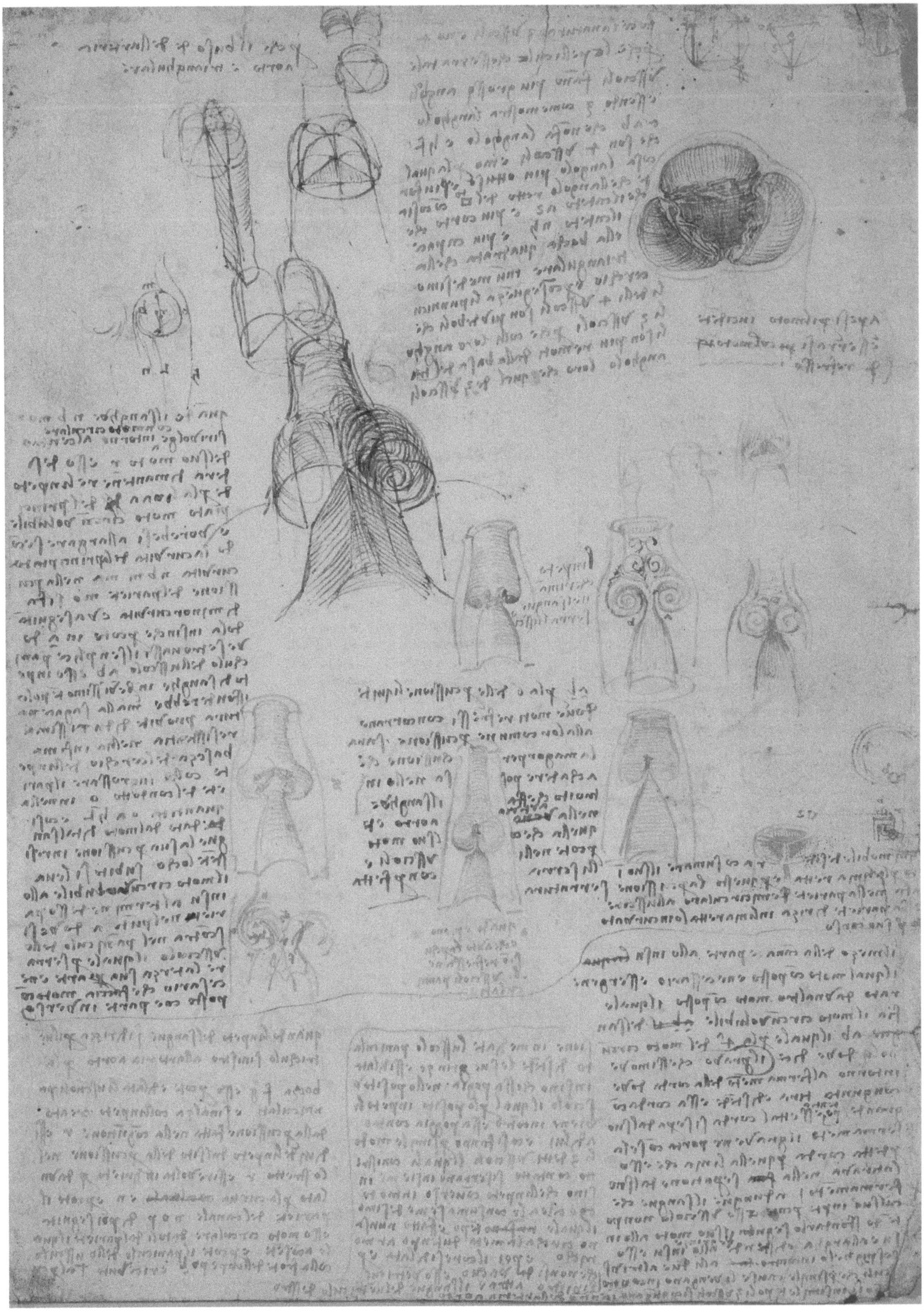

RL 19116 verso and RL 19117 verso, Blood flow through the aortic valve. Leonardo da Vinci (Vinci 1452-Amboise 1519). c.1512–13. Pen and ink, over some black chalk (Lent by Her Majesty The Queen. Royal Collection © Her Majesty Queen Elizabeth II)

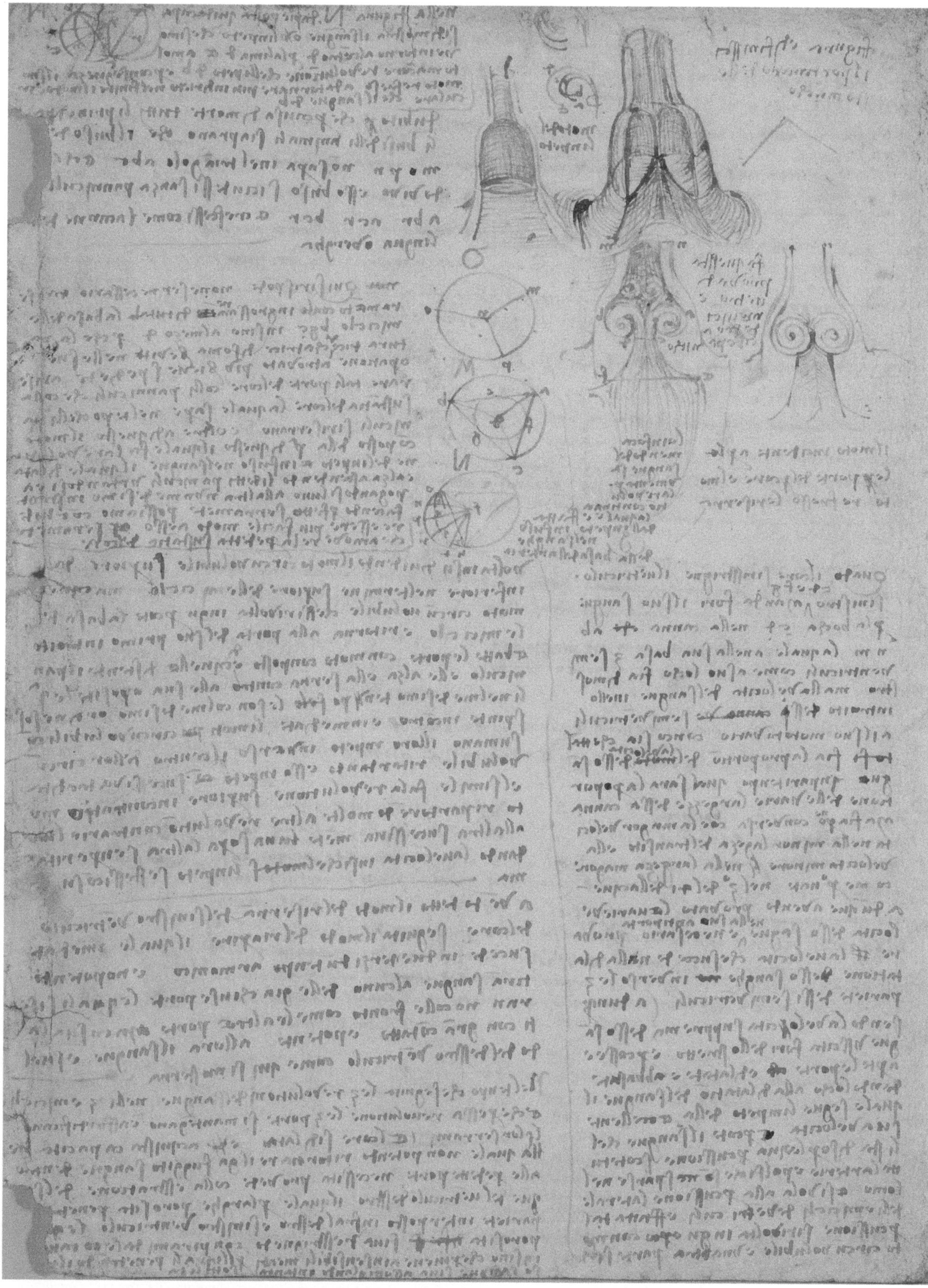

RL 19116 verso and RL 19117 verso (continued)

<table>
<tr><td colspan="2">19116–19117v (C IV 11v and CIV 12r). Pen and brown ink (two shades) over traces of black chalk.</td></tr>
<tr><td>[Fig. 1]</td><td>[Fig. 1]</td></tr>
<tr><td>[Fig. 2] a b c s</td><td>[Fig. 2] a b c s</td></tr>
<tr><td>[Fig. 3] e f (g) h n</td><td>[Fig. 3] e f (g) h n</td></tr>
<tr><td>[I] perche il buso de dellarteria | aorto e trianghulare</td><td>[I] Why the orifice of the aortic artery is triangular.</td></tr>
<tr><td>[II] fece la natura 3 vsscoli e nō 4 | perche le pellicule chesserra tale | vsscioli fāno piu grossj angh^o^lj | essendo 3 come mostra langholo | c a b che nō fa langholo e h f | che son 4 vssicolj e mo [sic: mostra] per lacqual | cosa langolo piu ottuso e piu for|te chellangolo retto del cō co sia | che il cateto a s e piu corto che | il cateto n h e piu capace | ella bocha quadrata chella | triangulare in ū medesimo | cerchio e per cōseguēza lipannicu|li delli 4 vsscoli son piv deboli che | li 3 vsscioli perche colli loro angho|li son piu remoti dalla basa del tri|angholo loro che quel de 3 vssciolj</td><td>[II] Nature made three little doors [cusps] and not four because the pellicles which shut these cusps make greater angles being three, as the angle c a b shows, than the angle e h f which is made by four cusps. And it shows for what reason the more obtuse angle is stronger than the right angle of the square since the cathete a s is shorter than the cathete n h; and the square orifice is more capacious than a triangular one within the same circle. Consequently the membranes of four valve-cusps are weaker than those of three valve-cusps because their [central] angles are more remote from the bases of their triangles than those of the three valve-cusps.</td></tr>
<tr><td>[Fig. 4]</td><td>[Fig. 4]</td></tr>
<tr><td>[III] ∂ apresi per il moto incidēte | e sserrasi (per i) col moto (īcj | de) refresso ©</td><td>[III] It is opened by the incident motion and closed by the reflected motion.</td></tr>
<tr><td>[Fig. 5, 6, 7, 8 9]</td><td>[Fig. 5, 6, 7, 8, 9]</td></tr>
<tr><td>[Fig. 10] m o r b a h L n</td><td>[Fig. 10] m o r b a h L n</td></tr>
<tr><td>[IV] quādo il sanghve n b m o a | si rivolge ^con moto circulare^ intorno alcentro | delsuo moto r esso desi|dera dj mantenere linpeto | de per la linaa [sic: linia] (de) del princj|piato moto circūvolubile | e vorebesi allargare secō|do la curvita delaprincipiata | curvita n b m ma nella percu|ssione delpariete m o si fa | dj mjnor curvita e vaseguitā|dola insin che percote in a do|ve setrouassi il senplice panj|chulo dellusscolo a b esso inpe|to dj sanghe in brevissimo tēpolo | isfonderebbe ma lla sagace na|tura provide dj durissima | resisstentia nella infima | baseza delcerchio dellinpe|to collo ingrossare ilpari|ete del condotto o in nella | quantita o a h L e cosi | (da) dato dalmoto djtalsan|gue la sua percussione in resi|stēte locho subito si leua | il moto circuvolubile allo | insu al termjne desso pa|riete nelpūto a dove si | scōtra nel pānjculo dello | vssciolo il quale per serra|re la terza sua parte e ne|cesario che faccia moto cō|posto coe parte inverso [:–:] il mezo della cāna e parte allo insu (la qua) | ilqual moto cōposto e necessario essergene|rato da vnaltro moto cōposto ilquale | fia il moto circūvolubile (a b il) del san|gue a b il quale per la 4ª del moto circū|volu<bile>dove dice ∂ il gravo che ssimove | intorno alfermamēto della corda dove | congunto tira e distēde essa corda cō | grande po^tentia^ e ssettal corda sisepara dal suo | fermamēto il grave ne porta cō se la | predetta corda per quella linja che esso | latirava nella (sua) separatione dalsuo | fermamēto © adunque il sangue che | colsuo inprete [sic: inpeto] percote esso vssciolo non po|tēdo sfondarlo seguita ilsuo moto allo in|su e allargha edjstende ^le grinze^ allo insu esso | sospignelo incontro (la e) alli due altri vs|culi che per simjle cause li vengano incontro | e cosi in simjle tēpo li 3 vsscolj sicōgūgano īcentro dellarteria aorto</td><td>[IV] When the blood n b m o a revolves back with circular motion around its centre of motion r, it desires to maintain its impetus along the line of the commenced circumvolving motion and would widen according to the curvature of the commenced curvature n b m. But in the percussion of the wall m o it makes a smaller curvature and it goes on following it until it percusses at a where it would find the simple membrane of the valve-cusp at a b, and this impetus of the blood would in a very short time rupture it. But the wisdom of Nature provides a very hard resistance at the lowest base of the circle of impetus by thickening the wall of the conduit o [aorta] into the quantity of o a h L. Thus when the motion of this blood has given its percussion to this resistant place, the circumvolving motion rises immediately upwards at the end of this wall at the point a where it meets the membrane of the valve-cusp, which in order to close its third part [of the whole aortic orifice] must necessarily make a composite motion which is partly towards the middle of the pipe and partly upwards. This composite motion is necessarily generated by another composite motion which consists of the circumvolving motion of the blood a b, which according to the fourth [book] On Circumvolving Motion states: 'The weight which moves around the fixed point of a cord where it is joined, pulls and stretches this cord with great power, and if such a cord is severed from its fixed point.' Therefore the blood which with its impetus percusses this valve-cusp, not being able to rupture it, continues its upward motion, enlarges and distends the folds [of the cusp] upwards, and presses it against the other two valve-cusps which for similar reasons come to meet it. Thus the three valve-cusps are joined together at the same time in the centre of the aorta.</td></tr>
<tr><td>[Fig. 11] a b c e f g n o p r</td><td>[Fig. 11] a b c e f g n o p r</td></tr>
<tr><td>[V] quando linpeto delsangue sidjrizza per il uē|trichulo sinistro allarteria aorto per la | bocha f g esso percote e djlata li uscuoli pa|nnjculati e sinalza collinpeto · creato · | dalla percussione fatta nella cōgūtione r essi djudjde linpeto dal sito della percussione nel|lo stretto r essirevolta in djrieto (per) da vn | lato per la curua (cocleale) e n e percote il | pariete del canale n o p djpoi seguita | esso moto circulare datoli dal pariete il qua|le acōsēte e percote il paniculo dello ussculo | colla frote [sic: fronte] dellinpepo [sic: inpeto] b e ricevuto Tal percu[:–:]sione inmedjate lusscolo panicula|to djstēde le sue grinze e ssidjlata | insino che ssapogia · nelloposito v|sciolo il qual per loposito inpeto li | viene incōtro essapoggia contro | a dj lui e cosi fanno per simjle modo | li 3 detti vsscioli li quali con istre|tto contatto siserrano insieme in|sino che llinpeto cōuerso in moto | chocleale consuma se medesimo | il quale (n ū sano) tēpo e fatto nun sa|no circha la meta dū tenpo armo|njcho e poi il core si djlata e perche no si da vacuo esso vētriculo | sinistro attrae il sangue del uētriculo desstro</td><td>[V] When the impetus of the blood is directed by the left ventricle into the aortic artery through the orifice f g, it percusses and dilates the membranous valve-cusps and it rises upwards with the impetus created by the percussion made at the junction r. And the impetus is divided at the narrow site of its percussion, r, and it revolves back along the curve e n and percusses the wall of the channel n o p [aortic sinus]. It then follows the circular motion imparted to it by the concurrent wall and percusses the membrane of the valve-cusp with the front of its impetus b. And having received such percussion, the membranous valve-cusp immediately extends its folds and dilates until it leans against the opposite valve-cusp which, by the opposite impetus, comes to meet it and leans against it. And thus in a similar way the three said valve-cusps shut together in close contact until the impetus, converted into cochlear motion, consumes itself. This time in a healthy person comes to about half a harmonic tempo; and then the heart dilates, and because a vacuum cannot exist, the left ventricle attracts the blood from the right ventricle.</td></tr>
<tr><td>[Fig. 12, 13, 14, 15]</td><td>[Fig. 12, 13, 14, 15]</td></tr>
<tr><td>[Fig. 16] a b</td><td>[Fig. 16] a b</td></tr>
</table>

19116–19117v (C IV 11v and CIV 12r). Pen and brown ink (two shades) over traces of black chalk.

[VI] a b per la 6 delle percussione liqujde | doue moti refressi concorrano | alla lor comune percussione sara | la magor per chussione che | achadere pos sa nello in|troito che ffa il sanghve | nella (*vena*) ^arteria^ aorto ed e | quella che co l suo moto | percote nellj vsscioli e | lli serra con perfetta | serratura –	[VI] *a b* according to the sixth [book] On the Percussion of Liquids, where [it says] reflected motions run together to their common point of percussion there will be the greatest percussion which can take place. This the blood makes on its entry into the aorta, and this it is which by its motion percusses on the valve-cusps and shuts them with perfect closure.
[Fig. 17]	[Fig. 17]
[VII] linpeto | che rimã | nelsangue | serra lussci^o^	[VII] The impetus which remains in the blood shuts the valve.
[Fig. 18, 19]	[Fig. 18, 19]
[VIII] quãdo e pieno | e che a<a>vto la perchu|siõ refressa ne|lli vsscioli pannj|culati –	[VIII] When it [the aortic sinus] is full and the reflected percussion has supported the membranous valve-cusps.
[Fig. 20, 21]	[Fig. 20, 21]
[Fig. 22] c b d	[Fig. 22] *c b d*
[Fig. 23] b retto	[Fig. 23] *b* straight
[IX] ognj mobile deside ra cõsumare il suo ĩ|[pe]to per laljnia retta e per questo la percussione | [d]ata nella pariete semjcirculare allusscire | [d]essa pariete djriza inlinja retto lo incurvato | p° suo corso –	[IX] Each mobile object desires to consume its impetus along a straight line. For this reason the percussion giving in issuing forth on to the semi-circular wall is made straight on to the wall in a straight line before its curved course.
[Fig. 24]	[Fig. 24]
[Fig. 25, 26]	[Fig. 25, 26]
[X] fa questa | prova dj | uetro e | moujci | dẽtro a|cqua e panicho	[X] Make this test in the glass [model] and move inside it water and panic grass seed.
[Fig. 27]	[Fig. 27]
[XI] il moto incidente apre le | le (*per*) porte del (*p*) core el mo|to refresso le riserra	[XI] The incident motion opens the gates [valve-cusps] of the heart and the reflected motion closes them again.
[Fig. 28] n m a b c d e f g	[Fig. 28] *n m a b c d e f g*
[XII] loinfoca|mento del | sangue sa|vmenta per | la revolu|tiõ continua | laquale e ffatta | dallinpeto infuso | nelsanghe | della basa dellarteria	[XII] The heating of the blood is augmented by the continuous revolution which is made by the impetus infused into the blood at the base of the artery [aorta].
[XIII] Quãdo ilcore sirisstrigne iluẽtriculo · | sinistro ^c d e f g^ manda fori ilsuo sangue | per la bocha c d nella canna (*che*) a b | n m la quale a nella sua basa 3 semj|ventriculj come a suo locho fia djmos|stro malla velocita del sangue inello | introito dessj (*canna ve*) semjvẽtriculj | a il suo mototo vario con co sia che ttal | (*to si*) fia la proportiõ del(*moto*) ^la velocita^ desso sã|gue Jnpari tenpo qual sara la propor|tione delle varie largezze dessa canna | Ma fia pro^{ne} conversa coe la magor veloci|ta nella mjnor largeza del transito ella | velocita mjnore (*l*) nella largeza magore | co me prouato nel 3° del 41 dellacque –	[XII] When the heart contracts the left ventricle *c d e f g* sends out its blood through the mouth *c d* into the pipe *a b n m* [aorta] which has at its base three semiventricles [aortic sinuses], as will be demonstrated in its place. But the velocity of the blood at the entrance to these semiventricles has its motion varied, for the proportional velocity of the blood at the same time will be proportional to the different sizes of the pipe, but that proportion will be inverse; that is to say the greater velocity will occur in the smaller-sized passage and the less velocity in the greater-sized passage, as is proved in the 3rd [proposition] of the 41st [book] On Water.
Adũque avendo provato leurarie ve|locita desso sãgue ^nella sua antiporta^ e necesario prova|re (*il*) la uelocita che succede nella djla|tatione desso sanghe (*ne*) inverso le 3 | pariete dessi semjvẽtriculj ∂ adunque | sendo la velo(*s*)cita suprema desso sã|gue vsscita fori dello stretto e percosse e | aperte le porte (*a d*) e djlatate e abbasate | dando locho alla djlatatiõ del sangue il | quale segue limpeto della eccellente | sua velocita e percote il sangue chel|li sta dj sopra laqua<1>percussione scote tu|tte larterie e possi che sõ (*ne*) sparse nel|lomo e si vola alla percussione laterale | delli emjciclj devẽtriculj e ffatta tal | percussione sirivolta in gu (*e per co*) con mo|to circũvolubile e vnalltra parte si rj[:–:]volta in su djuidendo ilmoto circũvolubile superiore dallo | inferiore neltermjne superiore dellemjciclo ma cque[sto] | moto circũuolubile che ssirivolta ingu percote la basa del|lemjciclo e ritorna alla porta del suo primo introito | e batte le porte con moto conposto e ^dj^ cquelle djstende ilpan|njculo elle alza ella serra contro alle sua oposite le qua|li nel medesimo ten(*v*)po (*se le*) le son colmedesimo ordjne sos|spinte incõtro e inmedjate li moto (*pr*) circũvolubili cõ|sumano illoro inpeto inverso ilcentro dellor circũ|volubile ritardano esso inpeto (*con*) sucessiva tradjta | el simjle fa la revolutione superiore in contrarjo mo|to ripartorẽdo molte altre revolutiõ contrarie luna | allalltra sucessiua mẽte luna sopra laltra senpre ritar|dando lauelocita insi<n>chel (*moto f*) linpeto se stessi cõsu|ma –	Therefore, having tested the different velocities of the blood in the aortic vestibule, it is necessary to test the velocity which succeeds this in the dilatation of this blood towards the three walls of the semiventricles [aortic sinuses]. Therefore when the greatest velocity of the blood issued out of the narrow [aortic vestibule] has percussed and opened the gates [valve-cusps] and dilated and lowered them, giving room for the dilatation of the blood which follows the impetus of this extraordinary velocity, and has percussed the blood resting above it, this percussion jolts all the arteries and pulses distributed throughout man and then it turns to the lateral percussion of the hemicycles of the ventricles [aortic sinuses]. Having made this percussion, it revolves downwards with a circumvolving movement. And another part [of the blood] revolves upwards dividing the upper circumvolving movement from the lower one at the upper edge of the hemicycle. But the circumvolving movement which revolves downwards percusses the base of the hemicycle and returns to the gate –valve-cusp] of its first entrance, and it beats the gates [valve-cusps] with a composite motion with which it distends the membrane [of the valves] and raises them up, shutting them against those opposite, which during the same time by the same arrangement are driven against each other. Without anything intervening the circumvolving movements consume their impetus towards the centre of their circumvolution, the impetus progressively slowing down. And the upper revolution does the same in its opposite movement [upwards], giving birth to many other revolutions in opposite directions to one another successively one above the other, their velocity always becoming lower until their impetus itself is consumed.

19116–19117v (C IV 11v and CIV 12r). Pen and brown ink (two shades) over traces of black chalk.

avēdo detto il modo del riserrā del sinjstro vētriculo | delcore seguita il modo del riaprire ilquale imedjate | sucede in due terzi dū tenpo armonjco e nopotendo | tirā sangue alcuno delle gia chiuse porte lequali si se|ran no colle fronte come le altre porte ma cosua la|ti con grā cōtatto e potente allora il sangue e suel|do del desstro vētriculo come quj si mosterra –	Having described the way the left ventricle of the heart closes, there follows the manner in which it reopens. This immediately succeeds in two-thirds of a harmonic tempo. [The left ventricle] not being able to draw back any blood from the previously closed [aortic] valve-cusps which are closed with their surfaces like the other valve-cusps, with their sides in great and powerful contact, the blood is at that time torn out of the right ventricle, as will be here demonstrated.
Nel tenpo che seguita le 3 revolutionj del sangue nelli 3 emjciclj | e che per essa reuolutione le 3 porte si mantēgano e affortificano | lj lor serramj ∂ El core sidjlata e acqujsta capacita ne|lla quale non potendo ritornare il ga fuggito sangue dentro | alle predette porte necessita provede colla esstratione del s[an]|gue del uētriculo desstro il quale per larghe porosita penetr[a il] | pariete interposto infral destro e sinjstro ventriculo le qu[ali] | porosita (*dj si s*) siuā resstrignēdo con piramjdale cōcauui[ta] | insino che peruienne a insensibili meatj per lliquali penetrādo il v[isco]|so sangue siua assottiglando in tanta sottilita	During the time which follows the three revolutions of the blood in the three hemicycles [aortic sinuses], and as by these three revolutions the three valve-cusps are maintained and strengthened in their closure, the heart dilates and acquires capacity. The previously expelled blood not being able to return into it inside the aforementioned valve-cusps, necessity provides for the extraction of blood from the right ventricle, which penetrates the wall interposed between the right and left ventricle through wide porosities. These porosities go on narrowing into pyramidal hollows [channels] until they become imperceptible ducts through which the viscous blood penetrates and proceeds to thin down to great fineness.
[Fig. 29] O m n o p	[Fig. 29] *O m n o p*
[Fig. 30 M a b c d e f g	[Fig. 30] *M a b c d e f g*
[Fig. 31] N f n o o p q q r r s t v	[Fig. 31] *N f n o o p q q r r s t v*
[Fig. 32] N a b c d e n n p r	[Fig. 32] *N a b c d e n n p r*
[XIV] Nella figure N· da pie posta qui da capo | sidjmostra ilsangue over linpeto che simo|ve intorno al cētro d per lalinia d e a mol|to magore revolutione chellīpeto d b e per conseguēza il suo | moto refreso a da tornare piu in djrieto nel finjre il moto cir|culare che il sangue d b	[XIV] In the figure *N* at the foot [Fig. 31] placed here at the head [Fig. 32] it is demonstrated that the blood, or in truth, the impetus which moves it round the centre *d* along the line *d e*, has a much greater revolution than the impetus *d b*. Consequently its reflected motion has to turn back further in finishing its circular motion than the blood *d b*.
[XV] dubito (*p*) che per causa dj morte tuttj li princjpa|li busi delli anjmali siaprano che ilbuso de[l] | m o p n nō sapra inel triāgolo a b c e che[ssen]|do vivo esso buso siciudessi sanza pannjculi | a b r a c r b c r e crescessi come la carne della | lingua o vergha –	[XV] I doubt whether all the principal orifices of animals are opened because of death, and whether the orifice *m o p n* [Fig. 29] does not open into the triangle *a b c* [Fig. 30], and whether in the living this orifice would shut itself without the membranes *a b r, a c r, b c r* [aortic cusps] increasing like the flesh of the tongue or penis.
[XVI] (*noi*) Qui sirispōde noneser necessario tal se|ramēto conlo in grossam(*eto*)^re^ di tuta (*b*) la basa delle|mjciclo b g c insino al mezo d perche la na|tura rice^r^chatrice dj soma brevita nelle sue operatione a trovato piv brieue spedjēti a rise|rare tali porte del core collj panniculi che colla sustātia del core laquale sapre neltēpo che lli pā|njculi si riserrano e oltre a dj questo il moto | cōposto della p^{a} dj questo ilquale fa la revolutio|ne delinpeto (*n*) infuso nel sangue il quale djlata | e alza astendendo li detti panjculi vrtandosi e a|pogandosi luno allaltra nun medesimo instante | facendo perfetto serramēto possiamo cōclude|re essere piu facile modo a esso (*as*) seramēto | che a movere la predetta sustātia del core –	[XVI] Here one replies: such shutting with increase of the whole base of the hemicycle *b g c* [Fig. 30] as far as its centre *d* is not necessary, because Nature, searcher for the greatest brevity in her operations, has found briefer expedients for shutting these valves of the heart with the membranes rather than with the substance of the heart, which is opened [in diastole] during the time that the membranes are closed. Besides this, the compound motion of the former makes the revolution of the impetus infused into the blood, which dilates and raises up whilst stretching the said membranes, throwing them against each other and apposing them at one and the same instant, making a perfect closure. We can conclude this to be an easier way for this closure to occur than by moving the above mentioned substance of the heart.
[Fig. 33]	[Fig. 33]
[Fig. 34] a b c d e	[Fig. 34] *a b c d e*
[XVII] moto del|linpeto	[XVII] Motion of the impetus
[XVIII] figura e djfinjsscj | il portinare dello | stōmacho	[XVIII] Draw and define the pylorus of the stomach.
[Fig. 35]	[Fig. 35]

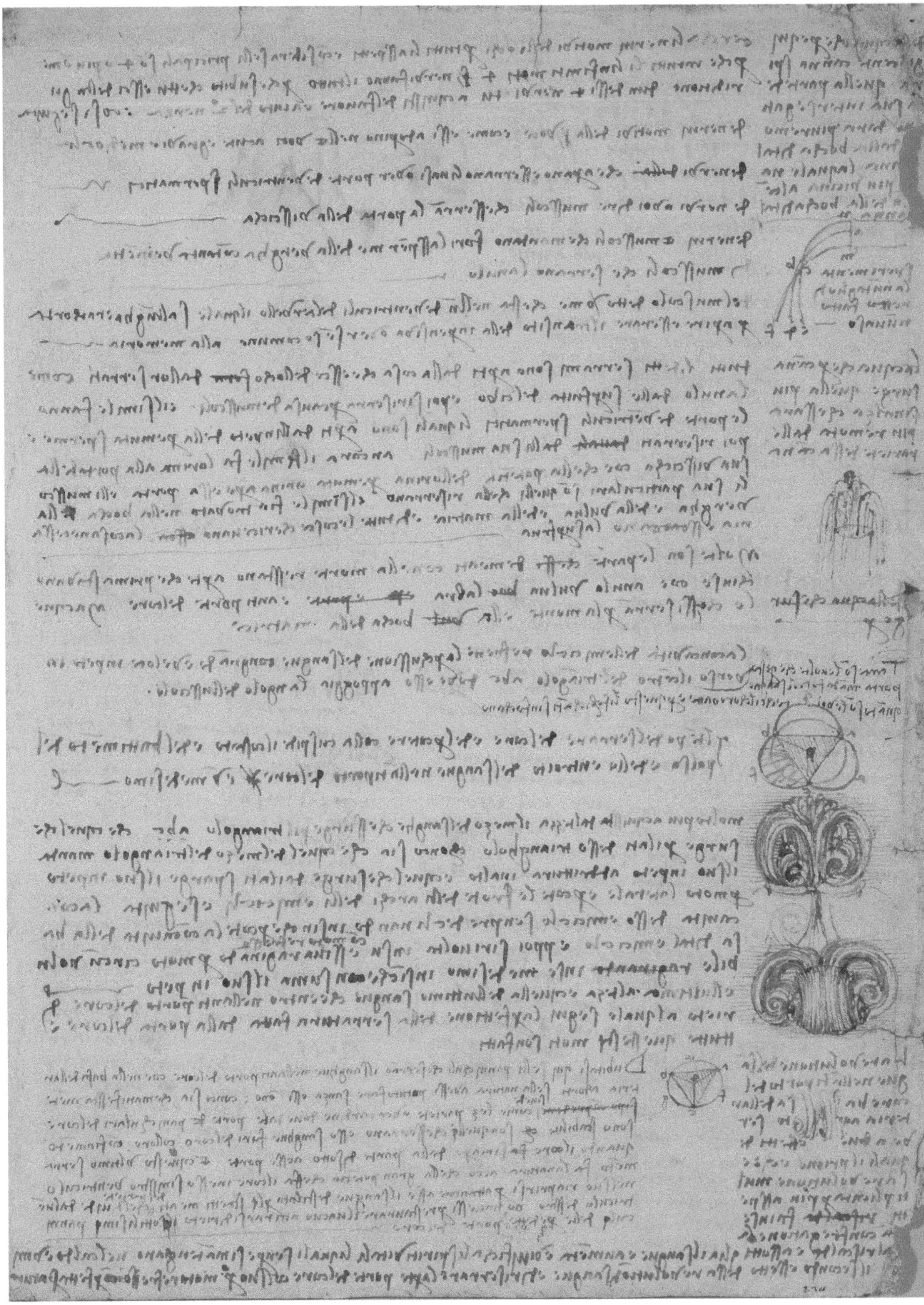

RL 19117 recto and RL 19116 recto, Studies of the nerves and muscles, the heart, and optics. Leonardo da Vinci (Vinci 1452-Amboise 1519). c.1512–13. Pen and ink (Lent by Her Majesty The Queen. Royal Collection © Her Majesty Queen Elizabeth II)

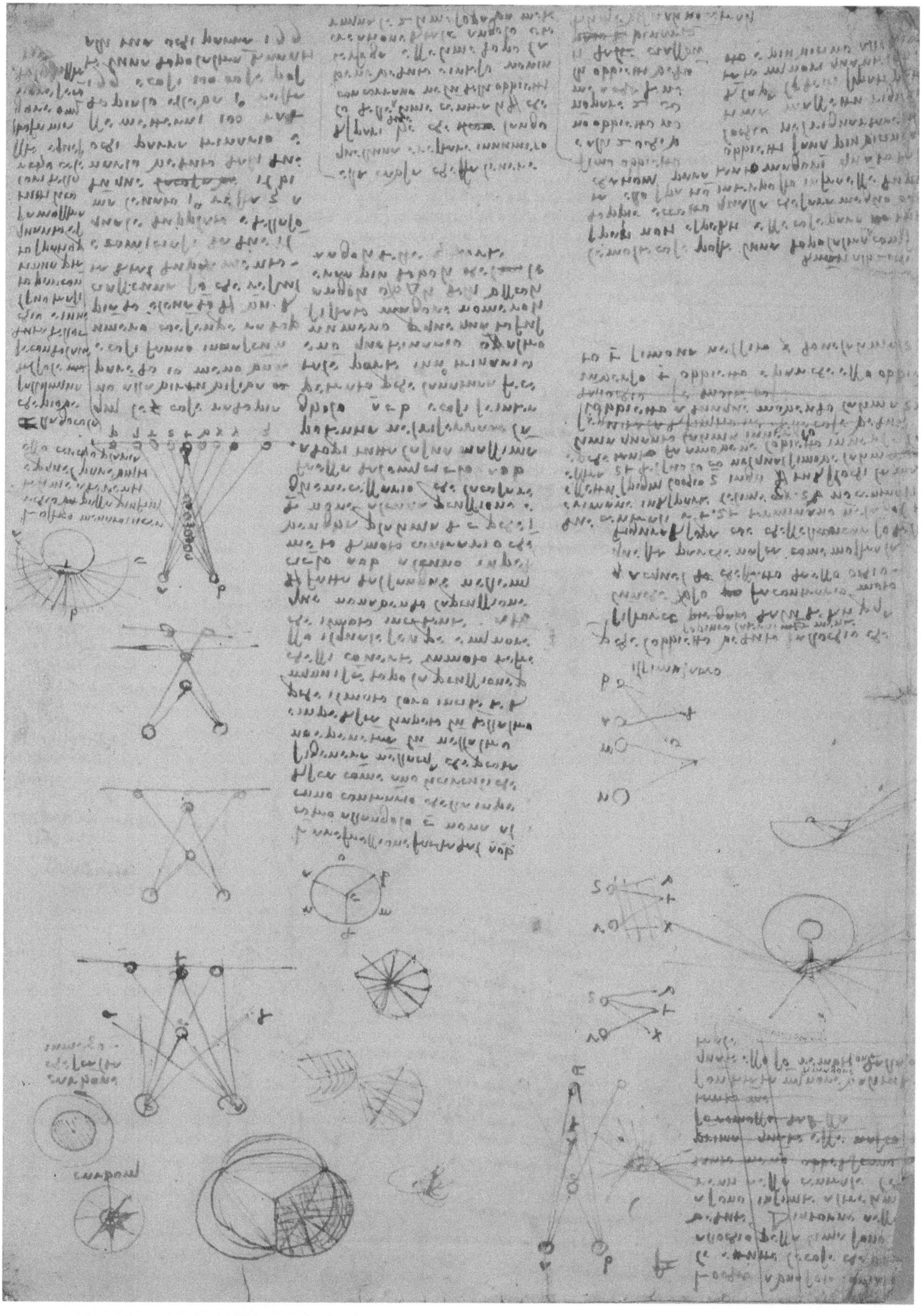

RL 19117 recto and RL 19116 recto (continued)

19116–19117r (C IV 12v and CIV 11r). Pen and brown ink. 307×438 mm.

[I] cercha li neruj motivi delli ochi per tutti li aspettj e cōsidera selli prīcipali sō 4 o piu o mē | perche intutti li linfiniti motj 4 (*f*) nervi fanno il tutto perche subito chettu essci della gu|ridjtione dun dessi 4 nervi tu acqujsti delfauore e aiuto del 2° neruo e cosi segujta de neruj motivi della (*j*) voce e come essi adoperino nelle voci acute e gravi e medjocle denervi (*della*) che aprano esserrano li uasi over porte deventriculi spermaticj – de nervi o voi djre musscolj chesserrā la porta della visscicha – de neruj e musscoli che mandrano fori lasspērme della vergha cō tanta vemētia dj musscoli che serrano lanulo – del muscolo detto verme chesta nellū deventriculi delcervello ilquale sallūgha e rachorta | per aprire esserare iltransito della inprensiva over sēso comune alla memoria – tutti il detti serramj sono aperti dalla cosa che essce dellocho (*sera*) dallor serrati come | lanulo dalle superf<r>uita delcibo e poj siriserra per causa demusscolj eilsimile fanno | le porte devētriculi spermaticj li quali sono aperti dallinpeto della premuta sperme e | poi riserrati (*da ttali*) dalli sua musscolj ancora il sjmjle fa lorina alla porta della | sua visscicha coe chella potētia dellorina premuta orina apre essa porta elli mussco|li sua particulari sō quellj chella riserrano elsimjle fia trovato nella bocha della | vergha e della vulua e della matrice edj tutte le cosce chericeuano (*essca*) lacosanecessa|ria e sscaccano la superfrua – Molte son le parte (*che ssi*) demeati c<h>enella morte resstano aperte che prima stavano | chiuse coe anulo vulua (*boc*) labra (*e p e porte*) e antiporte delcore Ma cque|le chessi serrā per la morte ella (*vul*) bocha della matrice	[I] Search for the motor nerves of the eyes from all aspects and consider whether there are four principal ones, or more, or less, because in all its infinite movements four nerves may accomplish all. For as soon as you leave the jurisdiction of one of the four nerves you acquire the benefit and assistance of the second nerves, and so it continues. On the motor nerves of the voice and how they act in high, low and medium-pitched voices. On the nerves which open and shut the vessels or gates of the spermatic ventricles [seminal vesicles]. On the nerves, or you should say muscles, which close the gateway of the bladder. On the nerves and muscles which eject sperm from the penis with such violence. On the muscles which close the anus. On the muscle called worm which is situated in one of the ventricles of the brain which lengthens and shortens to open and close the passage of the 'imprensiva' or 'senso comune' to the memory. All the said closures are opened by the thing which leaves the place shut by them, as the anus is by the superfluity of food, and then it closes again because of its muscles. And the gates of the spermatic ventricles do the same; they are opened by the impetus of the compressed sperm and then closed again by their muscles. Moreover the urine does the same thing at the gateway of the bladder, that is, the power of the compressed urine opens this gate and its particular muscles are those which shut it again. And the same is found at the mouth of the penis, of the vulva, and of the womb and of all parts which receive necessary things and expel the superfluous. There are many parts of ducts which remain open at death which previously stayed shut, that is the anus, vulva, lips and valves of the heart. But that which is shut at death is the mouth of the womb.
[Fig. 1] a b c d e f m n	[Fig. 1] *a b c d e f m n*
[II] dellacqua che per equj|giacente canna spi|ra quella parte de | la sua intersegati|one dara piuremo|[ta] dalla bocha djtal | [ca]nna laquale na|[sce] piu vicina alcē|[tr]o della bochadjtal | [c]anna	[II] Of the water which pours out of a horizontal pipe, that part of its interface which originates nearer the centre of the mouth of the pipe will go farther away from the mouth of the pipe.
[III] sperimenta|la nū taglio dj|retto fatto | nū uaso –	[III] Experiment on it with a straight cut made in a vessel.
[IV] lacqua che per cāna | surge quella piu | sinalza che ssara | piu remota dalle | pariete dessa cāna	[IV] Water that rises up through a pipe: that rises higher which is most removed from the walls of the pipe.
[Fig. 2]	[Fig. 2]
[V] (*dellaqua che sur|ge per*[…])	[V] (*Of the water which surges up through it …*)
[Fig. 3] a b c d e f	[Fig. 3] *a b c d e f*
[VI] la concavita delemjciclo refrette la perchussione delsangue congrāde e veloce inpeto in|verso incētro del triāgolo a b c dove esso appoggia langolo dellussciolo ·	[VI] The concavity of the hemicycle [aortic sinus] reflects the percussion of the blood with great and swift impetus towards the centre of the triangle *a b c* [aortic orifice] where it leans upon the angle of the valve-cusp.
[VII] Il tēpo delserrare del core e del percotere colla cuspide il costato e del battimēto del | polso e dello entroito delsangue nellātiporto del core (*v*) e £ medesimo –	[VII] The time of the closing of the heart [systole], of the percussing by its apex of the rib-cage, of the beating of the pulse, and of the entry of blood into the front gateway of the heart [aortic orifice], is one and the same.
[VIII] Tante sō leuolte che questa | porta māda fori ilsangue | quāto sō le uol te che ilchore batte e per questo li febrichāti sinfochano	[VIII] As many times as this valve sends out blood, so many times does the heart beat; therefore those who are feverish become inflamed.
[Fig. 4, 5]	[Fig. 4, 5]
[IX] molto piu acqujssta daltezza il mezo del sanghe che ssurge per il triangolo <u>a b c</u> che cquel che | surge per ilati desso triangholo chon co sia che cquel delmezo del triangolo manda | ilsuo inpeto addrittura inalto e cquel chesurge dailati sparge ilsuo inpeto | per moto laterale e percote le fro<n>te delli archi delli emjciclj e segujta la con|caujta desso emjciclo senpre declinando insin che percote la cōcaujta della ba|sa dj tal emjciclo e ppoi si riuolta insu ^cō moto refresso^ e ssiua ragirādo per moto circūvolu|bile (*ragirando*) inse medesimo insīche consuma il suo inpeto – e llultima alteza e cquella dellultimo sangue che entro nellantiporto delcore dj|rieto alquale seguj la perfettione della serratura fatta dalla porta del core e | ttutte questestj moti son fatti	[IX] The middle of the blood that spouts up through the triangle *a b c* [aortic orifice] acquires much more height than that which rises up along the sides of the triangle, because that in the middle of the triangle sends its impetus straight upwards, and that which rises up at the sides spreads its impetus by lateral motion and percusses the fronts of the arches of the hemicycles [aortic sinuses] and follows the concavity of the hemicycles, always going downwards until it percusses the concavity at the base of the hemicycle; then it turns upwards with reflected motion and goes on revolving round through a circumvolving (*gyrating*) motion on itself until it consumes its impetus. And the last elevation is that of the last blood which followed the perfect closure made by the valve of the heart. And these motions all fit together.

<table>
<tr><td colspan="2">19116--19117r (C IV 12v and CIV 11r). Pen and brown ink. 307×438 mm.</td></tr>
<tr><td>[Fig. 6] a b c d e f g</td><td>[Fig. 6] a b c d e f g</td></tr>
<tr><td>[X] Dubitasi quj selli pannjchuli che serrão il sanghue nellantiporto del core coe nella basa dellar|teria adorti sella natura avessi potuto fare sanza essi ōno: conco sia che manifesstamēte | (sipo cōprēdere) ^si uede^ come le 3 pariete over cardjne doue tale porte (de) panjchulari delcore | sono stabilitie (ch) sonquellj chesserrano esso sanghue fori del coro [sic: core] colloro cōfiamēto | quando il core (sa) siriapre dalla parte djsotto aesse porte E cquesto vltimo serra|mēto fa la natura a cco chella gran potētia cheffa il core in esso sinjsstro ventriculo | nelsuo riaprirsi per attrarre asse ilsangue destilato per lj stretti meatj ^del pariete^ chel djujde dal uē|triculo desstro nō hauessi per restaurare il uacuo attjrarsi djrieto j sottilisimj pannj|culj delle predette porte del core –</td><td>[X] Here it is doubted whether the membranes in the aortic vestibule of the heart which shut off the blood, that is to say at the base of the aorta – whether Nature could have done without them or not, because one may clearly see (one can understand) how the three walls or hinges where the membranous gates [valve-cusps] of the heart are stabilised. They are those which shut out the blood from the heart by their opposition when the heart reopens on the side below these valve-cusps. And Nature makes this final closure in order that the great power which the heart exerts in the left ventricle as it reopens to attract into itself blood distilled through the narrow pores of the wall which separates it from the right ventricle [interventricular septum], should not have to refill the vacuum by drawing backwards the very thin membranes of the aforesaid valve-cusps of the heart.</td></tr>
<tr><td>[Fig. 7]</td><td>[Fig. 7]</td></tr>
<tr><td>[XI] Larevolutione delsā|gue nellātiporto del | core ba sa dellar|teria aorto ser|ve a due effetti de | quali primo e che e|sa revolutione mul|tiplicata per piu asspe|tti (riscalda) fa in se | [gr]ā confregatione la | [qu]al riscalda e assottiglia il sangue e aumēta e viujficha li spiriti vitali liquali senpre si mātenghano nel caldo e vmj[do] | il secondo effetto dessa revolutiō dj sangue e dj riserrare laperte porte del core col suo primo moto refresso cō perfetta seratura</td><td>[XI] The revolution of the blood in the vestibule of the heart at the base of the aorta [Fig. 7] serves two purposes, of which the first is that this revolution, multiplied on most aspects, makes within itself great friction which heats and thins the blood and augments and vivifies the vital spirits which are continually preserved in the heat and moisture. The second effect of this revolution of the blood is to shut again the opened gates [valve-cusps] of the heart with their first motion reflected in perfect closure.</td></tr>
<tr><td>[Fig. 8, 9, 10, 11, 12]</td><td>[Fig. 8, 9, 10, 11, 12]</td></tr>
<tr><td>[Fig. 13] a b c d m n o</td><td>[Fig. 13] a b c d m n o</td></tr>
<tr><td>[XII] La refressione facta dal a o b | cōtro allangolo c nona al|cuno contrario chella inpe|djsca come āno li circuli che | sigenerā nellacqua che percota|no e penetrā lū nellaltro | e inpedjscā linpeto lū dellaltro | perche ilmoto loro incidēte dj|mjnuisce dopo la percussione per|che ssi cōuerte in moto refre|sso ilquale senpre e mjnore | che ilmoto incidente · adū|que nonavendo lapercussione | (de) fatta dal sangue nellemj|ciclo a o b alcuno inpedj|mēto di moto contrario che | vengha per linja d c perche ī | d non e alcuna percussione e|gli e necessario che la cosare|fressa dalemjciclo a o b | adoperi tutta la sua massima | potentia nelriserrare lā|gholo a c b e cosi se inter|petrato perche lanatura fece | tale porte in numero trinario | e nō queteinario o (p) altro | numero perque quāto fus|si stato magore nomero li | angoli over Δ^{li} dellj vsscoli | eran piu deboli chel (tri) le | angoli delle porte</td><td>[XII] The reflection made from a o b against the angle c has no opposition which impedes it as have the circles which are generated in water which percuss and penetrate one another and impede the impetus of one another, since their incident motion diminishes after percussion because it is converted into reflected motion, which is always less than incident motion. Therefore as the percussion made by the blood in the hemicycle a o b has no impediment from any contrary motion which might come along the line d c, because at d there is no percussion, necessarily the thing reflected from the hemicycle a o b exerts the whole of its utmost power in shutting again the angle a c b. Thus one has interpreted why Nature made these valve-cusps threefold in number, and not fourfold or in any other number. For if there had been a greater number the angles or triangles of the valve-cusps would be weaker than the angles of the three valve-cusps.</td></tr>
<tr><td>[Fig. 14]</td><td>[Fig. 14]</td></tr>
<tr><td>[XII] carbonj</td><td>[XII] coals</td></tr>
<tr><td>[Fig. 15]</td><td>[Fig. 15]</td></tr>
<tr><td>[XIV] carbone | che scalda | in mezo ·</td><td>[XIV] coal which is heated at the centre</td></tr>
<tr><td>[Fig. 16, 17, 18]</td><td>[Fig. 16, 17, 18]</td></tr>
<tr><td>[XV] (se) [–] Lochio a vna sola linja c[entra]|le e ttutte le cose che ven[gono] | allochio per essa linia sono [bene] | vedute Dintorno a ess[a lini]|a sono infinite altre lini[e ade]|renti a essa centrale le [quali] | (tanto meno obbedjscano all[a] | prima quāto esse nasconno | son remosse da essa | tanto me) | son di tāta mjnore valitudjn[e] | quāto esse sō ^di magore^ remotjone dalla ce[n]|trale – [–]</td><td>[XV] [–] The eye has only one central line and all things that come along this line to the eye are [clearly] seen. Around this line are infinite other lines adherent to the central line which have so much less validity as they are of greater remoteness from the central [line] [–]</td></tr>
<tr><td>[Fig. 19] r s t v x</td><td>[Fig. 19] r s t v x</td></tr>
<tr><td>[Fig. 20] r s t v x</td><td>[Fig. 20] r s t v x</td></tr>
<tr><td>[Fig. 21] n m</td><td>[Fig. 21] n m</td></tr>
<tr><td>[fig. 22] a b d</td><td>[fig. 22] a b d</td></tr>
<tr><td>[XVI] il simulacro
perche lobbietto veduto dallochio che | sistorce (piegato) ^so<s>pinto lateral(m)mente^ da lū de djti per qua|lunche verso (po) fa contrario moto | (q) a cquel (de) che ffatto da esso ochio – | questa par che nasca come mostra la | figura dj sopra coe chessell concor so dell[i] | due centrali r t · s t terminano nelāgol t | e rimane indjsparte lelinje s x · s v nō centralj | e ssettu spigni lochio s in gu (d) tu djslochi la ma|esstra s t del suo loco nel qual simove la linia s x | e perche tanto fa amovere lobietto inuerso la | (seguita la djfinjtione delle cose vedutj | dallochio se molte cose) | linia quanto la linia inuerso | lobbietto adunque movendo la linia s x | inverso t obbietto epar che esso obbie[t]|to t si moua nelsito x doue la linia s x</td><td>[XVI] The image
Why the object seen by the eye which is turned by being pushed sideways by a finger in any direction makes a movement contrary to that made by the eye. This seems to arise, as the figure above [Fig. 19] shows, that is, if the concourse of the two central [lines] r t, s t terminate in the angle t, and the non-central lines s x and s v remain apart, and if you push the eye s downwards, you displace the master-line s t from its place into which moves the line s x. And because the object moves as much towards the line as the line towards the object, therefore moving the line s x towards t the object, it appears that this object t moves to the position x where the line s x is.</td></tr>
</table>

19116--19117r (C IV 12v and CIV 11r). Pen and brown ink. 307×438 mm.	
[Fig. 23] a b c d e f	[Fig. 23] *a b c d e f*
[Fig. 24, 25]	[Fig. 24, 25]
[XVII] le molte cose poste luna dopo laltra ^djnãti alli 2 ochi^ con j s|spazi noti e spedjti esse cose parã *(do)* tutt[e] | doppie eccetto qualla che sara meglio ved[u]|ta ello spatio interposto infra esse dupl[i]|chationj parã tanto magori quãto tal[i] | obbietti sarã piu vicini a | lochio nel riguardare lul|tima masse ttu riguard|dj la p^a lj detti spati par[an] | tãto mjnori quanto lobb[ie]|tto e piu vicino allochio	[XVII] Many objects placed one behind another before the two eyes at known and open spaces, these objects all appear double except that one which is seen best. And the spaces interposed between these duplications appear so much larger as the objects are nearer to the eye in looking at the farthest one [Fig. 24]. But if you look at the first one, the said spaces appear so much the less as the object is nearer to the eye [Fig. 25].
[Fig. 26] a b c d e f	[Fig. 26] *a b c d e f*
[Fig. 27] a b c m n o p q r s t v x y z	[Fig. 27] *a b c m n o p q r s t v x y z*
[XVIII] quj le *(r)* cose radopia|no alla virtu visiva *(co)* | parẽdo 10 meno vna | e cosi fanno inciascũ n|umero coe senpre radop|piãdo e leuãdo *(de)* vn dj | ciasscuna sõ [sic: somna] che resul|ta di tal duplamento – | e comjciasi da due il | quale dupplato e della sõ|ma leuato 1° resta 3 a|dũque *(la cosa ve)* il bi|nario veduto dalli due | ochi parra trinario e | sse metterai 100 rad|dopialo elleva 1° resta | 199 e cosi 100 cose pos|te luna dopo laltra djnanti | alli tua ochi parrã 199 [:–:] ella cavsa cheffa levere [sic: levare] | quelluna e restare in numero djspari he che *(il con)* lango|lo delle ^due^ linie centrali *(dj)* che | concorrano nelũ delli obbiettj | bene veduto e inteso non in|tersega esse linie dopo la | creatione dj tale angolo e te|rmina le 2 linie sopra vn mede[:–:]simo obbietto | e alli 2 ochi v|no obbietto 10 | nõpare 2 co|me achade ne|lli obbietti vedu|ti da*(lle)* ciasscũ | *(paio di)* binario | djlinje visuali nõ cẽtralj	[XVIII] Here the objects are doubled to the visual power [virtu visiva] appearing as 10 minus 1. And they do this for every number, i.e. always redoubling and leaving out one from every total sum which results from such doubling. And one begins with 2, which doubled and 1 taken away from the total leaves 3; therefore the twofold seen by the two eyes will appear threefold. And if you put out 100, double it and take away 1, there remains 199; and thus 100 objects being placed behind the other in front of your eyes will appear as 199. And the cause which gets one taken away leaving an odd number is that the angles where the two central lines meet together on one of the objects is well seen and comprehended, and these lines do not intersect after the creation of such an angle. And the two lines end on one and the same object; and to the two eyes one object does not appear as two, as happens in objects seen by each of the twofold visual lines which are not central.
[Fig. 28] a b c	[Fig. 28] *a b c*
[XIX] Lostizo menato incer|rchio *(va)* passa per infinj|te ljnje aderenti | e per quel pare vnjto | esso cercho per laria	[XIX] The fire-brand moved in a circle passes through infinite adherent lines and through that circle appears united in the air.
[XX] E lla gocola | che piove | sallumjna | dal sole *(me)* | secondo la ue|duta delloc|chio e inne|l suo trãsi|to parcon|tinua per tã|to spatio per | quanto es|sa mosstra | tutti li co|lori della|rcho cele|sste eques|stofa ma|gore o mj|nore secõ|do la disstã|tia	[X] And the drop which rains down is illuminated by the sun according to the sight of the eye; and in its passage it appears continuous through as much space as shows all the colours of the rainbow. And this [rainbow] is larger or smaller according to the distance.

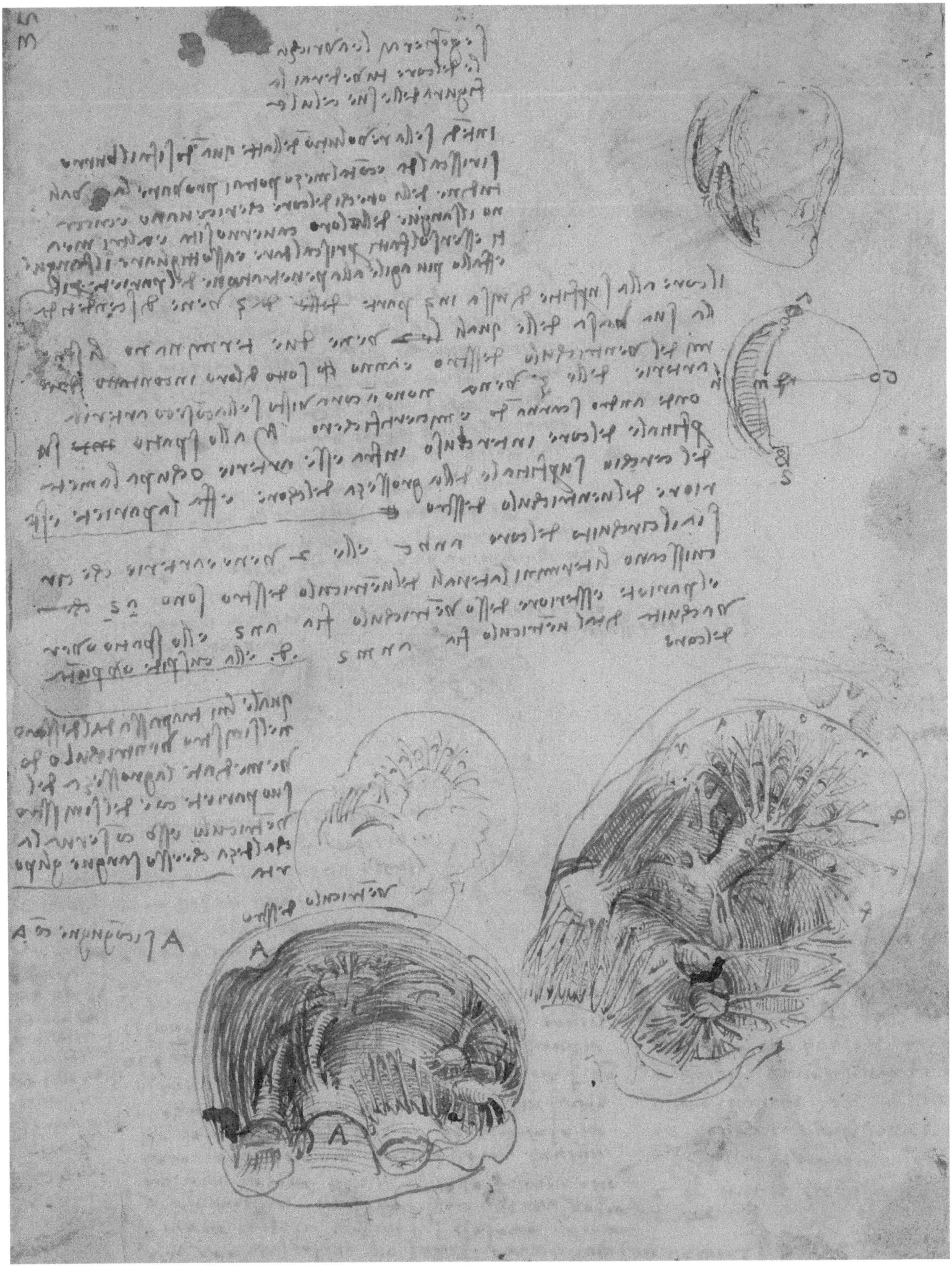

RL 19118 verso and RL 19119 verso The ventricles, valves and papillary muscles. Leonardo da Vinci (Vinci 1452-Amboise 1519). c. 1511–13. Pen and ink (Lent by Her Majesty The Queen. Royal Collection © Her Majesty Queen Elizabeth II)

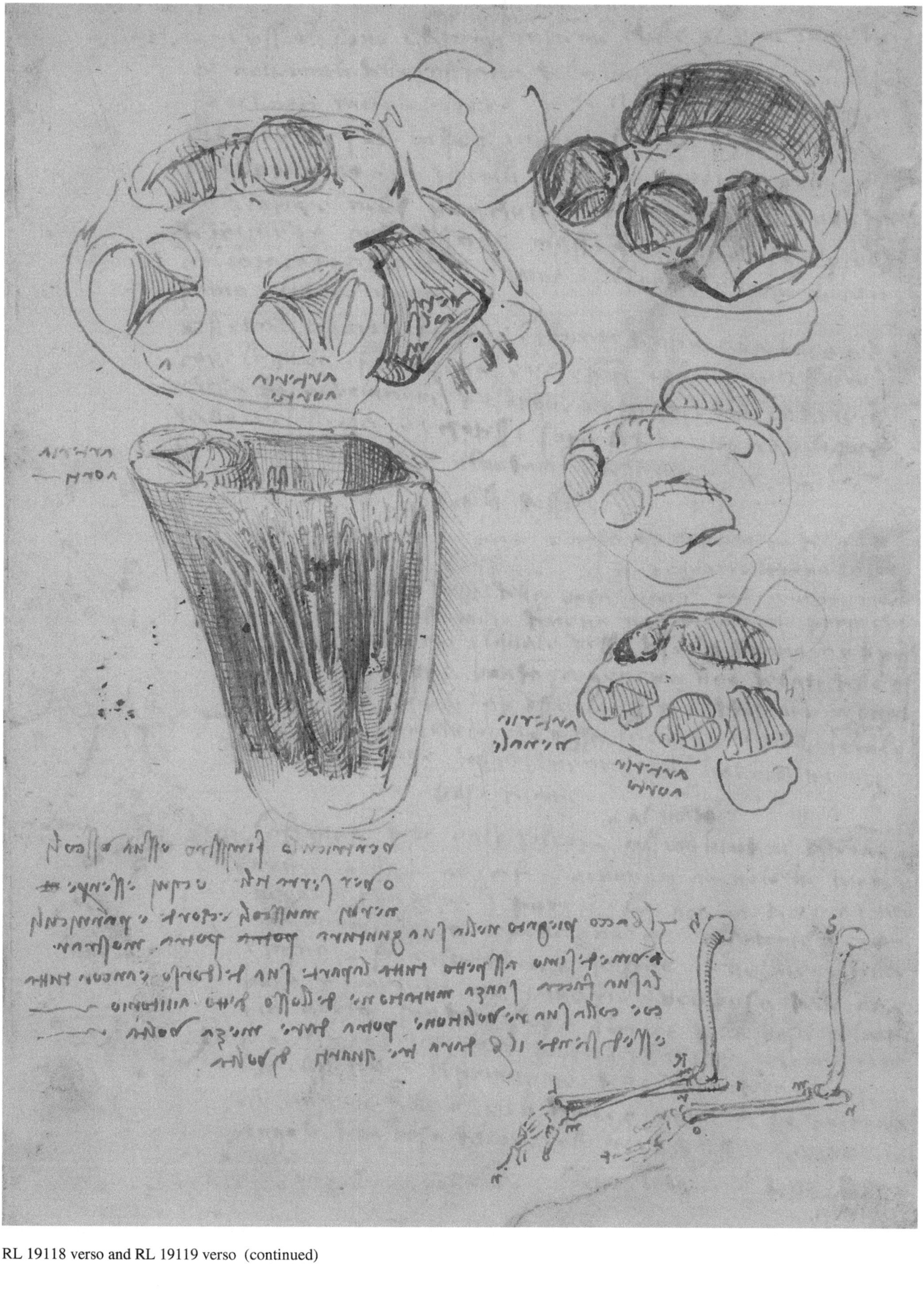

RL 19118 verso and RL 19119 verso (continued)

<table>
<tr><th colspan="2">19118–19119v (C IV 13v and CIV 14r). Pen and brown ink and wash.</th></tr>
<tr><td>[Fig. 1]</td><td>[Fig. 1]</td></tr>
<tr><td>[Fig. 2] vētri|col si|nj|stro arteria aorto a</td><td>[Fig. 2 left ventricle aortic artery a</td></tr>
<tr><td>[Fig. 3, 4]</td><td>[Fig. 3, 4]</td></tr>
<tr><td>[I] arteria | aorto</td><td>[I] Aortic artery</td></tr>
<tr><td>[II] arteria | venale</td><td>[II] Vein-like artery [pulmonary artery]</td></tr>
<tr><td>[Fig. 5]</td><td>[Fig. 5]</td></tr>
<tr><td>[III] arteria | aorti –</td><td>[III] Aortic artery</td></tr>
<tr><td>[IV] vētriculo sinjsstro e ssua vsscoli | over serratili e cquj e ssenpre (m) | nervj musscolj echorde e pannjculj</td><td>[IV] Left ventricle and its valve-cusps or shutters; and here there are always sinews, muscles, chordae tendineae and membranes [valve-cusps].</td></tr>
<tr><td>[Fig. 6] S m n o r f</td><td>[Fig. 6] S m n o r f</td></tr>
<tr><td>[Fig. 7] h i K l m n</td><td>[Fig. 7] h i K l m n</td></tr>
<tr><td>[V] Jl bracco piegato nella sua guntura (potra) potra mostrar|ti £medesimo asspetto tutta laparte sua deldorso e ancor tutta | lasua facca sanza mutatione dellosso detto aiutorio – | coe colla sua revolutione potra dare meza volta – | esse djstende il braccio dara tre quarti dj volta</td><td>[V] The arm bent at its [elbow] joint will be able to show you from one and the same aspect the whole of its dorsal side, and furthermore, the whole of its front side, without any change [in position] of the bone called humerus – that is to say, it will by revolving be able to give half a turn – and if you extend the arm it will give three-quarters of a turn.</td></tr>
<tr><td>[VI] se gōfieraj le avrichu|le delcore tu vederai la | figura delle sue celule</td><td>[VI If you inflate the auricles of the heart you will see the shape of its cells.</td></tr>
<tr><td>[VII] intēdj sella revolutiō dellatte quādo sifa ilburro | sirisscalda e cō tal mezo potraj provare la vali|tudjne delli orechi del core che riceuano e cacca|no ilsangue delleloro cauernosita e altri mea|tj essersol fatti per riscaldare e assottigliare il sangue | e ffallo piu agile alla penetratione del pariete per il [:–:] quale lui trapassa dal desstro | nel sinjstro ventrichulo do|ve medja<n>te la grosseza del | suo pariete coe del sinjsstro | vētriculo esso cōserua la | chaldeza cheesso sangue glipo|rta</td><td>[VII] Observe whether the revolution of milk when butter is made, heats it. And by such means you will be able to test the strength of the auricles of the heart which receive and expel the blood from their cavities and other passages; these being made only in order to heat and refine the blood and make it more quickly penetrate the wall through which is passes from the right into the left ventricle where through the thickness of its wall, that is, of the left ventricle, it conserves the heat which the blood carries to it.</td></tr>
<tr><td>[Fig. 8]</td><td>[Fig. 8]</td></tr>
<tr><td>[VIII] il core alla superfitie djujsa in 3 parte (delle) da 3 vene djscendēti da|lla sua basa delle quali (le 2) vene due termjnano li stre|mj del ventrichulo desstro e anno (co) sotto djloro incontatto due | arterie della 3ª vena nono ācora visto sella cō seco arteria | onde andro scarnādo e mj ciertifichero Ma llo spatio (inte) su|perfitiale del core intercluso infra esse arterie ochupa la meta | del cerchio superfitiale della grosseza delchore e ffa la pariete este|riore del uentrichulo desstro e<c>–</td><td>[VIII] The heart has its surface divided into three parts by three veins descending from its base, of which two delimit the edges of the right ventricle [great and small cardiac veins]. And they have below in contact with them two arteries [anterior descending branch of the left coronary artery and the right coronary artery]. As for the third vein I have not yet seen if it has an artery with it, so I am going to peel off the flesh to ascertain this. But the superficial area of the heart enclosed between these arteries occupies half of the superficial circular size of the heart and forms the external wall of the right ventricle.</td></tr>
<tr><td>[Fig. 9] a b c d m n s</td><td>[Fig. 9] a b c d m n s</td></tr>
<tr><td>[IX] sia il circhuito delchore a n b c elle 2 vene e arterie che cir|cuisscano li termjni laterali deluētriculo desstro sono a s (che) | el pariete essteriore desso vētrichulo fia a n s ello spatio over | vachuita djtal uētriculo fia a n m s·d·ella cuspide over pūta | delcore</td><td>[IX] Let a n c b be the circumference of the heart, and the two veins and arteries which encircle the lateral borders of the right ventricle are a s. And let the external wall of this ventricle be a n s and the area or empty space of this ventricle be a n m s. d is the apex or point of the heart.</td></tr>
<tr><td>[Fig. 10]</td><td>[Fig. 10]</td></tr>
<tr><td>[Fig. 11] a b c d e f n m o p q r</td><td>[Fig. 11] a b c d e f n m o p q r</td></tr>
<tr><td>[Fig. 12] A A</td><td>[Fig. 12] AA</td></tr>
<tr><td>[X] vētriculo desstro</td><td>[X] Right ventricle</td></tr>
<tr><td>[XI] A si cōgugne cō A</td><td>[XI] A is joined to A</td></tr>
</table>

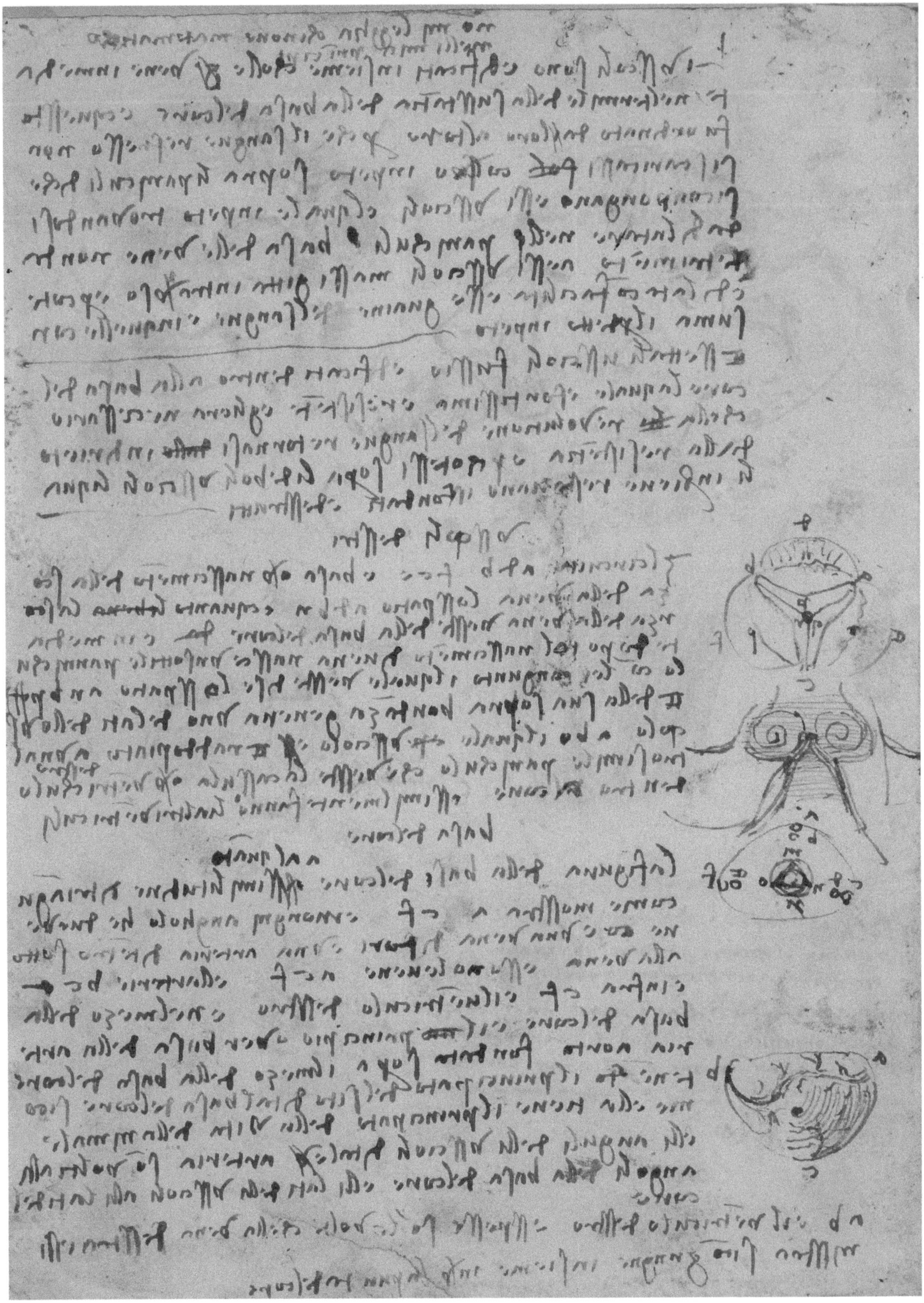

RL 19118 recto and RL 19119 recto, The right ventricle and the valves of the heart. Leonardo da Vinci (Vinci 1452-Amboise 1519). c. 1511–13. Pen and ink (Lent by Her Majesty The Queen. Royal Collection © Her Majesty Queen Elizabeth II)

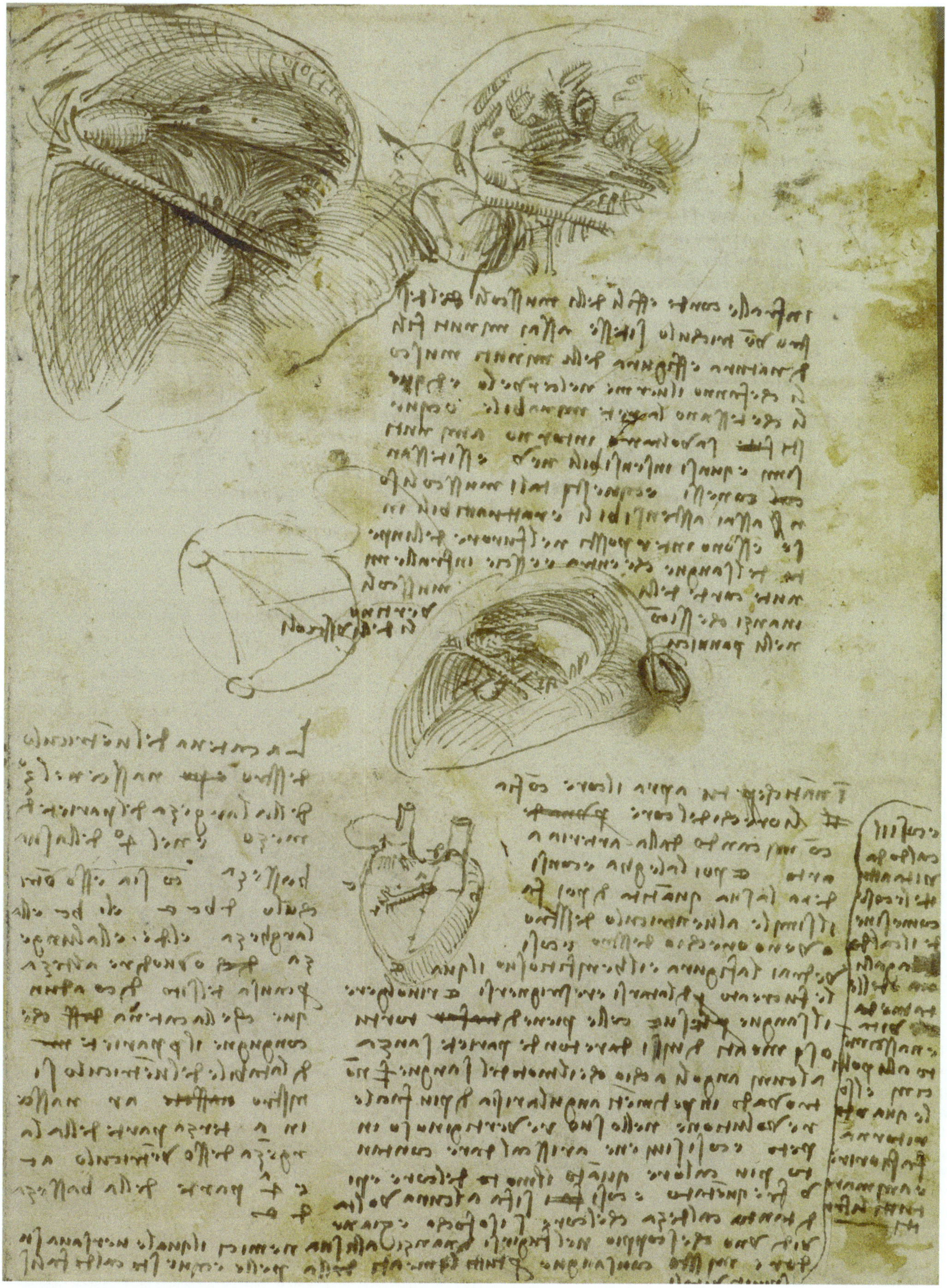

RL 19118 recto and RL 19119 recto (continued)

19118–19119r (C IV 14v and CIV 13r). Pen and brown ink. 221×312 mm.

[Fig. 1, 2]	[Fig. 1, 2]
[I] infralle corde e ffili delli musscoli del des|stro vētrichulo sitesse assaj mjnuti fili | dj natura e ffigura delli mjnuti musco|li chefanno iluerme nelcervelo e dj que|li tessano la rete mjrabile e cque|sti (*si te*) savoltano intorno amjnuti|simj e quasi insensibili ne<r>v<i>e ssitessan | (*col*) conessi e cquestj tali musscoli so|n (*s*) assai asstensibili e rattrattibili in | se essono interpossti nel furore dellinpe|to del sangue che entra e esscie infralle mj|nute corde delli musscoli | inanzi che ssicō vertino | nelli pannicu li delli vssciolj	[I] Between the cords and bands of the muscles of the right ventricle [trabeculae carneae] are woven very minute threads of the nature and shape of the minute muscles which form the worm in the brain [choroid plexus] and of those which weave the rete mirabile. And these are wound round the minutest and almost imperceptible tendons and are interwoven with them. And these muscles are themselves very extensible and retractile and they are interposed in the furor of the impetus of the blood which passes in and out between the minute tendons of the muscles [chordae tendineae] before they are converted into the membranes of the valves.
[Fig. 3, 4]	[Fig. 3, 4]
[II] īnāti che (*p*) tu apra core cōfia | (*il*) li orechi del core (*per vna de*) | cōmj<n>cando dalla arteria a|orto E poi lalegha e consi|dera la sua quātita dj poi fa | il simjle aluentriculo desstro | overo orechio desstro e cosi | vedrai lafigura e ilbenjfitio suo il qua|le fu creato per djlatarsi e retrignersi E riuolgere | il sangue per le sue celle piene dj (*trafor*) tortu|osj meati djujsi da retonde pariete sanza | alcunj angoli a chio che ilmoto del sangue (*s*) nō | trova<n>do inpedjmēti angulari sia dj piu facile | revolutione nello suo revertiginoso in|peto e cosi siujene a rissc aldare contan|to piu calore quāto ilmoto del core e pi|v frequētato e cosi (*fa i*) sifa alcuna volta | dj tanta caldeza chelcore si sofocha e gia ne | vidj vno che scoppio nel fugirsi djnanzi alli sua nemicj ilquale uersaua su|dore mjssto con sangue per tutti li meati della pelle e cquesto caldo fa li s|spiriti vitali [:–:] e cosi il | caldo da | vita a ttu|tte le cose | come si ue|de il caldo | della galli|na o delle | tac<i>ne da|re vita | e nassc imē|to alli polli|cinj el so|le quando | ritorna | fa fiorire | e anjmare | tuttj li fru|ttj –	[II] Before you open the heart, inflate the auricles of the heart, beginning from the aorta. Then tie it up and consider its size. Then do the same with the right ventricle or right auricle. Thus you will see its shape and the function of that which was created to be dilated and contracted and to revolve the blood throughout its cells full of tortuous passages separated by rounded walls without any angles so that the motion of the blood, not finding angular impediments, makes an easier revolution in its whirling impetus. And thus it comes to heat itself with as much more heat as the movement of the heart is more frequent [tachycardia]. Thus sometimes it creates so much heat that the heart is suffocated. And once I saw a man who burst in fleeing from his enemies: he poured out sweat mixed with blood through all the pores of his skin. And this heat makes the vital spirits. Thus heat gives life to all things, as one sees the heat of the hen or turkey give life and birth to chickens. And the sun when it returns makes all fruits blossom and produce seed.
[Fig. 5] a b c d e r	[Fig. 5] *a b c d e r*
[III] La catena del uētriculo desstro (*e po*) nassce nel 3° | della largeza del pariete dj | mezo e nel 4° della sua | basseza – cosi a esso vētri|chulo d b c e eli b c ella | largheza el·d e·e lla lunge|za (*dj ch*) o vuodjre alteza | per causa delsito djco adun|que che lla catena (*dess*) che | congugne il ppariete (*m*)| djlatabile deluētriculo si|njstro (*nasscie*) a r nassce | in a̲ terza parte della la|rgeza desso vētriculo a [sic: b] c | e 4ª parte della basseza | d e	[III] The chain of the right ventricle [moderator band: *a r*) arises at one-third of the width, from the middle wall and at a quarter of its depth [marked off *b c* and *d e* in the diagram]. Thus for this ventricle *d b c e*, *b c* is the width and *d e* is the length, or you may say height by reason of its location. Therefore I say that the chain which joins the dilatable wall from the left ventricle *a r*, arises at *a*, a third part of the width of this ventricle [*b*] *c*, and a fourth part of the depth *d e*.
[IV] nō mj leggha chi non e matematicho | nelli mja prīcipj	[IV] Let no one who is not a mathematician read my principles.
[V] Li vsscoli sono edjficatj insieme cholle (*g*) vene inmedja|te neltermjle della susstātia della basa delcore e cquessto | fu ordinato da(*i*) loro altore perche il sangue refresso non | siscaricassi (*sol*) colsvo inpeto sopra li panjculi dj che | si conpongano essi vssciolj el quale inpeto trovandosi | da djlatare nellj panjchuli basa delle vene non da | detrimēto aessi vssciolj ma ssi gitta intraverso e percote | e djlata cō facilita esse guaine delsangue e in quelle con|suma il predetto inpeto – E sse ttali ussciolj fussi<n>o edjficati dentro alla basa del | core la quale efortissima e resistēte egliera necessario | chella (*le*) revolutione delsangue retornasi (*dalle*) in djrieto | dalla resistētia e perchotessi sopra li deboli vsscioli liqua|li in brieue restauano isfondati e desstrutti –	[V] The valve-cusps are built up, together with the vessels, immediately at the termination of the substance of the base of the heart. This was ordained by their Author so that the reflected blood should not be discharged with its impetus on to the membranes of which the valve-cusps are composed. And this impetus, being devised for dilating the membranes at the base of the vessels, does not damage these valve-cusps but is thrown obliquely and percusses and dilates with ease these sheaths of the blood in which the aforesaid impetus is consumed. And if these valve-cusps had been built within the base of the heart which is both very strong and resistant, it would be necessary for the revolution of the blood to be returned from this resistance, and it would percuss on the weak valve-cusps which in a short time would be ruined and destroyed.
[Fig. 6] a b c d e f n o p q r r	[Fig. 6] *a b c d e f n o p q r r*
[VI] vsscjolj desstri Jlcircuito a d b f c e e basa over nasscimeto della sco<r>|za della vena losspatio a d b n e cquanto (*li bena*) la sco|rza della vena vesste della basa delcore (*da*) e inmedja|te dopo tal nasscimēto dj uena nassce vnsottile pannjchu|lo cō lei congunto ilquale vesste djse lo sspatio a n b p (*c f*) | E della sua soprabondāza genera vno delati dello vs|cjolo a b o ilquale (*ep*) vssciolo e (*s*) E raddopiato a vnal|tro simjle panjchulo che veste la cassula over vētrichulo ^destro^ | dentro al core essimjlmente fanno lialtri vētriculj	[VI] The right-sided valve-cusps [valve of the pulmonary artery] The circumference *a d b f c e* is the base or origin of the coat of the [arterial] vein [pulmonary artery]. The area *a d b n* is how much the coat of the [arterial] vein clothes the base of the heart and immediately beyond this origin of the vein there arises a fine membrane joined to it which clothes with itself the space *a n b p* [the cusp and sinus]. From its excess it generates one of the sides of the valve-cusps *a b o*, which valve-cusp is reduplicated by another membrane which is similar to that which lines the pericardium or the right ventricle inside the heart; and similarly the other ventricles.

19118–19119r (C IV 14v and CIV 13r). Pen and brown ink. 221 × 312 mm.

[Fig. 7]

[Fig. 7]

[Fig. 8] a b c d e f p m n o

[Fig. 8] *a b c d e f p m n o*

[VII] basa del core
lafigura della base delcore (*a*) ^a alquãto^ ssimjlitudjne dj triãgu<lo>| come mosstra a c f einongnj angholo he due ve|ne coe vna vena dj fori e vna arteria dj dẽtro sotto | alla vena essono leuene a c f ellarterie b c [sic: d] e | e infra c f e il uẽtriculo desstro e nelmezo della | basa del core e il (*na*) principio over basa della arte|ria aorto fondata sopra ilmezo della basa delcore | tenẽdo il principato del sito di tal basa delcore sico|me ella tiene il principato della vita dellanjmale | elli anguli delli vsscioli dj tale (*v*) arteria sõ volti alli | angolj della basa delcore elli lati delli vsscioli alli lati del | core

[VII] Base of the heart
The shape of the base of the heart is somewhat similar to a triangle, as *a c f* shows. And at each angle are two vessels, that is, a vein outside and an artery inside below the vein. *a c f* are the veins, and *b* [*d*] *e* the arteries. Between *c f* is the right ventricle. And at the centre of the base of the heart is the origin or base of the aortic artery [*o m n*], established on the middle of the base of the heart, keeping the principal position at the base of the heart, just as it keeps the principle of the life of the animal. And the angles of the valve-cusps of this artery are turned towards the angles of the base of the heart and the sides of the valve-cusps towards the sides of the heart.

[Fig. 9] a b c

[Fig. 9] *a b c*

[VIII] a b e il vẽtriculo desstro e sspesse sõ le volte chella vena desstra e ssi|njsstra sicõgungne insieme inver la punta delcore

[VIII] *a b* is the right ventricle and there are many times when the right and left vessels are joined together towards the apex of the heart.

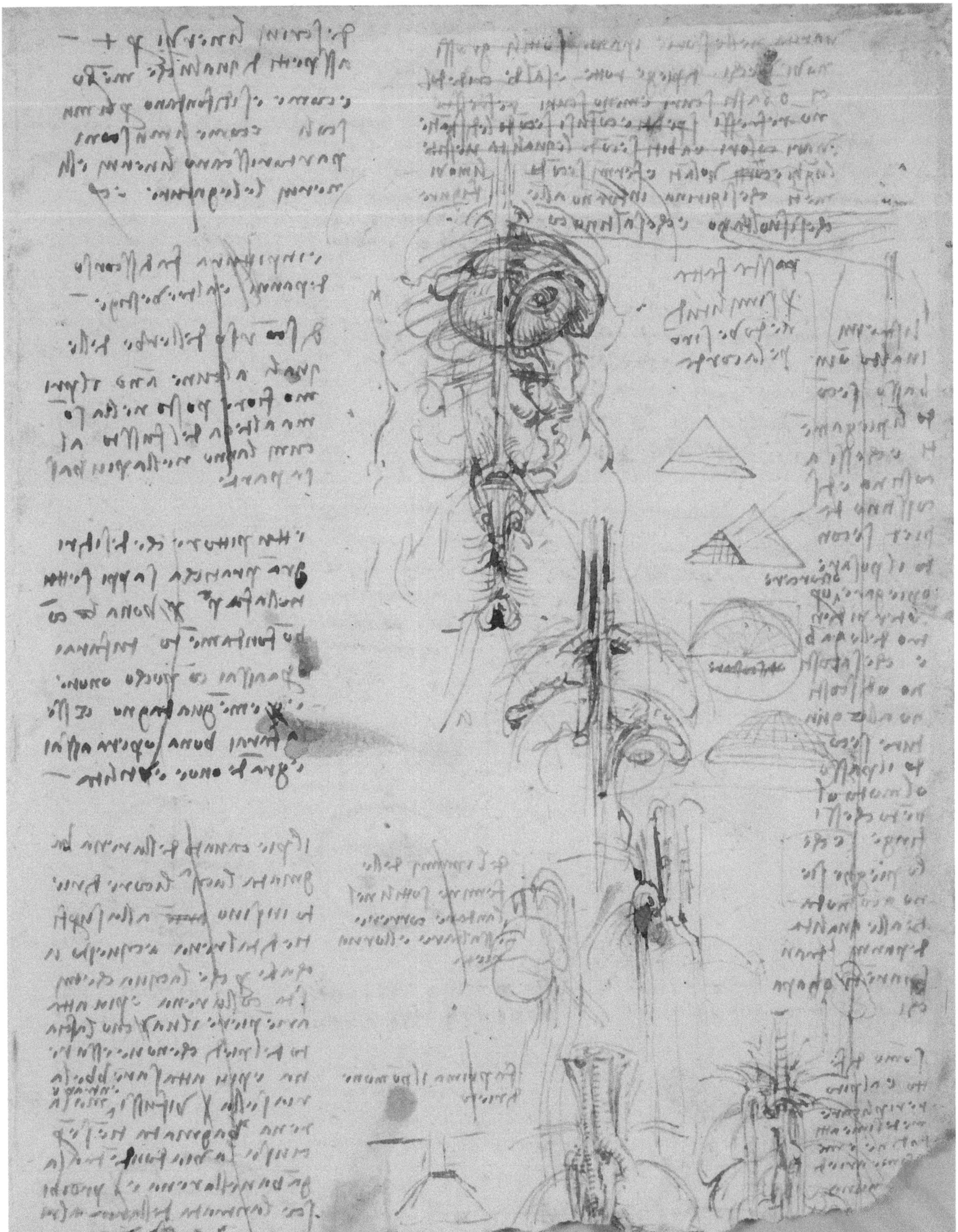

RL 19120 recto and RL 19121 recto, Studies of mechanics and anatomy, and notes on painting. Leonardo da Vinci (Vinci 1452-Amboise 1519). c. 1510–13. Pen and ink (Lent by Her Majesty The Queen. Royal Collection © Her Majesty Queen Elizabeth II)

19120–19121r (C IV 16 v and CIV 15 r). Pen and brown ink. 219×315 mm.

[I] [*–] varieta nelle storie i pannj sottilj grossi | novi vechi dj pjege rotte e salde cride dol|ci ōbrasti scuri e meno scuri refressi e | nō refressi spedjti e cōfusi secōdo le djstātie | e vari colori e abiti secōdo le qualita uestite [–*] | lūghi e cortj volāti e fermj secōdo li movi|mēti che si girino intorno alle figure | che sisuoltano e che saltino co [:–:] [*–] li stremj | in alto o in | basso secō|do li piegamē|ti e che ssi a|costino e djs|cosstino da | piedj secon|do il posare | o piegare ^o storcere^ o per|cotervi den|tro delle gāb|e che sacosti|no o djscosti|no alle gun|ture secō|do il passo | ol moto ol |vēto che ssi | finge e che | le pieghe sie|no acomoda|te alle qualita | depannj tran|sparēti o hopa|chi [–*]	[I] [*–] Variety in the depiction of events. Thin cloths, thick, new and old ones with broken and firm folds, pleasantly shaded, dark and not so dark, reflected and not reflected, clear cut or confused according to the distance and the different colours; and clothing according to the qualities of the suits [–*] long and short, floating about or still, according to the movements which twist around a figure, which turn round and leap up with [*–] edges high up or low down according to the folds. Those that cling close to or separate from the feet according to whether they are at rest or bending, twisting or being percussed inside by the legs. Those that keep close to, or move away from the joints according to their pace or movement, or the wind which is simulated. And let the folds be suited to the qualities of transparent or opaque cloths. [–*]
[II] [*–] sōmo djse|tto e al pito|re riplicare | medesime attj|tudjne e me|[desi]me arie di | […]n una | [–*]	[II] [*–] The biggest fault in a painter is to repeat the same postures and the same airs … in one … [–*]
[III] [*–] de li pannj delle | femjne sottili nel|landare correre | e ssaltare e llor ua|rieta [–*]	[III] [*–] On the thin clothes of women in walking, running and jumping, and their differences [–*]
[Fig. 1–7]	[Fig. 1–7]
[IV] [*–] fa prima il pōmone | djrieto [–*]	[IV] [*–] First do the lungs from behind [–*]
[V] [*–] descriuj li nervi per 4–| asspetti dj qualūche mēbro | e come e sidjfondano per li mu|scoli e come li musconi [sic: muscoli] | partorisscano li neruj ellj | neruj le legature &c [–*]	[V] [*–] Describe the tendons of any part from four aspects; and how they are diffused through the muscles, and how the muscles give rise to the tendons, and the tendons to ligaments, etc. [–*]
[VI] [*–] e inpittura fa djsscorso | depanni e altre vestige – | djscorso dellerbe delle | quali alcune anō ilpri|mo fiore posto nella sō|ma alteza del fussto al|cunj lanno nella piu bas|sa parte [–*]	[VI] [*–] And in [the book On] Painting make a discourse on cloths and other clothing. Discourse on plants of which some have their first flower placed at the very top of the stem and some have it placed at the lowest part. [–*]
[VII] [*–] e ttu pittore che desidjri | grā praticha sappi settu | nella fa p^a (*p*) bona (*lo*) cō | bō fondamēto tufarai | opera assai cō pocho onore | e (*g*) e mē guadagno e sse | la faraj bona opera assai | e grāde ono<r>e vtilita – [–*]	[VII] [*–] And you, painter who desire greatness in practice, know that if you should first develop yourself on a good foundation you will do much work with little honour and less reward; and if you do it well it will achieve much honour and be of great use. [–*]
[Fig. 8]	[Fig. 8]
[VIII] [*–] il pie cauato della rena ba|gniata lacqua li core djrie|to insino (*a riē*) alla superfj|tie dj talrena ecquesto a|chade perche lacqua che mj|sta colla rena e piu atta | a riēpiere il ua(*v*)cuo lascia|to delpiedj che non e essa re|na e piu atta sarebbe la|ria sellaj vi fussi ^entrare^ ma la | rena bagniata tiē sēpre | ciusa la via dondētra la | gāba nellarena e proibi|sce lantrata dellaria al ri|enpimēto desso vacuo – [–*]	[VIII] [*–] [When] a foot is dragged out of wet sand; water runs in behind it right up to the surface of the sand; and this happens because the water which is mixed with sand is more ready to fill up the vacuum left by the foot than is the san; air would be still more suitable if it could enter there, but the west sand always keeps closed the way through which the leg enters the sand, and prevents the entrance of the air to refill the vacuum. [–*]
[Fig. 9, 10]	[Fig. 9, 10]
[Fig. 11] (*e djscostate*)	[Fig. 11] (*has been put aside*)
[Fig. 12, 13]	[Fig. 12, 13]
[IX] passta fatta | per simjlitudj|ne dove sirō|pe la corda	[IX] Make a paste in the likeness of [the cone] where the cord cuts through.
[Fig. 14] p^a [ɪɪ]	[Fig. 14] First [figure] [ɪɪ]

19120–19121r (C IV 16 v and CIV 15 r). Pen and brown ink. 219×315 mm.

[X] confregatione [*–] Della confregatione depoli delle machine | sopra i loro sosstentaculj– ∂ multiplicha il braccio della lieua per tante volte 4 per quāte (*volte*) essoriceue | inse lacōtralieua e della soma leua vno e col rimanēte par|ti laconfregatione delpolo e llauenjmēto multiplica per 4 e cqe|sto e ilcontra peso equale alla confregatione ©– [–*] djreno adunque | n essere lagrosseza del polo n a lalungeza della lieua . n p . lalunge|za della cōtra lieue la qual sintende laconfregatiō dj tal polo colsuo | ricipiēte la qual dconfregatione senpre resisste pel 4° del peso cōfre|gato adunque io multiplico la lieua ^n a^ per tāte volte 4 quanto essa | riceue inse la cōtraliena n o che a presēte essendo la lieua equa|le alla contra lieua esse ricevano solo vna volta luna altra adū|que djreno vn vie 4 fa . 4 . // e della sōma leua 1° e colrimanē|te che sō 3 . (*le*) parti la cōfregatione del polo che 1° (coe la 4ª | parte del peso delpolo che 4) che ne viene 1/3 e questo 3° multi|plica per 4 (*e questo 4 e il uero cōtra peso* de) che sson 4/3 e | e cquessti quatro terzi. son vero cōtrapeso (*dj se e del pe*) della | confregatiō dj se insieme colla cōfregatiō del polo dove e 4/3 . | sisosstiene del quale tu torraj la quarta parte che 1/3 ello | gugnierai colla cōfregatione del polo che 1° coe la 4ª parte (*del*) | del peso del polo che 4 adunque givgnj insieme 1/3 (*e vn*) che | cōfregatiō delcontra peso e 3 terzi coe vno ītero che cōfrega|tione del polo e araj 4 terzi dj confregatiō dj tutto il peso equa|li a 4/3 del contrapeso e cquesta equalita dj dj cōfregatione dj | peso del polo col suo cōtra peso e nata perche (*il*) la lieva e equa|le alla contralieua – [–*]	[X] Friction [*–] On the friction of the axles of machines on their supports Multiply the arm of the lever by as many times four as that [lever] receives into itself the counter-lever, and subtract 1 from the total; and with the remainder divide the friction of the axle; and multiply the result by four–and this is the counter-weight equal to the friction. We shall say therefore, *n* is the size of the axle, *n a* the length of the lever, *n p* the length of the counter-lever which provides the friction of this axle with its recipient, which friction always resists by the fourth of the frictional weight. Therefore I multiply the lever *n a* by as many times 4 as it receives into itself the counter-lever *n p*. In the present instance since the lever equals the counter-lever these receive each other once only. Therefore we shall say 1 times 4 makes 4, and from the total take off 1, and with the remaining 3 divide the friction of the axle which is 1 (i.e. the fourth part of the weight of the axle which is 4) which comes to 1/3; and this 1/3 multiply by 4, which is 4/3. And these 4/3 are the true counter-weight of the friction between them together with the friction of the axle where the 4/3 are supported; from which you will take away the fourth part which is 1/3, and add it to the friction of the axle which is 1, that is the fourth part of the weight of the axle which is 4. Therefore you add together 1/3 which is the friction of the counter-weight, and 3/3, that is one unit, which is the friction of the axle, and you will have 4/3 of friction of the whole weight equal to 4/3 of the counter-weight. And this equality of friction of the weight of the axle with its counter-weight has been produced because the lever is equal to the counter-lever. [–*]
[XI] [*–] o modo secōdo che insignia procedure Tutti li pesi che ssi sosstēgano sopra il polo sanno acōnumerare | nel peso dj tal polo e ttuai acalculare senpre il peso del con|tra peso colla potentia della resistētia che fa la cōfregatione del | polo e nō tai a travagliare conaltro [–*]	[XI] [*–] A second way which teaches one how to proceed. All the weights supported by the axle are to be added to the weight of the axle itself. And you always have to calculate the weight of the counter-weight with the power of the resistance which the friction of the axle makes. And you have not got to trouble yourself with anything else. [–*]
[Fig. 15] 2ª [п]	[Fig. 15] Second [figure] [п]
[Fig. 16] 3ª [п]	[Fig. 16] Third [figure] [п]
[Fig. 17] 4ª [п]	[Fig. 17] Fourth [figure] [п]

4 Leonardo's Heart

Admirable instrument, invented by the Supreme Master[1]

Leonardo's heart studies from the years about 1513 to 1514 represent the pinnacle of his anatomical endeavours. They integrate structural form and dynamic function in a way that called upon his extensive experience and knowledge of hydrology, engineering, mathematics, and architectural design. From the evidence of the existing notes, he entered this period of study of the heart with a Galenic perspective of the movement of the blood. It is clear from the rhetorical nature of his commentary, however, that the only way that he could satisfy his desire to truly and accurately understand the detailed function of this complex organ was by using dynamic laws, and although there is no extant record of it, acknowledging the importance of mathematical proof wherever possible. This inquisitive and original approach inevitably brought him into territory that would conflict with Galen's concepts of the ebb and flow of the blood. Whilst Leonardo's extant notes fall short of a direct challenge to Galen's postulates, there are sections that clearly suggest different explanations. A good example of this is his conclusion that the air passages are not directly connected to the blood vessels in the lung, as had been suggested by Galen, and indeed by Erasistratus (304–250 B.C.E.) before him.[2] Galen's hypothesis explained how the vital spirits contained within the inhaled "pneuma" or air, could enter the heart. Leonardo carried out an experiment in which he inflated a pair of excised lungs and then tied off the principal airway (the trachea). In doing so, he found that there was no leakage of air from the lungs, thereby disproving Galen's hypothesis.[3] As far as we know, Leonardo did not offer an alternative solution to the transference of life-giving forces.

The dynamic nature of heart action and the ever-changing demands on its output and power make it an organ that can be understood only by integrating form with function at every level. This is the approach that Leonardo used in his descriptions of cardiac function. Even in the current era of advanced molecular biology, mathematics, and engineering, we struggle to fully decipher the complexities of the normal and diseased heart. Indeed, in the past 30 years the worldwide expenditure on cardiac research has been vast, and it continues to grow as the answer to each question throws up many new lines of enquiry.

Modern cardiologic investigation requires highly sophisticated techniques of dynamic imaging and measurement, all underpinned with sound experimental modelling and mathematical proof—a process of information gathering in the mould of Leonardo's ground-breaking studies. I hope to help the reader to appreciate the profound nature of some of Leonardo's observations and deductions by explaining the current understanding of some aspects of cardiac function and by using modern imaging technology wherever possible to illustrate some of his novel observations.

Whilst trying to elucidate the nature of his genius in this highly specialised subject, I have found that his use of rhetorical reasoning and purity of observation encourages an ability to see old problems in new ways. Indeed, concepts that he explored in depth, such as the nature of the flow of blood through the heart (as exemplified by his study of the vortices in the aortic root), have been largely neglected until the present day. Only now, with the advent of accessible technology such as advanced magnetic resonance imaging (MRI), are enquiries such as these being reinvigorated.

In the Quattrocento, scientific enquiry, as we understand it, hardly existed, but some of Leonardo's methods were of ground-breaking scientific nature. They arose from his unusually questioning mind, allied to a use of classical disciplines. In particular, he believed in the power of the forces of nature that caused everything to be in the form that it is for reasons that, once understood, would lead the enquirer to their full understanding. His imagination resulted in much

[1] RL 19029 recto [II].

[2] Erasistratus was a Greek philosopher and anatomist, who along with Herophilus founded a school of anatomy in Alexandria. He was the first to postulate a capillary link between the arteries and the veins, the basis of Harvey's description of the circulation of the blood. He also correctly began to describe the heart as a muscle pump and not the seat of the emotions. He also began to describe the valves of the heart and the normal rhythms and arrhythmias of the heart.

[3] RL 19071 recto.

F.C. Wells, *The Heart of Leonardo*,
DOI 10.1007/978-1-4471-4531-8_4,

cross-fertilisation of ideas. One can clearly see his experience in one discipline informing his understanding of others. This is particularly obvious in his studies of the valves of the heart, which utilised his extensive knowledge as a hydrodynamic engineer and his fascination with geometry, amongst other things.

The structure of the heart can be understood only by investigating the way in which it works. In the case of such an organ, the investigator must rapidly move on from simple observation and recording of the static anatomy into the dynamic function. This requirement was clearly appreciated by Leonardo. Although the cardiac notes are rich in images, they reveal a lot of thinking on paper through the use of many words. Leonardo stressed the dominance of the visual image for demonstrating structure, but he also accepted the need for words to satisfactorily explain function (physiology). Hence the density of text found on the pages of notes that address the heart.

In the time of Leonardo, two of the great challenges of natural philosophy (biological science today) were attempts to understand the way in which a new life is formed and then to explain its subsequent sustenance and growth. Any explanation of the growth of a newly formed life and its healthy existence until death requires an understanding of the access of all of the body's tissues to nutrition and the "vital spirits", inevitably leading to the need to understand how these vital forces are captured and distributed—and hence to the function of the heart, lungs, and intestines. Perhaps it should come as no surprise that the enquiring mind of Leonardo should engage in these challenges towards the end of his academic life. By that time, he had engaged in all of the necessary academic disciplines and he was uniquely placed to make new and exciting discoveries. Perhaps he had also accomplished all that he wanted to do in understanding the musculoskeletal aspects of the human body.

As we shall see, Leonardo had made considerable strides in the knowledge of the structure and the function of some of the subsystems of the heart; through a failure to publish, these were not turned into a significant and recognised contribution. Several of his ideas continue to sit well alongside our modern understanding of cardiac physiology, and some presage today's cutting-edge functional hypotheses. In judging his work, and in particular in judging its perceived deficits, it must be remembered that his pages on the heart represent a tiny fraction of his manuscripts that are still in existence, and their fragmentary nature suggests significant losses over the intervening years. The fragmentary nature of the extant notes makes it difficult to find an order to his work. Some parts have a degree of cohesion, perhaps following his preparation for publication. Some subjects are more finished than others, and some appear under headings; for example, RL19078 verso has at its head "Geography of the heart" and RL19074 recto is headed "Of the valves of the heart".

In an attempt to bring some kind of order to the work whilst respecting Leonardo's approach, I have arranged the sections in the following way: First, I have included a general description of the working heart, to enable readers who lack an anatomical or medical background to gain some insight into the specialist knowledge that might be assumed. The next section gives some general comments on the muscular nature of the heart, followed by the external appearance of the heart. We will then journey through the interstices of the heart, grouping together the internal subsystems following Leonardo's descriptions. Finally, I will attempt to put all of this work into the context of the movement of the blood as realised by Leonardo, and will conclude with a review of some of his references to the diseased heart.

The Normal Working Heart

The heart is a powerful example of the nature of evolutionary development. The progress of life forms from single-celled structures to multiple-organ individuals demands the emergence of a system to supply nutrition and remove effluent from all parts of the body. For the first days of human embryonic development, whilst embedded in the wall of the mother's uterus, oxygen and nutrients can reach all of the cells by simple diffusion from areas of high concentration to those of low concentration (along concentration gradients). Soon, however, the embryo reaches a size that requires a formal circulation, as the distances across which diffusion would be necessary become too great. Capillary networks form, through which the vital nutrients can reach the rapidly dividing cells. From the fourth to the sixth week of development, a group of cells form into a tube, which begins to differentiate into the chambers of the heart, beginning with the atria.[4] These cells possess the special property of contractility, and from a very early stage, cells can be seen to contract with a spiral wave beginning at the atrial end, in an action that resembles the wringing out of a towel. Interestingly, before the arteries and veins link together through the capillary networks, the first flow of blood within this tubular system is Galenic, with an ebb and flow pattern. It is only as the foetus grows and the capillary connections form that the formal circulation begins. As the tube of muscle cells grows, it twists and rotates in an extremely complex way, eventually establishing the relationships that we recognise in the adult heart.

The fully formed heart consists of four chambers arranged in two pairs. There are two weakly contractile receiving chambers called atria (from the Latin *atrium*, which means a

[4]This known fact in modern embryology makes sense of Leonardo's commentary that the heart is not the beginning of life: "The heart of itself is not the beginning of life but is a vessel made of dense muscle…." (RL 19050 verso).

hallway) and two powerful pumping chambers called ventricles. They are connected in series and they communicate through one-way, non-return valves. Between the right atrium and right ventricle is the tricuspid valve, and between the left atrium and left ventricle is the mitral valve. These valves allow blood to pass from the atria into the ventricles but oppose return flow into the atria. Similarly, two unidirectional valves control the outflow from the ventricles. The pulmonary valve prevents reflux of blood in the outflow from the right ventricle to the lungs, and the aortic valve performs the same function in the outflow from the left ventricle to the rest of the body.

The pairs of chambers are described as being either right or left. The concept of sidedness of the heart relates to their relationship to the lungs and not to their mature position within the chest. In the natural state within the body, the right atrium and right ventricle lie as much on the front of the heart as they do on the right side. The left atrium and ventricle are more towards the back of the chest than they are to the left side.

Blood leaves the right side of the heart to enter the lungs, from which it returns to the left-sided chambers, before departing for the rest of the body. Blood returning via the great veins from the many organs of the body to the right side of the heart is low in oxygen content and high in carbon dioxide. The reverse is true of the blood that returns to the left side of the heart from the lungs. The passage of blood through the lungs removes carbon dioxide and enriches it with oxygen.

A general rule states that arteries always carry blood away from the heart, and veins towards the heart. With the exception of the pulmonary artery, all arteries contain oxygenated blood. The pulmonary artery, which leaves the heart to carry blood to the lungs, carries the deoxygenated blood that has returned from the tissues of the body. Similarly, all veins other than the pulmonary vein contain blood rich in carbon dioxide.

Arteries branch like the branches of a tree, and veins receive tributaries, as does a major river. As we shall see, the cross-sectional area of the parent vessel is equal to the sum of the cross-sectional areas of its branches. Leonardo defined this relationship as a natural law.

All of the structures of the heart exist to provide the impetus, direction, and rhythmic flow of the blood. Therefore I will describe the structures of the heart in the order that the blood flowing through it will encounter them. Blood returns from the head and upper body via converging tributaries of veins that culminate in one large vein, the superior vena cava (SVC). This expands to form the upper part of the right atrium. Blood returning from the lower body via a myriad of converging venous tributaries enters another large venous channel called the inferior vena cava (IVC). The IVC then enters the inferior part of the right atrium. This chamber is in

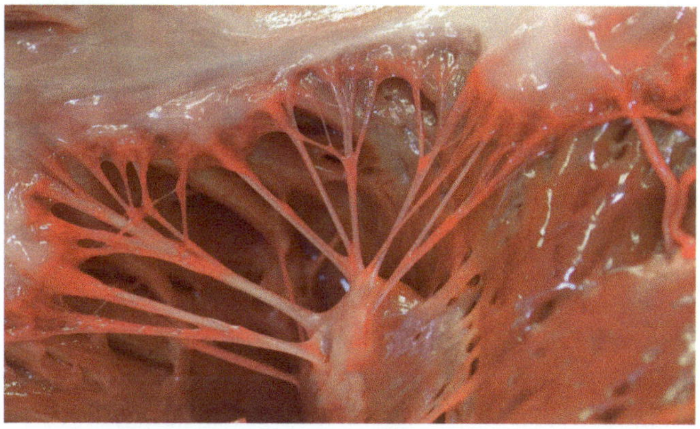

Fig. 4.1 Photo of a tricuspid valve. The chordae can be seen arising from the papillary muscles, reaching up to the leaflet of the valve. This arrangement in the mitral valve is very similar

two parts: The first is formed by the confluence of the superior and inferior vena cavae; it has a smooth internal lining and thus is called the sinus venosus part. The rest of the right atrium is formed as part of the heart tube and has a lining that is elevated in a mass of interwoven bars of muscle called trabeculae. This is where the true heart muscle begins. The junction between these two areas forms a ridge called the crista terminalis (terminal crest).

The blood then flows across the first of the atrioventricular valves within the heart, the tricuspid valve. This valve has a complicated relationship with the right ventricular muscle below. Fine, tendinous cords (which look rather like the strings on a parachute connecting the sheet to the person suspended below) connect the tricuspid valve to outcrops of muscle in the ventricle, called papillary muscles (Fig. 4.1). This valve, as with the other three valves, opens and closes under the force and direction of the flowing blood in response to the cyclical contraction and relaxation of the heart muscle. Other influences, such as vortex formation in the flowing blood, play an important part in the closure of the valves in such a way that no backward leakage of blood can occur. As we shall see, this idea was one of Leonardo's most accomplished pieces of work, not appreciated until the twentieth century.

When the ventricle is nearly full, the right atrium contracts to send the final portion of blood into the ventricle, whose cavity is then stretched open even further. Heart muscle has an elastic component, and this final stretch imparts extra energy to the ventricle. This elasticity was also recognised by Leonardo in relationship to the normal function of the aortic valve and root. The elastic recoil then powers the beginning of the ventricular contraction phase, which drives some of the blood up against the tricuspid valve, forcing it to close. The remainder of the blood forces open the pulmonary valve and flows onwards to the lungs. Whilst the ventricle is contracting and thereby emptying, blood continues to return to the atrium, filling it so that it is ready for the next phase of ventricular filling.

The main pulmonary artery divides into two, one part for each lung. These then divide again and again as the branches pass into the depths of the lung tissue. At their terminations, they become very small and are termed arterioles. At their ends, they merge into the capillaries. The capillaries are tubes that are only a single cell thick, forming a very weak barrier across which gas and nutrition can easily diffuse into the surrounding tissues.

The red cells of the blood pass through the fine capillary network of the lungs. They contain complex molecules called haemoglobin, whose special property is to loosely bind molecules of oxygen to themselves. Thus oxygen is carried onward to the tissues of the rest of the body. Oxygen is the active ingredient of the mystical vital spirit "pneuma". The transfer of oxygen, carbon dioxide, and the nutritious elements of the blood is passive. That is, the transfer occurs with the expenditure of minimal energy, along concentration gradients. In other words, the blood entering the lungs has given up a lot of its oxygen and is entering an environment where the oxygen concentration is high. Oxygen molecules therefore move from the lung air spaces into the blood. The reverse is true for carbon dioxide.

At the end of the capillary networks, the tiny vessels coalesce to form small veins (venules), which themselves coalesce into the larger veins, and ultimately these all merge into the pulmonary veins, which leave the lungs and enter the left atrium of the heart. There are two pairs of pulmonary veins (one pair from each lung), which enter either side of the left atrium. From the left atrium, blood then flows across the second atrioventricular valve, the mitral valve, located at the mouth of the left ventricle, the powerhouse of the systemic circulation. When this ventricle contracts, the blood is ejected under high pressure out through the aortic valve and onwards throughout the body via the aorta. The contraction of the left ventricle simultaneously causes the mitral valve to close.

The closure mechanism of the aortic and pulmonary valves is sophisticated and involves complex hydrodynamics that were first described by Leonardo. We will explore this discovery in detail later.

The contraction of the chambers of the heart is instigated by nerve impulses, which arise first in a specialised group of heart cells collectively called the sinoatrial node. Found at the junction between the superior vena cava and the right atrium, this is a concentration of electrically active cells that fire together at a rate that is controlled by the demands of the level of exercise or stress of the individual; increased demands for oxygen and energy cause chemicals in the blood to change, making the cells fire more rapidly. The nerve impulses then flow throughout the atrial muscle, causing a wave of contraction. They then coalesce at another clump of specialised heart muscle cells called the atrioventricular node. This lies in the crest of the septum (the dividing wall) between the two ventricles. The electrical wave then spreads throughout the two ventricles, causing them to contract. The two atria and the two ventricles contract together and in sequence, atrial contraction preceding contraction of the ventricles. As the atria contract at the end of the ventricular filling phase, the ventricles are relaxed. When the ventricular filling phase is ended, the ventricles contract, and at this time the atria relax to refill. The phase of ventricular contraction in the cardiac cycle is referred to as systole and the phase of relaxation as diastole. As we shall see, Leonardo commented upon the neuromuscular activity of the heart.

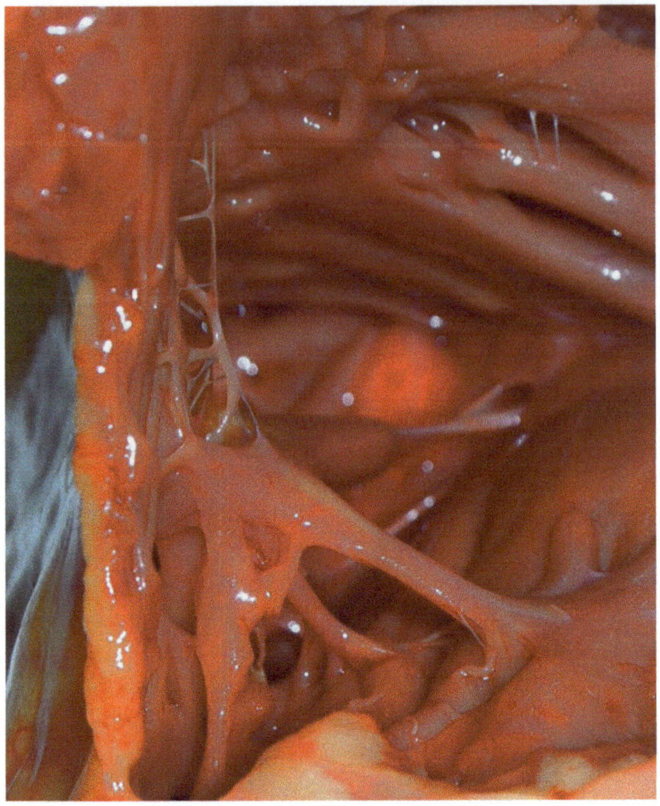

Fig. 4.2 Photograph of the moderator band in a human heart. Although quite well developed in this example, it is often very indistinct

The internal structures of the heart are complex both in their overall form and in their content. The right ventricle operates under lower pressures than the left, and consequently it has a thinner muscular wall. Its external wall is relatively unsupported, and as a result in some animal species it has a muscular bar stretching from the internal (septal) wall to the external wall. This is the structure first recorded and named by Leonardo as the "moderator band." Its presence prevents over-distension of the thin ventricular free wall (Fig. 4.2). The internal aspect of the ventricular walls is characterised by the protrusion of muscular bands, which form many nooks and crannies between them. These protrusions coalesce into large muscular projections called papillary muscles. From these arise the tendinous cords that support the ventricular side of the tricuspid and mitral valves. These

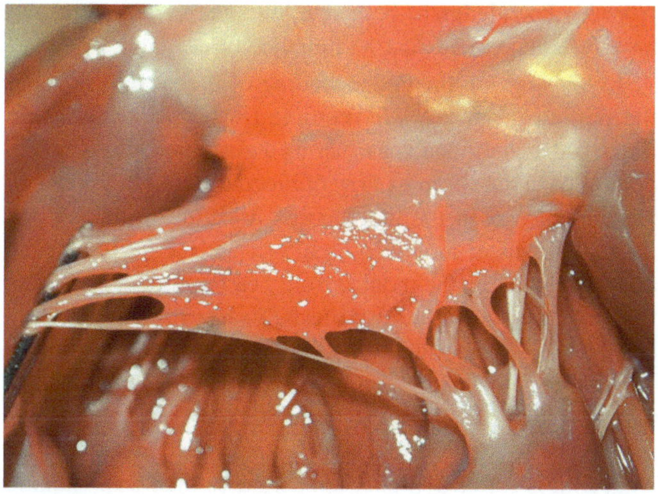

Fig. 4.3 Photograph of the underside (ventricular surface) of the mitral valve. The leaflet is being drawn up to show the arcades of chordae tendinae

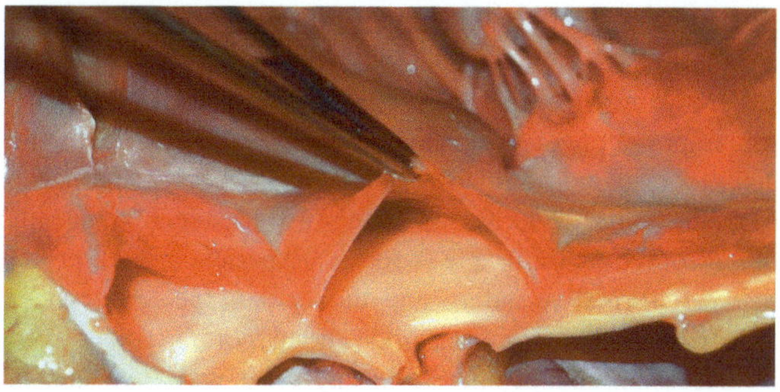

Fig. 4.4 Photograph of human aortic leaflets

cords are arranged in three layers, referred to as primary, secondary, and tertiary cords. The primary cords insert into the leading edges of the leaflets and act as guy ropes bringing the leaflet edges into line as they close in systole. The secondary cords, which are sturdier, reach up to the underside of the valve, where they fan out to spread like the fan vaulting of a church roof (Fig. 4.3). This arrangement allows allow the load placed upon them at maximum pressure of the ventricular contraction to be equally dissipated. The third group of tendinous cords, the tertiary cords, extend to the base of the valve just next to its origin from the ventricular wall. These cords are under tension throughout the cardiac cycle and act like "tie-rods" in an engineering sense, supporting the ventricular wall. These architectural relationships are important for the normal functioning of the ventricles. If it is not respected and preserved during valve replacement surgery, the heart will progressively distort from its normal conical shape to a sphere. This altered geometry causes the heart to fail over a number of years. An example of this type of cord is drawn by Leonardo, but without commentary.

The two arterial valves, pulmonary and aortic, are made up of three semilunar leaflets (so called because of their resemblance to a half moon). These "lunules" (valve leaflets) must support each other at their closure line, as no tendinous cords are attached to them (Fig. 4.4). They close under a lower pressure than the tricuspid and mitral valves, which must close in the powerful contraction phase of the ventricles as they generate their maximal pressure. The left ventricle generates a much higher pressure than the right ventricle, and thus the mitral valve has to close under the highest pressure of all of the valves.

The heart has its own blood supply. This is delivered to the heart by two main coronary arteries, the right and the left. They are the first branches of the aorta (the large artery that leaves the left side of the heart to supply blood to the whole body other than the lungs). The coronary arteries arise from hemispherical expansions in the root of the aorta, (the sinuses of Valsalva), which begin immediately above the leaflets of the aortic valve.[5] The coronary arteries give their names to the particular aortic sinuses from which they arise. Hence, there is a right, a left, and a noncoronary sinus. The latter (as its name implies) is the sinus of the aorta, which does not give rise to a coronary artery. As the blood exiting the ventricle enters the sinuses, vortices form within them as the blood is channelled back upon itself towards the upper side of the base of the leaflets. This motion begins to make the semilunar leaflets drift towards each other, initiating the first phase of valve closure. It also drives blood into the coronary arteries, enhancing perfusion of the heart.

The coronary arteries divide repeatedly and delve down into the heart muscle, providing its nourishment. Maximum flow in these vessels occurs during the relaxation phase of the cardiac cycle. Blood returns from the heart muscle via venous tributaries to enter the large vein, the coronary sinus, which ultimately flows back into the right atrium of the heart.

All of the structures of the heart are forged by the dynamic function that they are called upon to carry out. Therefore a teleological approach to their understanding is a laudable way to proceed. Thus Leonardo's method of using the observed effect to define its cause yielded fascinating results. In his own words: "O marvellous necessity, thou with supreme reason constrainest all effects to be the direct result of their causes, and by a supreme and irrevocable law every natural action obeys thee by the shortest possible process".[6]

[5]Antonii Mariae Valsalva was an Italian anatomist who was born in Imola, Romanga, in 1666 and died in Bologna in 1723. He described the origin of the coronary arteries from the root of the aorta in the sinuses that have since borne his name. The description was published by Giovanni Battista Morgagni in a collection of his works entitled *De vita et scriptis Antonii Mariae Valsalve commentariolum* (Venice 1740). The entry is on one page in one paragraph; there is a collection of small engravings and no discussion of physiological function.

[6]Codex Atlanticus 345 verso.

Leonardo's Methodology

Before we consider in detail Leonardo's research of the heart, it is important for us to recognise the novel methods that he devised to enable him to reach his conclusions. Leonardo took the natural world and man's experience of it as his touchstone for truth. He wrote, "Experience, interpreter between formative nature and the human species, teaches that, that which this nature works among mortals, constrained by necessity, cannot operate in any other way than that in which reason, which is its rudder, teaches it to work".[7]

His method consisted of extracting "experiences" by close scrutiny of the effects of phenomena from the greater world around him and distilling from them the questions of causation that he wanted to answer. These questions were then turned into the first examples of "experimentation" within a controlled environment. This personal experiential philosophy he described thus:

> First I shall test by experiment before I proceed farther, because my intention is to consult experience first and then with reasoning show why such experience is bound to operate in such a way. And this is the true rule by which those who analyse the effects of nature must proceed: and although nature begins with the cause and ends with the experience, we must follow the opposite course, namely begin with the experience, and by means of it investigate the cause.[8]

Seeing bodily organs in their functioning state, as if in life, was a sine qua non for Leonardo. Nature made everything in the form that it did for a specific purpose ("In nature there is no effect without cause...."[9]), so it must be examined in a state as close to its natural form as possible. Leonardo abhorred vivisection and therefore denied himself the easier route of observing the heart in its normal working state. Instead, he developed methods of demonstrating the chambers of the heart and the ventricles of the brain in their normal distended state by injecting them with wax. In the case of the brain, it was simple to remove the outer brain to reveal the solidified wax, which had then taken the form of the internal ventricles.[10] Comparison of Leonardo's images with those of an actual cast reveal the accuracy of his work (Figs. 4.5 and 4.6). Undoubtedly these are methods borrowed from his knowledge of both sculpture and engineering, and have reference to the lost wax method of casting.

In his work on the understanding of the aortic valve and the outflow from the left ventricle of the heart, Leonardo described a method of constructing a glass model in which to demonstrate the complex flow of blood through this channel. Here he also used the molten wax injection technique. He wrote, "A plaster mould to be blown with thin glass inside and then break it (the mould) from head to foot at 'a n'. But first pour wax into this valve of a bull's heart so that you may see the true shape of this valve" (Fig. 4.7).

A similar technique is used today by some valve manufacturing companies when modelling designs for replacement valves. Modern technology uses silicone rather than wax, as it is stable at higher temperatures than wax, making it easier to work with (Fig. 4.8).

Some authors (e.g., O'Malley and Saunders) have expressed doubt over whether Leonardo actually made and used this proposed model.[11] When they wrote, in 1952, no one had properly investigated the problem of how the aortic and pulmonary arteries actually worked. There was no published description of the true mechanism in the medical or scientific literature. Therefore those authors could not have known of the accuracy of Leonardo's work nor of the complexity that was entailed in its demonstration. Indeed, the first scientific examination of the problem that is reported in the literature is a paper by B. J. Bellhouse and F. H. Bellhouse from the Department of Engineering Science in Oxford University, which was published as a communication in *Nature* in 1968.[12] The only reference it contains is to Kenneth Keele's publication[13] describing Leonardo's original work, which of course was never published by its original author.

The work was further highlighted in an article by Professors Gharib, Kemp, and others, in which they further amplify the accuracy of the original observations[14] (Fig. 4.9). Professor Kemp retains a healthy scepticism with regard to whether Leonardo really proved his hypothesis with the use of the model.[15] He asks whether the model would have been sophisticated enough, even with the seeds dispersed within it, to visually demonstrate the detail that Leonardo reveals in his multiple drawings. I suspect that Leonardo's knowledge

[7]C.A. 86 recto.

[8]Paris Manuscript E. 55 recto.

[9]C.A. 147 verso.

[10]"Cast in wax through the hole 'n', at the bottom of the base of the cranium before the cranium is sawn through", and later on the same page, "Make 2 vents in the horns of the great ventricles and inject melted wax with a syringe, making a hole in the ventricle of the memoria, [IVth ventricle], and through such a hole fill the three ventricles of the brain. Then when the wax has set take away the brain [substance], and you will see the shape of the ventricles perfectly. But first put narrow tubes into the vents so that the air which is in these ventricles can escape and make room for the wax which enters into the ventricles."

[11]Charles D. O'Malley and J. B. de C.M. Saunders, *Leonardo da Vinci on the Human Body.* (New York, Henry Schuman, 1952; reprint, New York, Dover, 1983), 264.

[12]B.J. Bellhouse and F.H. Bellhouse, Mechanism of closure of the aortic valve. *Nature* 217(5123):66–7. 1968.

[13]Kenneth Keele, *Leonardo da Vinci on Movement of the Heart and Blood.* (London, Harvey and Blythe, 1952), 81.

[14]M. Gharib, D. Kremers, M.M. Koochesfahani, M. Kemp. Leonardo's vision of flow visualisation. *Experiments in Fluids.* 33 (2002): 219–23.

[15]Martin Kemp, *Leonardo da Vinci. Experience, Experiment, Design.* (London, V&A Publications, 2007), 80.

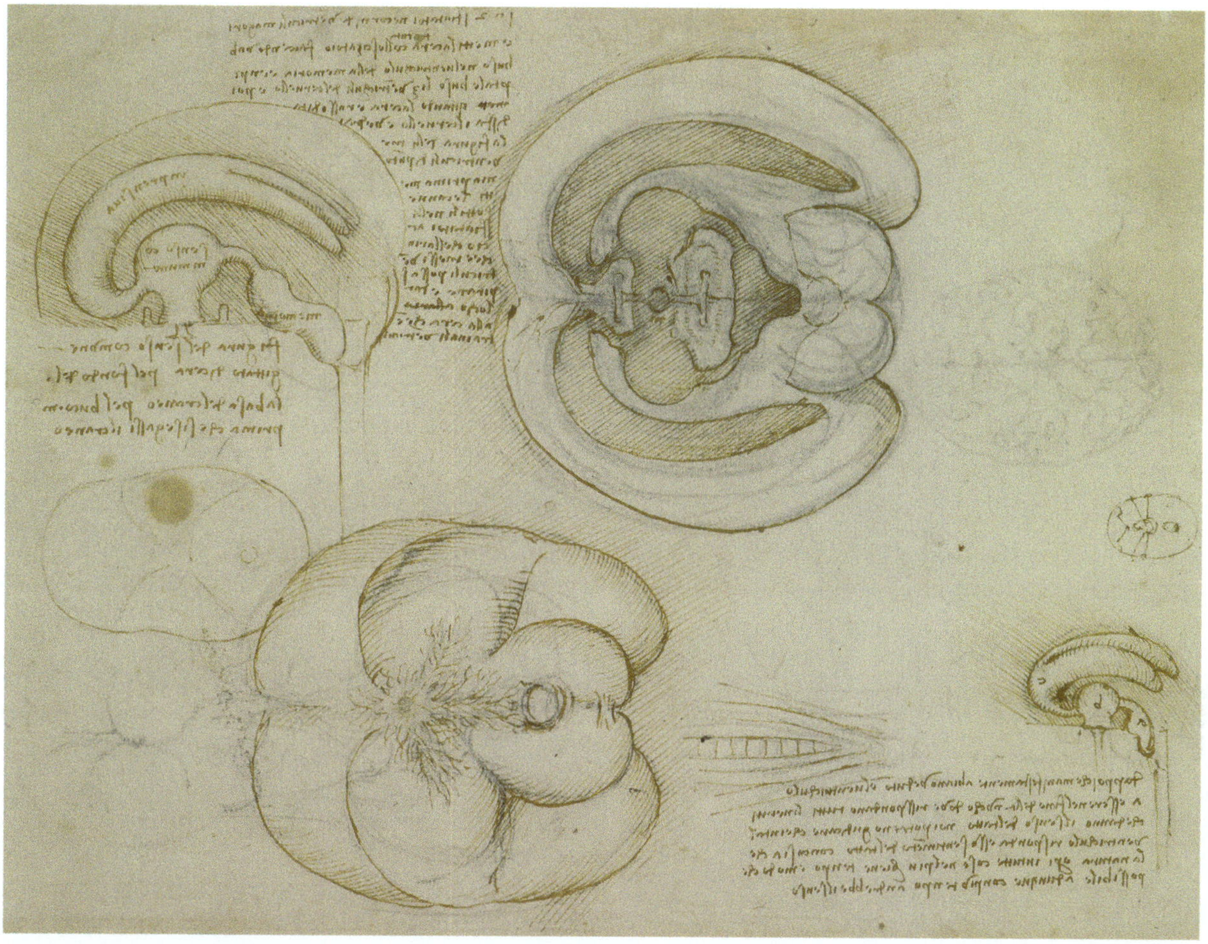

Fig. 4.5 Drawing to show the cerebral ventricles. RL 19127 recto, The brain. Leonardo da Vinci. c.1508-9. Pen and ink over black chalk (Lent by Her Majesty The Queen. Royal Collection © Her Majesty Queen Elizabeth II)

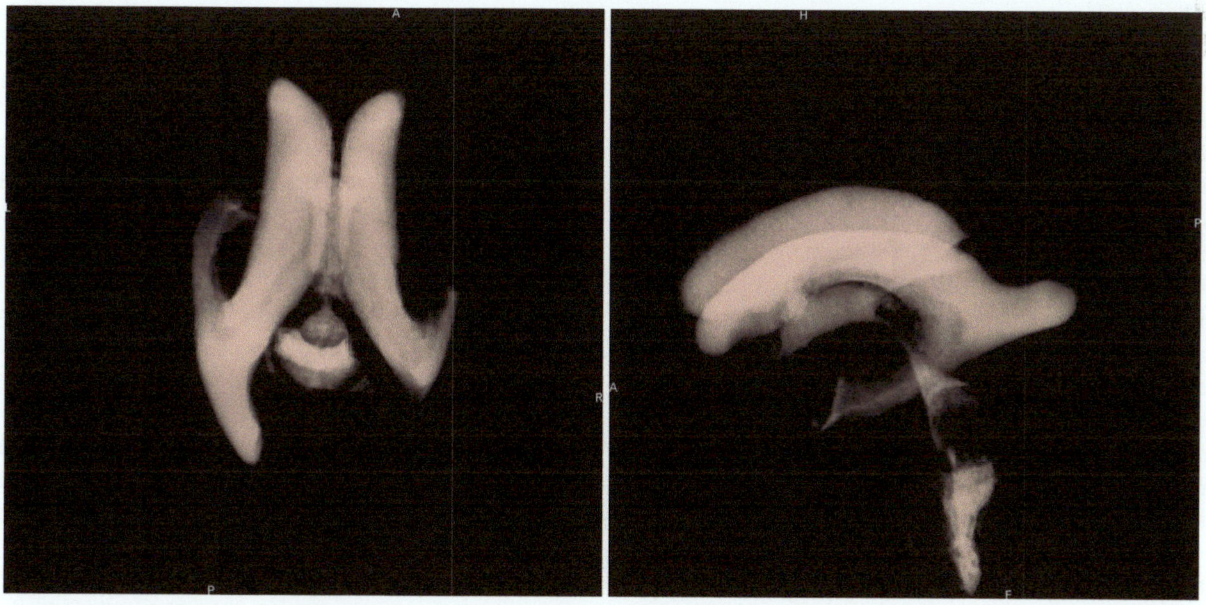

Fig. 4.6 (**a**, **b**) Modern images of the cerebral ventricles. Casts of the cerebral ventricles can be made by injecting hot liquid wax or silicone into them and then removing the brain tissue once the wax or silicone has set, revealing a cast of the ventricles. These can then be photographed and stored for teaching purposes. These days it is possible to render the shapes with CT or MR scanning and subtracting the data from around them, revealing their true shape, which can then be digitally imaged alone

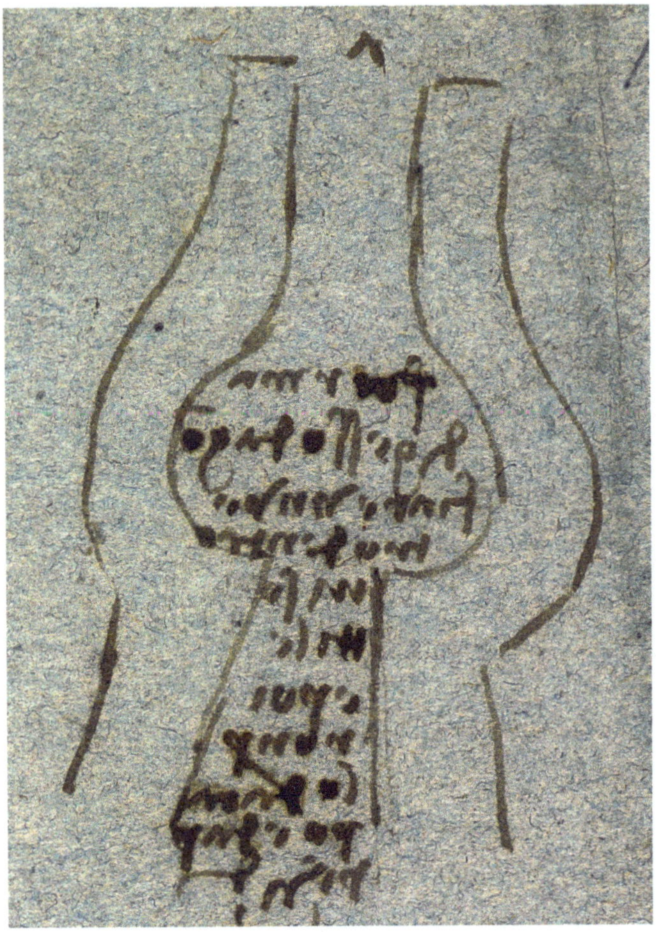

Fig. 4.7 The proposed plaster mould. Detail from RL 19082 recto, The aortic valve. Leonardo da Vinci. c. 1512–13. Pen and ink on blue paper (Lent by Her Majesty The Queen. Royal Collection © Her Majesty Queen Elizabeth II)

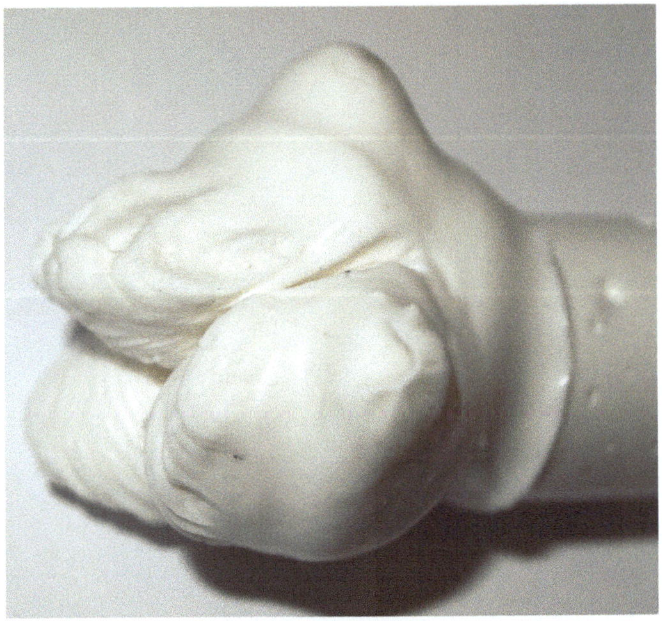

Fig. 4.8 A silicone mould of the aortic root of a pig. This gives a negative cast of the aortic root (in other words, the form of the blood within it). From this type of mould, Leonardo could deduce the tensioned shape of the structure (Photograph courtesy of Medtronic)

of hydrodynamics and his powerful deductive reasoning would have made him convinced in his own mind that his hypothesis was correct. Even the most subtle confirmation of his prior deductions in the model would have been enough to affirm his thoughts, although I suspect that having had the idea, he would have wanted to see it through. After all, the model was not complex to make, and the hard work had been done in the development of the idea. Although he deduced the expected findings from his other hydrodynamic work, his drawings of his final solution are so compelling as to suggest that he witnessed it. The reality is so complex that it was only been proven beyond doubt in the second half of the twentieth century and was only demonstrated in the live, working human heart as recently as the past few years.

In his descriptions of the morphology of the internal aspects of the auricles (atria) of the heart, Leonardo describes yet another method by which he was able to estimate the volume of the chamber: "Before you open the heart inflate the auricles of the heart beginning from the aorta. Then tie it off and examine its volume. Then do likewise to the right ventricle or right auricle. Thus you will see its shape and its function."

In trying to understand the mechanism of the aortic valve, Leonardo becomes a truly modern scientist. As with all good science, he identified the correct questions to ask from closely studying the shape of the distended aortic root of the bull's heart, and he then set about reconstructing the complex geometry and its relationship to the valve leaflets by designing the circuit in which to place the glass model of the aortic root. He went further by drawing, and probably making, his own form of a trileaflet valve with semilunar leaflets, accurately reproducing the natural valve complex (Fig. 4.10). This piece of work antedates modern synthetic valve design.

Thus, when Leonardo freed himself of the pre-existing concepts of his time and became immersed in concentrated analysis of his anatomical observations, he was at his most inventive. His resolution of problems remains apposite and accurate even through the perspective of modern knowledge. Indeed, his creativity in thinking through problems is at times quite breathtaking.

Accurate dissection manuals now guide the student of anatomy in the right direction, but of course in Leonardo's time there were no such manuals. He had to determine how to open the heart to optimise the view of the internal structures. In a geometric sense, the normal heart is a complex cone—that is, a cone with a twist. This twist distorts the internal architecture around the spiral, so that the number of cuts to be able to see the majority of the internal structures is limited. Leonardo appears to have worked out this problem

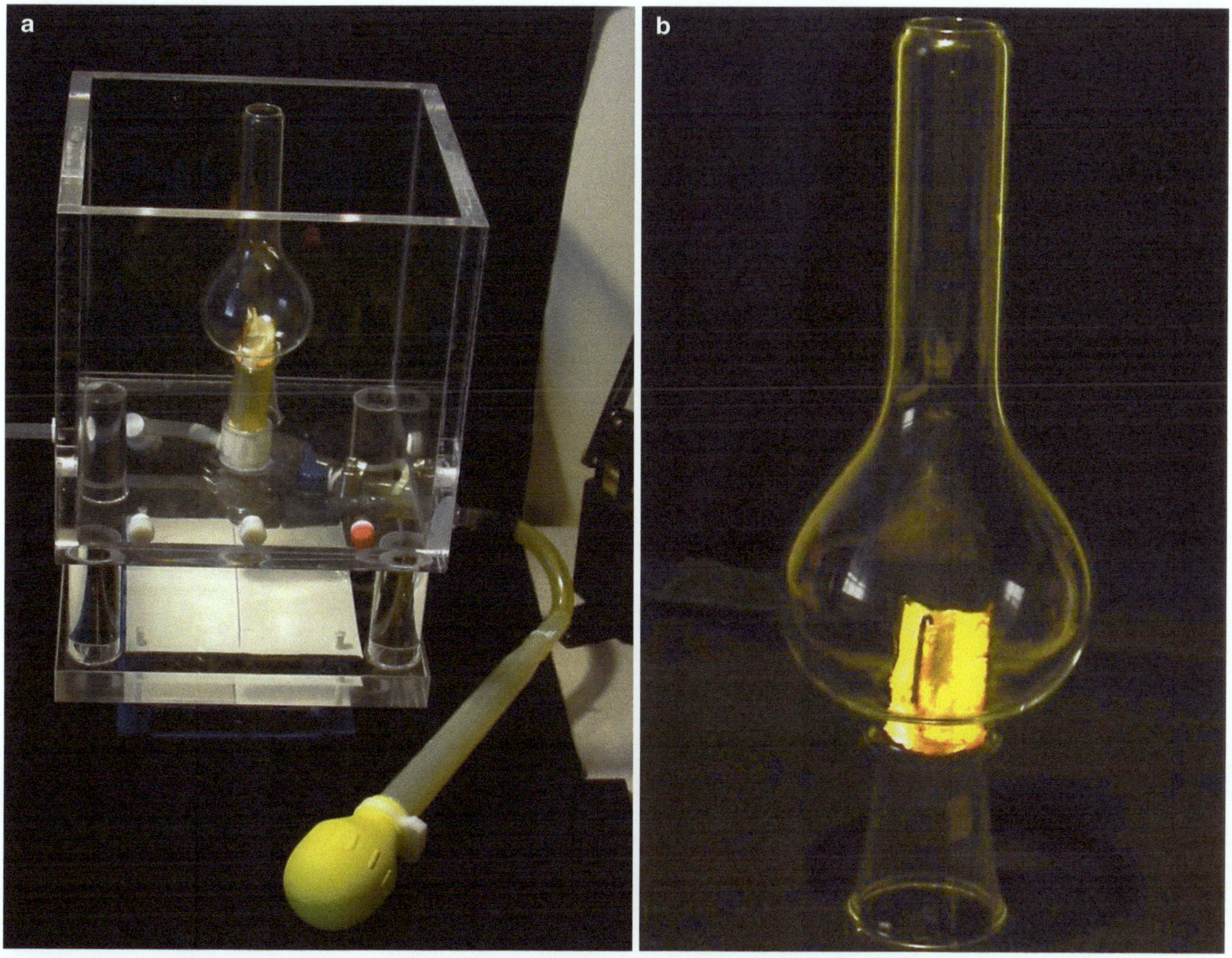

Fig. 4.9 (**a**, **b**) Gharib and Kemp's model of the aortic flow experiment (Photographs courtesy of Professor Morteza Gharib, Professor of Bioinspired Engineering, California Institute of Technology, Pasadena, California, USA)

and displays his incision lines in his drawings (Fig. 4.11). Contemporary cardiac pathologists inform me that Leonardo's chosen incisions are correct and similar to those used today[16] (Fig. 4.12).

Leonardo's task in trying to understand the heart was huge, but with great originality he advanced his knowledge through careful observation and experimentation. The great majority of Leonardo's cardiac research work is based on the bull's heart, although one drawing[17] appears to represent the interior of the right ventricle of a human heart (Fig. 4.13). The most obvious differences between the hearts of oxen and humans is that ox hearts are much bigger and more conical (Fig. 4.14). Also, the atria in the two species differ in their shape and in their relationship to the ventricles.

Although there are significant interspecies differences in cardiac anatomy, the principles involved in the function of both hearts remain the same. Both have the same number of chambers associated in the same way and separated by the same type of valves. Other than the presence of the moderator band in the ox, the internal components of the ventricular chambers are essentially the same, but they are arranged differently to accommodate the shape of the heart in a quadruped, where it is suspended from its base rather than sitting on the diaphragm, as in humans. Therefore it is unlikely that Leonardo's deductions about the function of the parts suffered significantly from this choice of material for his studies. Indeed, the exaggerated proportions of some of the anatomical features found in the ox heart probably inspired observations that allowed him to draw analogies from other disciplines such as hydrodynamic engineering.

[16]Personal communication from Dr. Susan Stewart and Dr. Martin Goddard, Papworth Hospital, Cambridge, England, 2011.

[17]RL 19119 verso is very suggestive of a human right ventricle, especially with the absence of an obvious moderator band (see below).

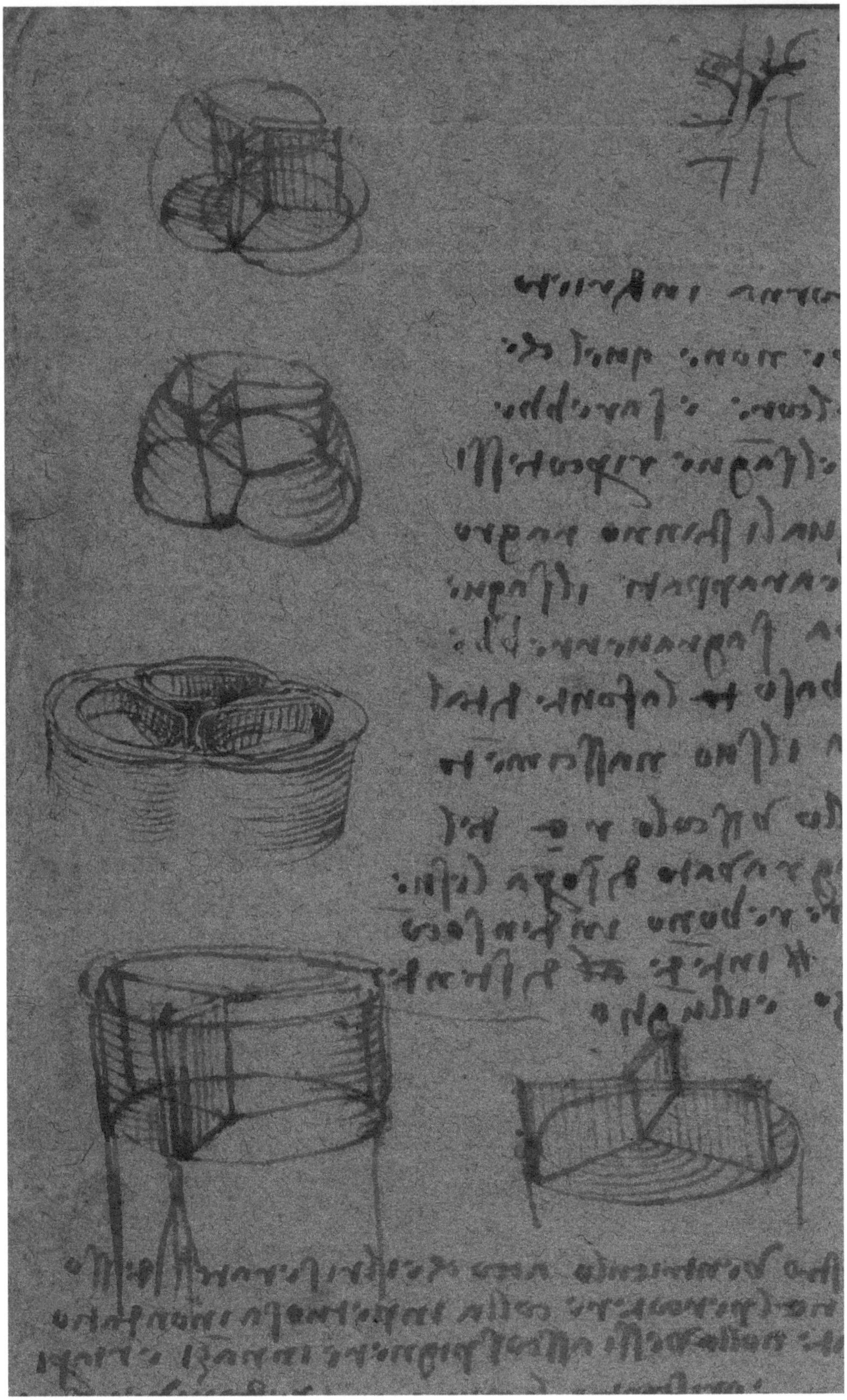

Fig. 4.10 Leonardo's drawings of the proposed design for the aortic/pulmonary valve. Detail from RL 19082 recto, The aortic valve. Leonardo da Vinci. c.1512–13. Pen and ink on blue paper (Lent by Her Majesty The Queen. Royal Collection © Her Majesty Queen Elizabeth II)

The choice of the ox heart for study was also probably stimulated by the ready availability of animal parts, unlike the regulations that surrounded the human body. Slaughterhouses, though perhaps repugnant to Leonardo, would have made access to specimens for dissection much easier.

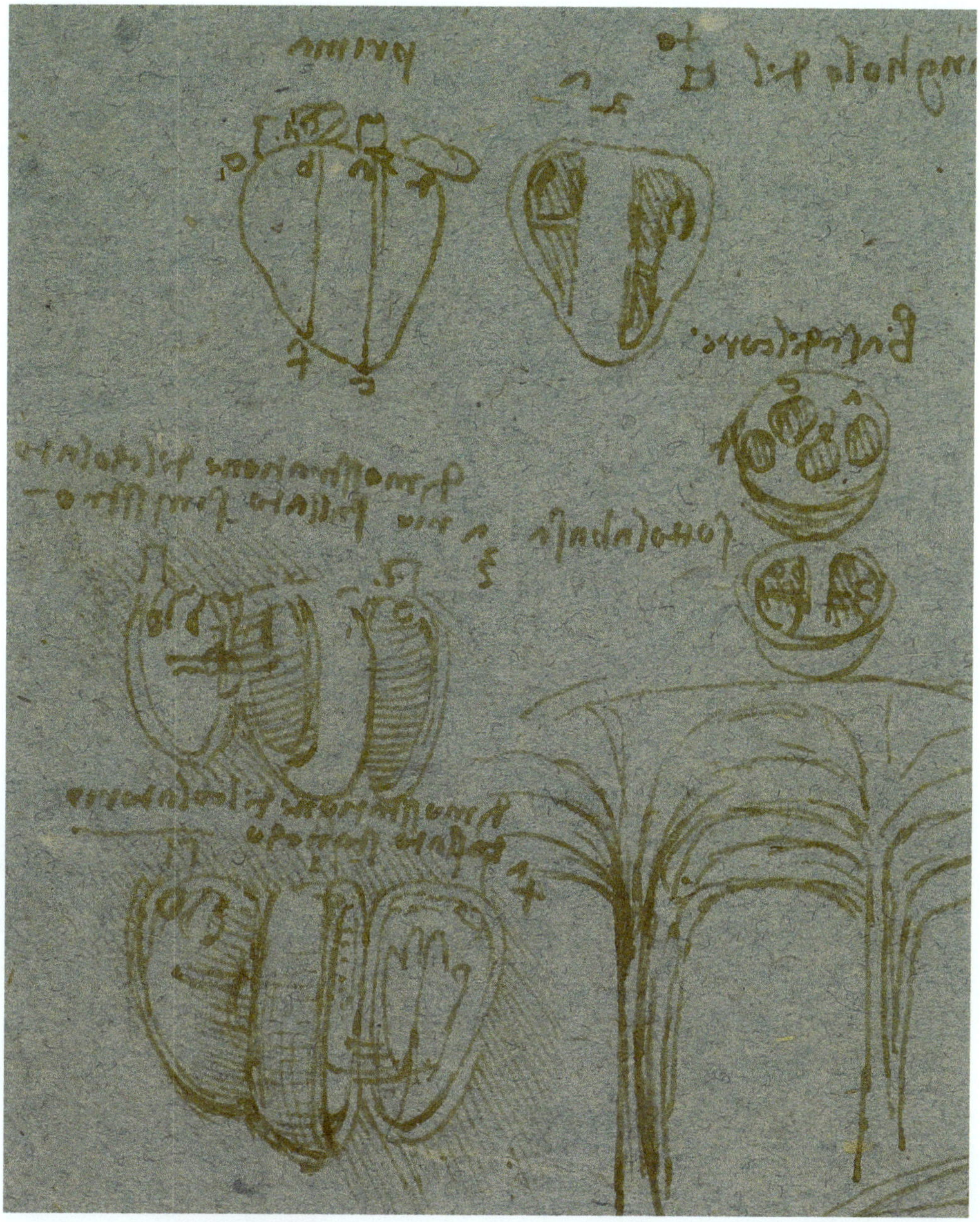

Fig. 4.11 Diagram of the incisions that Leonardo used in some of his dissection of the heart. Detail from RL 19080 recto. The left ventricle and mitral valve. Leonardo da Vinci. c.1512–13. Pen and ink on blue paper (Lent by Her Majesty The Queen. Royal Collection © Her Majesty Queen Elizabeth II)

The Centenarian

In Florence, during the winter of 1507–1508[18] Leonardo reported a dissection that he carried out on an old man, who shortly before he died told Leonardo that he was 100 years old. During the dissection of the heart, he discovered that the coronary arteries were diseased, and Leonardo documented the first known description of coronary artery disease. Regarding this incident, he wrote, "And this old man, a few hours before his death, told me that he was over a 100 years old and that he felt nothing wrong with his body other than weakness. And thus, while sitting on a bed in the Hospital Santa Maria Nuova in Florence, without any movement or sign of any mishap, he passed out of this life. And I made an anatomy of him to see the cause of so sweet a death."[19]

In this passage, Leonardo clearly states that the reason for the dissection was to discover the cause of death. Despite this man's very advanced age, the manner of his death clearly aroused Leonardo's curiosity. Therefore, it appears from his description that this was perhaps more of an autopsy than an

[18]Kenneth Clark. *Leonardo da Vinc,. Anatomical Drawings at Windsor Castle*. Vol. III, 2nd edition. (Oxford: Phaidon, 1969), RL 19027 verso, 13.

[19]RL 19027 verso.

anatomical dissection. Autopsy or post mortem examination to discover a cause of death was not unusual, even at that time. A physician contemporary with Leonardo, Antonio Benivieni, recorded a series of autopsies in a manuscript entitled *De abditis*.[20]

A recently published extensive review of the records of the hospitals in Renaissance Italy has shed considerable light on the practices of those institutions at that time.[21] That review reveals that although the hospital Santa Maria Nuovo was among the largest in Florence, there are no

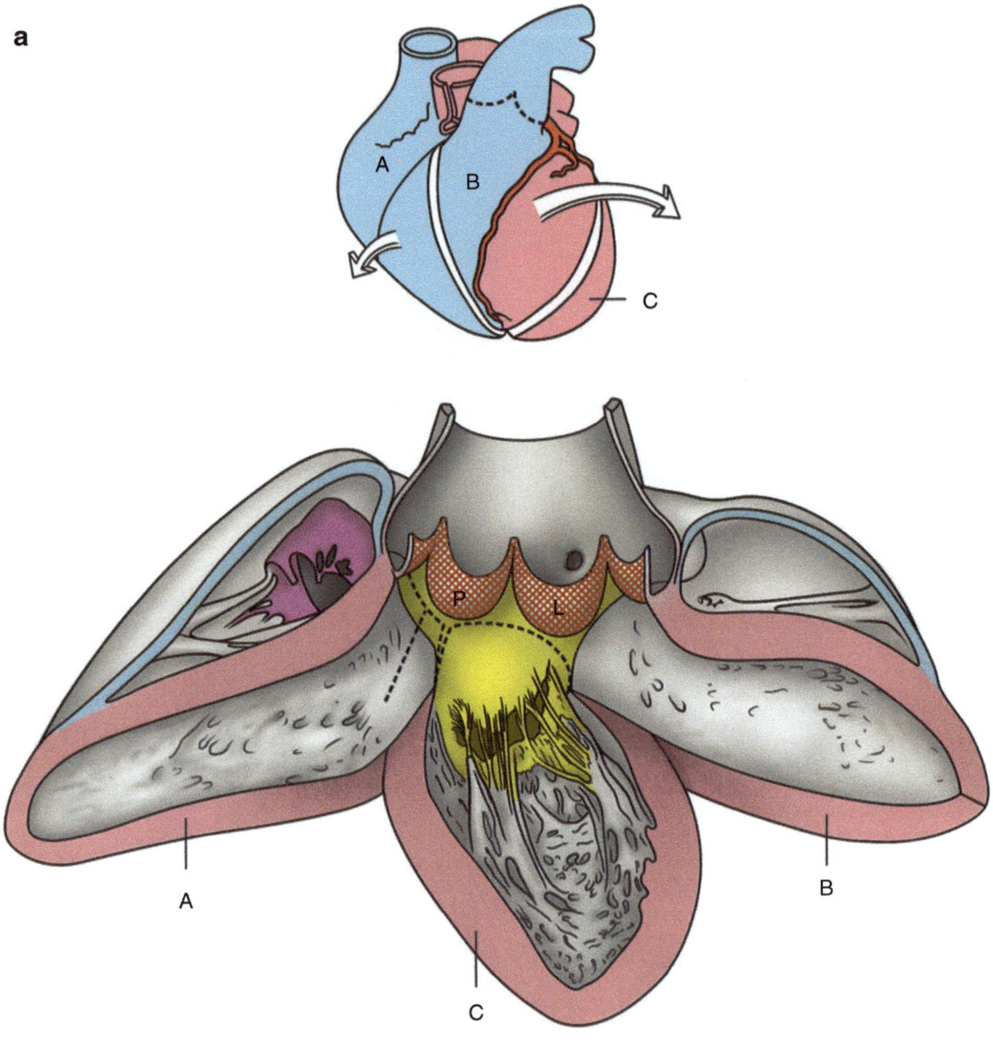

Fig. 4.12 (**a**) Illustration of modern incisions used to demonstrate the internal structures of the heart. The intact membrane may be seen to the best advantage by the use of the incisions demonstrated in the first drawing. The primary incision passes through the nadir of the right aortic annulus and is carried across the right ventricle between the segments *A* and *B*. These are distracted and two additional incisions are made to allow the interior of the specimen to be examined. The dotted line indicates the attachment of the chambers on the opposite side of the membrane. In the second drawing, this incision passes through the left anterior fibrous trigone and across the infundibulum of the right ventricle to the apex, whence it extends to the ostium of the left ventricle, separating the septal and posterior walls en route. The all-important ostium of the left ventricle is seen after the posterior papillary muscle is divided and then elevated. The left aortic annulus is separated 3 mm from the left ventricular ostium. This represents an extreme degree of the tendency of human hearts to demonstrate a membranous type of attachment of the aortic annuli to the ostium. The left, left anterior, and intervalvular fibrous trigones are in continuity. A heart with this type of configuration would be a prime candidate for the development of annular subvalvular left ventricular aneurysms, which are seen most often in the Bantu (From W.A. McAlpine. *Heart and Coronary Arteries.* (Berlin: Springer, 1975), 11). (**b**) Leonardo's same incision to show the septum (the sieve). RL 19074 verso, The heart and coronary vessels. Leonardo da Vinci, c. 1511–3. Pen and ink on blue paper (Lent by Her Majesty The Queen. Royal Collection © Her Majesty Queen Elizabeth II). (**c**) Leonardo's planes of section of the bovine heart

[20]A. Benivieni, *De abditis nonnullis ac mirandis morborum et sanationum causis.* Ed. G. Weber (Florence: L.S. Olschki, 1994).

[21]John Henderson, *The Renaissance Hospital. Healing the Body and Healing the Soul.* (New Haven, Yale University Press, 2006), 248.

b

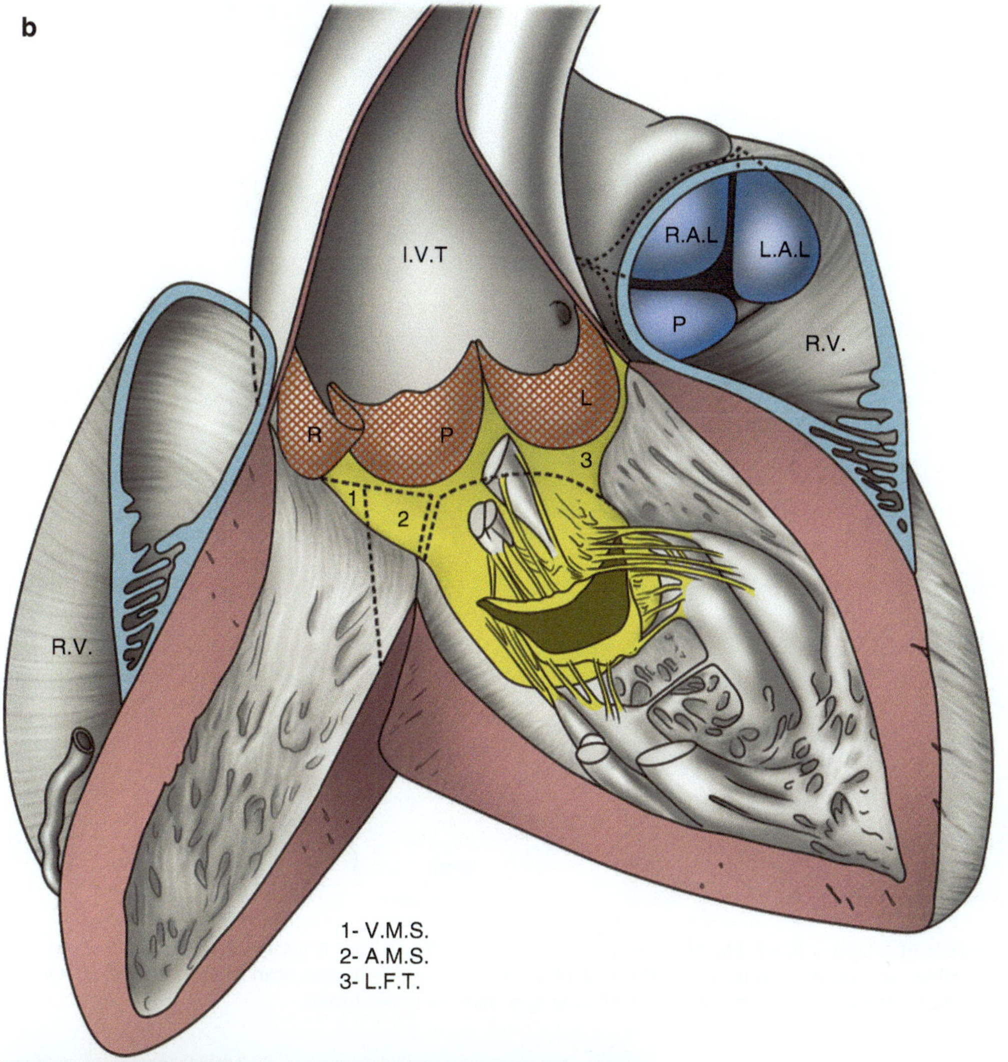

c

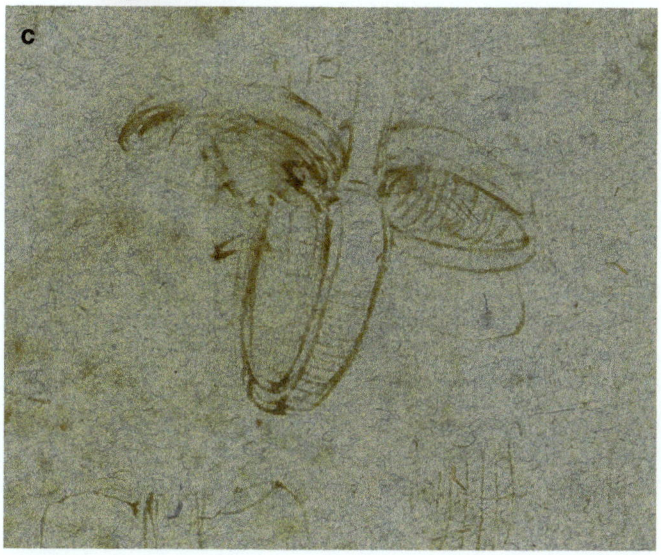

Fig. 4.12 (continued)

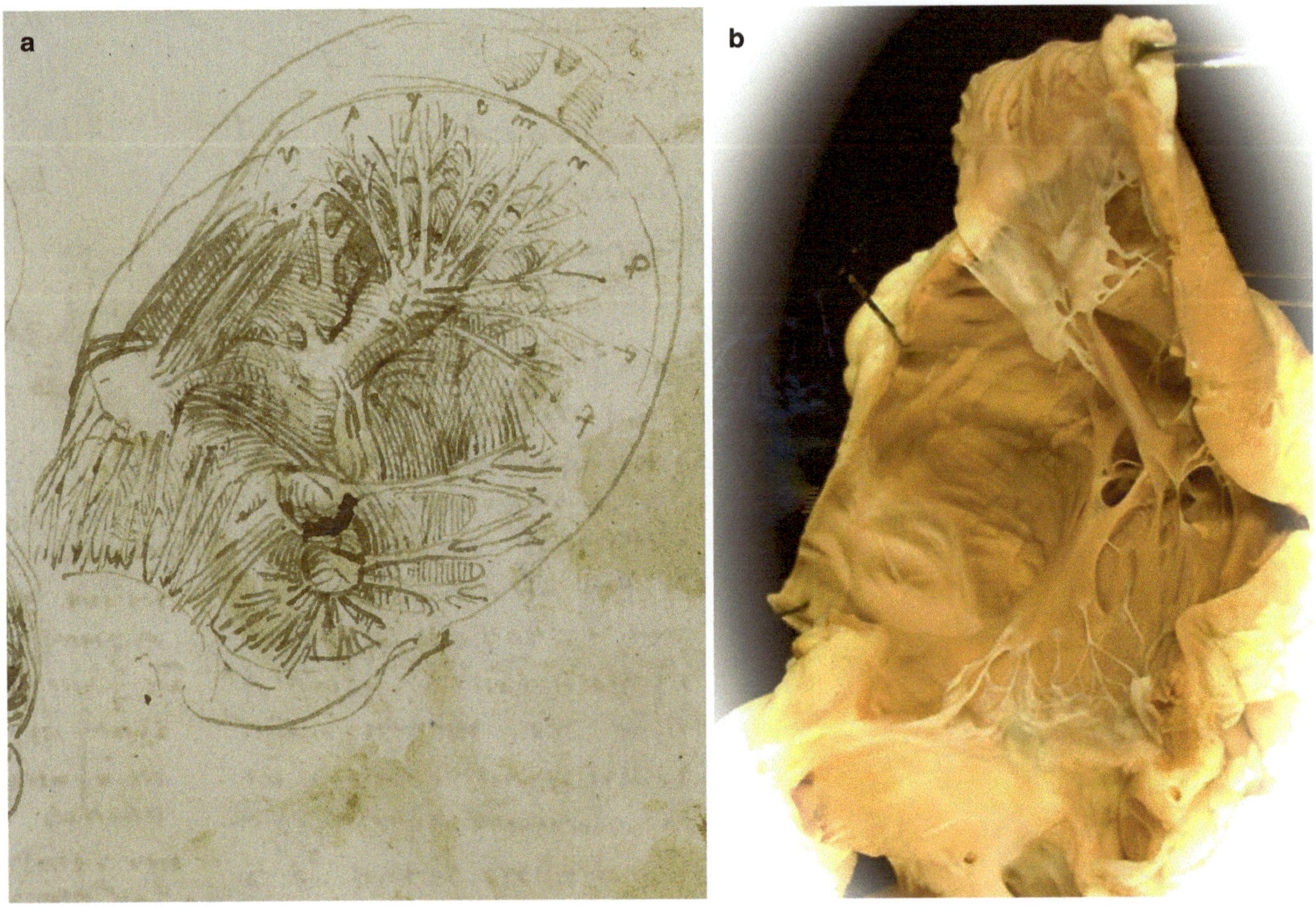

Fig. 4.13 (**a**) Interior of the right ventricle of a human heart. Detail from RL 19119 verso, The ventricles, valves and papillary muscles. Leonardo da Vinci. c. 1511–13. Pen and ink (Lent by Her Majesty The Queen. Royal Collection © Her Majesty Queen Elizabeth II). (**b**) Photograph of the interior of a human heart in the same orientation as (a) (Photograph courtesy of Medtronic)

records of its having a school of anatomy. Despite Benivieni's employment at Santa Maria Nuovo as a "house doctor" none of the dissections that he reported were carried out there. On the other hand, Leonardo's circumstances in association with that institution may have allowed him special privileges. Over the years, he had close ties to the institution and had used it as a depository for money and goods.[22] Whatever the circumstances of Leonardo's access to post mortem material, this dissection produced important and original observations.

It is here that we find the first record of arteriosclerosis, a disease that is recognised as the commonest cause of death in the modern western world. As we shall see, Leonardo compiled the three important features of the disease, which remain true today: the increased thickness in the wall of the vessels, the decreasing luminal diameter of the vessel (worst near its origin and exaggerated at divisions), and the increasing tortuosity and the resultant increasing length of the vessel.

From his dissection, he ascribed the old man's death to changes in the aorta and its branches (the first of which are the coronary arteries). In the continuation of the previous quote, he described the effect on the tissues supplied by these vessels: "And I made an anatomy of him in order to see the cause of so sweet a death. This I found to be a fainting away through the lack of blood to the artery which nourished the heart, and the other parts below it, which I found very dry, thin, and withered."[23]

[22]Martin Kemp, *Leonardo da Vinci. The Marvellous Works of Nature and Man*, Revised edition. (Oxford: Oxford University Press, 2006), 251.

[23]RL 19027 verso [III].

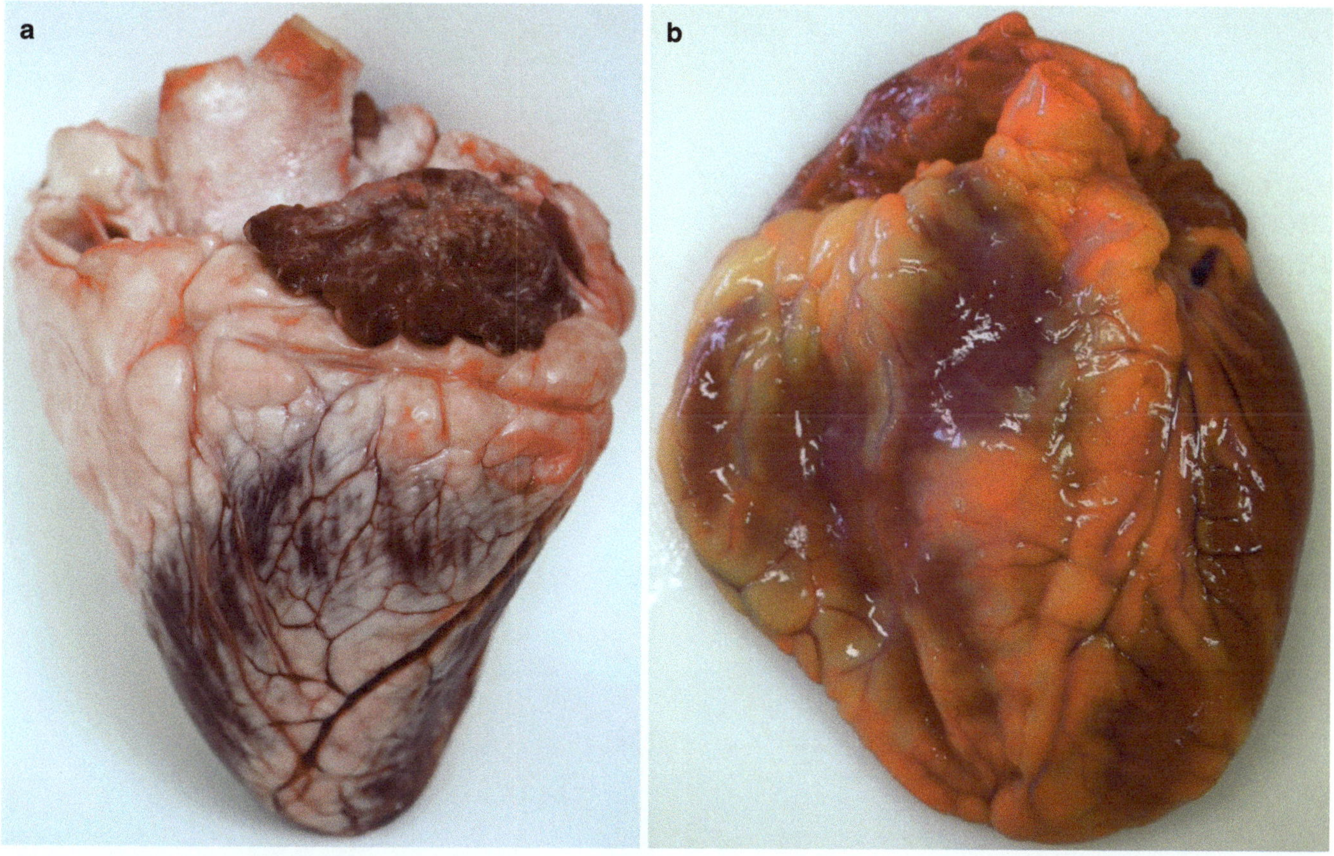

Fig. 4.14 (a) Photograph of an ox heart. (b) Photograph of a human heart

Leonardo also commented upon the ease of dissection in one so thin. He added with no reference to time or place that he had also opened a 2-year-old child. This comes as something of a surprise from a layperson at a time when dissection required specific permission and was generally permitted only in criminals.[24] The note continues, "This anatomy I described very diligently and with great ease because of the absence of fat and humours which greatly hinder the recognition of the parts. The other anatomy was on a child of 2 years, in which I found everything contrary to that of the old man."[25]

His line of enquiry with regard to the effect of aging on the blood vessels he lays out in the following statement: "Note whether the artery is bigger than the vein or the vein than the artery; and do the same in children, young and old, male and female, and animals of the earth, air and water."[26]

[24]Luca Landucci, *A Florentine. Diary from 1450 to 1516*. (London: J.M. Dent, 1927), 217. A commentary on the request for permission on 24 January 1506 to dissect a hanged criminal, from the medical students and doctors of the Studio, a well-known Florentine institution of Leonardo's time for the study of medicine, theology, and the arts.

[25]RL 19027 verso [III] continued.

[26]RL 19051 recto [V].

Of additional interest on this page are two sketches of the blood vessels of the left side of the neck and upper chest. Each is headed simply with the words "Of the old man".[27] They are notable for two reasons: First, they indicate the extent of dissection that Leonardo carried out on the "old man" as these structures require the removal of the anterior chest wall rather than simple opening of the chest, because of the inaccessibility of these structures as they pass behind the clavicle (collar bone) as they pass into the chest. Simply opening the centre of the chest could not reveal these areas. Second, along with the drawing on RL 19050 recto, the lower drawing of the two on RL 19050 recto is the earliest rendering of the internal mammary artery, the most important vessel used today in coronary artery bypass graft surgery (Fig. 4.15).

Evidence of Leonardo's natural use of comparative anatomy appears more than once in this work. Next to a significant drawing of the veins of the arm on RL 19027 recto is a good example, consisting of two small drawings of an artery, one of an old man and the other of a youth, both labelled as such (Fig. 4.16).

[27]RL 19051 recto.

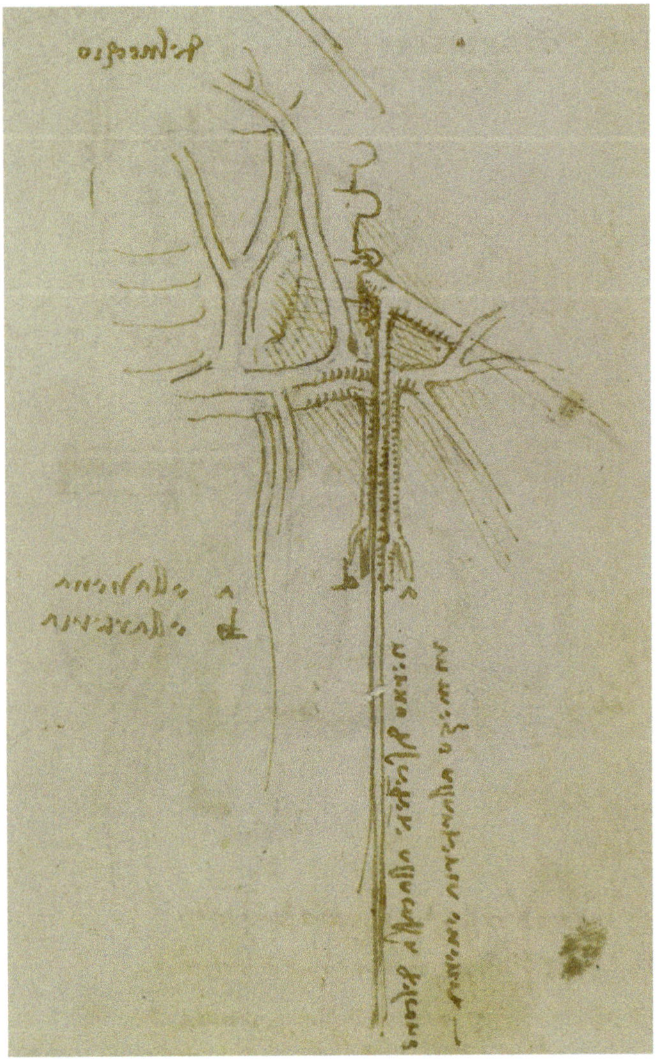

Fig. 4.15 Leonardo's drawing of the internal mammary artery. Detail from RL 19051 recto, The Vessels and Nerves of the Neck. Leonardo da Vinci. c.1508. Pen and ink (Lent by Her Majesty The Queen. Royal Collection © Her Majesty Queen Elizabeth II)

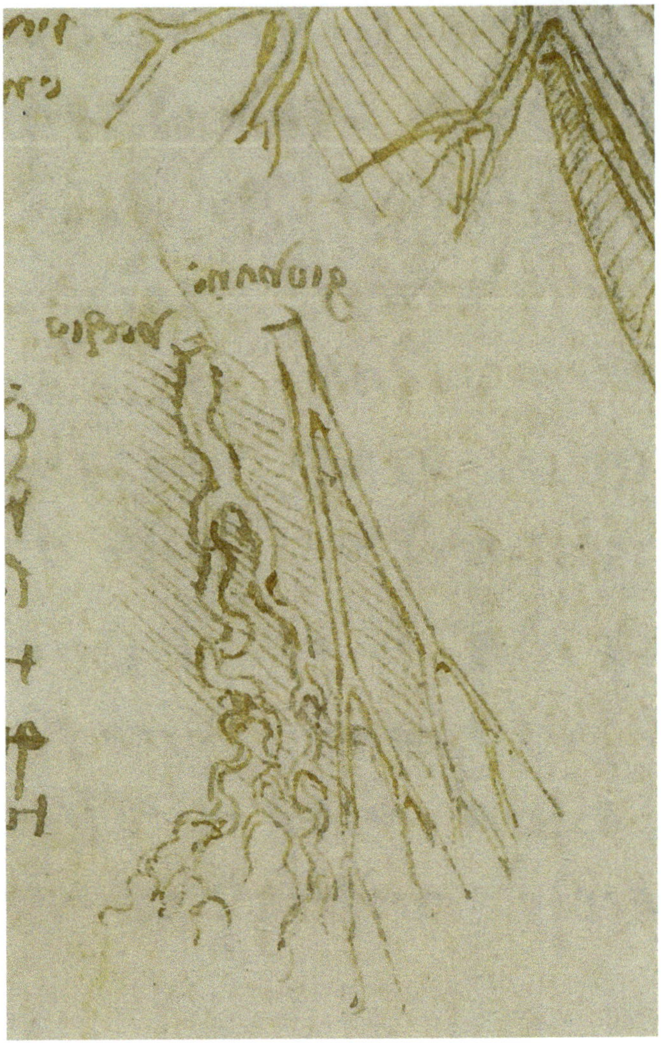

Fig. 4.16 The arteries of the old and the young, demonstrating the tortuosity that comes with old age. Detail from RL 19027 recto, The veins of the arm. Leonardo da Vinci. Pen and ink over traces of black chalk (Lent by Her Majesty The Queen. Royal Collection © Her Majesty Queen Elizabeth II)

It is from this comparative work on the child and the old man that Leonardo discovered the fact that aging brings about an increase in the length of vessels, and along with that an increasing tortuosity: "I also found in a decrepit old man the mesenteric veins [*the vessels to the bowel*], obstructing the passage of blood and doubled in length."[28] The only accompanying drawing on this page of extensive notes is a very small one that illustrates the tortuosity of vessels that arise from the abdominal aorta and supply the upper small bowel, liver, and spleen of the old man (Fig. 4.17). A similar but clearer illustration of the same part is to be found on RL 19028 verso (Fig. 4.18).

Leonardo describes the effect of aging on the thickness of the blood vessel walls in the following fashion: "The old who live in good health die through lack of nourishment. And this occurs because the passage of the mesenteric veins [*those that drain the bowel and liver*] is continually constricted by the thickening of the coats of these veins [*arteries?*], which are the first to be completely closed."[29]

It is likely that Leonardo is here referring to the arteries, as the same change does not occur in the veins of this or any other region. The narrowing of arteries to the heart, brain, and bowel often proceeds hand in hand. An infrequent but well recognised cause of serious problems and even death in some patients

[28]RL 19027 recto [III] [end of paragraph].

[29]RL 19027 verso [IV].

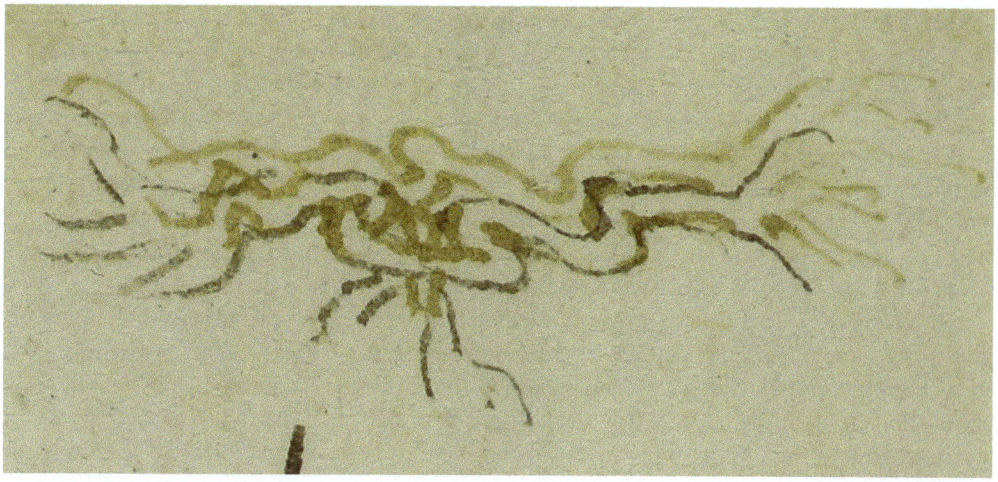

Fig. 4.17 A drawing illustrating the tortuosity of the abdominal vessels that can occur with age. Detail from RL 19027 verso, Notes on the death of a centenarian. Leonardo da Vinci, c. 1508. Pen and ink over traces of black chalk (Lent by Her Majesty The Queen. Royal Collection © Her Majesty Queen Elizabeth II)

Fig. 4.18 The tortuous abdominal vessels in old age. Detail from RL 19028 verso, The vessels of liver, spleen and kidneys. Leonardo da Vinci, c. 1508. Pen and ink over black chalk (Lent by Her Majesty The Queen. Royal Collection © Her Majesty Queen Elizabeth II)

undergoing coronary artery surgery is death of parts of the bowel as a result of reduced blood flow through the narrowed vessels.

In trying to find an explanation for the increasing thickness of vessels (worst near to their origin) and their tortuosity, Leonardo offers the following explanation:

> One asks why the veins in the aged acquire great length and those which were formerly straight become folded and their covering becomes so thick that they occlude and prevent the motion of the blood. From this arises the death of the old without disease.
>
> I judge that a structure that is nearer grows the more; and for this reason these veins being the sheath for the blood which nourishes the body, it nourishes the veins in proportion to their proximity to the blood.[30]

This observation is correct, as it is the earlier branches of the arteries that develop narrowing in the region of their branching. Then drawing on a natural analogy, Leonardo reasoned that the cause of death was narrowed and hardened arteries, not the blood thickening with age, as it is continuously destroyed and then replenished. In fact, the red cells in the blood have a life of only 40 days, or just over 1 month, and the white cells, even less. His deduction is astonishing as there was no way of verifying this with certainty: "And this coat on the vessels acts in man as it does in oranges, in which, in which as the peel thickens so the pulp diminishes the older they become. And if you say that it is the thickened blood which does not flow through the vessels, this is not true, because blood does not thicken in the vessel since it continually dies and is renewed."[31]

In describing the results of these changes, Leonardo wrote, "Vessels which [*in the elderly*] through the thickening of their tunics restrict the transit of the blood and, owing to this lack of nourishment, the aged failing little by little, destroy their life with a slow death without any fever."[32] The

[30]RL 19028 verso, translation from O'Malley and Saunders, *Leonardo on the Human Body*, 302.

[31]RL 19027 verso [IV].

[32]RL 19028 verso, translation from O'Malley and Saunders, *Leonardo on the Human Body*.

reference to dying "without any fever" gives an indication of the high death toll wrought by infection in our history. Death through infection would virtually always be accompanied by a significant fever. His commentary continues with a very modern health warning: "And this occurs through the lack of exercise since the blood is not warmed."[33]

On the reverse side of this manuscript sheet (RL 19027 recto), next to a drawing of the blood vessels of the arm, Leonardo wrote probably the first description of a false aneurysm[34] of an artery, in this case probably the brachial artery. This type of aneurysm is not actually an aneurysm, using the correct definition. It occurs when a hole is made in one side of an artery, usually by trauma. The extravasating blood is localised around the injury by the local tissues. The "aneurysm" is pulsatile and will grow progressively in size if not "bound up". If it is controlled, the blood eventually will clot and the injury will stabilise. The situation described by Leonardo occurs following an injury to the antecubital fossa (the soft tissue area of flexion of the arm opposite to the elbow): "Veins are extensible and dilatable. And testimony of this is given by my having seen a man who by chance was wounded on the common vein [*the median cubital vein or the brachial artery*], and it was immediately bound up with a tight bandage. In the space of a few days, there grew up a [*red*] sanguineous abscess as large as a goose's egg, full of blood; and so it remained for several years."[35]

As with the rest of Leonardo's anatomical work, this original contribution to medical knowledge remained hidden until long after others had begun to probe the same questions several centuries later. Among the first full descriptions was the one by John Hunter FRS in his casebooks.[36]

The Position and Status of the Heart

The special nature of the heart has been recognised throughout the history of anatomical exploration. The spiritual, emotional, and physical roles of the heart have always been interwoven. Although there is a predominance of discussion around the more subjective aspects, the physical (scientific) properties have been considered from earliest times. Aristotle, Hippocrates, and Galen all declared their beliefs with regard to the role of the heart. Its physical and spiritual place in the body was bound up together; its motion and its corporeal centrality vested it with importance.

Leonardo's comment upon the central position of the heart in the body is sparse and terse. He placed it midway between the seat of the soul and the source of life, between the brain and the testis. He wrote, "The heart is placed exactly in the middle between the brain and the testicles."[37]

He also spent time considering its position within the chest, describing it as lying obliquely, towards the left of centre. This he explained from the point of view of balance within the body. He stated that together with the spleen beneath, it counterbalanced the mass of the liver on the opposite side of the body.[38] Elsewhere he ascribed the obliquity of the heart to the difference in weight between the right and left ventricles: "The right ventricle was made heavier than the left ventricle in order that the heart might be placed obliquely."[39]

He rhetorically discussed this concept in another passage, in which he argued against the comparatively heavier weight of the thicker-walled left ventricle acting as a counterweight to the right, with its greater volume of blood. In this passage he introduced the concept of the support of the heart by its great vessels. He wrote,

> And if you say that the left exterior wall was made thick in order that it should acquire greater weight to form a counter weight to the right ventricle which has a greater weight of blood, you have not considered that such a balancing was unnecessary inasmuch all terrestrial animals except man carry their hearts lying horizontally. And likewise the heart of man lies horizontally when he lies in bed. But you would not be a good balancer because the heart has two suspensoria descending from the pit of the throat, and according to the fourth [proposition] of *On Weights*, the heart cannot be balanced unless it has above it one suspensorium only. These two suspensoria are the aortic artery and the vena cava. Furthermore, when the heart is deprived of the weight of blood on its contraction and gives it in deposit to the upper ventricles, the centre of gravity of the heart would then be on the right side of the heart and so the left side would be made lighter. But such weighting is not true, because as stated above, animals which lie down or stand on 4 ft have their hearts lying down [*horizontally*], and no such weighting of the heart is to be looked for. And if you say that the testicles are made only to open or close the spermatic ducts, you are mistaken, because they would only be necessary to rams or bulls which have very big [*testicles*]. And when these testicles had re-entered the body on account of the cold then coitus could not be performed. And the bat which sleeps and always places itself upside-down, how does it balance the right and left ventricles of its heart?[40]

The concept of balance is illustrated by Leonardo in the small sketch of a man holding a long object in front of himself, with his feet carefully placed to maintain balance and his head and back also positioned as to maintain a balanced posture.[41] This drawing is accompanied by a series of other

[33] RL 19028 verso.

[34] An aneurysm is an abnormal swelling or enlargement of an artery. They are described as "saccular" when they tend to bulge on one side only or "fusiform" if the dilatation is uniform around the whole vessel. As it enlarges, it is likely to rupture.

[35] RL 19027 recto [III].

[36] John Hunter, *The Case Books of John Hunter FRS*. Ed. Elizabeth Allen, J.L. Turk. (London, CRC Press, 1993), 36.

[37] RL 19104 verso [IV].

[38] RL 19085 recto.

[39] RL 19087 verso [II].

[40] RL 19087 recto [IV].

[41] RL 19087 recto. Small drawing at bottom right of page.

small ones showing the heart attached to the great vessels and one in which the ventricles are dilated and contracted (Fig. 4.19a). The small diagrams at the top of RL 19084 recto also show the heart in relation to the great vessels in a similar way (Fig. 4.19b). It would appear from these diagrams that he correctly concluded that the heart is maintained in its position by its attachment to the great vessels and that the proposed balancing effects of the respective weights of the two ventricles were not a factor. The importance of the anchoring effect of the great vessels can be seen in victims of rapid deceleration injuries (such as head-on collisions), in whom the weight of the heart and its contained blood is thrown forward within the chest. If the force is great enough, the aorta will tear just before the point where it is anchored to the posterior chest wall. This injury, which always occurs just beyond where the vessels to the head and neck arise, is commonly fatal.

Galen declared that the liver was the principal organ of the body, being the first to be formed, and as such was the source of all the blood vessels. Aristotle held another view, suggesting that the heart was the principal organ. More than a millennium later, Leonardo entered into this debate and concluded on the side of Aristotle that the heart had pride of place over the liver as the origin of the blood vessels. This he argued from a wider natural perspective using visual and verbal metaphors. He likened the heart to a seed or a nut, and wrote, "The heart is the nut which generates the tree of the veins; which veins have their roots in the manure, that is the meseraic [*mesenteric*] veins go on to deposit the blood they have acquired in the liver, whence then the upper veins of the liver [*hepatic veins*] are nourished."[42] Accompanying this text are two simple drawings, one with the word "nut" (*noccollo*) written at its centre and the other with the word "heart" (*core*) in the same place (Fig. 4.20). Near to them are the words, "A plant never arises from its branches, for the plant exists before the branches and the heart exists before the veins."[43] In the drawing labeled "heart" can be seen a simple representation of the vessels to the liver joining the inferior vena cava below the heart. In the parallel drawing, the roots and branches are shown directly emanating from the side of the seed. The visual analogy is clear.

Beside these small drawings is a more substantial one illustrating this important relationship of veins and arteries to the heart, the sovereign power. The major text on that sheet affirms in no uncertain terms this relationship:

> All the veins and arteries arise from the heart; and the reason is that the biggest veins and arteries are found at their conjunction with the heart. And the more they are removed from the heart the finer they become, and they divide into very small branches. And if you say that the veins arise in the gibbosity of the liver because they have their ramifications in this gibbosity, just as the roots of plants have in the earth, the reply to this comparison is that plants do not have their origin in the roots, but the roots and other ramifications take origin from that lower part of the plant which is situated between the air and the earth. And all the lower and upper parts of a plant are always less than this part which

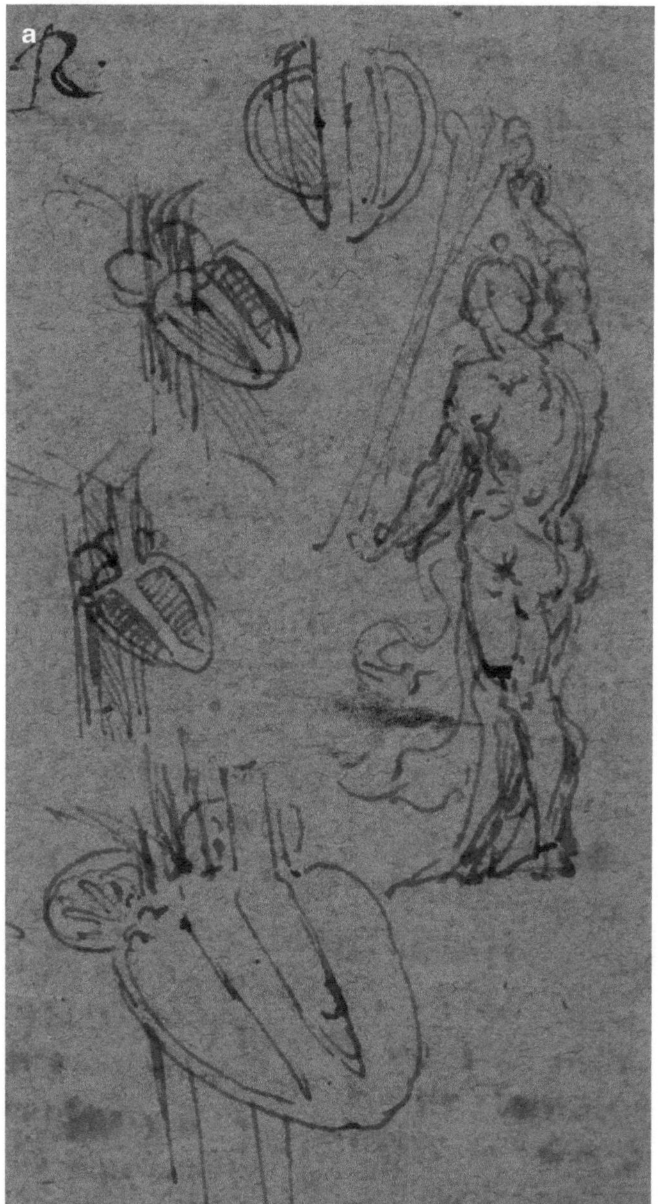

Fig. 4.19 The heart in relation to the great vessels. (**a**) Detail from RL 19087 recto, Studies of the heart, notes on light and shade, and a standing figure. Leonardo da Vinci. c. 1511–13. Pen and ink on blue paper (Lent by Her Majesty The Queen. Royal Collection © Her Majesty Queen Elizabeth II). (**b**) Detail from RL 19084 Recto, Studies of the heart, and notes on the vices of men. Leonardo da Vinci. c. 1511–13. Pen and ink on blue paper (Lent by Her Majesty The Queen. Royal Collection © Her Majesty Queen Elizabeth II)

[42] RL 19028 recto [VI].

[43] RL 19028 recto [IV].

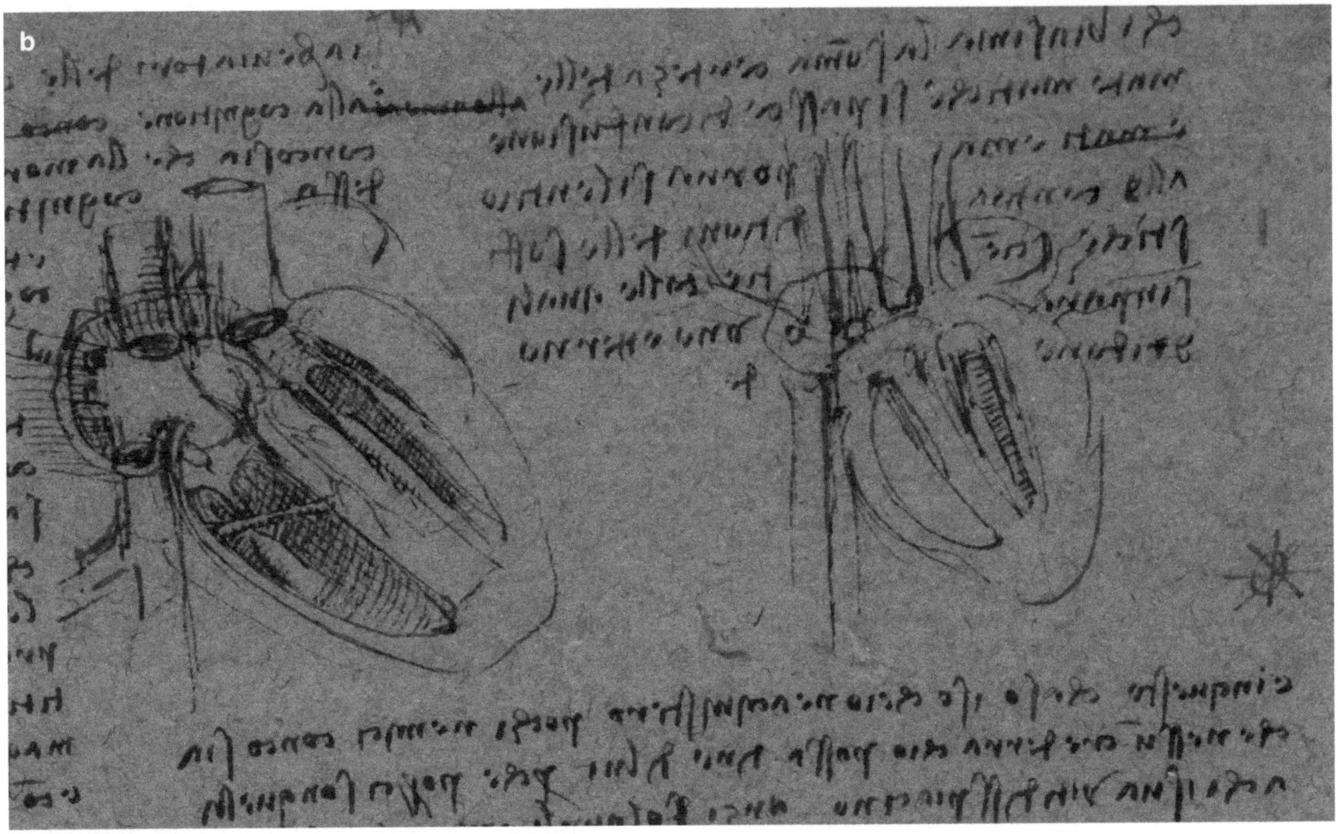

Fig. 4.19 (continued)

adjoins the earth. Therefore it is evident that the whole plant takes origin from such a size and in consequence the veins take their origin from the heart where they are biggest. Never does one find a plant which has its origin from the tips of its roots or other branches; and this observed by experience in the sprouting of a peach which arises from its nut as is shown above at a b and a c.[44] *(The labels are clearly seen on the accompanying small drawings.)*

Thus, in the opinion of Leonardo, the heart has pride of place within the body and is the source of all of the blood vessels.

The Nature of the Heart

As the heart is the only continuously moving part of the body, it is natural that investigators should look to it as a vital force in life. Its responsiveness to emotional change led to an extended debate as to its role in generating the emotional state as well as the viability of the whole animal. With the benefit of evolving investigative technologies, we have been able to demystify this crucial organ to the extent that we recognise the heart as a highly responsive, self-regulating muscle pump, which is the engine of the circulation. This statement seems obvious to us, but in the time of Leonardo, even the fact that it is made of muscle had not been established. This was Leonardo's first significant contribution to the practical understanding of the heart.

Modern biology recognizes three distinct forms of muscle: (1) Skeletal muscle, the most abundant form, which is responsible for the mobility of the animal. It has the microscopic characteristic of stripes running horizontally across its fibers, giving it the alternative name of striated muscle. (2) Smooth muscle, so called because it lacks the microscopic stripes of skeletal muscle. This type is found in places such as the eye and the bowel. It responds to chemical and autonomic nerve stimulation with such actions as the focusing of the eye and the movement of the gut. (3) Cardiac muscle, the unique muscle of the heart. It too is striated under the microscope, but the muscle fibers have branched endings, which interdigitate with the surrounding muscle fibers, forming a syncytium.[45] It has the property of continuous contraction and relaxation throughout life. It responds to chemical and nervous stimuli, driven in turn by exercise and emotional change.

There was no way of distinguishing these forms of muscle until the development of the microscope in the late sixteenth century, when It became possible to discern their different

[44]RL 19028 recto [III].

[45]A syncytium is a body of tissue in which multiple cell nuclei exist in one branching mass of cellular cytoplasm.

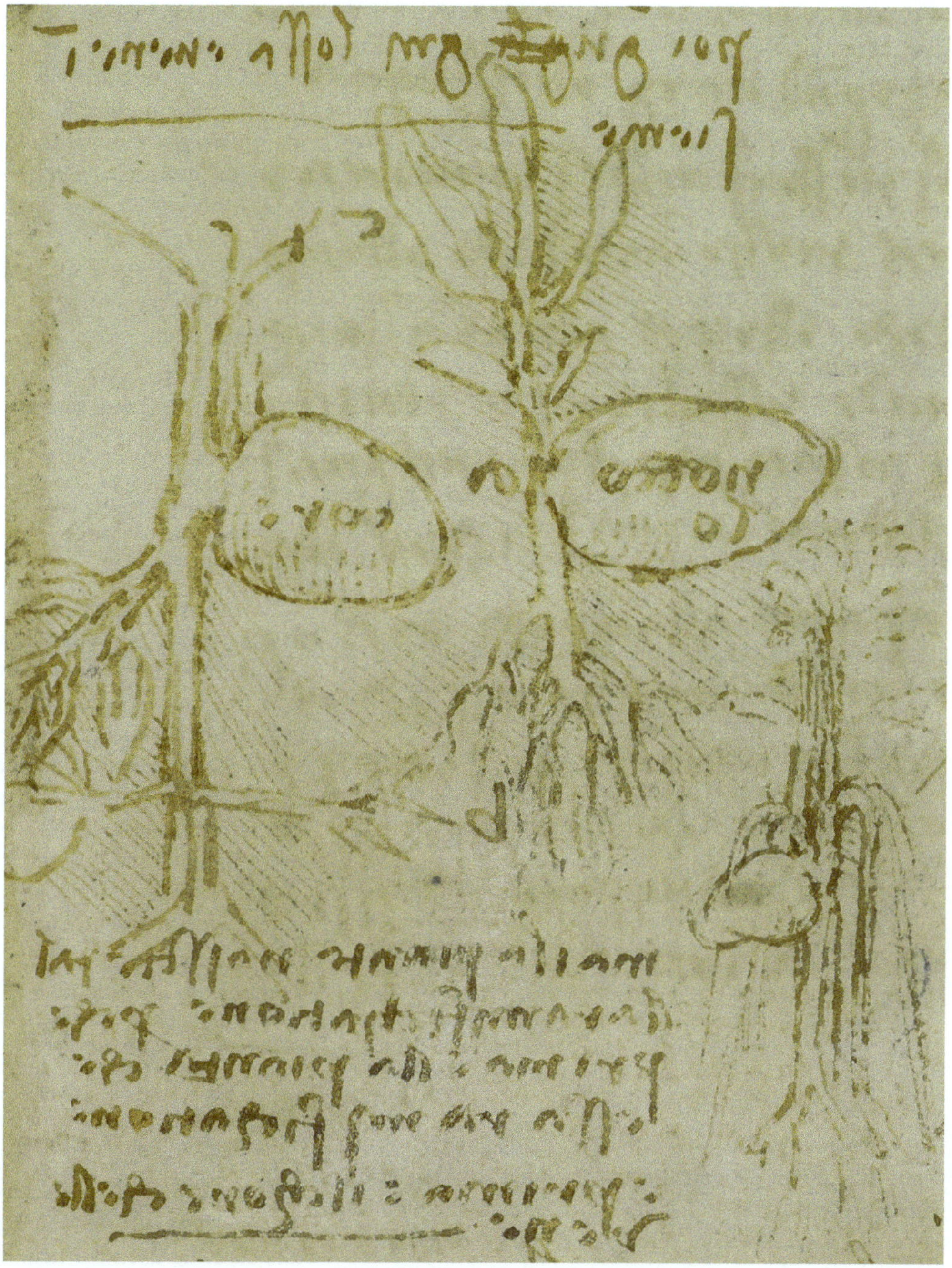

Fig. 4.20 The metaphor of the seed for the heart in an explanation of why the heart must be the source of the great vessels. Detail from RL 19028 recto, The heart compared to a seed. Leonardo da Vinci. c.1508. Pen and ink over black chalk (Lent by Her Majesty The Queen. Royal Collection © Her Majesty Queen Elizabeth II)

characteristics. Until then, differences could be determined only by examination of gross form and function. Galen recognised that the tissue of the heart was different and described it as having a special "hard" flesh. Whilst recognising that it was constructed of fibres resembling muscle, both he and Avicenna a millennium later said that it was in reality quite a different material.[46] In contrast to this position, Leonardo clearly stated that the heart was made of muscle. He recognised its difference from skeletal muscle in describing it as being of greater density. In support of his claim that it was a muscle, Leonardo pointed out that it possessed its own arterial blood supply and venous drainage: "The heart of itself is not the beginning of life but is a vessel made of dense muscle vivified and nourished by an artery and a vein as are the other muscles. It is true that the blood and the artery which purges itself in it are the life and nourishment of the other muscles."[47]

[46]K.D. Keele, *Leonardo da Vinci on Movement of the Heart and Blood.* (London, Harvey and Blythe, 1952), 54.

[47]RL 19050 verso. The remainder of this section emphasises the resistance of the heart muscle to heat.

In Galenic terms, the heart was perceived as having three specific roles: the movement of blood to the lungs and the periphery; drawing air from the lungs into the heart during the relaxation and filling phase of the ventricles; and generating the heat of the body, through its often-violent movement. Leonardo held to that philosophy, although, as we shall see, he began to question the direct connection between the trachea and the heart. A considerable amount of Leonardo's writing on the heart is spent in considering its role in the generation of innate heat. He perceived that several of the components and structural relationships in the heart are defined by their contribution to this vital function.

The presence of bodily heat in an animal implies life, and Leonardo stated that assumption as follows: "Where there is life there is heat, and where vital heat is, there is movement of vapour." [48] The last phrase of this quotation refers to the creation of convection currents within fluids, be they in the liquid or gaseous phase. Hot air and hot water rise, whilst the opposite is true for cold liquids.

In keeping with the beliefs of his time, Leonardo accepted the four elements of the universe—earth, water, air, and fire—and he saw the constitution of man as a reflection of the world around him and as such, as a part of the microcosm/macrocosm continuum:

> By the ancients man was termed a lesser world and certainly the use of this name is well bestowed, because, in that man is composed of water, earth, air and fire, his body is an analogue for the world: just as man has in himself bones, the supports and armature of his flesh, the world has the rocks; just as man has in himself the lake of the blood, in which the lungs increase and decrease during breathing, so the body of the earth has its oceanic seas which likewise increase and decrease every 6 h with the breathing of the world; just as in that lake of blood the veins originate, which make ramifications throughout the human body, similarly the oceanic sea fills the body of the earth with infinite veins of water. The nerves are lacking in the body of the earth.... But in all other things they are similar.[49]

Therefore it was natural for him to use examples from the wider world in his explanations of the inner workings of the human body. At work within the four elements were the four powers of nature: weight, force, movement, and percussion. The resistance between bodies, of which at least one was moving, creates friction.[50] Simple friction creates heat, and this Leonardo reflected in two statements on the production of heat within the heart: "And the revolvings made by the blood whirling round in itself in different eddies, and the friction that it makes with the walls and the percussions in the recesses [*the spaces between the pectinate muscle bundles of the right atrium*], are the cause of the heating of the blood, and of making the thick viscous blood fine...."[51] and "On The Cause Of The Heat Of The Blood: The heat is generated by the movement of the heart; and this is evident because the more quickly the heart moves the more the heat multiplies, as the pulse of febrile persons moved by the beating of the heart teaches us." [52]

A further use of natural analogy is found in other comments, in which he uses a river and a lake as similes for the blood flow and the heart. Here he is restating his belief that the ebb and flow motion of the blood is essential for the generation of heat:

> And such motion of the blood would behave like a lake through which a river flows, which acquires as much water from one side as it loses at the other. However, the only difference is that the movement of the blood is discontinuous and that of the river flowing through the lake continuous and from this lack of flux and reflux the blood would not be heated and consequently the vital spirits could not be generated and for this reason life would be destroyed.[53]

Leonardo's concept of the heart as a source of renewable heat is evidenced in the following words: "As natural warmth spread through the human limbs is driven back by the surrounding cold which is its opposite and enemy flowing back to the lake of the heart and the liver fortifies itself there, making of these its fortress and defence."[54]

Within Leonardo's interpretation of the necessity of the generation of heat by the movements of the heart was his insistence that the heating of the blood caused it to be less viscous and thereby to pass more easily through the fine pores in the interventricular septum, as described by Galen:

> Because the heat of the heart is generated by the swift and continuous movement which the blood makes by its friction within itself through its eddyings, also by the friction it makes with the pitted walls of the right upper ventricle into which it continually enters and escapes with impetus, these frictions made by the velocity of the viscous blood heat it, subtilise it, and make it penetrate through the fine meati [*pores*], giving life and spirit to all the organs into which it is infused."[55]

Leonardo used another example at another place in his notes to prove that continuous movement produces heat. In this case, he used the making of butter by the churning of milk as an example of heat production through motion. A difference with this argument is that in this process, the liquid milk becomes more viscous, whereas he argues that blood becomes "thinned". This observation recognises the difference in response to the motion of different liquids, which is predicted by the content of the liquid. The extreme

[48] Paris manuscript A 55 v.

[49] Paris manuscript A 55 verso.

[50] Forster II 131 verso.

[51] RL 19063 verso [I], "On the heart".

[52] RL 19029 recto [I].

[53] RL 19063 verso [I].

[54] Codex Atlanticus 212 verso/468 recto.

[55] RL 19063 recto [V].

example is that of a thixotropic liquid, one that is extremely viscous at rest (more like a gel), so that it will not flow. The energy imparted to it on violent agitation makes it much less viscous and allows it to flow freely for a while, but the as the energy is expended, the liquid returns to a gel state[56]:

> Observe whether the revolution of milk when butter is made, heats it. And by such means you will be able to test the strength of the auricles of the heart which receive and expel the blood from their cavities and other passages; these being made only in order to heat and refine the blood and make it more quickly penetrate the wall through which it passes from the right into the left ventricle[57] where through the thickness of its wall, that is, of the left ventricle, it conserves the heat which the blood carries to it.[58]

The analogy of the heart as a stove or a furnace is to be found on a page in the Codex Arundel.[59] There Leonardo illustrates a stove with an inlet and an outlet valve, which he labels "m" and "n". Writing about this stove, he regards the heart as the furnace and the trachea as the chimney. The air is drawn in through one opening labelled "n" and is blown out in the other direction through an opening labelled "m". Leonardo likens this to the cardiorespiratory system: "Short and frequent breathing suffocates the person, short because each breath does not change all of the air in the lung that has been heated. But a part of it remains there so greatly heated that it would do great damage to an animal unless big and long breaths were to drive it out of the lung."[60]

This analogy is carried over into an early drawing in the Windsor series, in which he labelled the two tubular connections to the lung in the same way "n" and "m".[61] In one part of the drawing, the labelled connections can be seen intact; another part shows the same view in cross section. Leonardo explains that they are the inlet and outlet for the products of combustion in the heart. The style and substance of this drawing suggest a date earlier than the final heart series of 1513–1514. Clark[62] and O'Malley and Saunders[63] date this and a similar drawing (RL 19112 recto) as 1504–1507.

Having established that the heart was in this way a "furnace", Leonardo had to adopt another property for heart muscle, namely its ability to withstand heat. He wrote, "The heart is of such density that fire can scarcely damage it. This is seen in the case of men who have been burnt, in whom, after their bones are in ashes, the heart is still bloody internally. Nature has made this great resistance to heat so that it can resist the great heat which is generated in the left side of the heart by means of the arterial blood which is subtelised in this ventricle."[64]

In reality, contracting muscle does generate heat. The obvious example is skeletal muscle on extreme exercise. Similarly, the heart produces more heat with increasing exercise, and this heat is dissipated in the blood flowing through it via the coronary arteries and the main stream of blood. The heat is then dissipated in the periphery by sweating and through the lungs in breathing. All metabolic processes in the body operate under the control of enzymes (biochemical catalysts), and these operate within limited temperature bands. Heat control is therefore very important for normal muscle function, albeit as part of a very different concept than that imagined by Leonardo.

The actions of the heart when an individual is at rest are barely perceptible. As exercise levels increase, however, its motion becomes increasingly forceful and obvious. At extremes of exercise, its movement can feel almost violent, producing a pounding in the chest and head and a bounding peripheral pulse. These enormous and sustainable reserves of power were recognised by Leonardo in the following words: "The heart is a principal muscle of force, and it is much more powerful than the other muscles."[65] As one of the four great powers, Leonardo's use of the term "force" was specific. His definition was as follows: "a spiritual energy, an invisible power which is created and imparted, through violence from without, by animated bodies to inanimate bodies, giving to these the similarity of life, and this life works in a marvellous way, constraining and transforming in place and shape all created things. It speeds in fury to its undoing and continues to modify according to the occasion."[66]

Galen's heavy reliance on the "vital spirits" for the satisfactory explanation of the viability of an animal led to his perceived importance of the relationship between the heart and the lungs for the extraction of the vital spirit, "pneuma" As we shall see later, access of "pneuma" to the heart was supposedly through the trachea. During the ventricular relaxation phase of the cardiac cycle, air is drawn from the trachea into the ventricle and was supposed to mix with the blood, re-entering the right ventricle after contraction of the atria.

[56]A good example of a thixotropic liquid is tomato ketchup!

[57]The lack of awareness of the capillary connections between artery and vein in the lungs made the proposition of the pores in the ventricular septum a necessity for the supply of blood to the rest of the body. These fictitious pores, suggested by Galen, were surprisingly easily accepted by Leonardo.

[58]RL 19119 verso [VII].

[59]Codex Arundel (British Museum) 24 recto.

[60]Ibid.

[61]K.D. Keele, *Leonardo da Vinci's Elements of the Science of Man.* (New York: Academic Press, 1983), 305.

[62]K. Clark, *Leonardo da Vinci. Anatomical Drawings at Windsor Castle.* (Oxford, Phaidon, 1969), 40, 43.

[63]C.D. O'Malley, J.B. de C.M. Saunders, *Leonardo da Vinci on the Human Body.* (New York: Greenwich House, 1982), 392.

[64]RL 19050 verso.

[65]Paris manuscript G. 1 verso.

[66]Codex Atlanticus 302.

On the next right ventricular contraction, part of the blood (now enriched with pneuma) would be forced through the imaginary septal pores and depart through the aortic valve on its way to the brain. Unlike Galen, Leonardo's writing lacks frequent reference to the "vital spirits" and places much heavier emphasis on the four powers of nature,[67] but he does spend some effort in trying to demonstrate this fictional relationship.

Although much of Leonardo's text on the function of the heart relates to the flux and efflux of pulmonary blood flow, it is clear from these lines that he was well aware of the importance of the heart in providing the flow of blood to the periphery. He stated as much in a section describing the way that an animal sustains itself by the consumption of nutrition: "And the same thing happens in the bodies of animals by means of the beating of the heart which generates a wave of blood through all the vessels which continually dilate and contract. And dilatation occurs on the reception of superabundant blood and diminution occurs on the departure of the superabundance of the blood received. This, the beating of the pulse teaches us when we touch the aforesaid vessels with our fingers in any part of the living body."[68]

Therefore, to Leonardo the heart was a muscle pump of considerable power, capable of withstanding great heat. He perceived it as the source of propulsion in the ebb and flow movement of the blood, as well as the generator of heat.[69]

The Movement of the Heart

> "Of the heart: This moves of itself and does not stop unless forever."[70]

The lifelong continuous contraction and relaxation of the heart sets it apart from all of the other organs in the body. It follows. therefore, that having proved to his satisfaction that it was made of a sturdy and most powerful form of muscle, Leonardo should ponder on the cause of this continuous motion. He asked himself, "And which part of the heart it is which is the cause of its movement; and whether it is inside or outside the heart".[71] We know that this continuous movement of the heart is activated by the spontaneous generation of a wave of cellular depolarisation that originates in the sinoatrial node and spreads throughout the heart. At rest, this occurs about 70 times a minute. It changes in response to exercise and emotional challenges. This change is moderated in part by a nerve originating in the brain called the vagus nerve (cranial nerve X). Leonardo was aware of this nerve and referred to it as the reversive nerve. It was given this name by earlier anatomists because on the left side of the body, it reverses its direction in the chest. After arising from the brain, it runs from the base of the skull to the abdomen. In the chest, it divides as it passes around the aorta, and part of it returns to supply the vocal cords of the larynx. The rest of the nerve continues along the pericardial sac surrounding the heart and continues into the abdomen, where it provides important stimuli for the normal function of the gut. On the right side of the body, it divides and reverses its direction higher up in the neck, looping around the right subclavian artery. As it passes over the pericardium, branches penetrate the heart and their activation contributes to the control of heart rate. This nerve is clearly shown in Leonardo's drawing on RL 19050 verso (Fig. 4.21). In a personal record of his dissection of this nerve as part of his attempts to discover more about its anatomy and its role in the control of heart muscle contraction, Leonardo wrote, "The reversive nerves arise at *a b*, and *b f* is the reversive nerve descending to the pylorus of the stomach. And the left nerve, its companion, descends to the covering of the heart, and I believe that it may be the nerve which enters the heart."[72]

The following quotation forms part of Leonardo's investigation of the opposing Aristotelian and Galenic positions on the seat of the soul using the role of the vagus as the arbiter:

> Do not fail to follow up the reversive nerves [*vagi*] as far as the heart; and see whether these [*nerves*] give movement to the heart, or whether the heart moves by itself. And if such movement comes from the reversive nerves, which have their origin in the brain, then you will make it clear how the soul has its seat in the ventricles of the brain, and the vital spirits have their origin in the left ventricle of the heart. And if this movement of the heart arises from the heart itself, then you will say that the seat of the soul is in the heart, and likewise to the other nerves because the motion of all muscles arises from these nerves which their ramifications are infused into the muscles".[73]

Though couched in the terms of his time, this discourse by Leonardo illustrates the use of rhetorical argument in an attempt to resolve a problem. The activity of the brain does control the activity of the vagus nerve, which in turn has an identifiable effect on the heart. Our heart rate quickens and slows in response to our emotional state. In this approach, Leonardo further allied himself with the view that the seat of the soul is the brain and not the heart. It is interesting to note that the movement of the heart was not understood by Harvey, who claimed that the movement of the heart was not

[67]K.D. Keele, *Leonardo da Vinci on Movement of the Heart and Blood.* (London, Harvey and Blythe, 1952), 60.

[68]RL 19045 recto [IV].

[69]RL19045 recto [IV].

[70]RL 19030 recto.

[71]RL19128 recto.

[72]RL 19050 verso.

[73]RL 19112 recto [V].

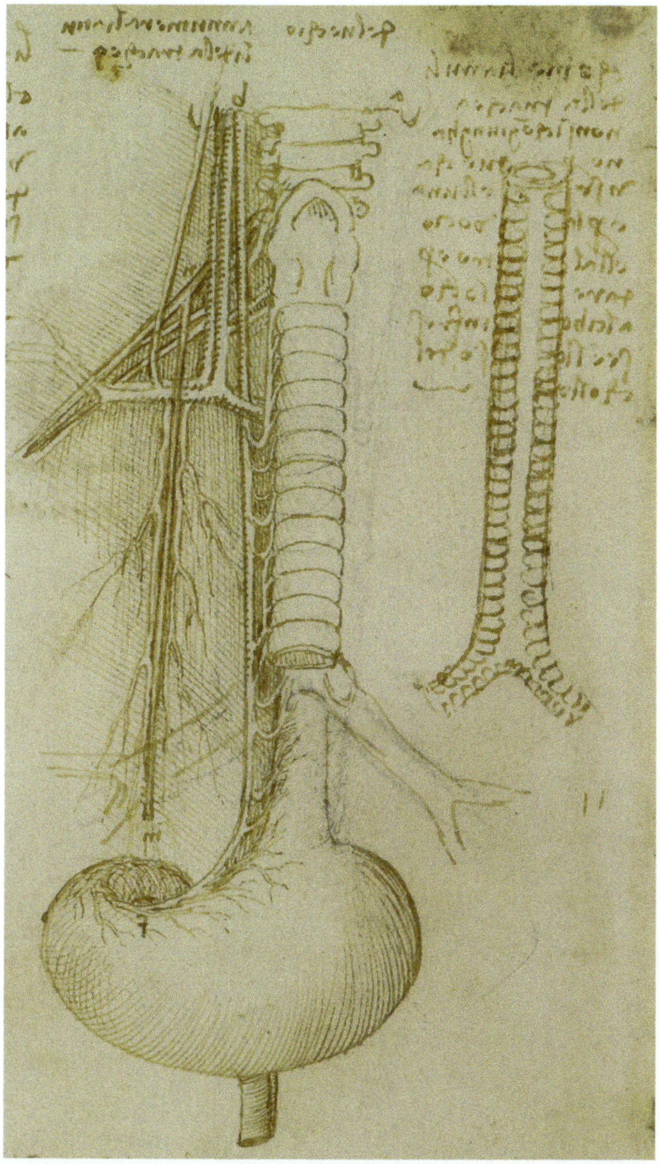

Fig. 4.21 The vagus (reversive) nerve and its branches. Detail from RL 19050 verso, The distribution of the right vagus and right phrenic nerves. Leonardo da Vinci. c.1508. Pen and ink over black chalk (Lent by Her Majesty The Queen. Royal Collection © Her Majesty Queen Elizabeth II)

spontaneous. He suggested that the vital spirits contained within the blood were the source of the heart's movement.[74]

In addition to the control of heart movement, Leonardo was interested in the movement of the heart in life and the form that it took after death. His most important description in this respect was that the atria and the ventricles move independently and separately. As the atria contract, the ventricles relax to fill with blood, and then as the ventricles contract, the atria relax. The atrial contraction occurs towards the end of ventricular diastolic filling and is called the "atrial kick" The muscular walls of the ventricles are significantly elastic and obey Hooke's law of elasticity: Up to a point (the elastic limit), the greater degree of stretch applied, the greater the force of recoil when released. The injection of blood into the ventricles engendered by the late atrial contraction further expands the elastic, muscular heart. This expansion significantly increases the power of recoil of the ventricular muscle and is an important contribution to cardiac efficiency. Leonardo illustrated this effect is his only known heart drawing housed outside the Windsor collection, found in Manuscript G in Paris, on the opening page[75] (Fig. 4.22). In this drawing, the dynamic nature of this motion is clearly visualised. On Windsor sheet RL 19087 recto are found a further series of sketches of the changes in shape of the atria and ventricles through the cardiac cycle (see Fig. 4.19a). Having established that the atria and the ventricles contract separately and consecutively, it is interesting to hear Leonardo's full description of the cycle of contraction of the cardiac chambers. First he makes the point that the atria are more distensible than the ventricles:

> ...And since the upper ventricles [*atria*] are more suited for the driving out of themselves the blood which dilates them than pulling it into themselves, Nature has so made it that by the closure of the lower ventricles (which close on their own) the blood which escapes from them is that which dilates the upper ventricles. These through being composed of muscles and fleshy membranes are suitable for dilation and for receiving as much blood as is pushed into them; they are also suited by their powerful muscles for contracting with impetus and driving out of themselves the blood into the lower ventricles, of which when one opens the other closes. And the upper ventricles [*atria*] do the same thing in such a way that when the right lower ventricle dilates the left upper ventricle contracts, and when the left lower ventricle opens, the right upper one closes."[76]

The sequential nature of atrial and ventricular contraction is correct, but the suggestion that the ventricles contract sequentially and not synchronously is not. The contraction of the right ventricle followed by the left ventricle is the opposite of what actually happens, when both ventricles contract in unison, as do both atria. This statement makes it clear that either at that time Leonardo had not had the opportunity to observe the beating heart in situ, or he had simply misinterpreted the observed facts. I suspect that his keen powers of observation would not have allowed the latter, as so much of the rest of his observations of anatomical fact are so accurate. Interestingly, on another sheet he declares that the ventricles contract in unison: "ON THE CONTRACTION AND

[74]Jonathon Miller, "The anatomised heart. Understanding the pump". In: James Peto (editor), *The Heart (Wellcome Trust Collection).* (New Haven: Yale University Press, 2007), 62.

[75]Paris manuscript G 1v.

[76]RL 19062 recto [I].

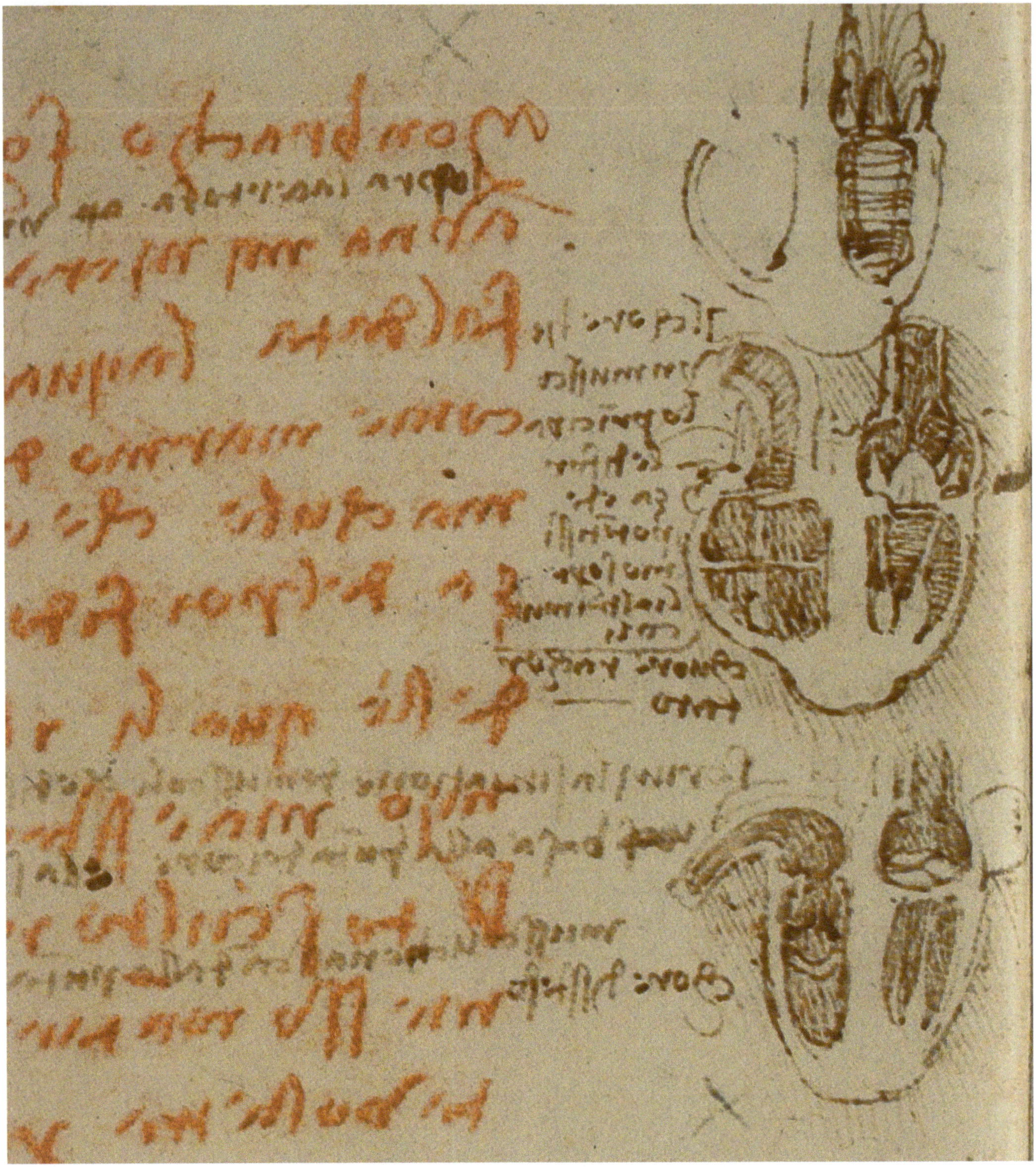

Fig. 4.22 Heart showing separate atrial and ventricular contraction. Leonardo da Vinci, Anatomy (heart). Manuscript G 1510–1516 (© RMN / René-Gabriel Ojeda)

DILATATION OF THE TWO VENTRICLES OF THE HEART: Of the two lower ventricles arising from the roots of the heart, their dilatation and contraction are made at one and the same time, following the first by reflux from the upper ventricles arising on top of the roots of the heart."[77] Clearly this was work in progress.

[77]RL 19073 recto [III].

Leonardo's correct, albeit inconsistent, assumption about the cyclical motion of the heart gave rise to his error in calculation of the number of contractions of the heart per minute. From the normal resting heart rate of 70 beats per minute, it is easy to calculate the number of beats per hour. The correct solution is of course $70 \times 60 = 4{,}200$. Below is Leonardo's solution. The main purpose of this calculation is to estimate the amount of new blood that is needed from the liver to maintain the total volume: "How much blood is the liver able to deliver owing to the opening [*diastolic relaxation*] of the heart? It restores to it as much as it consumes, that is a very small part. Since in 1 h about 2,000 openings of the heart occur, it is a great weight.[78]

Since Leonardo assumed that the ventricles opened sequentially, he divided the normal heart rate by two, giving the total of about 2,000 openings per hour. This is followed by a small calculation "25 × 12/300" followed by "7 oz".[79] It is likely that this small amount (7 oz) represents the amount of blood derived from the liver, although where this sum came from is impossible to be sure. Multiplying 25 by 12 does indeed amount to 300, and if this is divided by 60 for the number of minutes in an hour, 1 arrives at the total of 5; in other words 5 oz/h to be replenished by the liver. We cannot say whether this actually equates to Leonardo's workings.

Leonardo continued his discussion of the movement of the heart in musical terms of tempo: "…The time interposed between two percussions of the pulse is half one musical temp."[80] And later in the same passage, "Therefore in every harmonic or, as you might say, musical tempo, the heart has three movements which are contained below. Of these tempi, one hour contains one thousand and eighty. Therefore the heart moves three thousand five hundred and forty times during each hour in opening and shutting."[81]

What is more interesting is Leonardo's statement that the amount of blood expelled from the ventricles per hour is a "great weight". It was this very observation that led William Harvey to derive his theory of the circulation of the blood. One cannot help but wonder where Leonardo's mind was taken with this thought. Sadly there is no more commentary on it. We can only assume that he was satisfied with the idea that much of the blood continuously ebbed and flowed between the atria and the ventricles acting as a store for the endogenous heat of the body. This idea appears to be reinforced by the following words, in which he considers the capacity of the ventricles: "I say that the greatest receptive capacity of the ventricles is the quantity of blood which escapes from them when they contract. This is received on deposit by the upper ventricles called the auricles of the heart, and it is kept in these until it is returned on the succeeding dilatation of the heart, but with so much less quantity as was contributed to the nourishers of life [i.e., *the veins*]. This deficiency is compensated by the liver, generator of such blood.[82] From Leonardo's calculation this amount would appear to be 7 oz.

As mentioned earlier, the other important function of the heart's action was to deliver the vital spirits (pneuma) to the brain and the rest of the body. Leonardo's description of how this came about involved a novel but incorrect function for the pericardium (the fibrous sac around the heart):

> Air which takes the place occupied by the upper ventricle when it holds the blood deposited in it by the contraction of the lower ventricle, flows in to replace the upper ventricle when it contracts, and it comes from that in the capsule of the heart [*pericardium*]; and this capsule having given away its air to the place left by the contraction of the upper ventricle is restored by drawing more air into the lung through the trachea than usual; therefore the air drawn into the lung cannot always be equal.[83]

In other words, when the blood in the ventricle has emptied backward into the atrium, it is mixed with air from the lungs that has passed via the pericardium. That which has been lost from the pericardium to the heart is restored by air drawn in through the trachea (in other words, a breath). As we shall see, Leonardo is troubled by this assumption.

At that time, the ventricles were regarded as being formed from three different fibre groups: transverse, oblique, and longitudinal.[84] Whilst Leonardo used these terms,[85] he also simplified the subdivision of the muscle mass into intrinsic and extrinsic sets using the more customary classification: "And the intrinsic muscles suitable for contracting are of one and the same nature in the four ventricles. But the extrinsic muscle exists only on the intrinsic ventricles [*ventricles*] of the heart. On the extrinsic ventricles [*atria*] there is only a continuous dilatable and contractible skin."[86]

This difference can be appreciated in the ox heart, in which the atrial wall is very much thinner than that of the ventricle. Although this classification seems very naive today, it must be remembered that Leonardo had no means of observation other than the naked eye.

In Leonardo's opinion, the contraction of the longitudinal muscles of the ventricular wall and the papillary muscles within the ventricle caused the heart to widen and in the

[78]RL 19087 verso [VIII].

[79]RL 19087 verso.

[80]RL 19081 recto [III].

[81]RL 19081 recto [III].

[82]RL 19087 verso [III].

[83]RL19062 verso [III].

[84]K.D. Keele, *Leonardo da Vinci on Movement of the Heart and Blood.* (London, Harvey and Blythe, 1952), 71.

[85]RL 19063 verso [VI].

[86]RL 19073 verso [I].

process to shorten from apex to base: "S N a hard muscle [*papillary muscle shown on the same page in the centre of the ventricle*], contracts and is the primary cause of the movement of the heart, and in contracting it enlarges, and in enlarging it shortens and draws back all the muscles below and above it, and closes the gate M [*mitral valve*] and shortens the distance interposed between the base and the apex of the heart, whence it comes to empty itself, and to attract to itself fresh air."[87] This process of ventricular shortening is indeed an important component in ventricular emptying and is measured on a regular basis in the assessment of ventricular performance. Leonardo describes the function of the transverse and the longitudinal muscles as follows: "And these movements are made by two sorts of muscle that is transverse and longitudinal. But the transverse are those which narrow the size of the heart and the longitudinal ones extend its length."[88]

Among the most important observational studies carried out by Leonardo was an effort to reveal normal cardiac motion. Nowhere in his notes does Leonardo give the impression of witnessing a beating working heart, so that many of his conclusions on the motion of the heart are deductions from observed structure. This experiment therefore is exceptional for him in that he was making observations from a living animal, albeit in the process of slaughter. He observed the slaughter of pigs by Tuscan farmers. The method that they used was to pierce the heart of the animal with a spike. As the point moved in the ventricular muscle, the reverse motion could be seen in the part of the spike that remained outside the chest. The section begins "Anatomy":

> Whether the heart changes its position at death or not: The change of the heart through its death is equal to the change that it makes in the expulsion of its blood, and somewhat less. This is shown when they kill pigs in Tuscany where they pierce the heart of the pigs with an instrument called a "spillo" with which wine is drawn from casks and so turning the pig on its back and fastening it down firmly they pierce its right side together with its heart with this "spillo" pushing it straight inside. And if this "spillo" pierces the heart when it is elongated, in its expulsion of blood the heart shortens and draws the wound upwards together with the point of the "spillo" and as much as is raised the point of the "spillo" inside so much it lowers the handle of the "spillo" outside. And then when the heart is distended and pushes this wound downwards then the part of this "spillo" outside makes a movement contrary to that part inside which moves together with the movement of the heart. And this it does many times in such a way that at the end of life the external part of the "spillo" remains in the middle between the two extremes at which were the last contrary movements of the heart when it was alive. And when the heart becomes quite cold it retracts by a very small fraction and is shortened by as much space as was occupied by its heat, because heat increases a body which it enters and diminishes it when it leaves. And this I have seen many times; and I have observed such measurements, and I have let such an instrument remain in the heart until the animal was cut up.[89]

[87]RL 19029 recto [III].

[88]RL 19081 recto [III], 3rd sentence.

[89]RL 19065 recto [I].

Whether the last comment from Leonardo means that he carried out such an experiment or simply directed the slaughter man is conjectural. The script is accompanied by small diagrams that illustrate the position of the "spillo" in the heart (Fig. 4.23).

Then follows a further lengthy discourse on how the position of the "spillo" in the heart affects its movement. He recorded the movement of the heart in systole and diastole by following the handle of the instrument, which moved in the opposite direction to the heart. From this work, Leonardo demonstrated that there is little or no movement at the middle of the ventricle, and this is shown in the lower of the three accompanying diagrams. This method of demonstrating cardiac motion was first reported in the literature by Jung of Basel in 1835.[90]

In 1891, Berry Haycraft developed the recording mechanical cardiograph by attaching a light, inked straw to a needle introduced into the heart. Later the two concepts, the electrical activity of the heart and the recording of cardiac motion, came together with the advent of the electrocardiogram.

The following quotation from a page of dense notes with no drawings suggests that Leonardo had observed the changes in shape of the heart at different times after death: "How the heart widens only at the death of a man without changing its position at its base. The heart widens and shortens at the death of a man because its transverse muscles extend and its longitudinal muscles are drawn up and thus only the lower part is raised and not the upper."[91]

This description by Leonardo from a human heart is unusual, although we shall see later that there is some visual evidence of his access to human hearts for dissection in at least one of his drawings.[92] This sheet was probably written during his last period of anatomy, whilst he was in Rome (circa 1513–14). This assumption is strengthened by a note in the margin that admonishes his detractors, "Make a discourse on the censure required by students who hinder anatomy, and abbreviators thereof."[93] It was at that time that Leonardo was in conflict with a German mirror maker with whom he shared rooms in the Vatican palace.

Thus the heart of Leonardo was a pump made of a special type of muscle with the property of continual movement. It could contract with great force and was responsible for the impetus to the blood to move it to and from the lungs and the periphery. It could be regulated by external impulses from the vagus nerves, was responsible for generating heat, and was capable of withstanding great heat. It was not the seat of the soul but was the quartermaster of the vital spirits (pneuma) for the brain. With the exception of the concept of a circulation of the blood rather than an ebb and flow movement, these concepts remain true today.

[90]C.D. O'Malley, J.B. de C.M. Saunders, *Leonardo da Vinci on the Human Body*. (New York: Greenwich House, 1982), 234.

[91]RL 19063 verso [III].

[92]RL 19119 verso, Drawing of the opened right ventricle.

[93]RL 19063 verso [V].

The External Appearance of the Heart and the Coronary Vessels

The existing collection of Leonardo's manuscript pages contains very few detailed drawings and little in the way of written notes on the surface anatomy of the heart, but that is not to say that he does not succeed in rendering a good external visual image of this structure. The main drawings of the overall perspective of the ox heart are so clear that few words are necessary other than to guide the student to the important structures. The most significant sheets are RL19074 verso and RL 19073 verso, which together represent perhaps Leonardo's most finished work on the subject, although even here there is a note on further work that needs to be done: "As for the third vein I have not yet seen if it has an artery within it, so I am going to peel off the flesh to ascertain this".[94]

The other significant images are RL 19071 recto and RL 19072 verso, which display the back of the ox heart and the arrangement of the arteries that supply blood to the airways. Only two others, RL 19118 recto and RL 19119 verso, have small drawings of the surface anatomy of the coronary arteries.

Much of the commentary relating to the surface anatomy of the heart is of the coronary arteries, probably because it is these structures that bound and define the chambers of the heart. Perhaps the earliest extant comment by Leonardo on his desire to illustrate the vessels on the surface of the heart is found on RL 19028 recto, where he states: "Draw the veins which are in the heart, and also the arteries, which vivify and nourish it."[95] O'Malley and Saunders date this sheet around 1504–1506.[96]

The external appearance of the ox heart and its coronary anatomy seen on RL 19074 verso and RL 19073 verso were originally together as one. On RL 19074 verso, the heart is shown from two viewpoints at approximately 45 degrees to one another, in keeping with Leonardo's demands of multiple views to better understand the structures in question. The heart is enveloped in a fibrous sac called the pericardium. Leonardo recognises its presence as a part of cardiac anatomy and notes: "First make [*draw?*] the capsule [*pericardium*]."[97]

[94]RL 19073 verso [VIII].

[95]RL 19028 recto [II].

[96]C.D. O'Malley, J.B. de C.M. Saunders, *Leonardo da Vinci on the Human Body*. (New York: Greenwich House, 1982), 282.

[97]RL 19072 verso [VI].

Fig. 4.23 Diagrams to show the rotation of spikes when placed in the heart of the pig. Detail from RL 19065 recto, The movement of the heart. Leonardo da Vinci. c.1511–12. Pen and ink (Lent by Her Majesty The Queen. Royal Collection © Her Majesty Queen Elizabeth II)

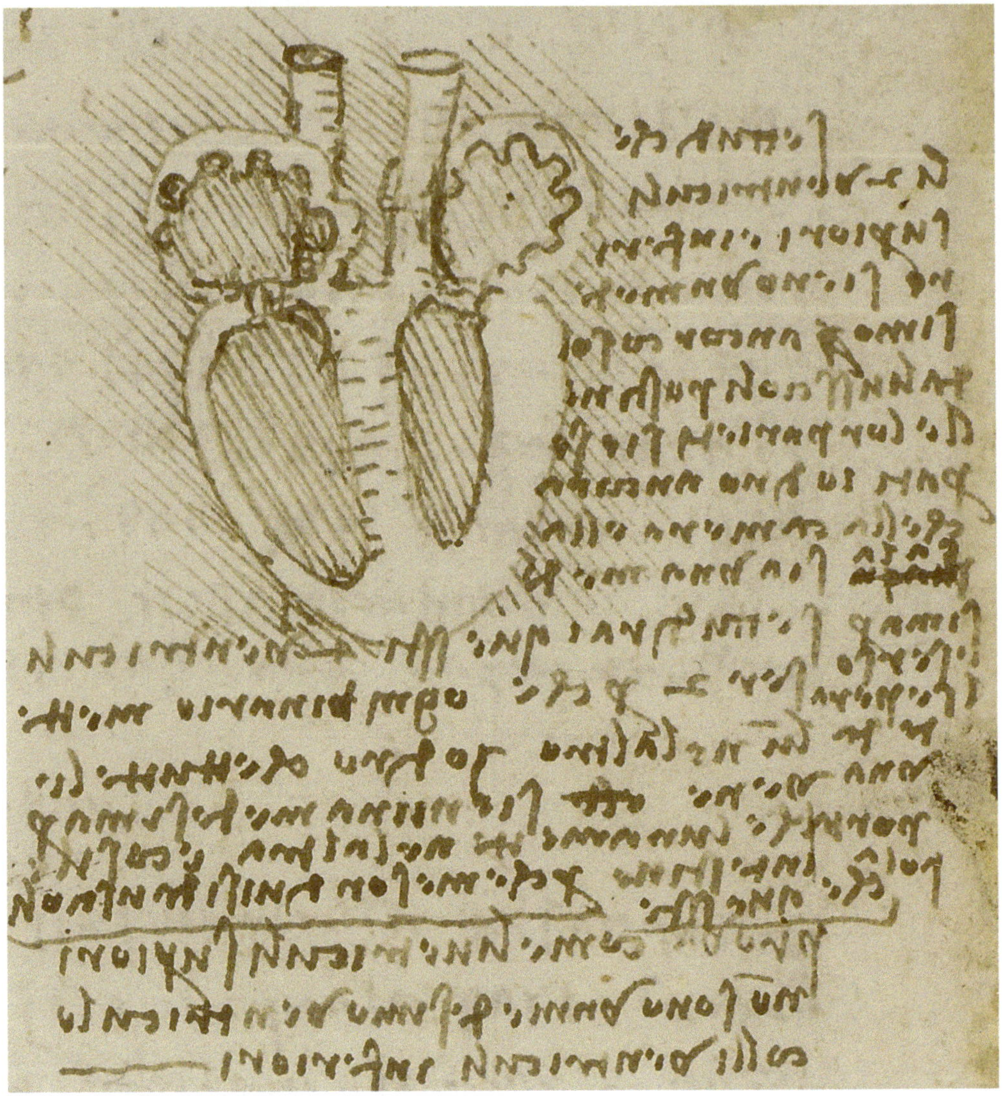

Fig. 4.24 Drawing illustrating the heart as a four-chambered organ. Detail from RL 19062 recto, The atria and ventricles of the heart. Leonardo da Vinci, c.1511–12. Pen and ink (Lent by Her Majesty The Queen. Royal Collection © Her Majesty Queen Elizabeth II)

Leonardo recorded the first two drawings that he should make to reveal the form of the heart: "First make the heart with its vessels from two aspects, that is from behind and in front."[98] He described the relative lengths of the right and left ventricles from the surface view and noted correctly that "the right ventricle extends to a depth of three-quarters of the length of the heart."[99]

The main discussion on this page (RL 19074 verso), as on so many others, is of a more general nature regarding the function (physiology) of the heart. In this case it is of "the subtilisation of the blood" by its percussion[100] between the lower "intrinsic ventricles" and the two upper "extrinsic ventricles" called "auricles".

From the time of Galen anatomists had regarded the heart as a two-chambered structure. These chambers were the intrinsic ventricles (lower or actual ventricles) of Leonardo. The atria were regarded merely as extensions of the great veins. In the following discourse Leonardo took issue with this view:

> Reply to the adversary against the number of ventricles saying that they are two not four because they are continuous and not united together, the two right by themselves making the same one, and the left doing likewise. Here one replies that if the right and the left ventricles are one single right and one single left ventricle only, it is necessary that during one and the same time they should perform one and the same function, and not opposite functions on the right side as is manifested by their flux and reflux. Furthermore, if it is one and the same, there is no need for the membranous doors [*valves*] which separate one from the other; and if it is one and the same [*ventricle*] there is no need for one part to open when the other closes.... Therefore it is concluded that the upper ventricles of the heart [*atria*] are different in their functions, in their substance, and in their nature from those below, and that they are separated by gristle and various substances interposed between one and the other, that is the sinewy membrane and much fat[101] (Fig. 4.24).

[98]RL 19072 verso [VII].

[99]RL 19074 verso.

[100]"Percussion" is defined by Leonardo as the end of the swift motion of bodies against resisting objects. It may be simple or complex. Paris manuscript B 27 verso.

[101]RL 19062 recto [I] 3rd paragraph.

It is worth quoting here another significant rebuttal of the two-chambered heart theory, as it was such an important departure from the current view of Leonardo's time. He began by using his perceived universal law of nature, that everything exists for a reason: "On the Heart: Whether Nature make the right ventricle larger and leave out its upper ventricle [*auricle*] or not. Nothing is superfluous and nothing is lacking in any species of animal or product of Nature, unless the defect comes from the means by which she produces it. Therefore it follows that the upper right ventricle was necessary for the flux and reflux of the blood which is generated through the help of this ventricle."[102] His emphatic rebuttal of the Galenist view on the number of cardiac chambers is irresistibly argued from observed anatomical features but is reinforced by the Galenic concept of heat production and his view of their physiological necessity for this production to occur.

On another sheet, after restating the heart has four ventricles, "…that is two upper called auricles of the heart, and two below these called right and left ventricle"[103], Leonardo began the inaccurate description of the weight of the right ventricle based upon supposition of its position in the chest, which was discussed earlier. This point does suggest that he may have been recounting the observed position in the human chest rather than that of the ox, where it lies more vertical.

The drawings on RL 19074 verso display the important relationship of the coronary arteries to the aorta, the pulmonary artery, and the atrial appendages. A note on this page correctly identifies the origin of the coronary arteries from the first part of the aorta after it has arisen from the left ventricle: "a b are two arterial vessels [coronary arteries] which arise in two openings outside the left ventricle."[104]

The right and left coronary arteries arising from the base of the aorta can be clearly seen in the right-hand drawing of the two main drawings on sheet RL 19074 verso (Fig. 4.25). In this drawing, the pulmonary artery has been removed just above its origin, giving a clear view of the base of the aorta. In the undisturbed anatomy, the pulmonary artery arises between these two coronary arteries, obscuring the view of their origin. This dissection has clearly been made to allow the inspection of this important relationship.

Whilst clearly and correctly identifying the pulmonary artery in the image, Leonardo appears to have made an error in his description of the origins of the two arteries. He stated that both of those great arteries arose from the right ventricle: "The right ventricle has two [*valvular*] orifices, one into the 'vena aorta'; they open from within outwards. The other orifice is in the venous artery [*pulmonary vein–left atrium*] which goes from the heart to the lung and has only one tunic and it is called artery (because of the subtelised blood) and vein because of the enlarged vessel [*atrium*]."[105] However,

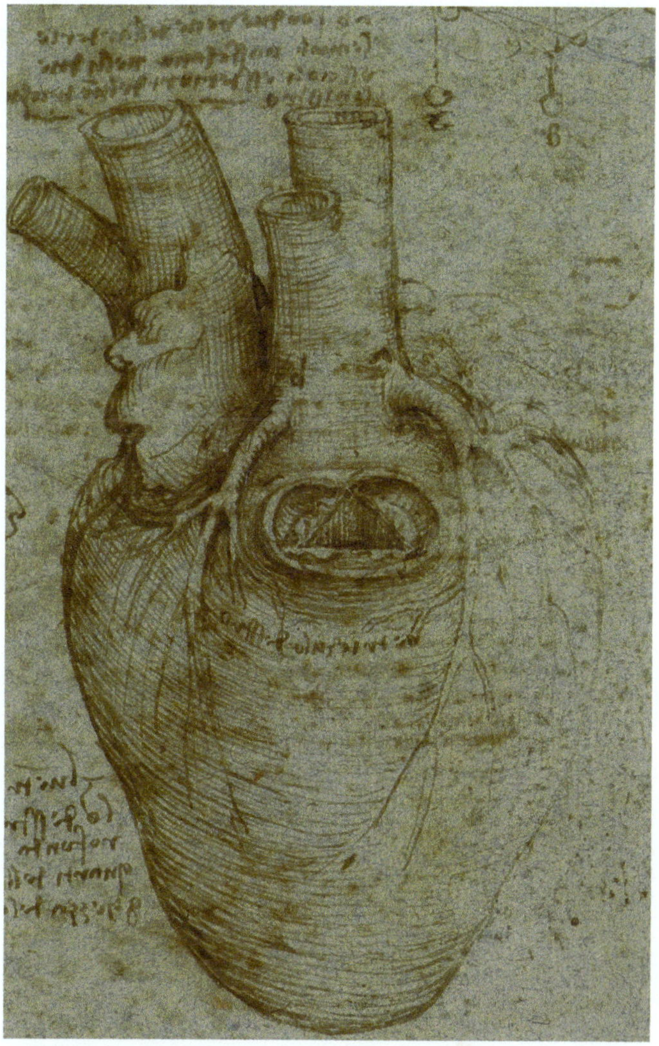

Fig. 4.25 Root of the aorta from behind with the pulmonary trunk removed. This reveals the origin of the two main coronary arteries labelled *a* and *b*. Detail from RL 19074 verso. The heart and coronary vessels. Leonardo da Vinci, c. 1511–13. Pen and ink on blue paper (Lent by Her Majesty The Queen. Royal Collection © Her Majesty Queen Elizabeth II)

the full text indicates that he was referring to a chamber that had an inflow from the lungs (the left atrium) and an outflow which he named the aorta. It is likely that this is a simple mistake, having written "right" when he meant "left" (a "lapsus calami" as suggested by Boruttau in 1913).[106]

The detailed anatomy of the coronary arteries is unique in that period. The drawing at the left side of RL 19074 verso demonstrates the left main coronary artery arising from the aorta and then dividing into its principal branches of left anterior descending and circumflex arteries. The circumflex vessel soon disappears under the left atrial appendage (auricle) and must be formally dissected out to be seen properly (Fig. 4.26).

[102]RL 19063 verso [I].

[103]RL 19087 verso [I].

[104]RL 19073 verso [II].

[105]RL 19072 verso [IV & V].

[106]C.D. O'Malley, J.B. de C.M. Saunders, *Leonardo da Vinci on the Human Body*. (New York: Greenwich House, 1982), 220.

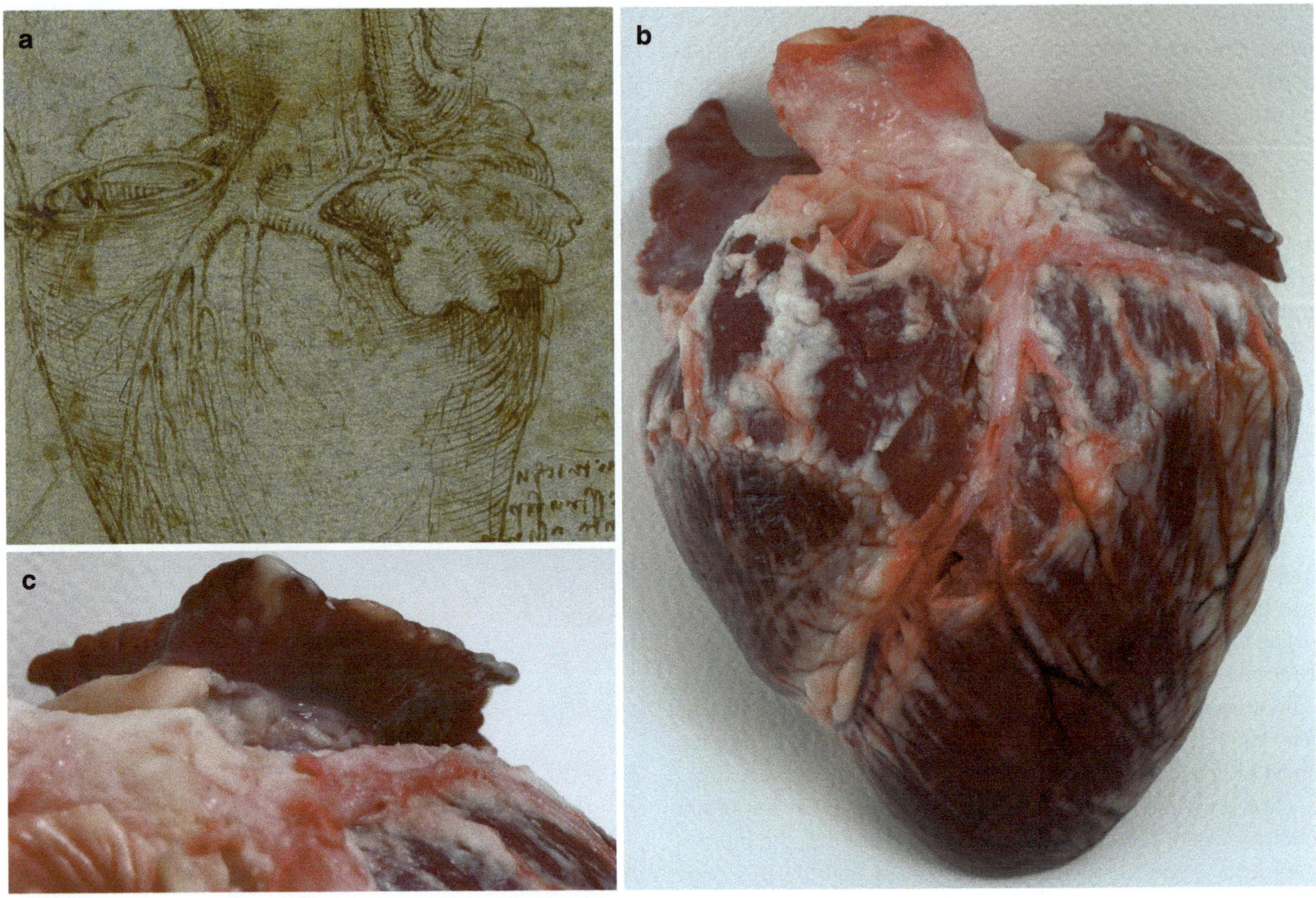

Fig. 4.26 (**a**) This drawing illustrates a dissection of the left main coronary artery with its major divisions. (**b**) Close-up photo of dissected heart to show the same detail. (**c**) Detail of the relationship between the coronary vessels and the base of the left atrial appendage. Detail from RL 19074 verso. The heart and coronary vessels. Leonardo da Vinci , c. 1511–13. Pen and ink on blue paper (Lent by Her Majesty The Queen. Royal Collection © Her Majesty Queen Elizabeth II)

Leonardo recognised the consistent position of the coronary arteries on the surface of the heart. On sheet RL 19119 verso is a description of the surface anatomy of the heart relating, correctly, how the arrangement of the coronary vessels demarcates the edges of the right ventricle. The following words relate to the small drawings at the top right-hand part of the sheet (Fig. 4.27):

> The heart has its surface divided into three parts by three veins descending from its base, of which two delineate the edges of the right ventricle [*great and small cardiac veins*]. And they have below in contact with them two arteries [*anterior descending branch of the left main coronary artery and the right coronary artery*]. As for the third vein, I have not yet seen if it has an artery with it, so I am going to peel off the flesh to ascertain this. But the superficial area of the heart enclosed between these arteries occupies half of the superficial circular size of the heart and forms the external wall of the right ventricle.[107]

This passage is enlivened by Leonardo's commentary on the progression of his dissection. This is clearly work in progress; though it is correct in its observations thus far, it is not a finished statement on the anatomy of the part.

On another sheet (RL 19073 verso), Leonardo correctly stated that the right coronary artery is the source of blood for the right ventricle, having again defined the margins of that ventricle by the position of the coronary arteries:

> The vessels a c [*anterior interventricular branch of the left coronary artery*] and b c [*posterior interventricular branch of the right coronary artery*] surround the edges of the right ventricle. And the arterial vessel [*aorta*] sends the branch o p [*right coronary artery*] to nourish the wall a b c which makes the complete covering of this right ventricle.[108]

The relationship of the coronary vessels to the surface of the heart is again addressed on RL 19119 verso (Fig. 4.19). Beneath the upper, more representational drawing is a

[107]RL 19119 verso [VIII].

[108]RL 19073 verso [X].

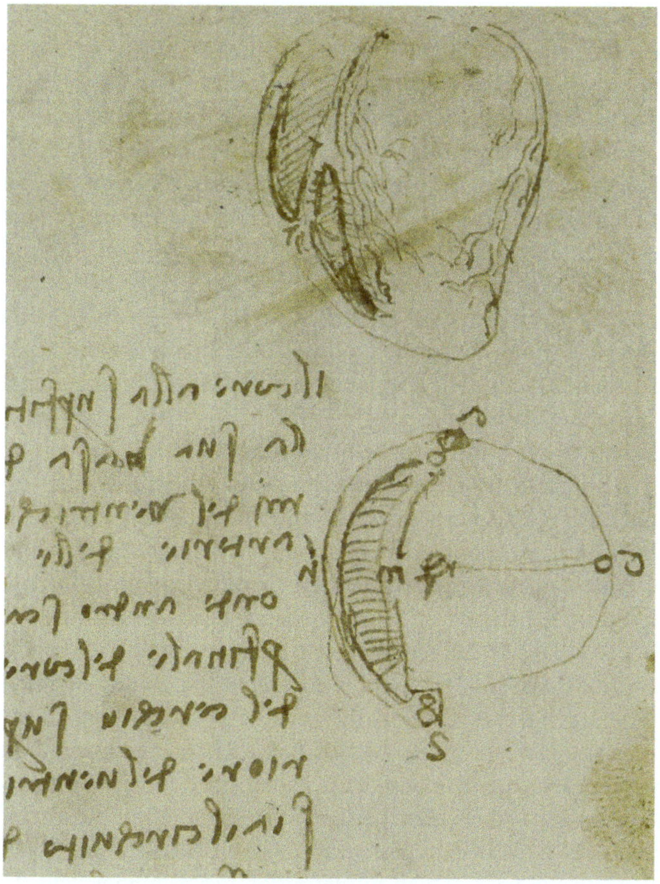

Fig. 4.27 Sketch to show the relationship of the coronary arteries to the outer aspect of the ventricles. Detail from RL 19119 verso. Leonardo da Vinci, c. 1511–13. Pen and ink (Lent by Her Majesty The Queen. Royal Collection © Her Majesty Queen Elizabeth II)

schematic one showing in cross-section the relationship of the coronary vessels to the surface of the right ventricle and incidentally, correctly displaying the crescentic shape of the right ventricle draped over the left ventricle, which remains as an empty circle.

The heart is an approximately pyramid-shaped organ and as such has a base and an apex. The base of the heart is the broader end to which the atria are attached and from which the great arteries arise. On manuscript sheet RL 19119 recto, in a short section headed by Leonardo "Base of the Heart", he likened the external shape of the heart to a triangle viewed from the base, again highlighting the relationship of the vessels on the surface of the heart:

> The shape of the base of the heart is somewhat similar to a triangle, as a c f shows. And at each angle are two vessels, that is, a vein outside and an artery inside below the vein. a c f are the veins, and b [d] e the arteries. Between c f is the right ventricle. And at the centre of the base of the heart is the origin or base of the aortic artery [o m n] established on the middle of the base of the heart, keeping the principal position at the base of the heart, just as it keeps the principle of the life of the animal. And the angles of the valve cusps of this artery are turned towards the angles of the base of the heart and the sides of the valve cusps towards the sides of the heart.[109]

A second note relates to the small drawing at the bottom of the page, which views the heart from the pointed apex looking upward toward the base: "a b, is the right ventricle, and there are many occasions in which the right and left ventricles are joined together in the region of the apex of the heart."[110] Here Leonardo describes the termination of the three principal coronary arteries and notes the interconnection between the finer branches of the right and left coronary arteries. These connections (collaterals) are important as alternative routes for blood to reach areas of muscle whose direct arterial supply may be obstructed by coronary artery disease (atherosclerosis). In that condition, the obstructions occur closer to the origins of the vessels than to their termination, and frequently occur at areas of division. Thus connection to other vessels toward their termination allows retrograde flow into the obstructed vessel, which can be lifesaving for some people. If the arterial occlusion progresses slowly, these interconnecting vessels enlarge and may maintain an adequate flow for moderate exercise for many years (Fig. 4.28).

Leonardo was the first to correctly record in detail the anatomy of the coronary veins. It is clear that he treated both the arteries and the veins with the same degree of importance. He described the relationship of the veins with the coronary arteries and the surface of the heart. He described the veins as lying on top of the arteries, except in some cases where ramifications of the arterial branches may cross the vein. On RL 19071 recto, he wrote, "The artery lies below the vein; it is true that some of its branches sometimes lie above the vein."[111]

In general, venous anatomy is more difficult to trace accurately, but Leonardo clearly identified the main venous tributaries, culminating in the coronary sinus. This is the terminal vein that enters the right atrium and is shown in the top two large drawings on RL 19073 verso. The left-hand drawing correctly identifies the coronary sinus "S G" accompanying the circumflex branch of the left coronary artery and entering the right atrium. The drawing to the right of it reveals the confluence of the great and middle cardiac veins as they pass under the left atrial appendage "N". These are complex relationships that are not easy to establish, as learned through my experience of dissection of the ox heart, even with prior knowledge of the anatomical arrangements. To correctly identify all of these connections, the vessels must be kept intact. The easiest way to accurately accomplish tracing them

---

[109]RL 19119 recto [VII].

[110]RL 19119 recto [VIII].

[111]RL 19071 recto [III].

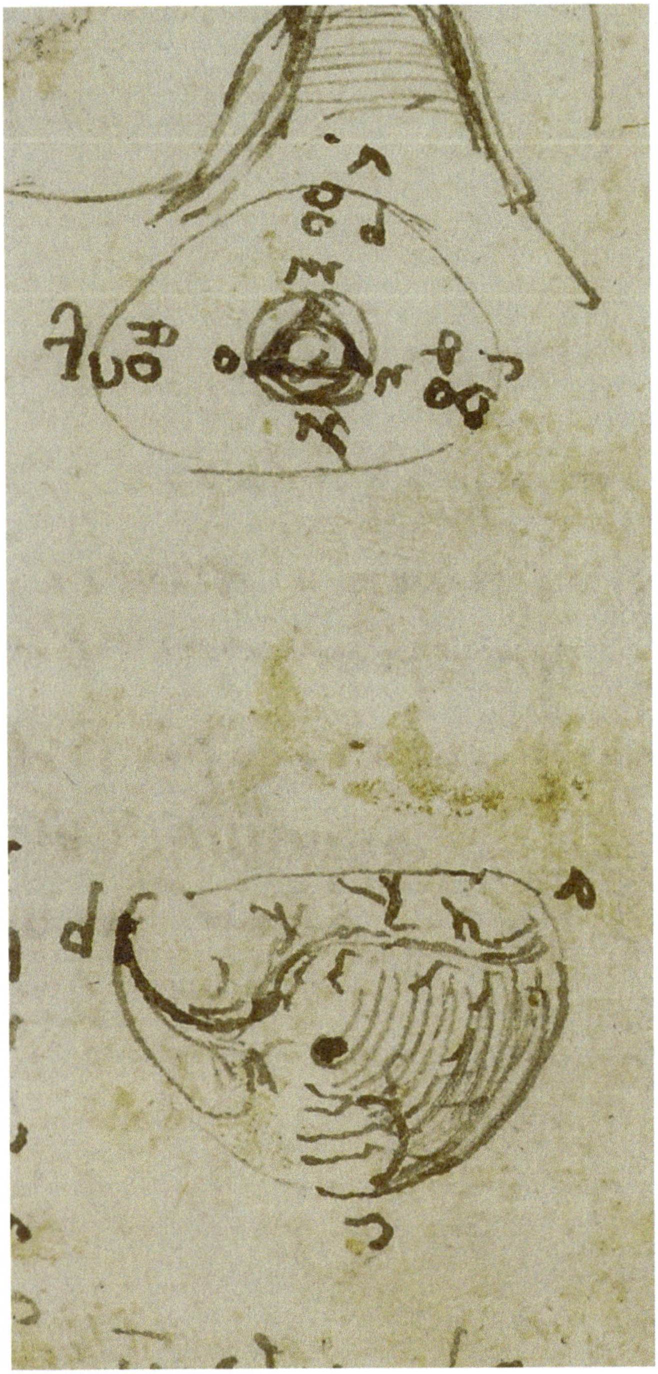

Fig. 4.28 Sketch to show the termination of the coronary arteries at the apex of the heart. Detail from RL 19118 recto, The right ventricle and the valves of the heart. Leonardo da Vinci, c. 1511–13. Pen and ink (Lent by Her Majesty The Queen. Royal Collection © Her Majesty Queen Elizabeth II)

is to pass a fine, malleable probe into their lumen and then dissect down onto them. There is no evidence that Leonardo used this technique.

On manuscript sheet RL 19073 verso, he noted that "the artery [*left coronary, H*] is always beneath the black vein [*great cardiac vein*]"[112] On sheet RL 19071 recto, he recorded another important fact with regard to the location of the coronary vessels. He stated, "On the veins that nourish the heart. They are always exposed on the heart; and their lateral parts are surrounded by tallowy fat, that is, the dense fat".[113] He continued in the next paragraph on the same sheet with the following observation: " The veins and the artery joined together have a membrane, finding this vein and artery apposed to the other flesh, clothes half of their thickness; the other half being clothed by the flesh to which they are applied. But if they are not applied to this flesh, then this membrane surrounds them, or is continued farther and is continued into space interposed between the bifurcation of the vessels and maximally in the substance of the lung".[114]

This factually correct description of the arrangement of the arteries and veins in relationship both to each other and to the structures immediately around them holds significance for the cardiac surgeon when carrying out coronary artery bypass graft surgery. First, the visceral pericardium must be incised. This is Leonardo's "outer membrane". If the back wall of these small vessels is accidentally partially incised, the surgeon may have to rely on the normally very close apposition of the back wall of the artery to the muscle to prevent leakage of blood. This is described by Leonardo as "the flesh to which they are applied". Quite often the artery that is in need of a bypass graft must be dissected out from beneath the vein, which by definition has a thinner wall. If not kept intact, troublesome bleeding may ensue.

The presence of the "dense fat"[115] always found around the coronary arteries, occasionally in sufficiently large amounts to completely obscure the vessels, frequently bars the easy discovery of the exact site of the vessel to be exposed. Sometimes in these dissections they can be impossible to find, bearing witness to the difficulties that Leonardo could have experienced.

Detailed drawings of the coronary anatomy are found on RL 19073 verso. In the lower two drawings, Leonardo shows them dissected out and elevated from the underlying muscle. This is difficult to accomplish and requires a great deal of patience. It can be achieved much easier in a specimen less than 6 h after it has been removed from the carcass of the freshly killed animal. After that time, the fat encasing the vessels begins to harden significantly. We cannot say whether this drawing represents an actual dissection or is purely schematic.

These drawings correctly identify the progressive narrowing of the arteries, noting that they diminish in diameter after

[112]RL 19073 verso [V].

[113]RL 19071 recto [I].

[114]RL 19071 recto [II].

[115]RL 19071 recto.

each division. Leonardo says this about the nature of the branching of these vessels: "The ramifications of the vessels are always that much larger as is the larger trunk from which they arise, that is the main vessels; it follows likewise in the ramifications of the ramifications right to the end."[116] Next to a small drawing at the bottom of this page, which depicts the dissected vessels alone, he wrote: "Vessels are always larger within the bifurcations of their trunks than they are beyond these trunks."[117]

Leonardo recognised that the branching of structures in nature followed a universal law: The total cross-sectional area of a branching structure (such as a tree trunk, a blood vessel, or the airways of the lung) is always equal to the cross-sectional area of the parent trunk at all points of division until the final one. He wrote, "All the branches of trees at every stage of their height, united together, are equal to the thickness of their trunk. All the ramifications of watercourses at any degree of their length, if they are of equal motion, are equal to the size of the parent stream."[118] This law was derived from his appreciation that the optimal flow of a fluid passing through a tube of any size is always proportional to the cross-sectional area of that tube. Hence, if flow is restricted but passed through a smaller or a larger tube, then it will become turbulent and thereby damage the walls of the tube constraining it. This inherently unstable situation would be contrary to the laws of nature and could not be sustained, as it would cause erosion or destruction of the tube.

The arrangement of the vessels on the back of the heart is less easy to define, as the arteries and vein interlace to some degree. Leonardo, in a note referring to the small group of four diagrams to the right hand side of sheet RL 19073 verso, described this in the following way: "The heart viewed from the left side will have its veins and arteries intersecting like someone crossing his arms. And they will have above them the left auricle and inside them the gateway with triplicated triangular doors [*aortic valve cusps*] and the opened gateway will remain triangular."[119] This arrangement is shown in the four small drawings on the right-hand side of the page. (See detail from RL 19073 verso.)

These manuscript pages reveal that his descriptions of the coronary vessels and their position on the heart were derived from accurate observation of careful and probably multiple dissections. The results were recorded using a clear drawing style allied to concise verbal descriptions, so that only refinements in detail of little practical importance can be added to the present day.

[116]RL 19074 verso [V].

[117]RL 19074 verso {VI].

[118]Paris manuscript I folio 12 verso [IV and V].

[119]RL 19073 verso [IX].

The Internal Structure of the Heart

All new research projects must begin with an awareness of what has gone before and the state of the art at the time. All researchers, no matter how great, operate in the climate of the times, and so it was for Leonardo. As we have seen so far, he was inevitably constrained by the academic context of the late middle ages. The more one studies the detail of his text and his drawings, however, the more apparent it becomes that he was willing to strive to see beyond Galenic dogma and to realise his own interpretations. Careful reading of the manuscript pages that relate to the heart and its mechanisms reveals the contrasting nature of the information within the text and the information conveyed in the images. Despite the polemic surrounding the function of the heart, Leonardo began to tease out significant new structural information that remains correct and relevant today. On superficial inspection, it is easy to be blinded to the originality of many of the thoughts he expressed. The forthcoming sections will show that even in the closely argued passages relating the function of the heart as a furnace, there are gems of deduction that impinge upon modern debates about heart mechanisms. In his rhetoric on the necessity of the heart to have two upper and two lower chambers, Leonardo began to unpick the details of the hydrodynamics of blood flow between the two chambers.

Much of Leonardo's work on the internal structures of the heart is original. In his work on the valves of the heart, the imagery created by his prose integrates with the accompanying drawings more than in many other areas of his anatomical work. The drawings are outstanding in both their accuracy and the level of information that they contain.

We have discussed the principal types of drawing that are to be found in Leonardo's work on the detail of the interior of the heart, predominantly the representational and the diagrammatic types. As one might expect of such a good draughtsman, both are used to good effect. In the sections on the aortic/pulmonary valve, most drawings are diagrammatic, showing a general scheme or outline of an object or its parts, whilst in the section on the atrioventricular valves, many drawings are accurately representational, seeking to portray the physical appearance of the object. The pages that deal with the overall function of the heart include some drawings that are more schematic in type (representing objects by symbols), such as the analogy of the seed (nut) as the heart.[120] These aid the verbal explanations by the use of metaphor.

The quality of Leonardo's drawing on these pages is defined more by their representational accuracy than by their finesse or their degree of completion. Clark described them as having a "deliberate carelessness of touch as though

[120]RL 19028 recto.

Leonardo was denying himself the comeliness of his earlier style."[121] This terseness of style is not reflected in his writing and therefore cannot be ascribed to any physical decay in Leonardo himself. It is my impression that this work was so challenging that these drawings represent more a form of visual note-keeping than finished illustrations.

The Atria and the Atrioventricular Relationship

In examining Leonardo's drawings of the atria (auricles), it is possible to follow his developing appreciation of their autonomy, connections, and structure. In perhaps his earliest drawing of the vascular tree,[122] dated c.1487 or even earlier,[123] the heart is almost secondary to the liver and is represented as a single-chambered structure (See RL 12597 recto); the same relationship is shown in cross-section on RL 19112 recto. The atria are subsumed into the veins. This is Leonardo's visual summary of his understanding of the Galenic circulation. In RL 19104 verso (c. 1504), the situation has not changed. The venae cavae enter the ventricles directly, and the pulmonary vein enters the left ventricle with no sign of a left atrium. The detail of this drawing is so similar to that of the earlier one that it is difficult to believe that it was done 20 years later. Perhaps in returning to his anatomy at that time, Leonardo was picking up from his previous thoughts as he gathered himself for a fresh and more autonomous assault. By perhaps 1505, as shown in RL 19112 recto, he had begun to establish visually the separation of the atrial chambers, as there is clear identification of a left atrial appendage, although the caval veins can be seen confluent with the pulmonary veins in the absence of a formal chamber (see RL 19112 recto).

By 1513, there is clear separation of the atria, as revealed in a small drawing on RL 19062 recto that accompanies a carefully worded argument for their existence. In the same period, other depictions of the right atrium (e.g., detail at the top of RL 19084 recto) are equally clear in establishing this new fact. RL 19062 recto even depicts the rough zone, the so-called trabecular portion of the atrium.

Leonardo's keen observation also noticed an abnormality in one of his dissections; he reported, "I have found from 'a', left ventricle (auricle), to 'b', right ventricle, a deep hole from a to b, which I note here in order to see whether this is so in the other auricles of other hearts."[124] (Fig. 4.29)

Despite the confusion in his terminology of the "ventricles" with the "auricles", this short passage is likely to be the first description of a significant congenital heart defect anywhere in the literature. We now refer to this as an atrial septal defect (in the vernacular, a type of "hole in the heart"), which is the most common congenital defect found in the human heart.

Leonardo integrated his anatomical observations of the atria within the polemic of his functional theory of the heart in the way that has already been described, but because of his novel assertion that there are two chambers connected in series on each side of the heart, it was necessary for him to generate original ideas on the way the two chambers would work together. Two significant stumbling blocks prevented a full realisation of their true interaction. The first was the absence of an appreciation of a continuous circulation, which made the fictitious ventricular pores a necessity. The second was a need to explain how the vital spirit contained in air gained access to the blood. In these two deficiencies, it is possible to see the carefully argued reasons behind his partially inaccurate descriptions of atrioventricular workings. The bulk of these arguments are to be found on RL 19062 recto and verso, RL 19063 recto and verso, and RL 19087 verso. He reasoned that the atria were separate chambers because if they were not, then there would be no need for "nature" to have placed the valves between them. "If you say that these four ventricles are two because each pair leads from one into the other, I will say that all veins are one and the same because one leads into the other, and likewise the intestines because they are divided up by little doors likewise."[125] Leonardo continued, "If you say that the two upper and lower ventricles are one and the same, although they are separated only by one such little door placed in their walls, I will further say that the chamber and the hall are one and the same because they are separated by a small doorway."[126] Leonardo then uses the analogy of the impossibility of bearing two conflicting emotions at the same time as an argument for the separateness of the chambers: "At one and the same time in one and the same cause two contrary movements cannot exist, that is repentance and desire. Therefore if the right upper and lower ventricles are one and the same it is necessary that at the same time the whole should produce one and the same effect and not two effects arising from directly contrary tendencies as one sees produced by the right upper ventricle with the lower, since when the lower contracts the upper one dilates and incorporates the blood which has been driven out of the lower ventricle."[127] Other sheets bear related passages, but these probably represent a short period of concentrated contemplation on the problem that consolidated the concept in Leonardo's mind.

[121] K. Clark, *Leonardo da Vinci*. Revised and introduced by Martin Kemp. (London: Penguin Books, 1988), 233.

[122] RL 12597.

[123] M. Kemp and D. Lorenzo, personal communication. These authorities suggest the date of 1482 for this drawing, 2009.

[124] RL 19081 recto [IV].

[125] RL 19062 recto [II].

[126] RL 19062 recto [III].

[127] RL 19062 recto [IV].

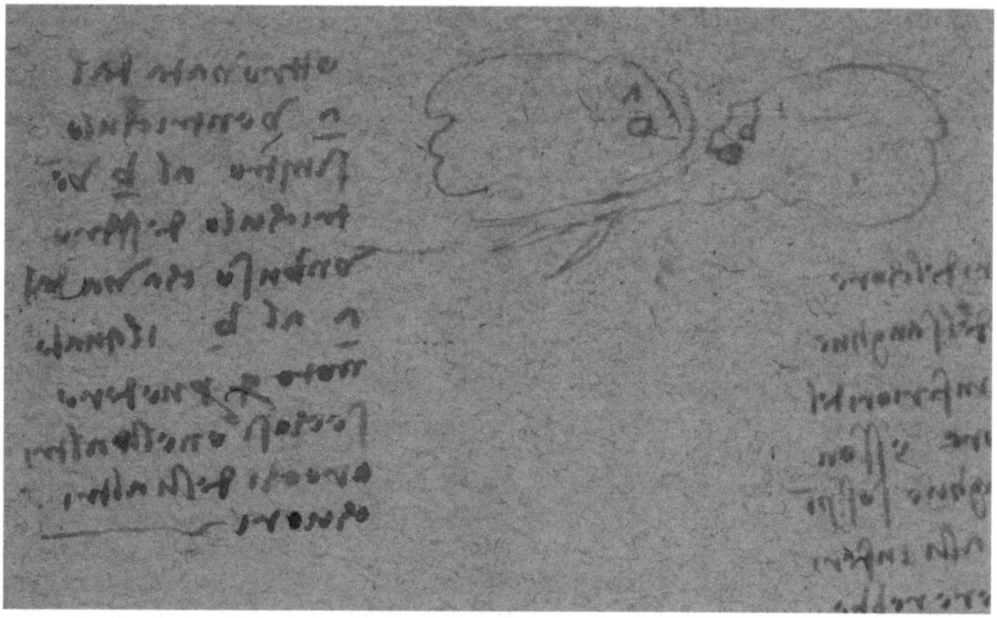

Fig. 4.29 Small sketch of the first described atrial septal defect. Detail from RL 19081 recto, Studies of the motion of the heart. Leonardo da Vinci, c. 1511–13. Pen and ink and red chalk on blue paper (Lent by Her Majesty The Queen. Royal Collection © Her Majesty Queen Elizabeth II)

Therefore, the atrium as described by Leonardo is a separate and important chamber. It has thinner walls, with only a single layer of muscle. Part of the wall is lined with irregular projections called pectinate bands in its trabecular portion. It is contractile and expansible. Two large veins (the venae cavae) connect it to the liver and the upper part of the body. This topographical description remains valid today.

Leonardo's description of the hydraulics of atrial and ventricular filling was in part true. However, except in the pathological state of tricuspid and mitral valve leakage, the ventricles cannot push blood backwards into the atria as he proposed, but his description of the mechanism of ventricular filling was quite accurate. The fact that the atria are distensible he stated thus: "These through being composed of muscles and fleshy membranes are suitable for dilation and for receiving as much blood as is pushed into them; they are also suited by their powerful muscles for contracting with impetus and driving out of themselves the blood into the lower ventricles."[128] This contractile ability is important for the generation of the atrial kick for final filling of the ventricles to maximise their elastic recoil. This he describes again elsewhere in his description of ventricular function: "And the capacity of the upper ventricle [*atrium*] diminishes by the same amount as the quantity of blood which issues from it and falls down into the lower ventricle because this [*upper*] ventricle contracts behind the blood which is pushed out by means of its muscles."[129]

[128]RL 19062 recto [I] Second paragraph, second sentence.

[129]RL 19063 verso [II].

The inspiratory phase of breathing produces a negative pressure inside the chest, and this draws more blood into the right side of the heart, an effect that increases with exercise. This effect is partly predicted theoretically by Leonardo in the following passage: "When it happens that at one and the same time the lung is dilated when the right lower ventricle also dilates, then the lung is increased inside the diaphragm and it squeezes and presses blood out of the vena cava which is closed between the lung and the dorsal spine. Some of this blood enters the right lower ventricle [via *the upper*] and just as much blood as this vein pours into that ventricle, so much less is drawn from the gibbosity of the liver."[130] Although the last phrase betrays his accuracy, as in reality blood drains from the whole of the venous system, not just the liver, it is remarkable that without being able to witness any of this either directly or experimentally, Leonardo should link cardiac and pulmonary physiology in this way.

The Interior of the Ventricles and the Atrioventricular Valves

These subjects have been grouped together because Leonardo recognised that they are mutually dependent. In modern terms, we describe this association as an "atrioventricular loop". This rather simplistic term refers to the fact that the atrioventricular valves arise from the margins of the atrioventricular junction and are linked by tendinous cords to muscular projections that arise from the walls of the ventricles, called papillary muscles. It is possible to trace an

[130]RL 19063 recto [II].

unbroken line over the surface of the atrioventricular valve leaflet, down the tendinous cords to the papillary muscle, and then along them to the ventricular wall and back up to the base of the leaflet in the atrioventricular junction, thereby forming a continuous loop. We will examine this relationship in more detail later.

Leonardo's discourse on the ventricles placed particular emphasis on the right ventricle rather than the left, as it is central to the Galenic concept of heart and lung function. It has direct access to the lungs and the liver, whereas (according to Galen) the left ventricle is connected to the right via the septal pores (the sieve[131]). The left ventricle then acts as delivery system to the periphery of the body in a pulsatile but ebb-and-flow model. Therefore we find much more detailed drawings of the tricuspid valve and the interior of the right ventricle than we do of the left-sided structures.

Leonardo used incisions described earlier by Mundinus to gain access to the interior of the heart.[132] He wrote, "Cut the heart first on the right side and begin from its point in such a way that you will not touch the opposite wall but cut along the side of the middle ventricle, and you will at once see that right ventricle and inside its two orifices."[133] Leonardo's use of these instructions can be seen in RL 19080 recto, top left.

The most realistic illustrations of the interior of the ventricles of the ox are cross-sectional representational drawings found on RL 19073 recto. Comparison of these drawings with recent photographs of the same site reveals their accuracy (Fig. 4.30). At first sight, the mass of the papillary muscle looks fanciful, but comparison with photographs reveals that they are a true representation. These very striking columns or pillars of muscles are also illustrated in longitudinal section on other pages, which address the mitral valve (Fig. 4.31) The papillary muscles in the right ventricle are less striking, which can be seen in illustrations on RL 19078 verso. These muscles are condensations of the trabecular portion of the ventricular wall. Leonardo concluded from the sheer size of these structures that they must have some role in preventing the complete closure of the ventricular cavity. He wrote, "The muscles interposed within the ventricular spaces are those which by their excessive density prevent complete closure of the right and left ventricles.... The aforesaid muscles interposed in the ventricles do not let these ventricles close completely. But they do what is drawn at a b c d; and their interposed muscles are a b c d."[134] Modern imaging has demonstrated that, whilst not actually preventing wall apposition in the ventricles, the papillary muscles do have a role in compartmentalisation of the ventricle in the different phases of the cardiac cycle.[135]

The normal ventricular wall has two distinct zones. The outer zone has dense layers of muscle with no spaces. Inside that is a layer of muscle with nooks and crannies, called the trabecular part of the ventricular wall and also referred to as the noncompacted layer because of this looser arrangement of muscle fibres. This arrangement can be seen well represented in a drawing by Leonardo of the tricuspid valve with a specialised band of muscle, the moderator band, arising from it (Fig. 4.32). The papillary muscles take their origin from this loose layer of muscle, and their relationship with the underlying compacted part of the ventricular muscle is only now coming to light. A paper by Leon Axel[136] highlights the quite unexpected somewhat tenuous association of the two. Leonardo examines this relationship to some extent in his drawing of the origin of the papillary muscle, found at the bottom of RL 19074 recto (Fig. 4.33).

Other points to note about the drawings on RL 19073 recto are the correct shape and disposition of the right ventricle, giving the appearance of being draped over the much thicker-walled left ventricle. The left ventricle forms a thick-walled, cylindrical tapering tube or cone, whereas the thinner-walled and more distensible right ventricle is wrapped around its anterior wall. The other correct detail is the depiction of the right ventricular side of the septum as having a somewhat wrinkled surface (Fig. 4.34). This crenellated appearance reveals the elastic reserve of the right ventricle to stretch to accommodate increasing blood volumes. It may also have instigated the Galenic idea that there were perforations in the septum. The pitted nature of its appearance in the collapsed heart does suggest the entrance to apertures in the muscle, although of course on probing no such apertures exist. Leonardo emphasised the appearance in a small drawing and accompanying note thus: "The dividing wall of the

[131]RL 19080 recto [VII] and [VIII].

[132]Mondino, *Anathomia*. Strasburg 1513. Ioannes Adelphus's edition.

[133]Ibid.

[134]RL 19074 recto [II].

[135]J.I. Fann, N.B. Ingels, C. Miller, "Pathophysiology of mitral valve disease and operative indications." In: L.H. Edmunds (ed.), *Cardiac Surgery in the Adult*. (New York: McGraw-Hill, 1997), 959–87.

[136]L. Axel, Papillary muscles do not attach directly to the solid heart wall. *Circulation*, 2004;109:3145–8.

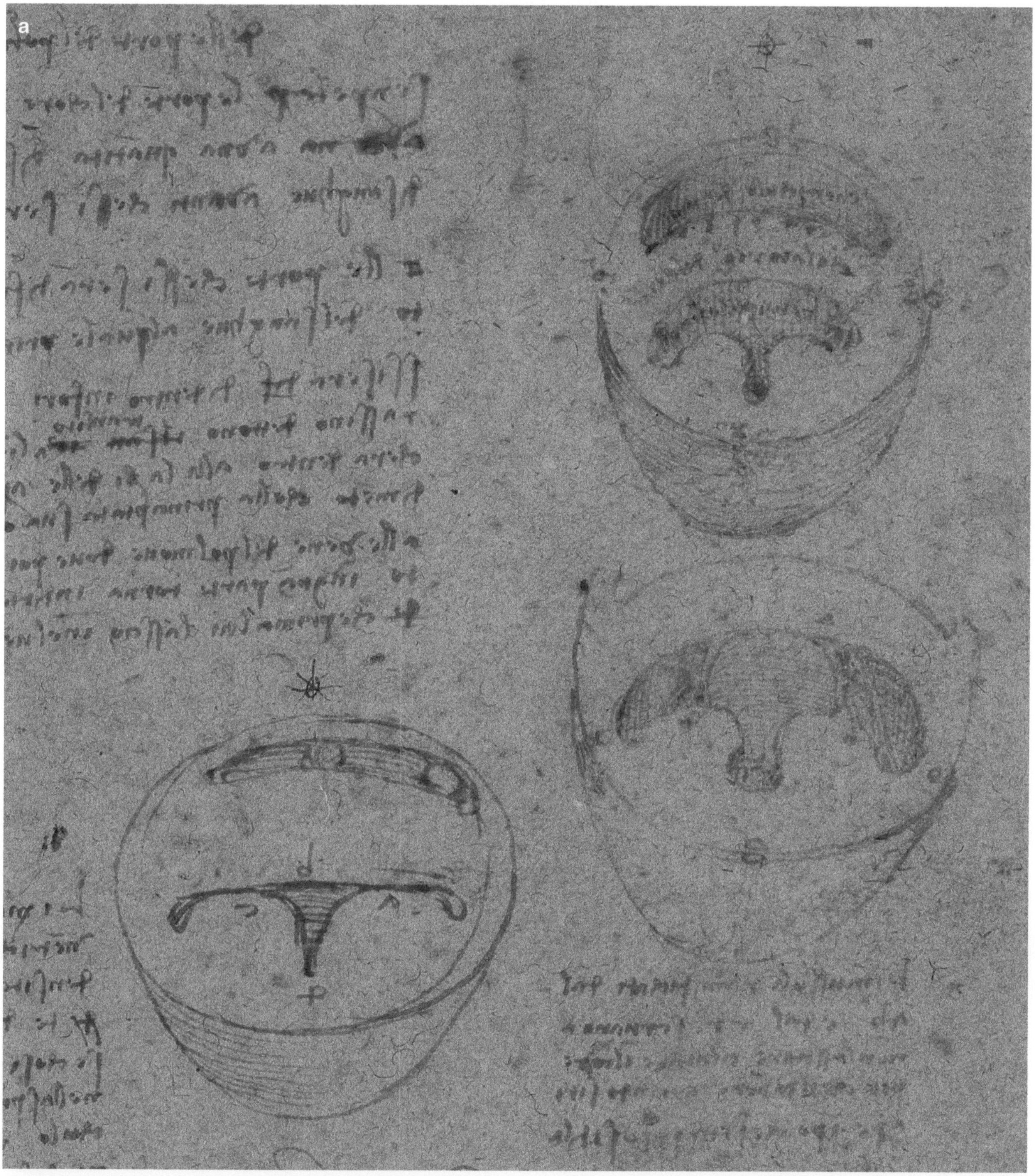

Fig. 4.30 (**a**) Drawings of cross-sections of the ox heart. Detail from RL 19073 recto, The ventricles, papillary muscles and tricuspid valve. Leonardo da Vinci, c. 1511–13. Pen and ink on blue paper (Lent by Her Majesty The Queen. Royal Collection © Her Majesty Queen Elizabeth II). (**b**) Photograph of cross-section of the ventricles of the ox heart. (**c**) Photograph of a lower cross-section of the ventricles of the ox

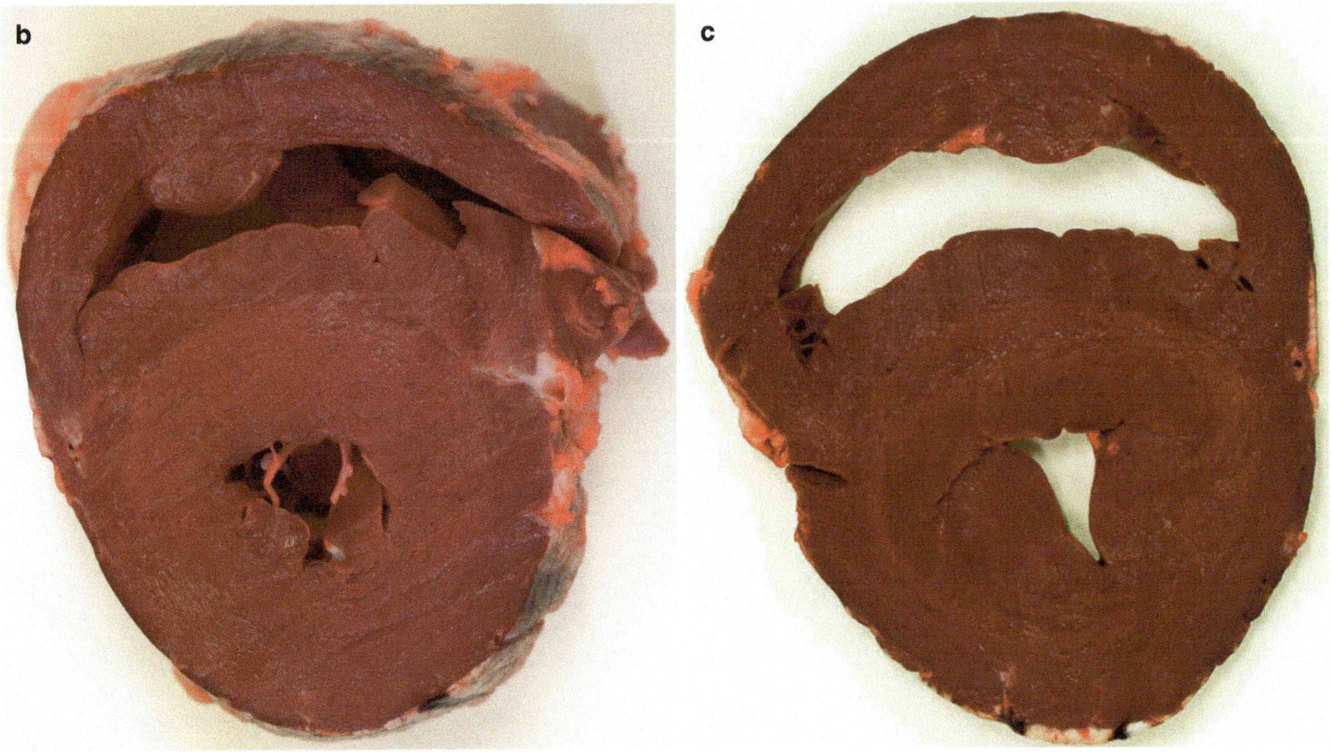

Fig. 4.30 (continued)

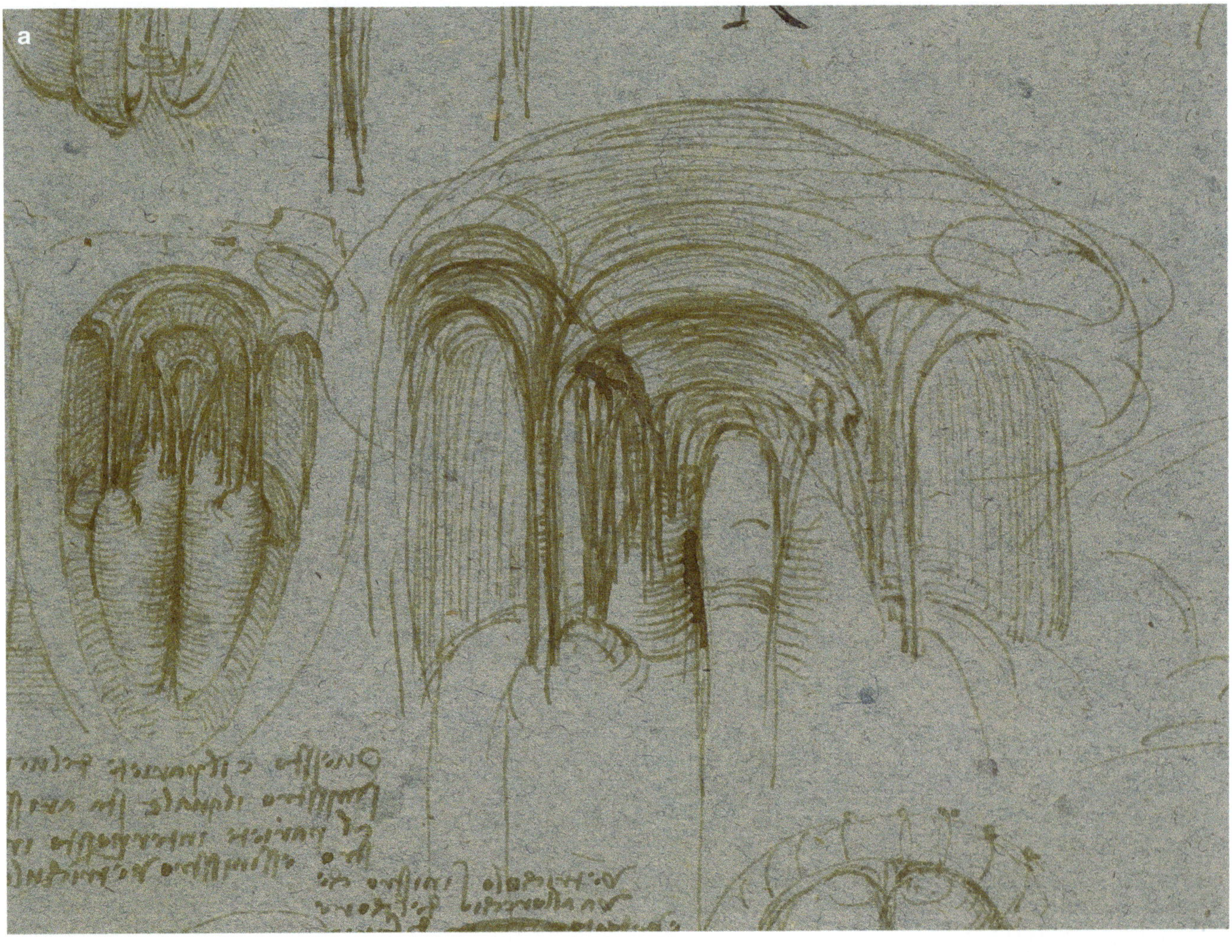

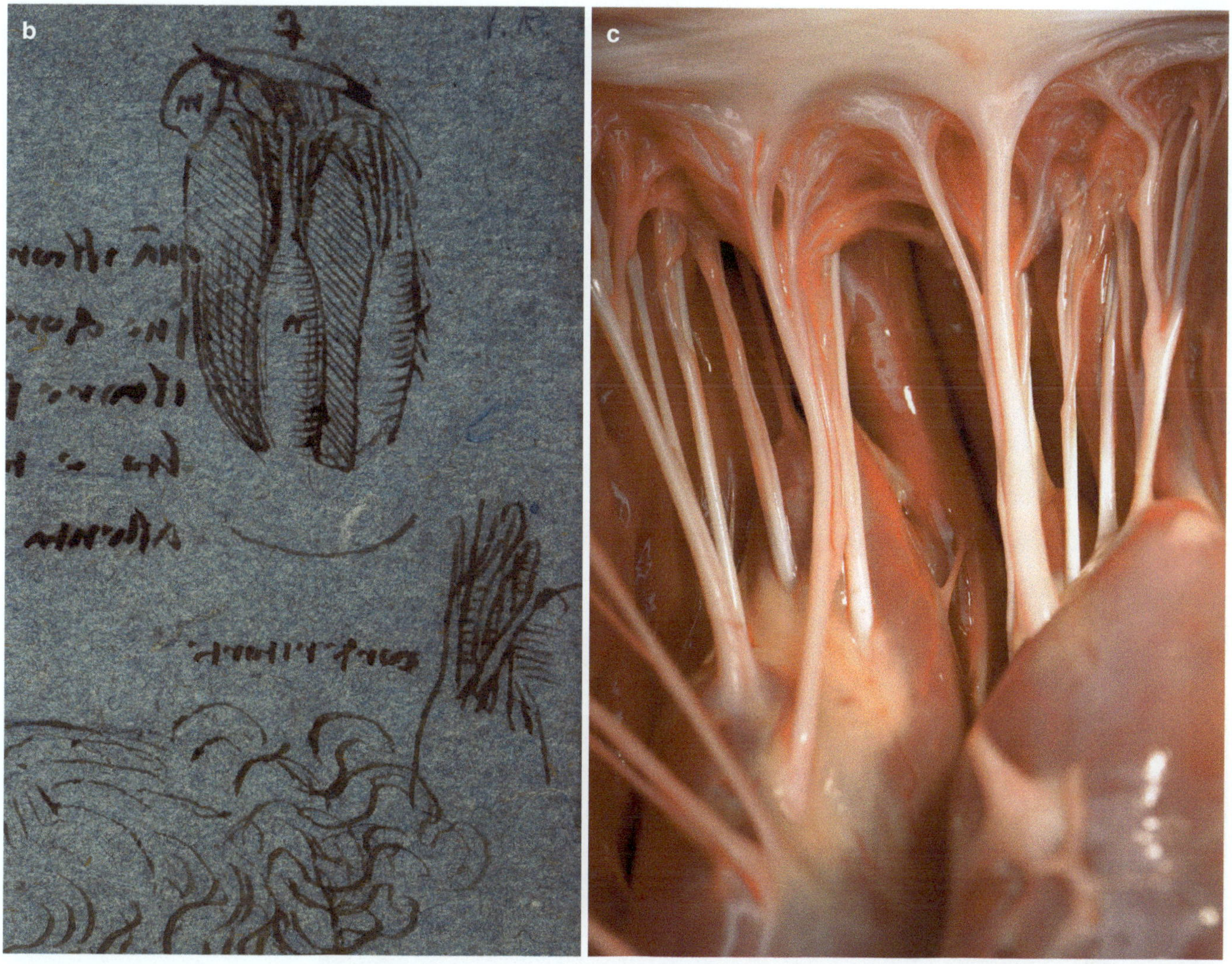

Fig. 4.31 (continued)

heart-and thus it must be drawn to make it understood."[137] (Fig. 4.35)

Whilst accepting the concept of septal pores, Leonardo did comment that the channels tapered down in a pyramidal fashion such that their opening on the left ventricular side was imperceptible. Thus he did record that he was not able to see them himself within the left ventricle: "…Necessity provides for the extraction of blood from the right ventricle, which penetrates the wall interposed between the right and left ventricles through wide porosities. These porosities go on narrowing into pyramidal hollows [channels] until they become imperceptible ducts through which the viscous

[137]RL 19074 recto [VII].

Fig. 4.31 (**a**) This main drawing by Leonardo shows the column-like papillary muscles of the mitral valve. Detail from RL 19080 recto, The left ventricle and mitral valve. Leonardo da Vinci , c.1512–13. Pen and ink on blue paper (Lent by Her Majesty The Queen. Royal Collection © Her Majesty Queen Elizabeth II). (**b**) This small drawing again shows the robust pillar-like papillary muscles that support the left ventricle and the chordae tendinae to the mitral valve. Detail from RL 19093 recto, Studies of the heart, and the head of a youth in profile. Leonardo da Vinci, c. 1511–13. Pen and ink on blue paper (Lent by Her Majesty The Queen. Royal Collection © Her Majesty Queen Elizabeth II). (**c**) This photograph reveals the accuracy of Leonardo's observations on the papillary muscles

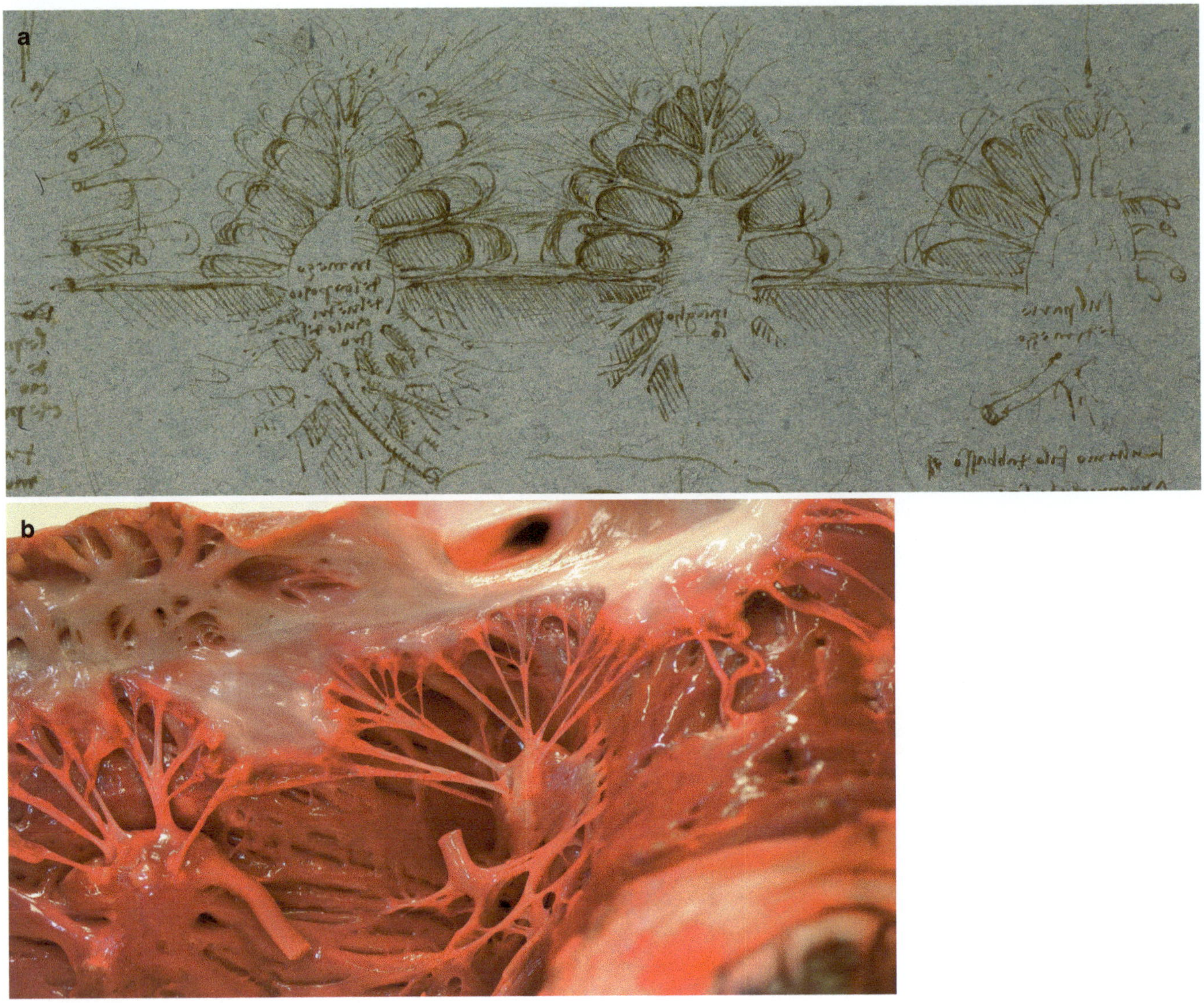

Fig. 4.32 (**a**) This drawing reveals the opened-out right ventricle with the moderator band divided; the trabecular portion of the right ventricular muscle can be seen drawn in at the left side. Detail from RL 19078, The right ventricle and tricuspid valve. Leonardo da Vinci, c.1512–13. Pen and ink on blue paper (Lent by Her Majesty The Queen. Royal Collection © Her Majesty Queen Elizabeth II) (**b**) Photograph of the same part in the ox dissection

blood penetrates and proceeds to thin down to great fineness."[138]

Close inspection of the second drawing down on manuscript page 19073 recto suggests that the tendinous cords stretch from the papillary muscles to the septal wall on the opposite side (Fig. 4.36). This is not the case, but when we carried out this dissection, we were surprised to find that the orientation of the tendinous cords was indeed in this disposition.

As described earlier, there are three sets of tendinous cords, the primary, secondary, and tertiary cords. The tertiary cords do indeed run from the base and posterior surface of the papillary muscles to the ventricular wall just beneath the attachment of the mitral and tricuspid valves, as shown by Leonardo. They are there to support the wall of the ventricle and maintain that relationship throughout the cardiac cycle. They act rather like tie-rods in an engineering sense, preventing overdistraction of the walls that they support. This is

[138]RL 19116 verso [XII].

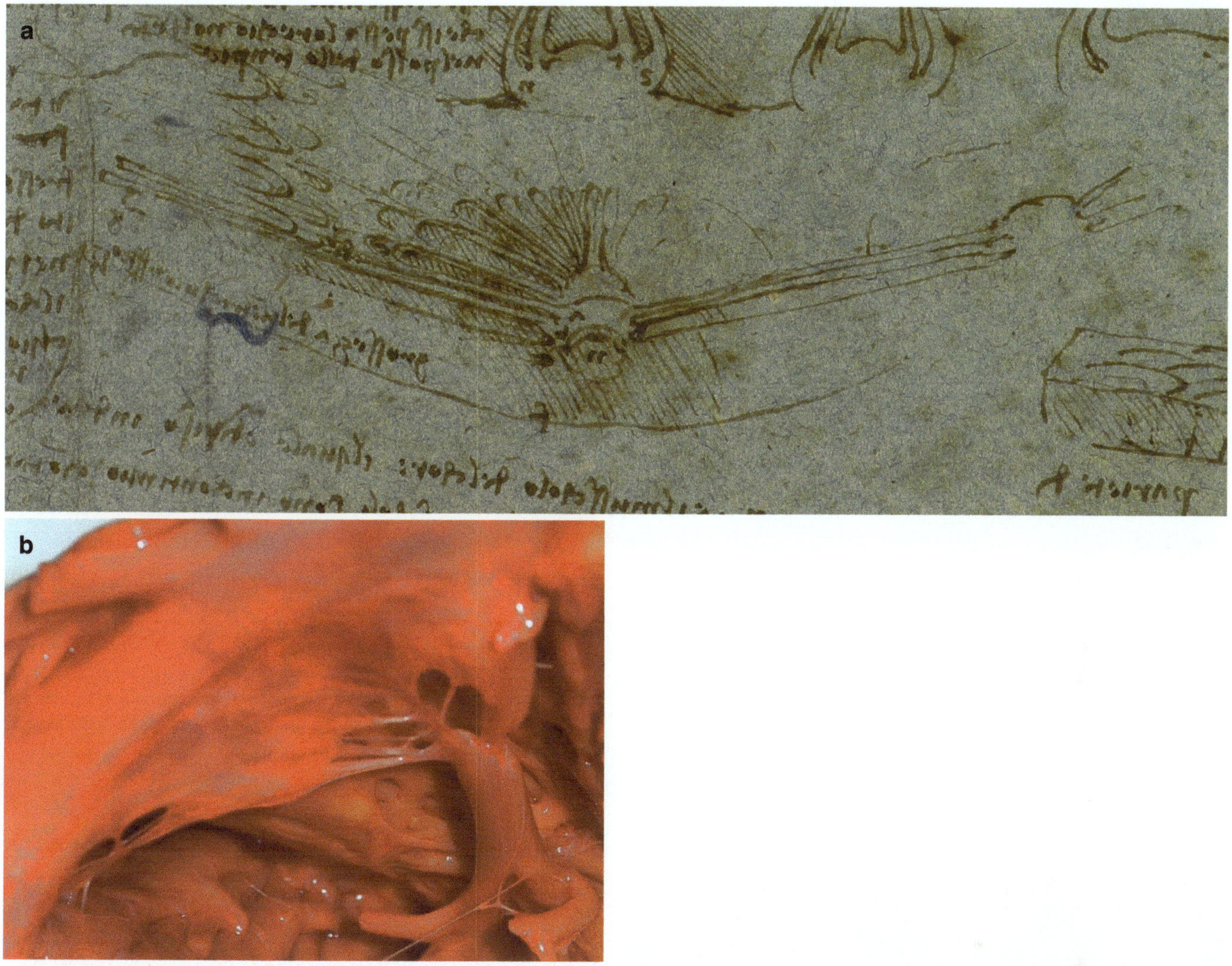

Fig. 4.33 (**a**) This drawing by Leonardo represents the origin of the papillary muscle from the right ventricle and the origin of the cords from it. Detail from RL 19074 recto, The ventricles, papillary muscles and tricuspid valve. Leonardo da Vinci, c. 1511–13. Pen and ink on blue paper (Lent by Her Majesty The Queen. Royal Collection © Her Majesty Queen Elizabeth II) (**b**) A clinical photograph of the base of a human papillary muscle

what Leonardo says about them on the same page: "The muscles ramifying from a b and from c d serve to prevent the heart from dilating more than it should when it reopens. For if it dilated too much it would have to draw too much blood back from the vessels into which it first pressed it."[139]

Leonardo continued to rationalise this relationship by suggesting that if this were not the arrangement, then the increased blood flow and its velocity allied to its density would be so great that it would cause the artery to catch fire. Clearly the reasoning is fanciful, but his almost architectural deduction that the cords prevent overdistraction of the ventricle is correct. The primary cords are like guy ropes guiding the leaflets into apposition under the force of the blood, which pushes back against the underside of the valve as the ventricle contracts. They prevent the leading edges of the leaflets from overshooting each other. The secondary cords are the ones that bear the greatest force on closure of the valve. This is borne out by their character and shape—hence the double layers in the leaflet in this area. Leonardo realised this and described their insertion into the leaflets in some detail. Indeed, the spreading cordal insertions strongly resemble the beams in a roof. It is unlikely that Leonardo

[139]RL 19073 recto [V].

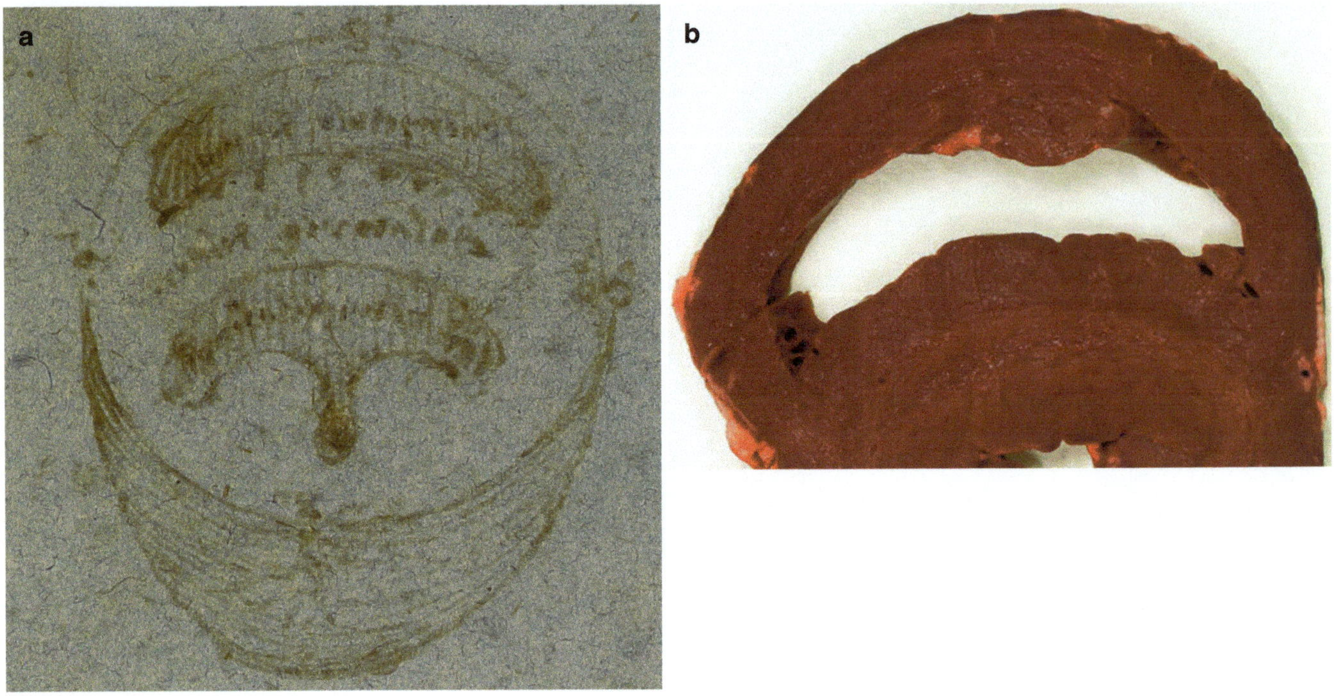

Fig. 4.34 (**a**) Drawing of a cross section of the ventricles. Detail from RL 19073 recto, The ventricles, papillary muscles and tricuspid valve. Leonardo da Vinci, c. 1511–13. Pen and ink on blue paper (Lent by Her Majesty The Queen. Royal Collection © Her Majesty Queen Elizabeth II). (**b**) Photograph of dissected ox heart in cross-section that reveals the correctness of Leonardo's drawing of the same part in the same aspect

Fig. 4.35 This little drawing reveals Leonardo's interpretation of the "Galenic pores" in the septum. They are most likely the pits seen in the surface of the post-mortem collapsed septal surface. Detail from RL 19074 recto, The ventricles, papillary muscles and tricuspid valve. Leonardo da Vinci, c. 1511–13. Pen and ink on blue paper (Lent by Her Majesty The Queen. Royal Collection © Her Majesty Queen Elizabeth II)

would have missed the structural resemblance between the drawing of the underside of the mitral valve on RL 19080 recto and his architectural drawings of the dome of a chapel.

Leonardo commented upon the timing of the contraction and relaxation of the papillary muscles in the following section: "When the heart broadens, n [papillary muscle] is shortened; drawing itself up towards f through its chordae, and this is the cause of shortening and opening the heart. And when the heart narrows [contracts] n elongates and is carried farther upwards, and through its elongation and being carried upwards relaxes the valve cusps, and the blood which percusses them from within shuts them."[140] (The lettering corresponds to RL 19093 recto top of page right.)

It may surprise the reader to note that the exact timing and detail of contraction and relaxation of the papillary muscles and the rest of the ventricular muscle mass remains unclear. In his review of 1952,[141] Kenneth Keele reflected that Leonardo was in error with his interpretation of ventricular muscle contraction, as cardiological opinion in the 1950s was that the whole of the cardiac muscle contracted en masse. Modern electrophysiologic measurement has shown, however, that this is not true: the muscle mass contracts and relaxes sequentially, more in the fashion that Leonardo described. The emptying of the ventricle is in fact somewhat like the wringing out of a towel with a synchronised,

[140]RL 19093 recto [I].

[141]Kenneth Keele, *Leonardo da Vinci on Movement of the Heart and Blood*. (London, Harvey and Blythe, 1952), 73.

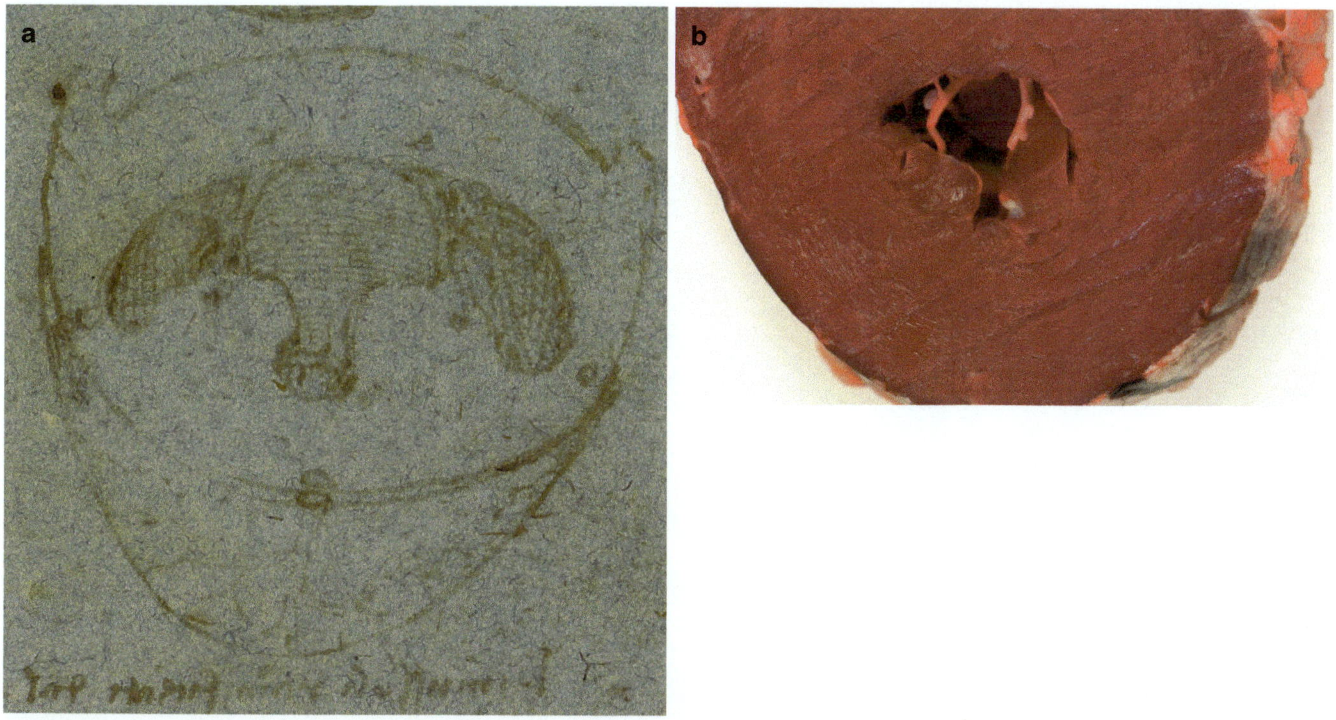

Fig. 4.36 (**a**) This drawing appears to show the chordae of the mitral valve running from papillary muscle to the opposite ventricular wall. This is a distortion by the plane of section but is exactly how it looks in the real transected heart in the same plane (see **b**). Detail from RL 19073 recto, The ventricles, papillary muscles and tricuspid valve. Leonardo da Vinci, c. 1511–13. Pen and ink on blue paper (Lent by Her Majesty The Queen. Royal Collection © Her Majesty Queen Elizabeth II). (**b**) Transected heart in the same plane

sequential twist. Indeed the loss of synchrony in ventricular muscle contraction is an important cause of heart failure. The contraction sequence can be restored towards what is understood to be normal with the use of pacemaker devices that resynchronise heart muscle contraction.

An unmistakable structure in the right ventricular cavity of the ox heart is a muscular band called the moderator band. Leonardo first identified this structure. He illustrates it several times, but perhaps it is best shown in the drawing to be found on RL 19119 recto. He named it the "catena" (chain). He correctly shows it stretching from the ventricular septal wall to the free wall of the right ventricle. These illustrations have clearly been aided by Leonardo's technique of distending the ventricles after death, probably using hot wax, which he allowed to set within their cavities. This would fix them in the same state as they were when filled with blood and working. The moderator band is there to prevent overdistension of the right ventricular chamber and was described by Leonardo thus: "The chain of the right ventricle [moderator band: *a r*] arises at one-third of the width, from the middle wall and at a quarter of its depth [marked off *b c* and *d e* in the diagram]. Thus for this ventricle *d b c e*, *b c* is the width and *d e* is the length, or you may say height by reason of its location. Therefore I say that the chain which joins the dilatable wall from the left ventricle *a r*, arises at *a*, a third part of the width of this ventricle [*b*] *c*, and a fourth part of the depth *d e*."[142] This disposition in the heart is illustrated in the small sketch in the bottom centre of RL 19119 recto (Fig. 4.37a).

The moderator band is illustrated on several sheets; RL 19084 recto, RL 19081 recto, and RL 19087 verso. Although present in the human heart, this structure is much less obvious. Indeed, the magnitude of its presence in the ox heart allows us to differentiate the internal structure of the right ventricle of the ox from that of the human heart (Fig. 4.37b).

In his earlier drawings, Leonardo shows bands crossing the chambers of both ventricles.[143] Aristotle proposed that a number of fine strands crossed the ventricles, but it was Leonardo who ultimately described the single significant muscular band in the right ventricle only. This structure was not properly described again until King in 1837.[144]

The lower drawings of the open right ventricle on manuscript sheet RL 19119 verso are remarkable. We know that these are drawings of the right ventricle for two reasons. First, Leonardo states that that is the case.[145] Second, the

[142]RL 19119 recto [III].

[143]RL 19112 recto.

[144]TW King, An essay on the safety valve function of the right ventricle. *Guy's Hosp Rep* 2:104, 1837.

[145]RL 19119 verso [X] "Right Ventricle".

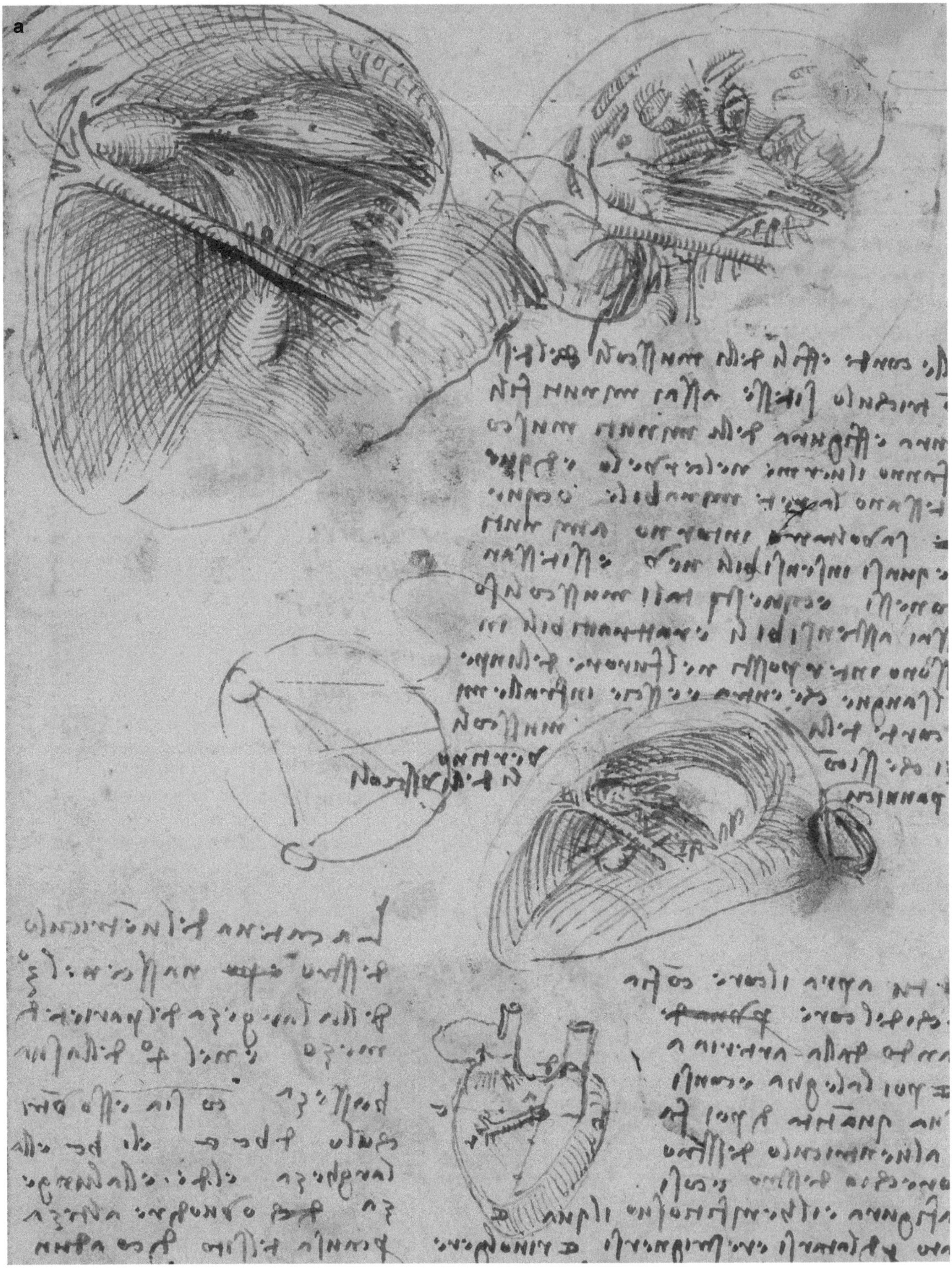

Fig. 4.37 (**a**) The moderator band. Detail from RL 19119 recto, The right ventricle and the valves of the heart. Leonardo da Vinci, c. 1511–13. Pen and ink (Lent by Her Majesty The Queen. Royal Collection © Her Majesty Queen Elizabeth II). (**b**) Photograph showing the moderator band in an ox (See also Fig. 4.2, showing the moderator band in a human)

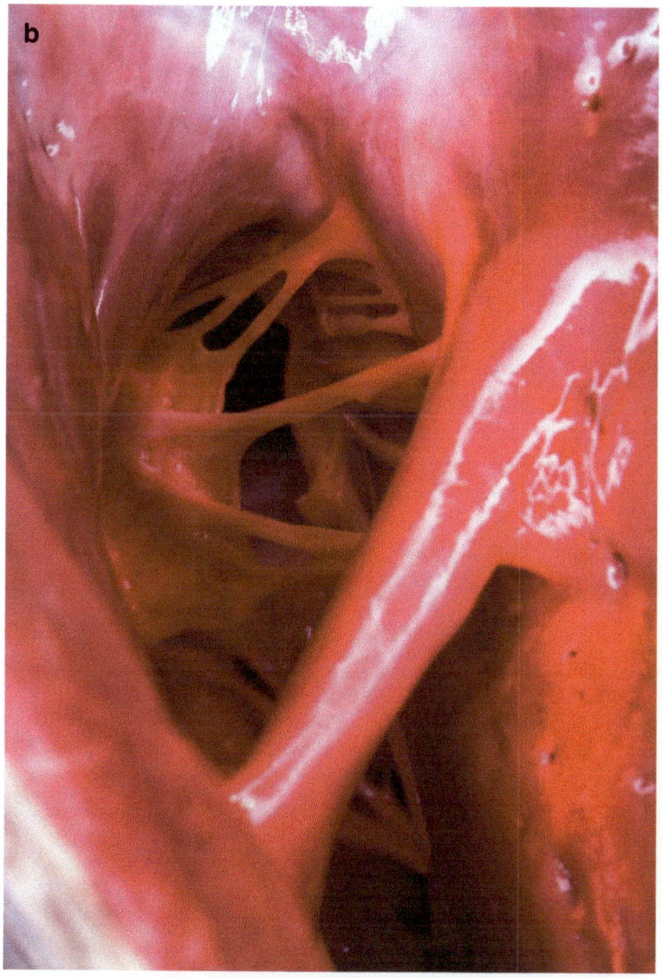

Fig. 4.37 (continued)

drawing reveals three valve leaflets whose attachment is to papillary muscles that are too fine to be from the left ventricle as drawn many times by Leonardo. They are remarkable because they bear a very strong resemblance to the interior of the right ventricle in a human heart. No visible moderator band is illustrated, and we know that Leonardo recognised this as a denominator of right ventricular identity. It is hard to imagine that he would complete a drawing of this structure from memory without including it, as he does in all of his other drawings. Also, the detail is such that it is a truly representational drawing, as can be seen from comparison with a photograph of an opened human right ventricle as shown in the accompanying photograph. Whilst the moderator band is found to a varying extent in human hearts, it is a much-reduced structure and in many is not visible as a discrete structure (Fig. 4.38).

The drawing is certainly one from Leonardo's Roman period, from 1513. The fact that it is very likely to be human suggests that it is drawn from a dissection that Leonardo carried out in the church hospital near to the Vatican. Indeed, such dissections may relate to the ensuing difficulties he had with access to dissection, as suggested in his protestations over the German mirror maker and his ultimate departure from Rome.

The Atrioventricular Valves

The role of these valves as gatekeepers to the entrance to the ventricles from the atria was certainly appreciated by Leonardo. In a section entitled "ON THE VENTRICLES", he wrote, "The upper [atria] are separated by certain little doors *(or gateways)* from the lower ventricles."[146] Later on the same page, pressing his case that the heart has four chambers, not two, he wrote, "Furthermore if it [*the atrium and ventricle*] is one and the same, there is no need for the membranous doors [*valves*] which separate one from the other."[147] Therefore, he had decided that the valves demarcated the two chambers from each other, but what of the way in which they worked? The flow of blood across the valves when the ventricle relaxes and dilates (diastole) is described thus: "Therefore, since in the time of dilating the lower ventricles the little doors [*valve cusps*] which provide the supporting base to the blood kept in the upper ventricle are opened, it is necessary that this lower ventricle should be filled first with the blood which is in contact with it rather than that which is enclosed within the liver and is distant from it."[148]

Further details on the way in which Leonardo perceived the operation of these valves is described in the following passage, which he titled "On the closing of the greater gates [valvular orifices] of the heart:"[149]

> The cusps of the greater valves of the heart [*mitral and tricuspid*] are closed by the percussion of the blood which escapes from the lower ventricles of the heart to the upper ventricle outside the heart [*atrium*]. They are reopened by the reflux of the blood pushed from the upper ventricles into the lower. And the vacuum which would be generated by the opening of the lower ventricles when they reopen is the cause of pulling back into themselves the blood of the upper ventricles when they are emptied.[150]

Whilst his use of the term "vacuum" may seem an exaggeration, it does approach reality in the sense that the flow of blood through the heart is along pressure gradients, with a higher pressure being present in the atria than in the ventricles as they begin to relax at the end of their systolic emptying phase.

The leaflets of both valves are supported from the ventricular side by tendinous cords that arise from the papillary muscles, part of the atrioventricular loop referred to earlier. Leonardo described their responsibilities in the following way: "The chords which arise from the muscles of the heart

[146] RL 19062 recto [I].

[147] RL 19062 recto [I].

[148] RL 19063 recto [III]. This subtlety of description distances Leonardo's thoughts from those of Galen, who describes the filling of the ventricles with blood directly from the liver where it is constantly formed. It also declares Leonardo's insistence that the heart has four chambers and that the blood has to pass through the atrium before reaching the ventricle, something not recognised by Galen and his followers. Thus this is a real statement of independent thought setting Leonardo apart from the academics of his day and potentially leading him into conflict with his detractors and the "abbreviators" that he refers to in his writing.

[149] RL 19081 recto [I].

[150] RL 19081 recto [I].

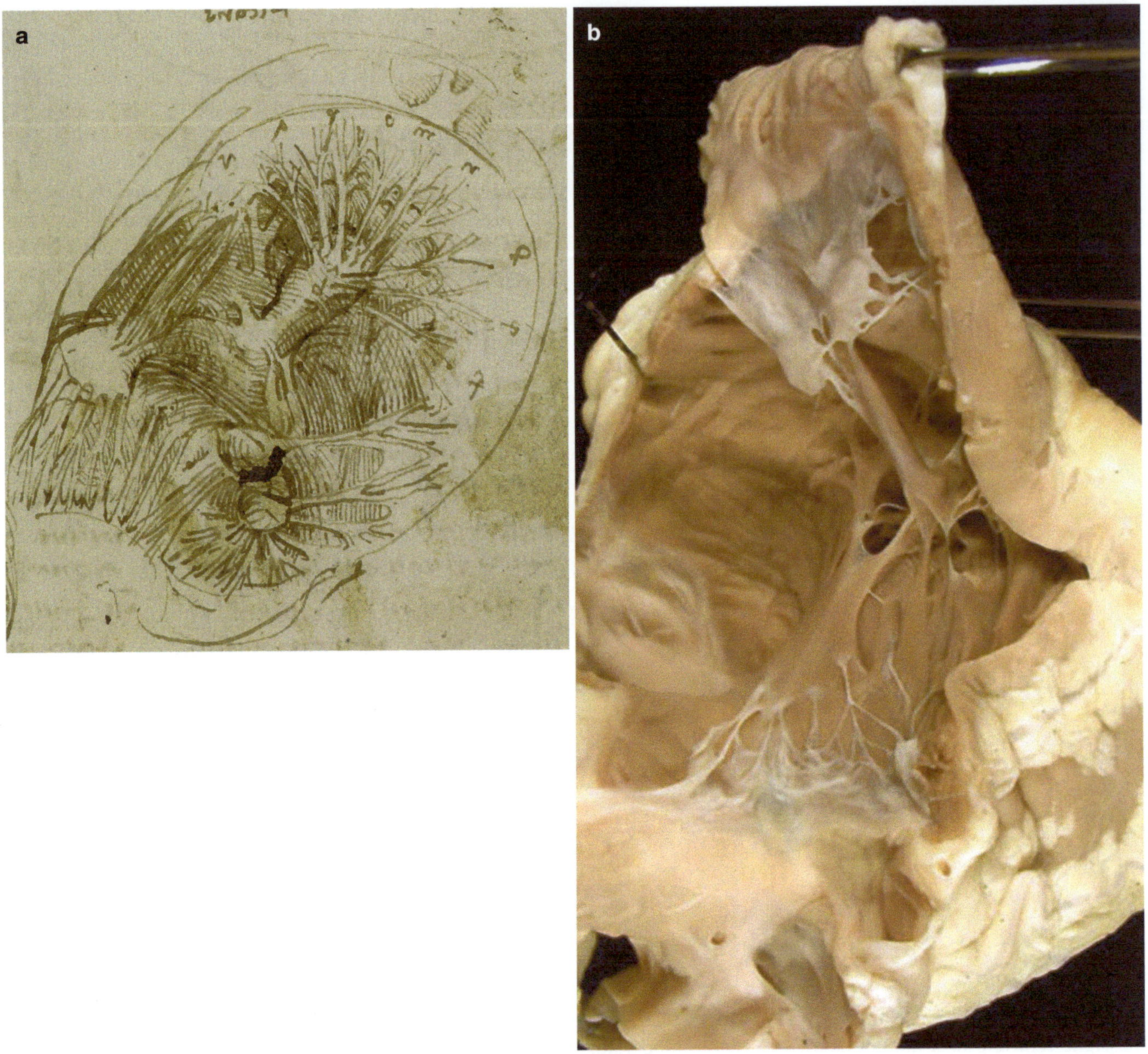

Fig. 4.38 (**a**) The opened right ventricle to show the arrangement of the papillary muscles and the cords and leaflets. Detail from RL 19119 verso, The ventricles, valves and papillary muscles, Leonardo da Vinci, c. 1511–13. Pen and ink (Lent by Her Majesty The Queen. Royal Collection © Her Majesty Queen Elizabeth II). (**b**) Photograph of the interior of the human right ventricle oriented in the same plane for comparison (Photograph with permission of Medtronic Inc.)

and are converted into the membranes which become the cusps of the great valves of the heart are those which hold the cusps of the valve so that they do not pass out of the opening, but with their extension they increase and are applied to each other making perfect closure."[151] Elsewhere he wrote, "In front [*atrial side*] the gates [*valves*] are found to be membranous and they are held from behind, that is, protected from within, by threads [*chordae tendinae*] which prevent their reversal."[152] His drawing of the "plait" of cords as they arise from the papillary muscle is convincingly lifelike (Fig. 4.39). He wrote next to this small drawing "plaited cords". This drawing accurately illustrates the arrangement of the chordae tendinae arising from a papillary muscle.

[151]RL 19081 recto [II].

[152]RL 19078 verso [III].

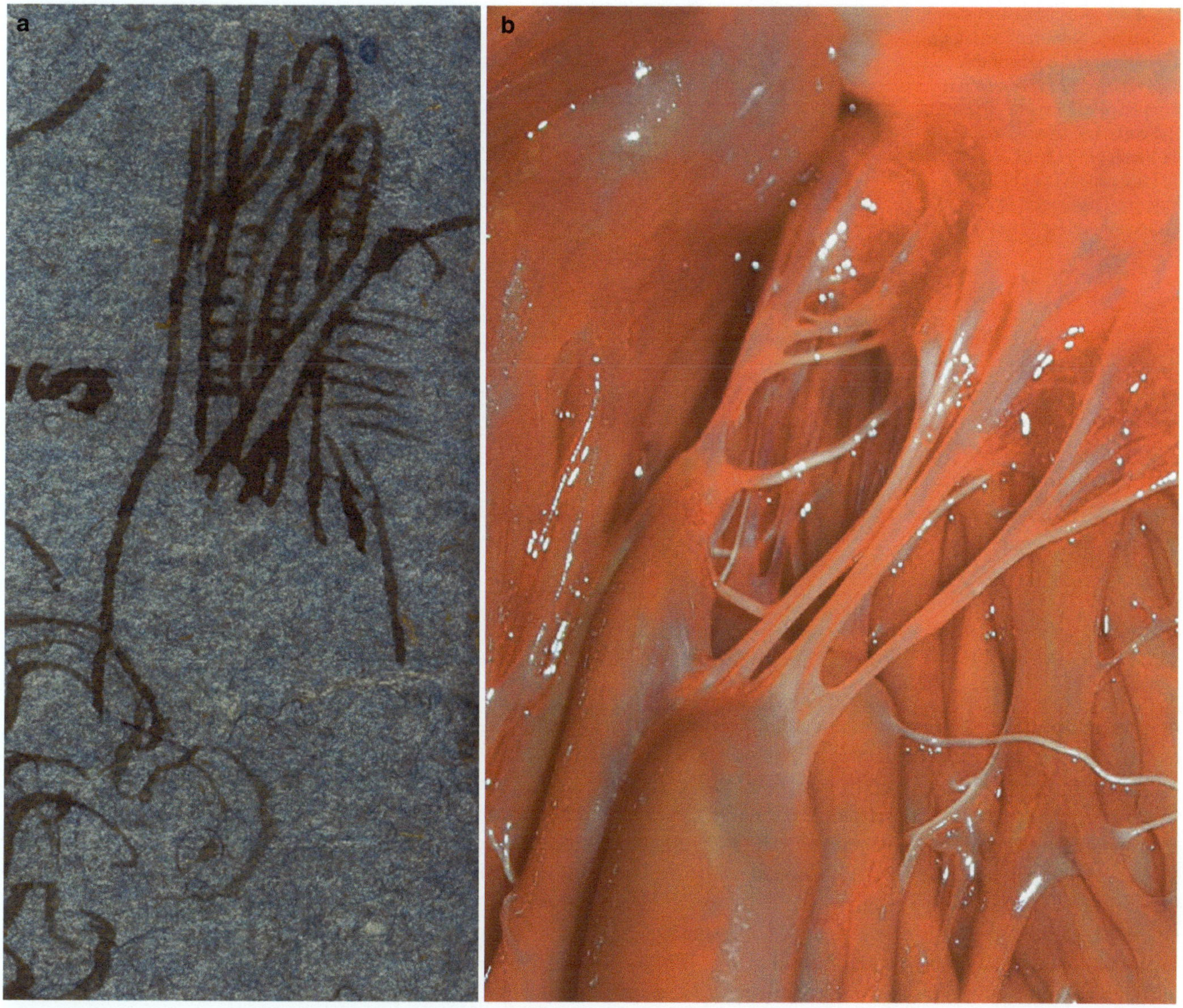

Fig. 4.39 (**a**) Small drawing reveals the plait-like arrangement of the cords to the atrioventricular valves. Detail from RL 19093 recto, Studies of the heart, and the head of a youth in profile. Leonardo da Vinci, c. 1511–13. Pen and ink on blue paper (Lent by Her Majesty The Queen. Royal Collection © Her Majesty Queen Elizabeth II). (**b**) Photograph of human chordae

An allusion to the geometry of an atrioventricular valve is to be found as a small faint sketch to the right of RL 19074 verso, which relates to the mitral valve (Fig. 4.40). Here, through the illusory use of ropes and pulleys, the observer can sense the idea of the tensioned relationship between the valve leaflets, the muscle walls of the ventricles, and the papillary muscles. Leonardo had transferred his observations of the anatomical relationships to an engineering/geometric setting. This is the first demonstration in the history of anatomical illustration (not even discussed for centuries to come) of the true nature of this complex relationship. The base of the triangle is supported by ropes passing over two pulleys to which weights are attached. This represents the tension in the ventricular wall transmitted to the leaflets. In the centre (at the apex of the triangle) is another rope to which is attached a further weight, representing the tension exerted by the papillary muscles. Leonardo thought that in the contraction phase of ventricular action, the pull on the lateral wall tensioned the leaflets and drew them up into apposition. In diastole (the relaxation phase), the central pull of the papillary muscle contraction caused the leaflets to open. Today this relationship

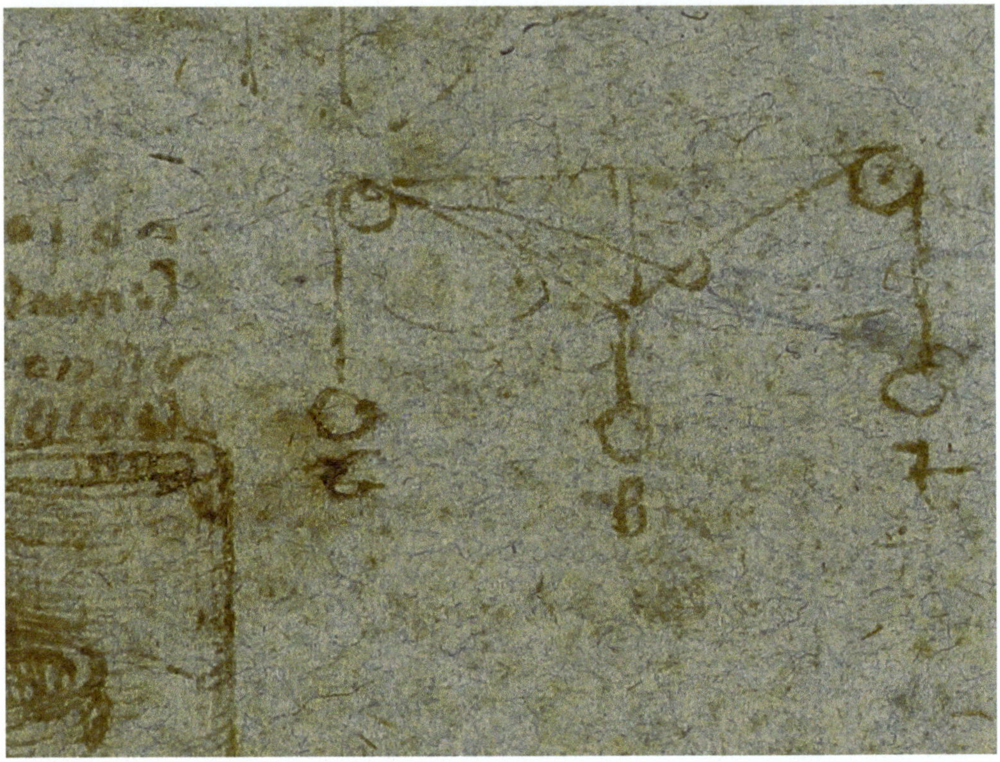

Fig. 4.40 A geometrical sketch to illustrate the forces acting on the leaflets and cords of the atrioventricular valves. Detail from RL 19074 verso, The heart and coronary vessels. Leonardo da Vinci, c. 1511–13. Pen and ink on blue paper (Lent by Her Majesty The Queen. Royal Collection © Her Majesty Queen Elizabeth II)

is referred to as the "atrioventricular loop" to remind the reader that this is a complex interrelationship and that for normal function of these valves all aspects of this geometrical arrangement must be functioning normally.

Leonardo perceived the valve leaflets to be of two layers. Where the cords spread out on the underside of the leaflets, they diminish in thickness. The atrial side of the leaflet has a smooth covering of uniform thickness. This distinction of layers in the leaflets is accurate and is used today. The cords and the leaflets blend into one another (rather as Leonardo described) as they develop harmoniously. The exact nature of the development of this relationship is still not well understood. Indeed, atrioventricular valve development has not yet been completely described. Leonardo's description of this relationship and his suggestions for the arrangement (as found on RL 19078 verso and RL 19074 recto) would not be out of place in more modern texts.

The three rows of cords, which he correctly illustrated, have different responsibilities. Those at the leading edge of the leaflets guide them into position as they close. The stronger ones behind (the secondary cords) are the ones that bear the force of closure, and the third row act like tie rods in an engineering sense between the ventricle and the base of the heart, as noted in the section on the normal working heart. The arcade-like arrangement of these groups of cords is displayed in a tiny drawing at the top of RL 19078 verso (Fig. 4.41). Drawings at the right-hand edge of the same manuscript page (RL 19078 verso) reveal the insertion of the chordae on the surface of the closed leaflets of the tricuspid valve.

Further important drawings at the top of both RL 19078 verso and RL 19074 recto reveal the form of the valve after the ventricle has been cut longitudinally and opened up (Fig. 4.42). Leonardo had designed the drawing in Fig. 4.42a to enable the teaching of this anatomy through the reconstruction of the valve and its orifice. He wrote beneath it (under the title "Geography of the heart") "Cut out these three muscles with their cords and membranes [*chordae tendinae and valve cusps*] and then join them together in the way in which they are placed when the right ventricle is closed. Then you will see the true shape of the [*valve*] membranes and what they do with their chordae [*tendinae*] when they are shut."[153] (This must be the first suggestion of a cut-out model in the teaching of anatomy. This form of modelling became a teaching tool only in the latter part of the seventeenth century.) The accuracy of these drawings can be appreciated by comparing them with the photographs of modern dissection specimens prepared imitating Leonardo's work.[154]

In closure of this valve, the contraction of the muscle at the base of the ventricle (that is, the part around the valve orifice), in conjunction with the rest of the ventricle, narrows

[153]RL 19078 verso [I].

[154]Dr. Martin Goddard, Consultant Cardiac Pathologist at Papworth Hospital, Cambridge. September 2007. Ongoing work.

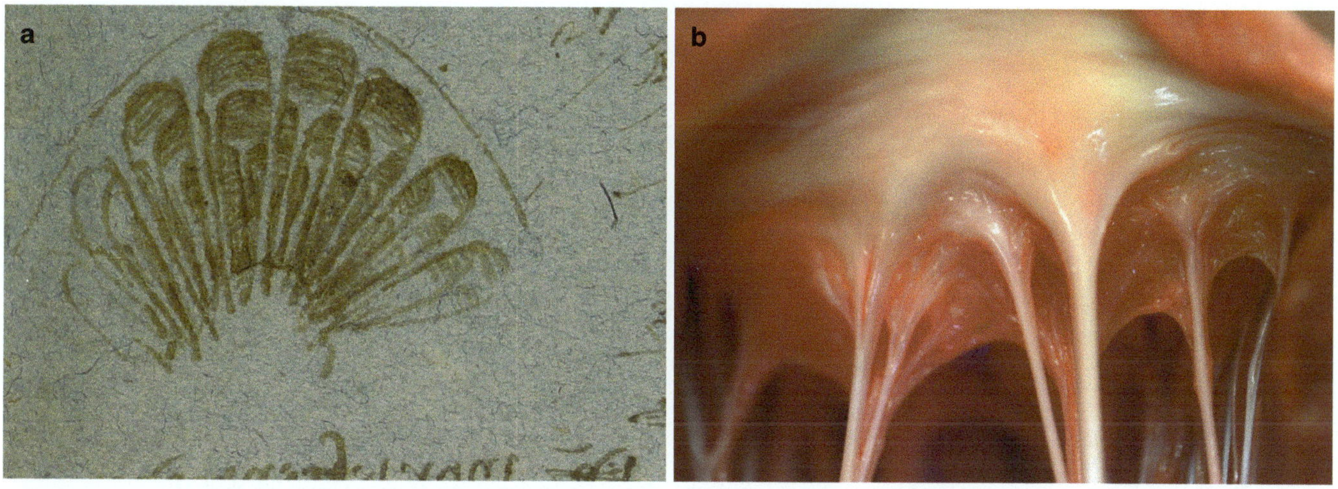

Fig. 4.41 (a) A small sketch to show the arrangement of the three layers of cords in the normal atrioventricular valve. Detail from RL 19078, The right ventricle and tricuspid valve. Leonardo da Vinci, c.1512–13. Pen and ink on blue paper (Lent by Her Majesty The Queen. Royal Collection © Her Majesty Queen Elizabeth II). (**b**) This dissection specimen reveals the accuracy reproduced in Leonardo's small sketch of this arrangement of cords

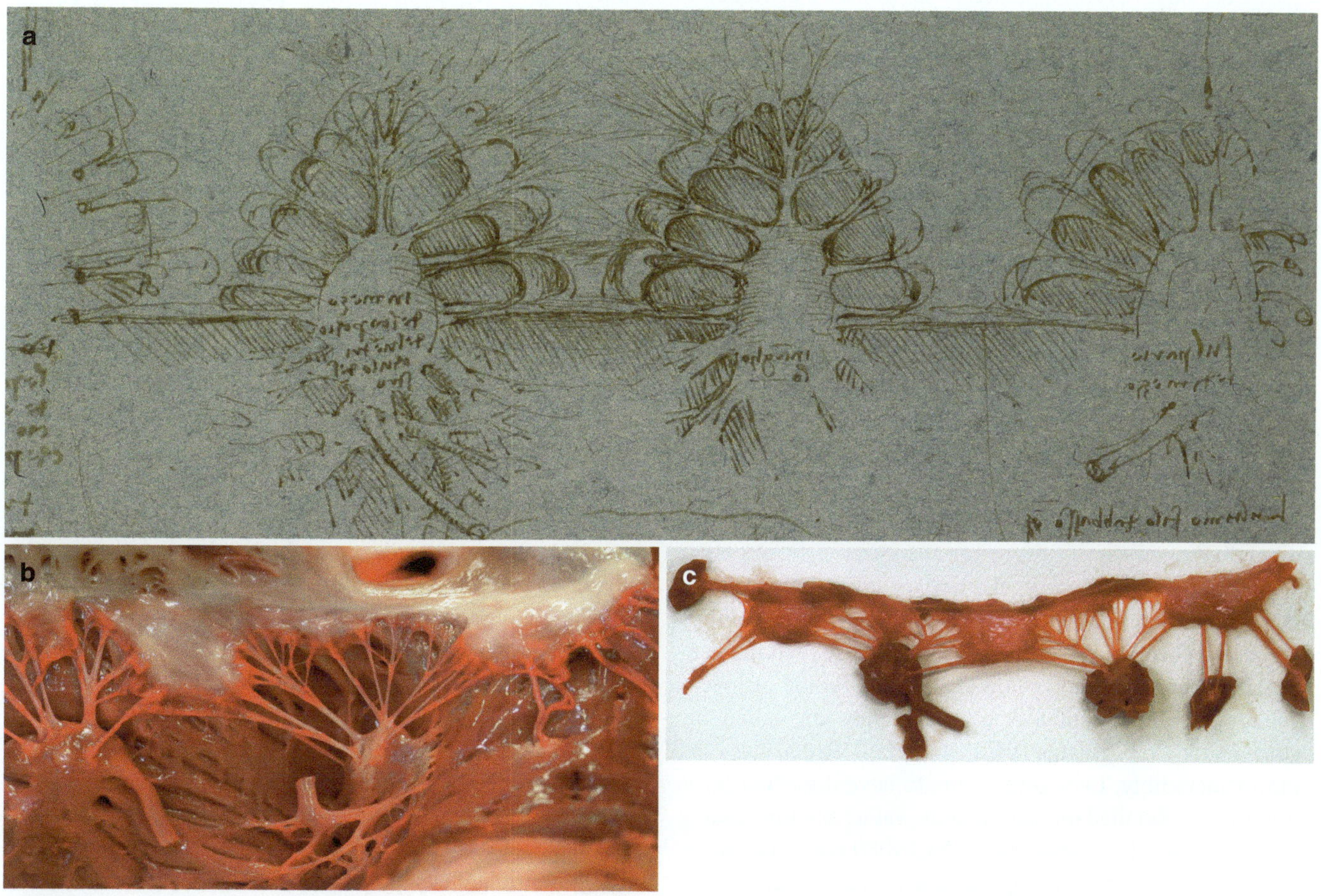

Fig. 4.42 (a) Drawing to show the opened tricuspid (right atrioventricular) valve with the divided moderator band, with the idea of making a paper model to demonstrate the valve function. Detail from RL 19078, The right ventricle and tricuspid valve. Leonardo da Vinci, c.1512–13. Pen and ink on blue paper (Lent by Her Majesty The Queen. Royal Collection © Her Majesty Queen Elizabeth II). (**b**) Photograph of a dissected tricuspid valve in situ, reproducing Leonardo's idea. (**c**) The valve excised

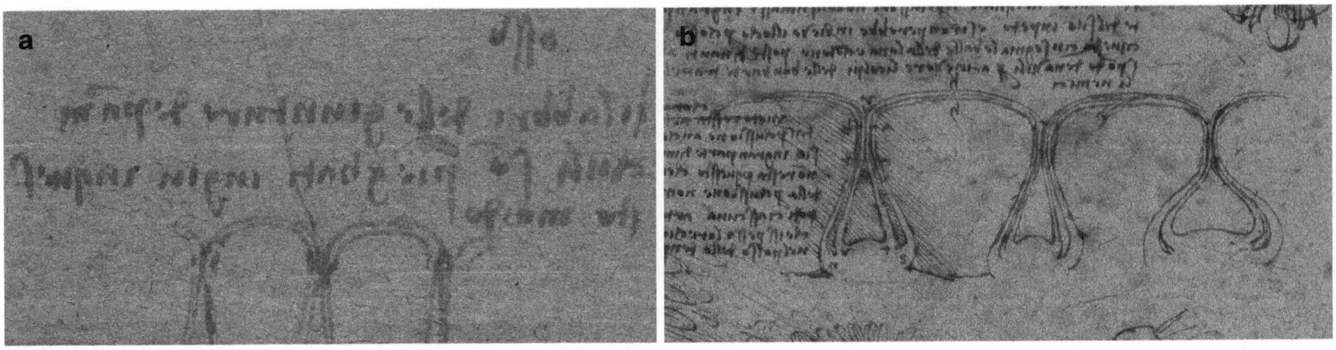

Fig. 4.43 (**a**) Small drawing to show the way that the leaflets of the mitral valve bend down to rest against each other where they meet. Detail from RL 19080 recto, The left ventricle and mitral valve. Leonardo da Vinci. c.1512–13. Pen and ink on blue paper (Lent by Her Majesty The Queen. Royal Collection © Her Majesty Queen Elizabeth II). (**b**) This drawing shows with great clarity the nature of the closing mechanism of the atrioventricular valves. Lengthy coaptation is required for each leaflet to support the other. Detail from RL 19074 recto, The ventricles, papillary muscles and tricuspid valve. Leonardo da Vinci, c. 1511–13. Pen and ink on blue paper (Lent by Her Majesty The Queen. Royal Collection © Her Majesty Queen Elizabeth II)

the valve orifice. Where the leaflets meet, they fold downwards so that a length of leaflet from each side comes to rest against the other over a distance of up to half a centimetre. This line of apposition is called coaptation and adds extra support to the valve leaflets by sharing the load between both over their surfaces. The coaptation surface has the effect of load distribution and intelligent use of forces, somewhat in the same way as the keystone in an archway does. Remove the keystone and the arch falls. Remove coaptation and the leaflets will prolapse and the valve will leak. The valve could not safely and completely work if it relied on a perfect edge-to-edge junction. Leonardo realised that this was how the valve closed, and he drew this vitally important relationship on more than one sheet (RL 19080 recto, at the very bottom; RL 19078 verso, lower middle left; and RL 19074 recto, small drawing at the top of the page and next to bottom central) (Fig. 4.43). His only known reference to it in his notes is in this simple statement: "The lips of the junctions of the membranes [*of the valve-cusps*] are bent down in this manner".[155]

The third layer of cords, the tertiary cords, are so unremarkable on first inspection as to be easily ignored, but they are very important in acting like tie-braces between the ventricular wall and the underside of the valve at the atrioventricular junction. In modern parlance, they are called the "strut cords" recognising their important structural role. Almost incredibly, Leonardo seems to have dissected these cords out in a detailed dissection of the valve, and this finding is recorded in a faint drawing on RL19080 recto. No reference to his thoughts in doing this dissection is found in his notes. Recent dissection of the ox heart has revealed the veracity of this observation (Fig. 4.44).

Further functional thoughts on the mitral valve can be found in sketches on manuscript page RL 19087 recto and verso (Fig. 4.45). Both of these pages seem to have been used to record Leonardo's thoughts on the muscular contractions and relaxation during the cardiac cycle. The tiny drawing at the bottom right-hand side of manuscript page RL 19087 versois accompanied by an arithmetical sum with accompanying words commenting upon the cardiac output of blood, concluding that this must be a great weight! No further comment is made on this potentially momentous observation. It will be remembered that it was just this realisation that became a crucial point in William Harvey's argument for a "circulation of the blood" rather than the Galenic "to-and-fro" motion, which Harvey disputed in his groundbreaking publication, *De Motu Cordis*.

These little drawings seem to be musings on the way that the valve leaflets and cords act during ventricular contraction, drawing the apex of the heart upwards towards its base at the atrioventricular ring. Leonardo states, "An object which is squeezed increases the distance which exists between its centre and its ends. In contracting, the walls increase upwards, drawing behind them the point of the heart."

In these observations, Leonardo stresses the "full and perfect closure" of the leaflets of these valves during systole, the strength of this closure being greater the more that the heart contracts. Clearly his concerns about the destination of the blood from the ventricles can be heard in his tortuous attempts to explain how his correct interpretation of valve function might still allow for the Galenic theory that the blood passes across the septum through the imaginary pores and must be replenished by the liver. Leaving this aside, his observations on valve function and the considerable importance of the contracting ventricular muscle in supporting the whole process is only now appreciated as being crucial. The dilatating, failing heart in ischaemic heart disease and cardiomyopathy fails to support the normal closure

[155]RL 19080 recto [XI].

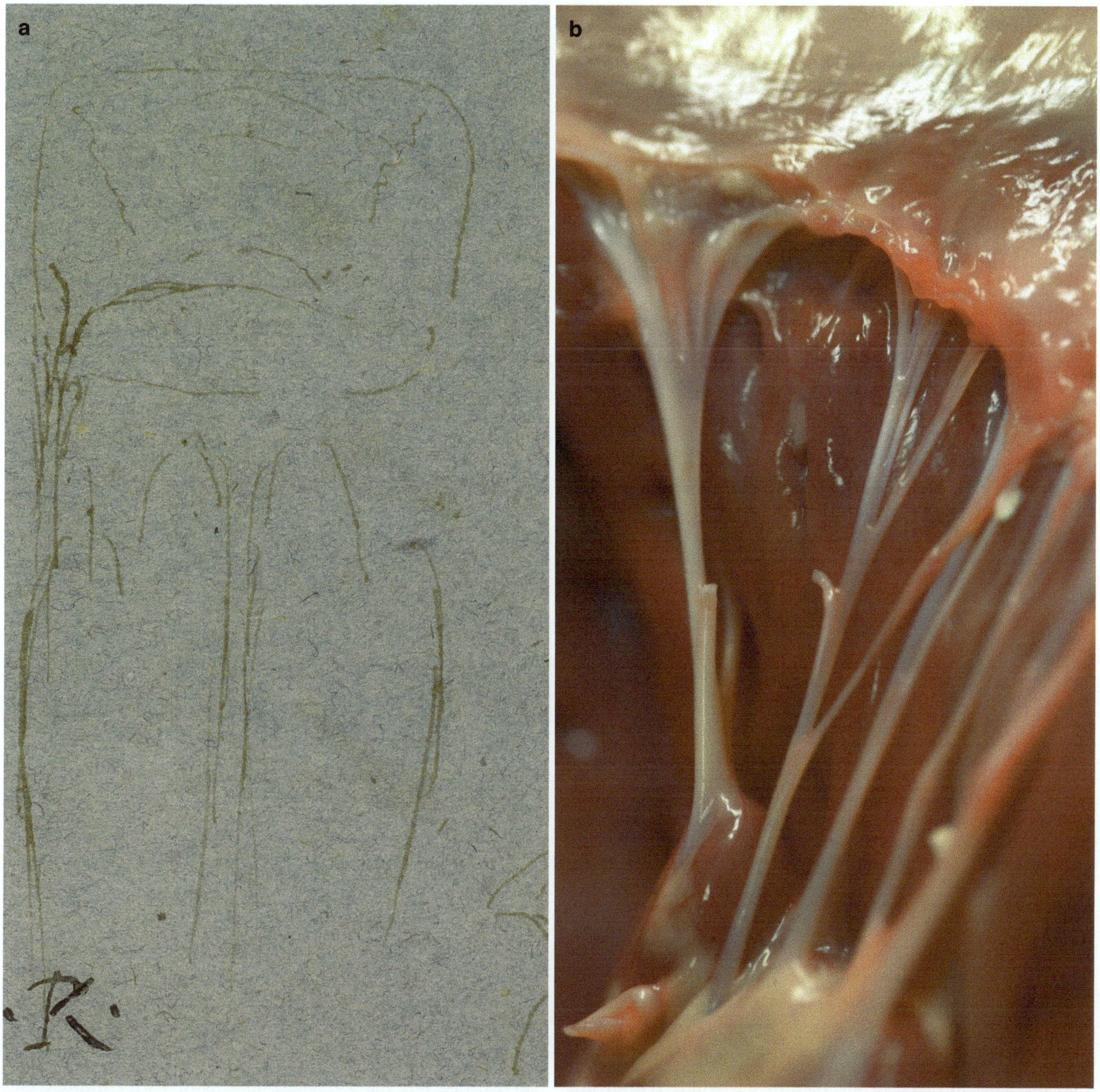

Fig. 4.44 (**a**) Leonardo's dissection of the tertiary (strut) cord. Detail from RL 19080 recto, The left ventricle and mitral valve. Leonardo da Vinci, c.1512–13. Pen and ink on blue paper (Lent by Her Majesty The Queen. Royal Collection © Her Majesty Queen Elizabeth II). (**b**) Photograph of the same structure on dissection of an ox heart

mechanism and draws the leaflets apart as the heart dilatates, leading to secondary mitral regurgitation. This form of progressive leaking of the valve is the most difficult to deal with surgically. It cannot be cured by simply addressing the valve failure, as the primary problem is the failing ventricular muscle. Although modern-day surgeons pay lip service to this fact, many continue to try to resolve the issue with surgical assaults on the mitral valve, and without fully addressing the failing ventricle, they have very limited success.

During the summer and into the winter of 1504, Leonardo spent time in Piombino as a military consultant to Jacopo Appiani. While there, he sailed between that town and the island of Elba.[156] Madrid manuscript II includes a number of drawings of ships' sails and rigging, and on Madrid MS II,

[156]K. Keele and C. Pedretti, *Leonardo da Vinci. Corpus of the Anatomical Studies in the Collection of Her Majesty the Queen at Windsor Castle* (New York, 1979).

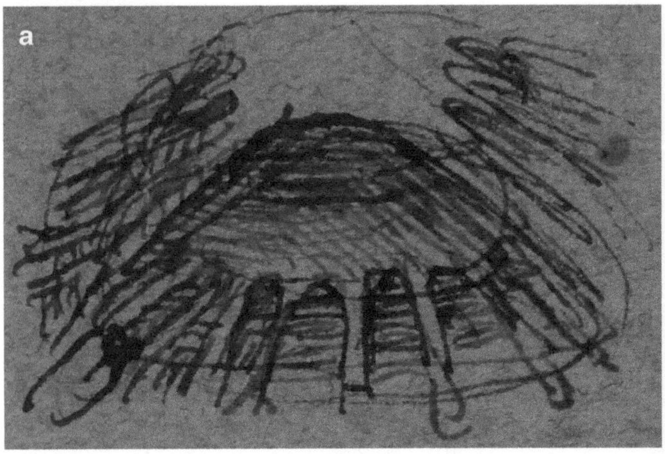
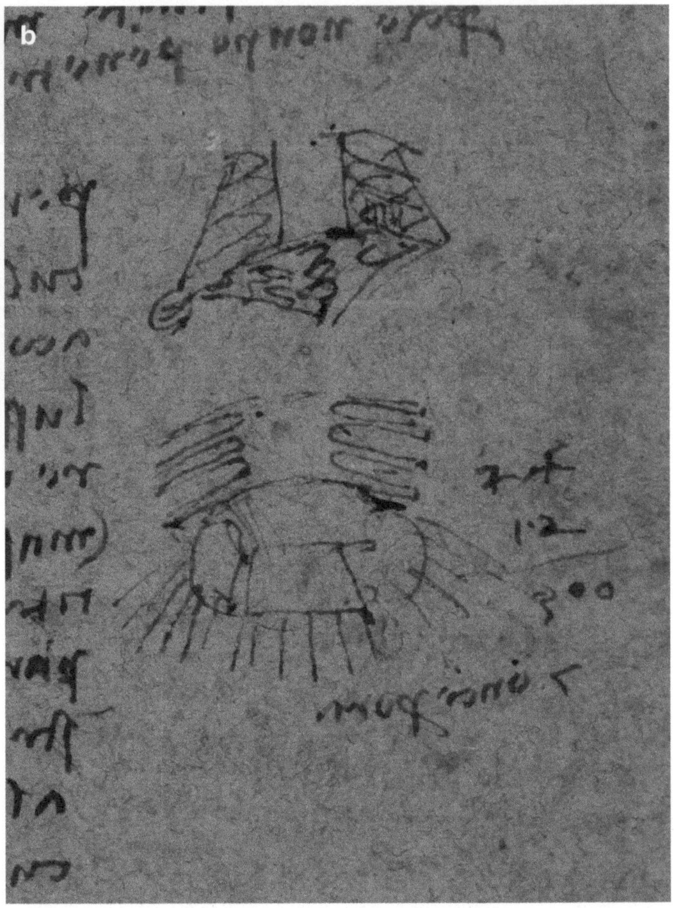

Fig. 4.45 (**a**, **b**) Sketches of the opening and closing of the atrioventricular valve apparatus in relation to ventricular function. (Part a) Detail from RL 19087 recto, Studies of the heart, notes on light and shade, and a standing figure. Leonardo da Vinci, c. 1511–13. Pen and ink on blue paper (Lent by Her Majesty The Queen. Royal Collection © Her Majesty Queen Elizabeth II). (Part b) Detail from RL 19087 verso, Studies of the motion of the heart. Leonardo da Vinci, c. 1511–13. Pen and ink on blue paper (Lent by Her Majesty The Queen. Royal Collection © Her Majesty Queen Elizabeth II)

f. 121 verso, he uses the terms that are again used in relationship to the valve. He used this information to declare that the tendinous cords of the valve are to be found on the underside; they are there to prevent the leaflets being blown inside out by the force of the blood in the same way that the ropes hold the sails in place in the face of the wind. Within this passage there lies the detail of the full development of the leaflets, like that of a sail filling with wind to gain maximum effect. Just as the wind unfurls the released sail, so does the blood unfurl the leaflets as they move into apposition with each other. Indeed, on another page Leonardo invokes picturesque nautical language in terminology for the parts of the valve: "Give the names to the chordae, which open and shut these two sails, that is, call the main one the bowline, topmast and the like"[157] (Fig. 4.46). Beneath this section he wrote, "Here are the membranes which close the gates of the right ventricle. In order that the re-shutting of this ventricle by the percussion of the wave of impetus on these membranous valves should not force them from behind, necessity provides them with hard and powerful chordae which can sustain the percussion of such impetus; and the muscles of these chordae are very hard almost like cartilage. And these valves, just like the left sided ones, shut with complete and perfect closure."[158]

This passage is notable for two statements. The first is that the atrioventricular valves shut with complete and perfect closure. This is at variance with other statements, where he suggests that the valves leak during closure to allow reflux of blood into the right ventricle as part of the churning mechanism to arouse heat. He wrote: "On the valves of the heart. On the shutting of the heart the valve cusps of the heart [*mitral and tricuspid*] always give passage first to a quantity of blood before they shut from within outwards."[159] This

[157] RL 19082 recto [IV].

[158] RL 19082 recto [V].

[159] RL 19074 recto [I].

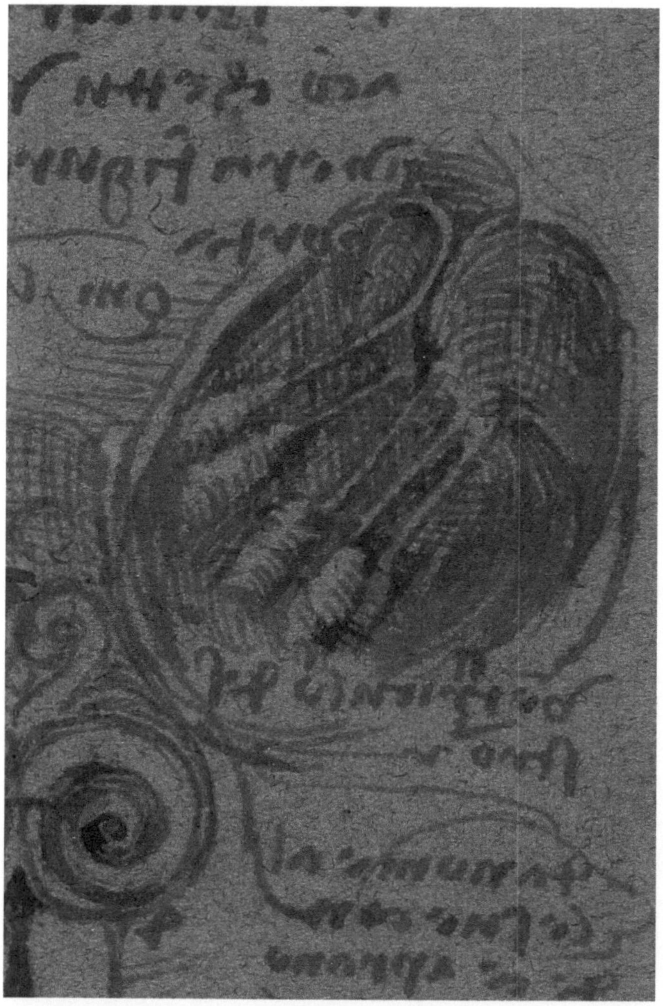

Fig. 4.46 Small drawing of the mitral valve. Detail from RL 19082 recto, The aortic valve. Leonardo da Vinci, c.1512–13. Pen and ink on blue paper (Lent by Her Majesty The Queen. Royal Collection © Her Majesty Queen Elizabeth II)

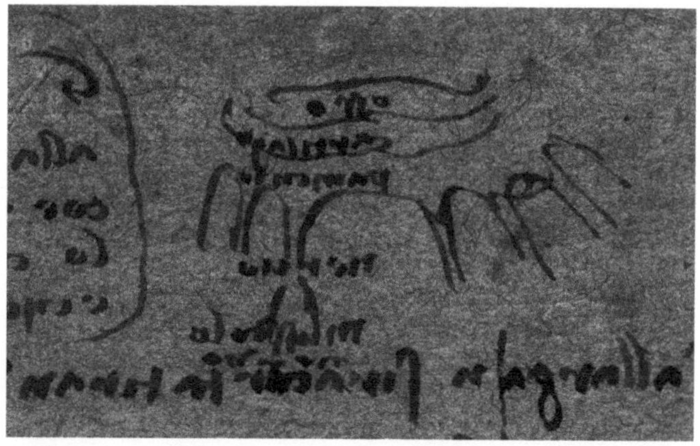

Fig. 4.47 This small drawing of a mitral valve leaflet shows a layer of calcification that occurs in mitral valve rheumatic disease. Detail from RL 19093 recto, Studies of the heart, and the head of a youth in profile. Leonardo da Vinci, c. 1511–13. Pen and ink on blue paper (Lent by Her Majesty The Queen. Royal Collection © Her Majesty Queen Elizabeth II)

statement is true. In the early closure phase of the valve, blood will pass backwards into the atrium. The duration of this phase is milliseconds. Secondly, Leonardo refers to the chordae in one report as being hard, almost like cartilage. In humans, this pathology is found in people who have suffered from rheumatic valve disease. It is important to know whether his subject was ox or human, as different explanations are likely for each species.

The ox naturally has bone formation at the base of the heart where the valve leaflets of all four valves meet. This is called the *os cordis* (bone of the heart). This explanation probably fits with another passage, in which Leonardo likens the valve unit (bony base of leaflet, leaflets, and cords) to a motor unit in the skeletal system. The latter consists of the bone from which muscles arise, the muscle itself that has to cross a joint to produce movement, and the tendon of insertion into the next bone. Next to a drawing of this arrangement at the top of page RL 19093 recto he wrote, "The things that go together in composing movements are six; that is, bone, cartilage, membrane, tendon, muscle, and nerve; and these six things are in the heart."[160]

If the accompanying illustration was of a human heart (something that we cannot know), then this may be referring to the condition of rheumatic fever, in which the leaflets thicken and then calcify over many years, particularly at their base of attachment to the ventricle. The leaflets stiffen and the chordae stiffen, calcify, and fuse (Fig. 4.47)

In a fairly lengthy description of the right ventricle, Leonardo adds the following note, which describes changes to the valve that could represent rheumatic fever,[161] although confused with post-mortem clot entwined within the chordae and the interstices of the right ventricle:

> And these fibres multiply like thick short tow[162] and lengthen and wrap themselves around the chords of the membranes [*chordae tendinae*] which close the right ventricle [*tricuspid valve*] in such a way that with the ageing of the animal the orifice [*of the valve*] cannot close well and a large part of the blood which ought to penetrate the narrow porosities of the middle wall into the left ventricle for the creation of the above-mentioned spirits, escapes through the imperfectly closed valve orifice into the right upper ventricle [*atrium*]. For this reason all the spirits are insufficient in the aged and they often die whilst speaking."[163]

---

[160]RL 19093 recto [IV].

[161]Rheumatic fever is an autoimmune response to a streptoccocal infection of the throat, the acute phase of which usually occurs in young children. The resulting inflammation in the valves of the heart goes on to cause calcification and fixity of the valve leaflets exactly as described by Leonardo, although the process is different from the one that he described, with the deposition of blood clot as the cause.

[162]Tow is the fibre of flax, hemp, or jute prepared for low-grade spinning and formed into a rope. (Oxford University Press: *The New Shorter Oxford English Dictionary*, 1993).

[163]RL 19062 verso [II].

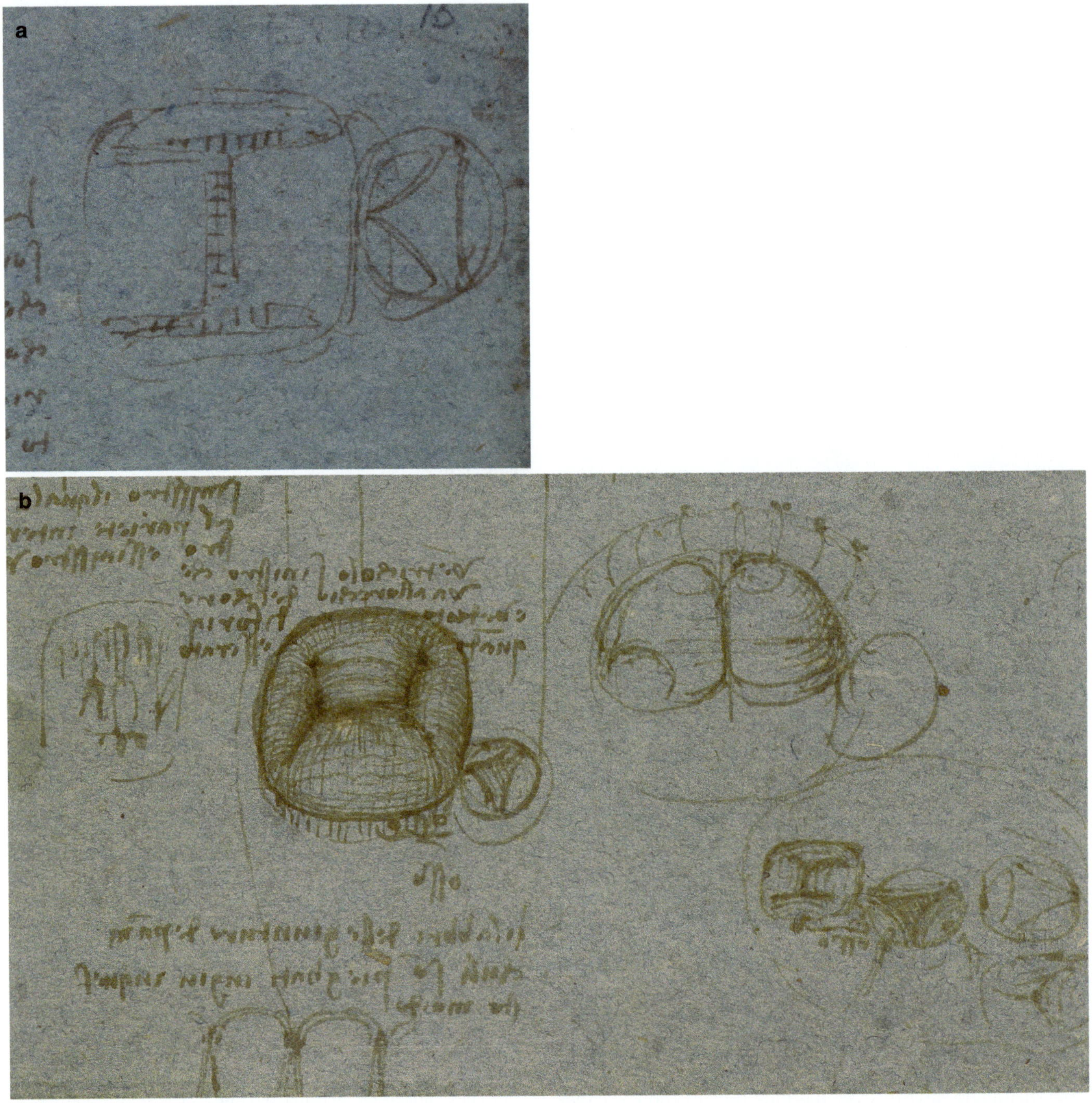

Fig. 4.48 (**a**) Faint sketch of the relationship between the aortic and mitral valve. Detail from RL 19081 recto, Studies of the motion of the heart. Leonardo da Vinci, c. 1511–13. Pen and ink and red chalk on blue paper (Lent by Her Majesty The Queen. Royal Collection © Her Majesty Queen Elizabeth II). (**b**) Further drawings of the relationship between the aortic and mitral valves. Detail from RL 19080 recto, The left ventricle and mitral valve. Leonardo da Vinci, c.1512–13. Pen and ink on blue paper (Lent by Her Majesty The Queen. Royal Collection © Her Majesty Queen Elizabeth II)

As discussed earlier, the mitral valve receives much less attention in the notes of Leonardo. The most interesting drawings of the valve appear on RL 19080 recto. (RL 19093 recto also focuses on the mitral valve.) It is interesting and important to note that in virtually all of his drawings of the mitral valve, Leonardo relates its structures to the correct relationship with the aortic valve. This is a very intimate relationship, with the two valves having a membranous continuation from the noncoronary leaflet of the aortic valve to the anterior/aortic leaflet of the mitral valve. Its intimacy is well relayed by Leonardo in a faint drawing at the top of RL 19081 recto (Fig. 4.48). All of the drawings on manuscript page RL 19118 verso are devoted to this very important relationship, the understanding of which is important for the cardiac surgeon, as either valve can be injured with surgery on the other.

The "Base" of the Heart

Leonardo recognised the conical shape of the heart, with the apex of the ventricles forming the apex of the cone and the atria forming the base of the cone (See RL 19074 verso, RL 19073 verso, RL 19072 verso, and RL 19112 recto). Modern anatomical description correctly places the mitral and tricuspid valves (the atrioventricular valves or inlet valves) and the aortic and pulmonary valves (the outlet valves) in the same plane within the base of the heart. This relationship between the atrioventricular valves and the aortic and pulmonary valves in the plane of the base of the heart is important both anatomically and surgically. Leonardo correctly illustrated this relationship for the first time in the anatomical record. Although his drawings of this relationship may appear to be a small and insignificant part of Leonardo's anatomical exploration of the heart, they do reveal the methodical and rigorous approach that he adopted in this work.

These relationships are well documented on the manuscript sheet RL 19118 verso. On this page, all but one of the drawings show this view. In a further drawing to the left of the centre of the page, Leonardo depicts the mitral valve in longitudinal section and its relationship to the aortic valve. To achieve this view, both atria, the aorta, and the pulmonary artery must be removed. This is the only way to establish accurately the relationship of all four valves of the heart, which lie approximately in this same plane. On RL 19081 recto, Leonardo poses two simple questions, which are almost certainly related to these relationships: "Which orifice is deepest at the base of the heart", and "Which orifice goes deepest in the base of the heart."[164] At the top of this page, there is a faint drawing of the relationship between the aortic and mitral valves. The detailed nature of this enquiry suggests that RL 19081 recto and RL 19118 verso are contemporaneous. By examining these relationships in such detail, Leonardo demonstrates his deep interest in the whole arrangement of the heart in a way that goes further than is required to understand Galenic teaching.

In many of the manuscript pages discussed in this section, his work appears contradictory. His drawings that are based upon observation are accurate and full of original insight, whereas his notes often reflect the Galenic philosophy of the time. His understanding of the anatomy of the atrioventricular valves, reflected in the accuracy of his drawings, are only occasionally backed up by the written word.

[164]RL 19081 recto [V] and [VI].

In reality, these valves are all about unidirectional flow and architectural support of the muscle pump. Leonardo cannot avoid the truth of the first of those elements, despite the need for him to be able to show that these valves leak on ventricular closing (contraction) to meet the perceived truth of heat generation. As we have seen, however, at one point he cannot but accept that these valves are so designed as to make "complete and perfect closure".[165] His words on valve ventricular connection and support are correct up to the point of his conclusions on the timing of papillary muscle contraction and relaxation, but even with modern imaging and measurement, the exact detail of these connections remains to be worked out. We are left to wonder about his final position on these fundamental questions of heart action, considering that other pages may have been lost.

The Aortic and Pulmonary Valves

As we have said, the valves of the heart exist to ensure that blood flows in one direction. The direction in which they open and close may at first seem to be all we need to know about them, but we have seen that the aortic and pulmonary valves have a configuration completely different from that of the atrioventricular valves. Whereas the mitral and tricuspid valves have cords to support them, the aortic and pulmonary valves rely entirely upon a less distinct support mechanism. Their configuration and their position within the narrowest part of a tubelike structure, (the exits from the right and left ventricles into the arteries beyond) suggests an important role for hydrodynamic effects. Apart from Leonardo's commentary on the subject, it is surprising that it was not properly investigated until the twentieth century.

Through modern eyes, therefore, Leonardo's detailed analysis of the mechanism of aortic and pulmonary valve function, which resulted in a very farsighted and correct explanation, must represent the apogee of his achievements in anatomy. The way in which he approached this study suggests significant intellectual comfort with the challenge, as he is able to make sense of his observations utilising his knowledge and experience gained in other spheres of interest. In many ways, the problems that he was grappling with are more complex than the idea of a theory of the circulation of the blood. After he had demonstrated the efficiency of function of these one-way valves, what did he think happened to all the blood that passed through them? Reading through his notes, there is the sense that he was drawn into the microcosm of his world of hydrodynamic engineering and did not see the consequences of his discovery. To have defined the closure mechanism was enough.

[165]RL 19082 recto [V].

In this work, Leonardo brought together careful and accurate observation with his acknowledgement of the influence of the powers of nature in shaping natural form. Using his unusual ability to exercise lateral thought and his knowledge of geometry and hydrodynamics, he suggested an experiment (whether or not he carried it out) that strengthened his belief in the complex solution that he described.

His work on this subject is summarised on six manuscript pages from a total of 30 pages on cardiac anatomy—in other words, one fifth of the extant important sheets on the heart. These six pages are RL 19116 recto, RL 19116 verso, RL 19117 verso, RL 19118 recto, RL 19082 recto, and RL 19083 verso. Four other sheets—RL 19045 recto, RL 19073 verso, RL 19079 verso, and RL 19080 recto—have visual reference to the aortic/pulmonary valve. It is difficult to place these sheets in any particular order, and it is very likely that Leonardo added his thoughts and observations to them at different times. Indeed, RL 19116 recto includes notes on the eye in addition to the notes on valve function.

Where did the inspiration for this detailed analysis come from? Of course we cannot know Leonardo's mind, but visual inspection of the anatomy of the outflow of the left ventricle of the ox heart reveals dramatic form. This complex comprises the valve leaflets and the hemispherical dilatations above the leaflets (the sinuses of Valsalva). It is likely that this complex, three-dimensional geometry fired his imagination. The conclusions of this work were hidden for centuries, and only recently has original scientific investigation confirmed his description in all of its original form.

All of the principal sheets are on blue paper. According to Clark and Kemp, these drawings were completed either in Vaprio after leaving Milan for the last time or during Leonardo's residence in the Vatican, about 1513–1514.[166] Another blue sheet is RL 19077 verso, which bears drawings of the diaphragm and the empty thorax of an ox, as well as three sketches revealing the internal organs of a man. Clearly, at that time Leonardo was thinking of (if not carrying out) dissections of the chest in dead oxen. That page is dated 9 January, 1513. Along with notes on the lungs and abdominal viscera, this page also mentions a chamber in a turret in Vaprio.[167] As mentioned earlier, the only dissection of the heart that appears to be from a human is on cream paper. Perhaps his work on the ox was done in Vaprio and his work on human material was done in Rome.

As if setting the tone for the student of his work, he wrote at the head of RL 19118 recto, "Let no one who is not a mathematician read my principles".[168] He begins this page with a brief but important discussion of the anatomical origin of the leaflets of the aortic and pulmonary valves: "The valves are built up, together with the vessels, immediately at the termination of the substance of the base of the heart."[169] He correctly claims that this arrangement is necessary to prevent the percussion of the refluxing blood from tearing the valve leaflets from their origins (during relaxation of the ventricle), and he alludes to the shock absorption supplied by the surrounding musculature: "This impetus, being devised for dilating the membranes at the base of the vessels, does not damage these valve cusps but is thrown obliquely and percusses and dilates with ease these sheaths of the blood in which the aforesaid impetus is consumed."[170]

This prescient statement is relevant in the design of modern tissue replacement valves. Ignorance of this important phenomenon in the early design of such valves led to unexpected early failure through the tearing of the leaflets from their housing, as the full shock of closure was borne by the leaflets alone. To a great extent, the realisation of these forces has led to the wider acceptance of replacement tissue valves that have less or no rigidly stented (supported) sewing rings, but simply a rim of supple material that allows shock to be dissipated more readily.

It is worth considering Leonardo's view of the basic form of the valve before developing his discourse on its dynamic function. From the beginning, he appears to have seen the valve as a functional unit comprising the muscle at the base of the aortic orifice, the leaflets, and the sinuses above to the point of origin of the tubular part of the aorta itself. This is a completely modern way of visualising this structure and the one that is operational today. His description is to be read in conjunction with the illustration (Fig. 4.49) that accompanies it:

> The right-sided valve cusps [*valve of the pulmonary artery?*]. The circumference ***a d b f c e*** is the base or origin of the coat of the [*arterial*] vein [*pulmonary artery*]. This area ***a d b n*** is how much the coat of the vein [*artery*] clothes the base of the heart and immediately beyond this origin of the vein there arises a fine membrane joined to it which clothes with itself the space ***a n b p*** [*the cusp and sinus*]. From its excess it generates one of the sides of the valve-cusps ***a b o***, which valve cusp is reduplicated by another membrane which is similar to that which covers the right ventricle inside the heart; and similarly the other ventricles [*the endocardium*]."[171]

The circular orifice of the mouth of the valve is defined by *a,d,b,f,c,* and *e; a,n,b,* and *p* define the sinuses and their junction with the valve cusps, which in turn are defined by *a, b,* and *o*. This passage would appear to describe the continuity of the valve leaflets with the endocardium, or inner lining of the ventricle. More specifically, Leonardo wrote of the origin of the valve leaflets in a passage that we will refer to again later on the

[166]K. Clark, *Leonardo da Vinci*. Revised and introduced by Martin Kemp. (London: Penguin Books, 1988), 230; and M. Kemp, *Leonardo da Vinci. The Marvellous Works of Nature and Man*, Revised edition. (Oxford: Oxford University Press, 2006), 285.

[167]K. Clark, *Leonardo da Vinci. Anatomical Drawings at Windsor Castle*. Vol. III, 2nd edition. (Oxford: Phaidon, 1969), 32.

[168]RL 19119 recto [IV].

[169]RL 19118 recto [V].

[170]RL 19118 recto [V cont'd].

[171]RL 19118 recto [VI].

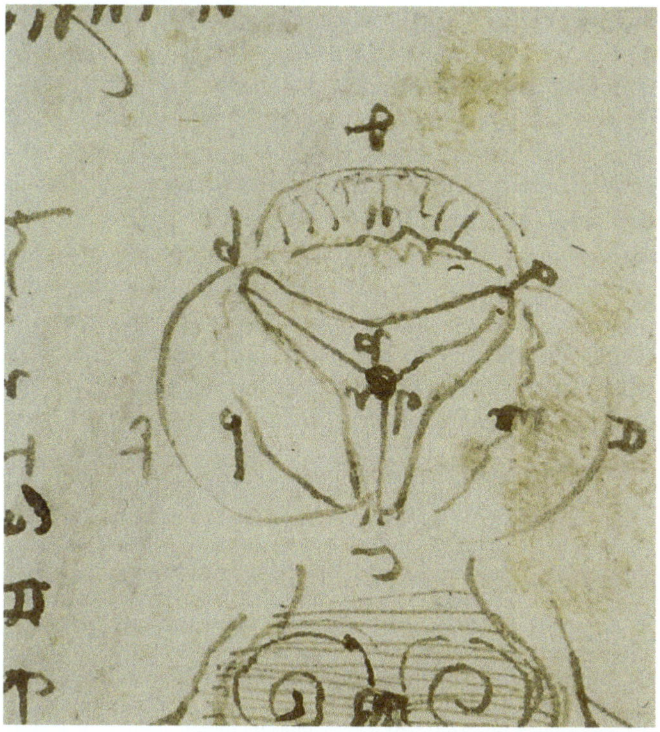

Fig. 4.49 Diagram of the arrangement of the leaflets of the aortic and pulmonary valves. Detail from RL 19118 recto, The right ventricle and the valves of the heart. Leonardo da Vinci, c. 1511–13. Pen and ink (Lent by Her Majesty The Queen. Royal Collection © Her Majesty Queen Elizabeth II)

elastic properties of the aortic root: "The valve-cusps are built up, together with the vessels, immediately at the termination of the substance of the base of the heart."[172] The site of the aorta is defined within the base of the heart in detailed relationship with coronary arteries and veins.[173] This level of detail with regard to the structure and relationships of the valve leaflets with the great vessels, the aorta and pulmonary arteries, and their lining tissue is extraordinary and reveals a level of detailed observation that goes beyond anything that could be of practical use for centuries to come. Again this reveals Leonardo's work as an inquiry within the sphere of understanding natural philosophy and nothing to do with anatomy that could be applied within the confines of medicine or surgery.

In the following words, he correctly associated ventricular contraction with the opening of the aortic valve and the presence of the peripheral pulse: "The time of closing of the heart [*systole*], of the percussing by its apex of the rib-cage, of the beating of the pulse, and of the entry of blood into the front gateway of the heart [*aortic orifice*], is one and the same."[174] This simple statement, so obvious today, is at the centre of the concept of cardiovascular integration.

[172]RL 19118 recto [V].

[173]RL 19118 recto [VII].

[174]RL 19116 recto [VII].

On this same manuscript page (RL 19118 recto), Leonardo began his discussion of the geometry of the leaflets. He pointed out that the orifice of the aorta takes up the centre of the base of the heart.[175] He correctly related the relationships of the coronary arteries and the cardiac veins to the surface anatomy of the heart, and their relationship to the three leaflets of the aortic valve.[176]

Using geometry, Leonardo explored the reason why nature provided three leaflets for the normal function of these valves:

> Nature made three little doors (leaflets) and not four because the pellicles which shut these cusps make greater angles being three, as the angle *c a b* shows, than the angle *e h f* which is made by four cusps. And it shows for what reason the more obtuse angle is stronger than the right angle of the square, since the cathete [i.e., *the altitude of a triangle*] *a s* is shorter than that cathete *n h*; and the square orifice is more capacious than a triangular one within the same circle. Consequently the membranes of the four valve cusps are weaker than those of the three valve cusps because their central angles are more remote from the bases of their triangles than those of the three valve cusps."[177]

This statement explains that the distance from the centre of a square to its base on any side is greater than the distance from the centre of a triangle to its base. That equates to a wider angle from the base to each side of the leaflet and hence greater strain, increasing the likelihood of shearing the leaflets from the wall of the aorta. Small geometric diagrams illustrating this geometry can be seen near the top of page RL 19117 verso.

These correct observations explain the cause of abnormal valve function when more or fewer leaflets have formed. Aortic valves with only two leaflets, as drawn by Leonardo at the very top right-hand corner of RL 19117 verso, are among the most common causes of aortic stenosis in adult humans, leading to the need for aortic valve replacement. The turbulent flow across the valve, which cannot open properly, causes fibrosis and ultimately calcification, with a reducing and fixed orifice. Aortic valves with four leaflets do occur in nature and also tend to fail early, requiring aortic valve replacement.

The other line of theoretical enquiry that Leonardo followed in this project was general hydrodynamics. He began by applying his prior knowledge derived from his studies of the flow of water through a pipe. These observations are to be found on manuscript RL 19116 recto (Fig. 4.50). Here he wrote, "Of the water which pours out of a horizontal pipe, that part of its interface which originates nearer the centre of the mouth of the pipe will go further away from the mouth of the pipe."[178] He also says, "Water that rises up through a pipe:

[175]RL 19118 recto [VI].

[176]RL 19118 recto [VII].

[177]RL 19117 verso [II].

[178]RL 19116 recto [II].

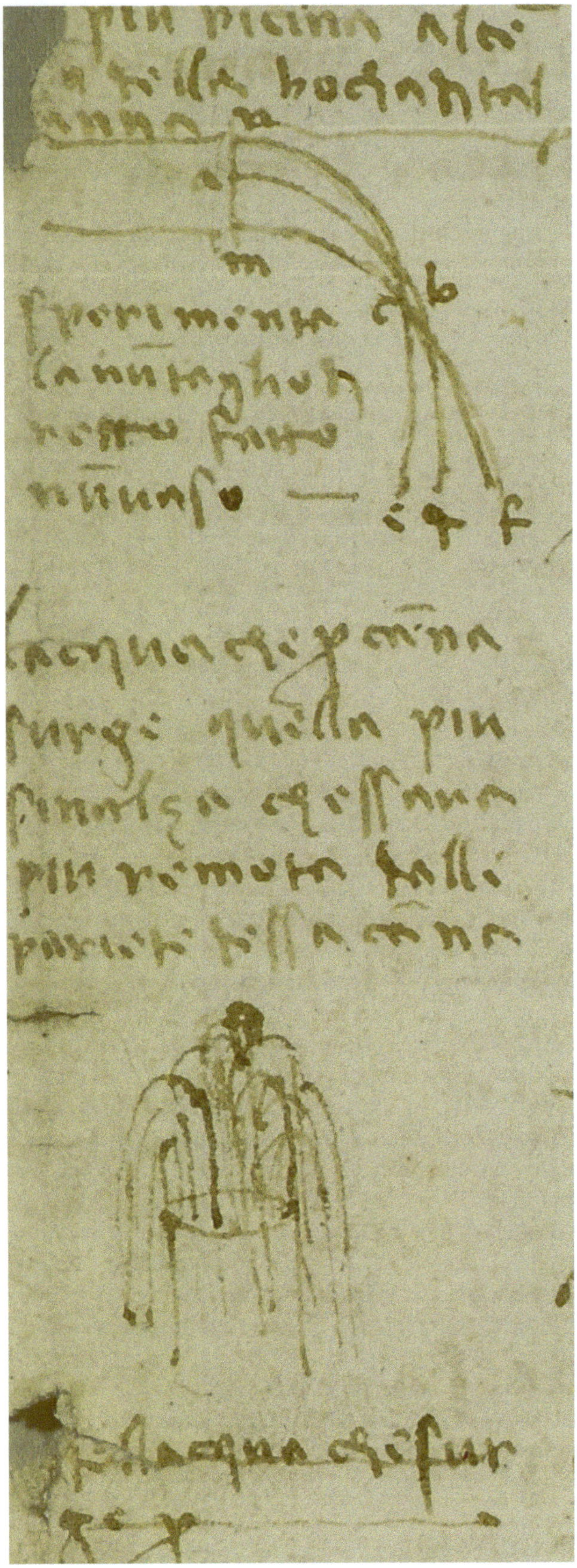

Fig. 4.50 Drawings to illustrate the principle of fluid drag. Detail from RL 19116 recto, Studies of the nerves and muscles, the heart, and optics. Leonardo da Vinci, c.1512–13. Pen and ink (Lent by Her Majesty The Queen. Royal Collection © Her Majesty Queen Elizabeth II)

that rises higher which is most removed from the walls of the pipe."[179] In other words, the velocity of flow is greatest in the middle of the column of fluid and least at the margins. This we know as "fluid drag" in a system of nonlaminar flow. In other words, the friction of the fluid against the wall of the pipe (fluid drag) will slow it down; the slower layer of water will reduce the rate of flow of that next to it and so on in a diminishing way until the centre of the column of water, where fluid drag will be least and velocity of flow highest. Leonardo then applied this observation to the setting of the aortic valve and the structure of the aorta just above the valve, where the dilatations of the wall of the aorta give rise to the so-called sinuses of Valsalva:

> The middle of the blood that spouts up through the triangle *a b c* [*aortic orifice*] acquires much more height than that which rises up along the sides of the triangle, because that in the middle of the triangle sends its impetus straight upwards, and that which rises up at the sides spreads its impetus by lateral motion and percusses the fronts of the arches of the hemicycles [*aortic sinuses*] and follows the concavity of the hemicycles, always going downwards until it percusses the concavity at the base of the hemicycle; then it turns upwards with reflected motion and then goes on revolving round through a circumvolving (gyrating) motion on itself until it consumes its impetus.[180]

Here Leonardo factors in the expansion of the column of blood flowing into the sinuses and its impact on flow direction and form. This statement is accompanied by three small drawings (Fig. 4.51). Leonardo had conducted other physical experiments on this phenomenon, and examples are recorded on Madrid I 134 verso and Paris Manuscript I, 89 verso.

Having established the relevance to this situation of the principle of fluid drag and laminar flow, Leonardo focused on the impact of a fluid flowing through the geometric shape that is the aortic root. Vortex formation in fluids (air and water) had occupied his mind on many occasions, and the representation of the vortex is found frequently in his notes. Drawings from his "deluge" series show very dramatic vortices in the clouds and in the water from a downpour of rain (Fig. 4.52). In Paris Manuscript A on sheet 60 r, Leonardo described the effect of water issuing from a narrow opening: "Water issuing swiftly from a narrow opening percusses the slow-moving water in front and is reflected, not in straight lines as are light and solid bodies, but in equivalent curved lines."

He illustrated this and other effects of various obstructions to flow through his drawing (Fig. 4.53). Perhaps his most beautiful rendering of this phenomenon is in a watercolour sketch he made to try to resolve the problem of riverbank erosion (Fig. 4.54). More specifically, in four simple sketches in RL 12666 recto (Fig. 4.55), he drew the formation of vortices as the effect of a fluid meeting a curved surface.

Leonardo transferred these arguments to the aortic root and described the result in the following words:

[179] RL 19116 recto [IV].

[180] RL 19116 recto [IX].

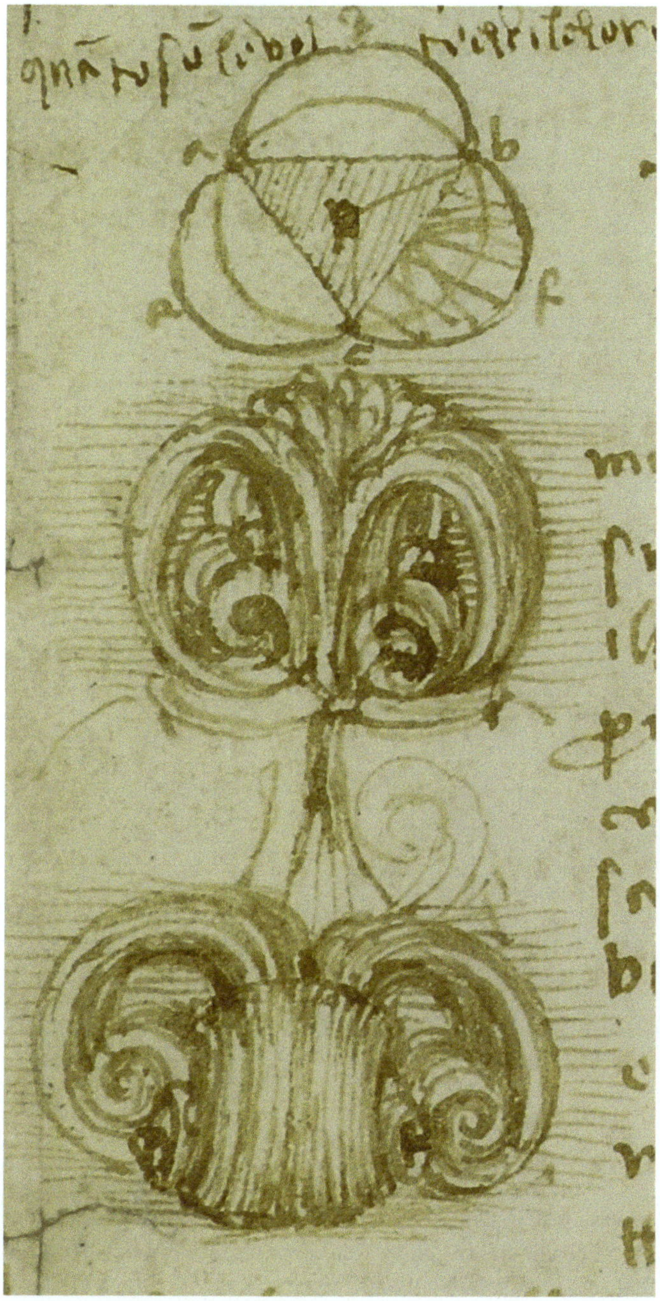

Fig. 4.51 Drawings to show the formation of the vortices in the sinuses above the aortic valve. Detail from RL 19116 recto, Studies of the nerves and muscles, the heart, and optics. Leonardo da Vinci, c.1512–13. Pen and ink (Lent by Her Majesty The Queen. Royal Collection © Her Majesty Queen Elizabeth II)

When the blood *n b m o a* revolves back with circular motion around its centre of motion *r*, it desires to maintain its impetus along the line of the commenced circumvolving motion and would widen according to the curvature of the commenced curvature *n b m*. But in the percussion of the wall *m o* it makes a smaller curvature and it goes on following it until it percusses at *a* where it would find the simple membrane of the valve cusp at *a b*, and this impetus of the blood would in a very short time rupture it. But the wisdom of nature provides a very hard resistance at the lowest base of the circle of impetus by thickening the wall of the conduit *o* [*aorta*] into the quantity of *o a h L*. Thus when the motion of this blood has given its percussion to this resistant place, the circumvolving motion rises immediately upwards at the end of this wall at the point where it meets the membrane of the valve cusp, which in order to close its third part [*of the whole aortic orifice*] must necessarily make a composite motion which is partly towards the middle of the pipe and partly upwards. This composite motion is necessarily generated by another composite motion which consists of the circumvolving motion of the blood *a b*, which according to the fourth book on Circumvolving Motion states: "The weight which moves around the fixed point of a chord where it is joined, pulls and stretches this chord with great power, and if such, a chord is severed from its fixed point". Therefore the blood which with its impetus percusses this valve-cusp, not being able to rupture it, continues its upward motion, enlarges and distends the folds [*of the cusp*] upwards and presses it against the other two valve cusps which for similar reasons come to meet it. Thus the three valve cusps are joined together at the same time in the centre of the aorta.[181]

This text accompanies a very simple diagram, to which the letters apply. The leaflet, which is best seen with the mirror image reversed, can be seen with the vortex unfurling it (Fig. 4.56).

This text explains what happens to a fluid entering a curved space with vorticeal motion. The description of a weight on a rope revolving around a fixed point encourages the student to grasp the outward force that this movement places on the curved upper surface of the leaflets, which does not just have the effect of pushing the leaflets closed but unfurls them and forces them to fully open out to meet the other leaflets. The power in a vortex is determined by the flow rate and the incident force on a surface. Leonardo explained this in the context of the sinuses of Valsalva as follows: "Each mobile object desires to consume its impetus along a straight line. For this reason the percussion giving in issuing forth on to the semicircular wall is made straight on to the wall in a straight line before its curved surface."[182] This is the expenditure of force that caused the erosion of the embankment of the river in Leonardo's watercolour painting referred to earlier.

As the flow of blood through the orifice of the aortic valve reaches its maximum velocity, it causes the edges of the leaflets to flutter vertically. This movement can be appreciated in the drawing at the top of RL 19117 verso (Fig. 4.57) and now can be seen with good-quality echocardiography (ultrasound).

There is a significant subtlety to the form of the aortic sinuses. The aortic wall that arises from the termination of the sinuses is about 10 % narrower than the orifice of the ventricle at the aortic valve level. (This junction is called the sinotubular junction.)[183] This means that the blood towards the periphery of the column leaving the ventricle will encounter these protrusions. In addition, the fluid drag at the edge of the column, along with the vortex formation predicted by the law that calls for vortex formation as a fluid passes from one chamber

[181]RL 19117 verso [IV].

[182]RL 19117 verso [XII].

[183]The diameter of the sinotubular junction is 10 % less than that of the orifice of the valve. (Dr. Tirone David, personal communication, Contemporary.)

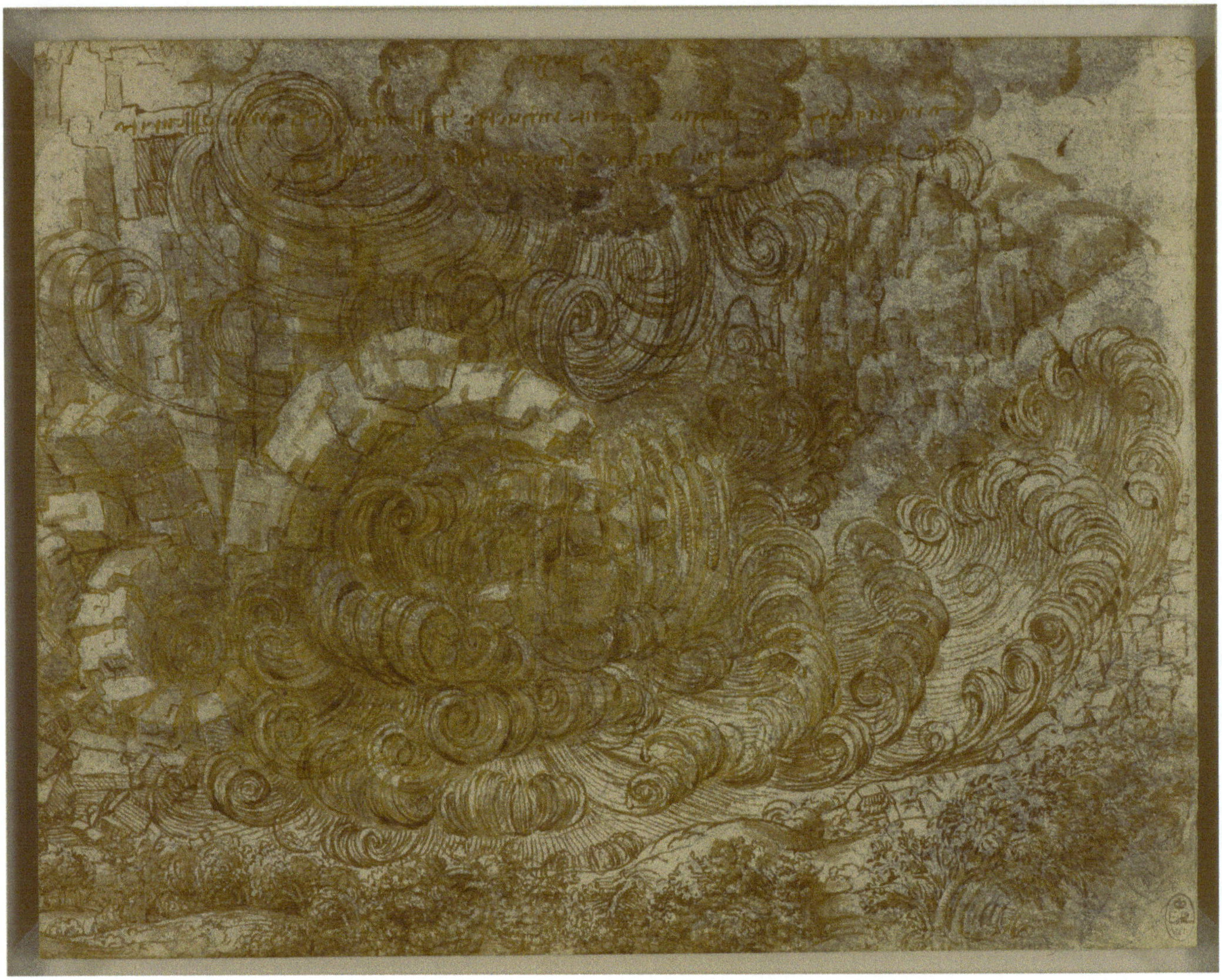

Fig. 4.52 One of the so-called deluge drawings with dramatic vorticeal formation. RL 12380, A deluge. Leonardo da Vinci, c.1517–18. Pen and black ink with wash (Lent by Her Majesty The Queen. Royal Collection © Her Majesty Queen Elizabeth II)

through a narrow opening into a wider chamber, means that this fluid will enter the upper part of the sinus and then be deflected downwards towards the base of the sinus in a semicircle, becoming the vortex described above by Leonardo.

We may summarise this information in the following way. Armed with all of this knowledge, Leonardo constructed his hypothesis that as the blood passed through the aortic valve into the wider chamber above, it had to form vortices in the aortic sinuses. Fluid drag and the anatomy of the aortic root would drive the peripheral blood at the edges of the emerging column into the sinuses. As the energy of the forward impetus of the blood in the central part of the fluid column begins to wane, the higher energy contained within the vortices begins to push back the leaflets of the valve towards each other prior to the end of the forward flow of the blood, which is in late systole. This motion is further amplified by the semicircular nature of the leaflets themselves. The effect of the swirling vortex on the curved leaflets is to open them out fully, as the wind does with a sail. In other words, the valve begins to close as the ventricle is in the final phases of emptying itself.

Simple logic might dictate that the aortic and pulmonary valves close in a much simpler way by the reflux of the ejected blood at the end of the systolic impetus. It may be suspected that this reversed flow would force the valve leaflets to close, similar to the way a strong wind will blow closed an open door. Indeed, this was the commonly accepted mechanism until recently. In a personal rhetorical argument, Leonardo made both cases thoroughly. In this process, he produced evidence as to why the latter option is not the one that occurs in nature:

> How the blood which turns back when the heart reopens [*diastole*] is not that which closes up the gates [*valves*] of the heart. This would be impossible, because if the blood percussed back on to the valve-cusps of the heart which stay curled, wrinkled and snatched up, the blood which pressed on them from above would weigh on them and press the surface of each membrane down on to its origin, as is shown by the valve-cusp *r o* which being weighed down from above its folds would shut up in solid contact. And Nature intends to stretch it out in height, breadth and width.[184]
> *and*

[184]RL 19082 recto [II].

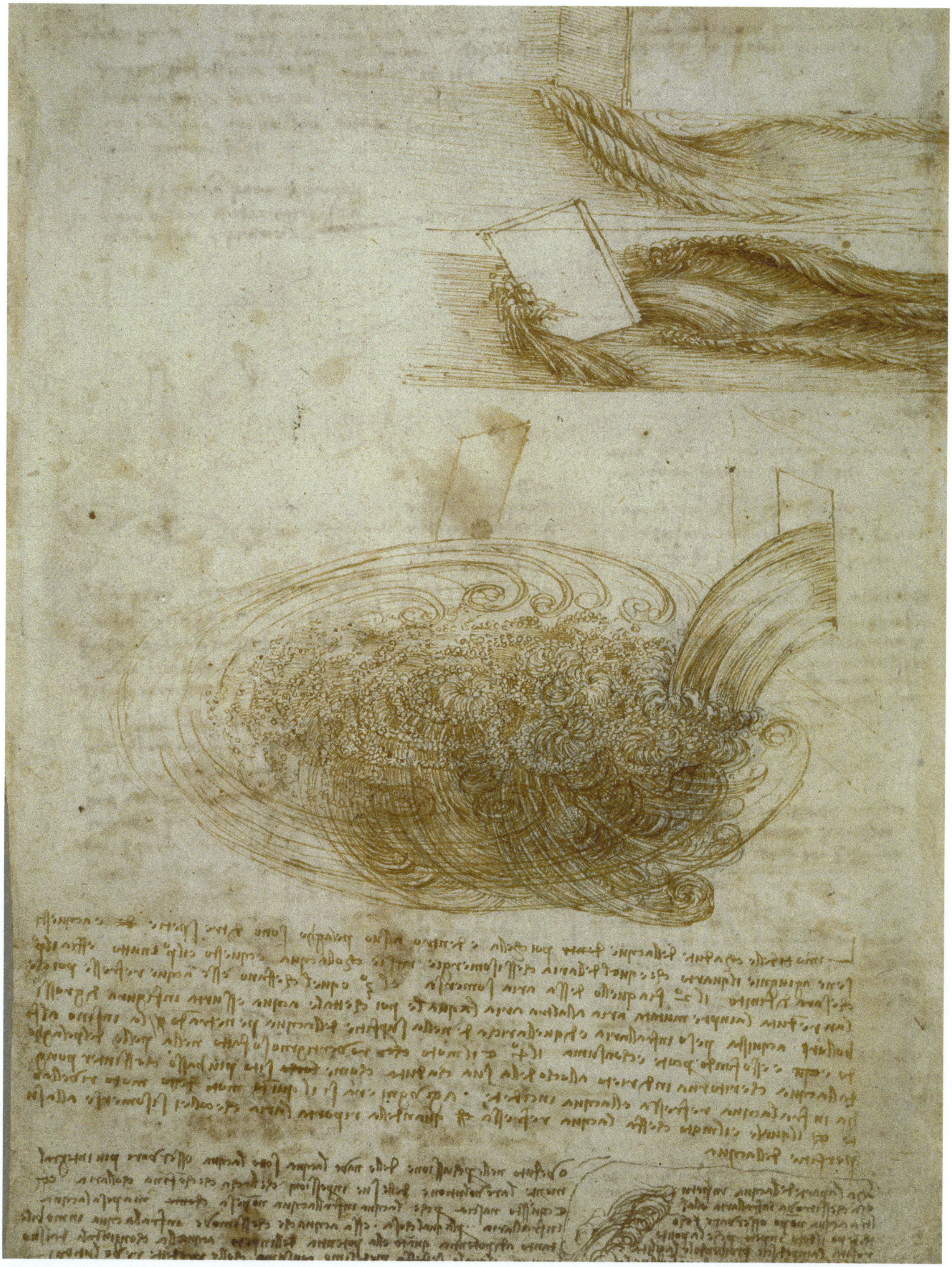

Fig. 4.53 Two sketches revealing the results of impediments to flow and the exhaust of water from a square opening. RL 12660 verso, Studies of flowing water, with notes. Leonardo da Vinci, c.1510–13. Pen and ink over traces of black chalk (Lent by Her Majesty The Queen. Royal Collection © Her Majesty Queen Elizabeth II)

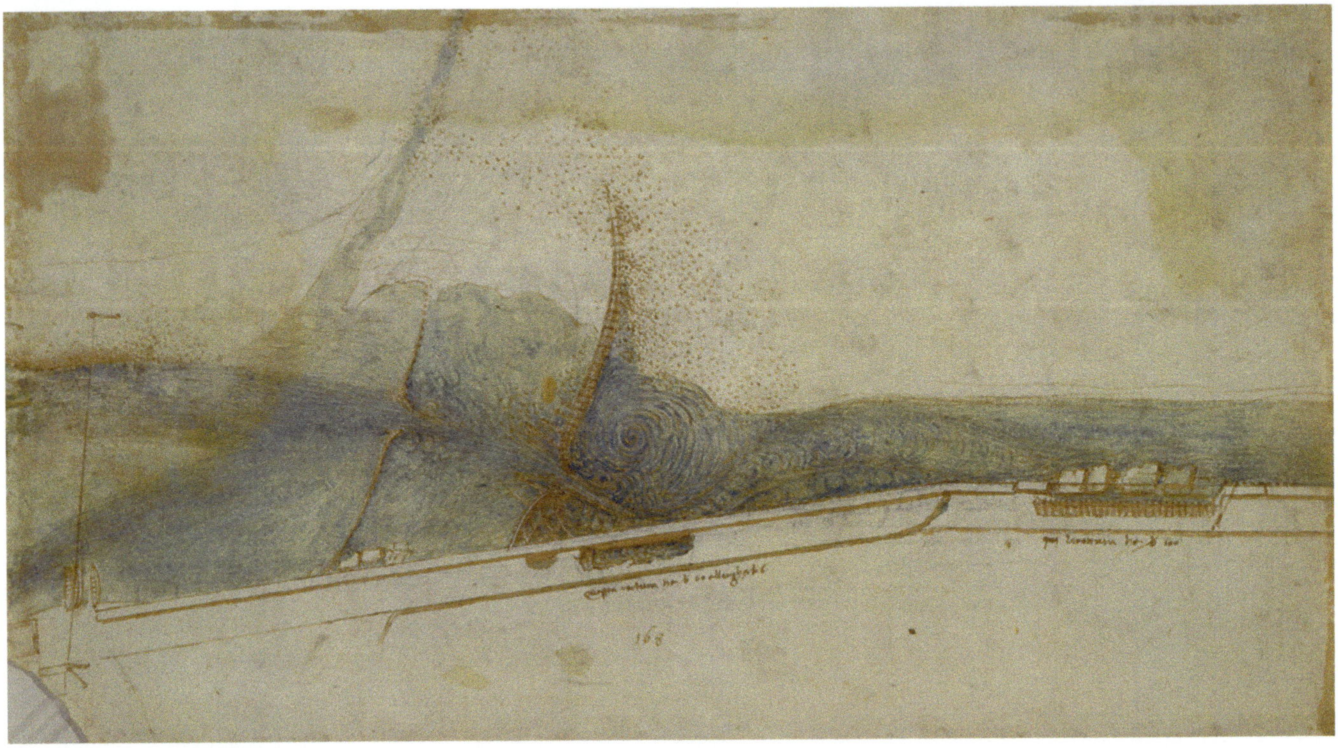

Fig. 4.54 Watercolour illustration of the erosion of a riverbank. RL 12680, A map of a weir and embankment on the Arno east of Florence. Leonardo da Vinci, c. 1504. Pen and ink and watercolour (Lent by Her Majesty The Queen. Royal Collection © Her Majesty Queen Elizabeth II)

> The adversary says that part of the left ventricle of the heart is necessarily shut when this ventricle reopens; and this happens because the blood which is above this valve descends on to the valve, percusses, weighs on it and shuts it. And this is denied by the drawing of the valve; it would be more readily trampled down than shut."[185]

This he illustrated with a small drawing at the top of page RL 19082 recto and an even smaller one on RL 19083 verso (Fig. 4.58).

The final stage of proof came with the idea of testing these concepts in a glass model. Whether or not this model was made and used continues to cause differences of opinion, but the level of detail that Leonardo discussed in his theory leads me to conclude that he did indeed test his ideas in this way. Within a diagram of the plaster mould for the model of the aortic root, Leonardo wrote, "A mould of gypsum for blowing up inside with thin glass; and then break it from head to foot at *a n*. But first pour wax into the gate of a bull's heart so that you may see the true shape of this gate"[186] (see Fig. 4.7).

On another page of drawings of the aortic root and the vortices within, he wrote: "Make this test in the glass [*model*] and move inside it water and 'panic' grass seed."[187] To accomplish this test, he would have to make a closed glass circuit within which he would mount his glass model of the aortic root, then fill it with water and instill grass seeds into the water. Then he must pump it around the circuit to observe the stream of seeds within the flow of water as it passes through the orifice of the modelled aortic root. Gharib and Kemp have shown that this model does work, using laser light to map the blood flow,[188] and we have used state-of-the-art MRI to show that the phenomenon is accurate and can be demonstrated in a live person (Fig. 4.59).

Leonardo's designs for the valve to be placed within the model are recorded on RL 19082 recto. They are quite extraordinary, as they antedate a current method of tissue valve production used by some biotechnology companies. As the three natural valve leaflets are always disproportionate, some producers have chosen to select from several valves to find leaflets that are of even size and shape; they separate them and recombine them in a new synthetic tissue valve in which all of the leaflets are identical. This reproduces the way in which Leonardo has seemed to place the leaflets together within some form of a cylindrical housing. (See Fig. 4.10 and RL 19082 recto.) Leonardo used this type of model in other situations with seeming success.[189] Perhaps even more

[185] RL 19083 verso [III].

[186] RL 19082 recto [I].

[187] RL 19116 verso.

[188] M.Gharib and M.Kemp. Personal communication, 2009.

[189] In Codex Atlanticus 81r, Leonardo records the use of millet seeds to show the movement of water. In Codex Atlanticus 170 vb, he records the use of two colours of seeds to compare complex flow patterns within the flow of water.

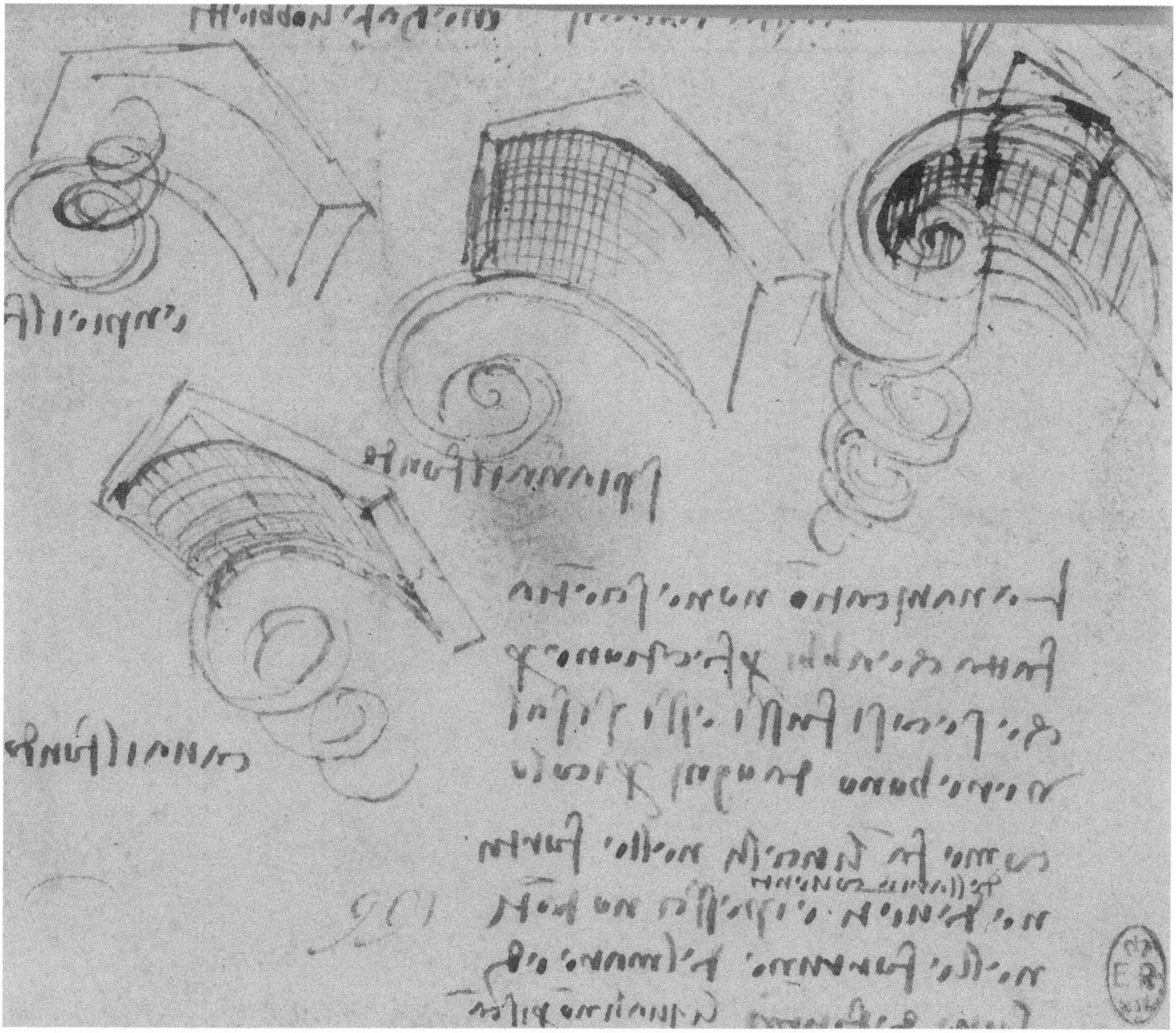

Fig. 4.55 Sketches illustrating the effect of water hitting a curved surface. RL 12666 recto, Studies of vortices. Leonardo da Vinci, c.1510–13. Pen and ink (Lent by Her Majesty The Queen. Royal Collection © Her Majesty Queen Elizabeth II)

extraordinary is that in his notes he records visually the presence of primary and secondary vortices in the column of blood in the aorta beyond the primary vortex above the valve, and in one drawing he suggests tertiary vortices. He could not have seen these with any ease at all in his described model, but once again modern science has proved him correct!

In 1968, two Oxford engineers reported their research on the mechanism of aortic valve closure. In the course of their detailed scientific and mathematical investigation, they had tested Leonardo's theory and proved it correct.[190] The extensive mathematical proof of this concept was spread over several pages, which in its necessity paid homage to Leonardo's demand for mathematical proficiency in anyone attempting to understand or challenge his work. As previously mentioned, testimony to the originality and prescience of Leonardo's work is that the only reference they cited was a description of that work.[191]

An interesting and important corollary to this work, which was suggested by Leonardo, was that the closed aortic valve leaflets are further supported by the sphincter muscular tone in the aortic outflow tract (the cylindrical tube of muscle in the heart immediately beneath the aortic valve).[192] Leonardo suggested that this area of muscle remains contracted,

[190]B.J. Bellhouse and F.H. Bellhouse, Mechanism of closure of the aortic valve. *Nature* 217(5123):66–7. 1968.

[191]Kenneth Keele, *Leonardo da Vinci on Movement of the Heart and Blood.* (London, Harvey and Blythe, 1952), 81.

[192]RL 19082 recto [VI].

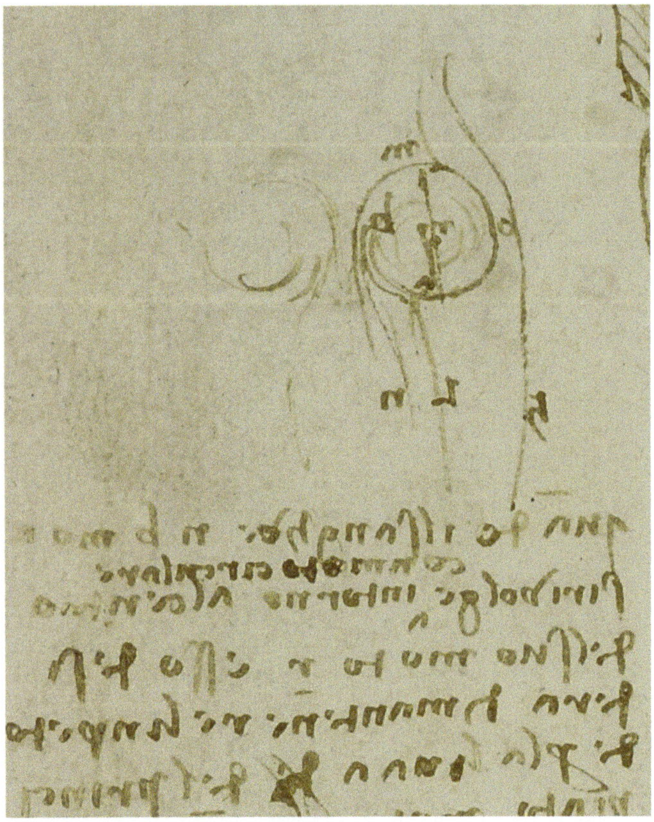

Fig. 4.56 Fluid (blood) being forced into a vortex by the curved sinus of the aorta. Detail from RL 19117 verso, Blood flow through the aortic valve. Leonardo da Vinci, c.1512–13. Pen and ink, over some black chalk (Lent by Her Majesty The Queen. Royal Collection © Her Majesty Queen Elizabeth II)

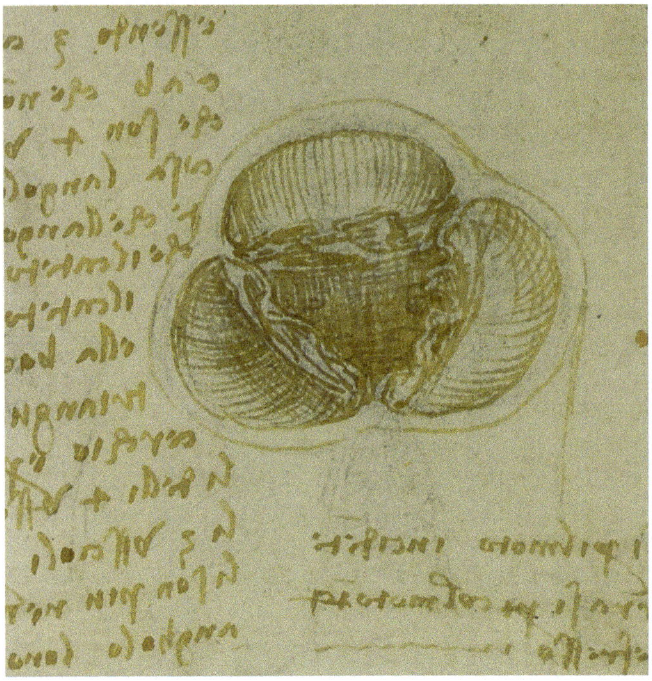

Fig. 4.57 A beautiful drawing showing the fluttering, everting leaflets of the aortic valve as the blood is forced through from the left ventricle. Detail from RL 19117 verso, Blood flow through the aortic valve. Leonardo da Vinci, c.1512–13. Pen and ink, over some black chalk (Lent by Her Majesty The Queen. Royal Collection © Her Majesty Queen Elizabeth II)

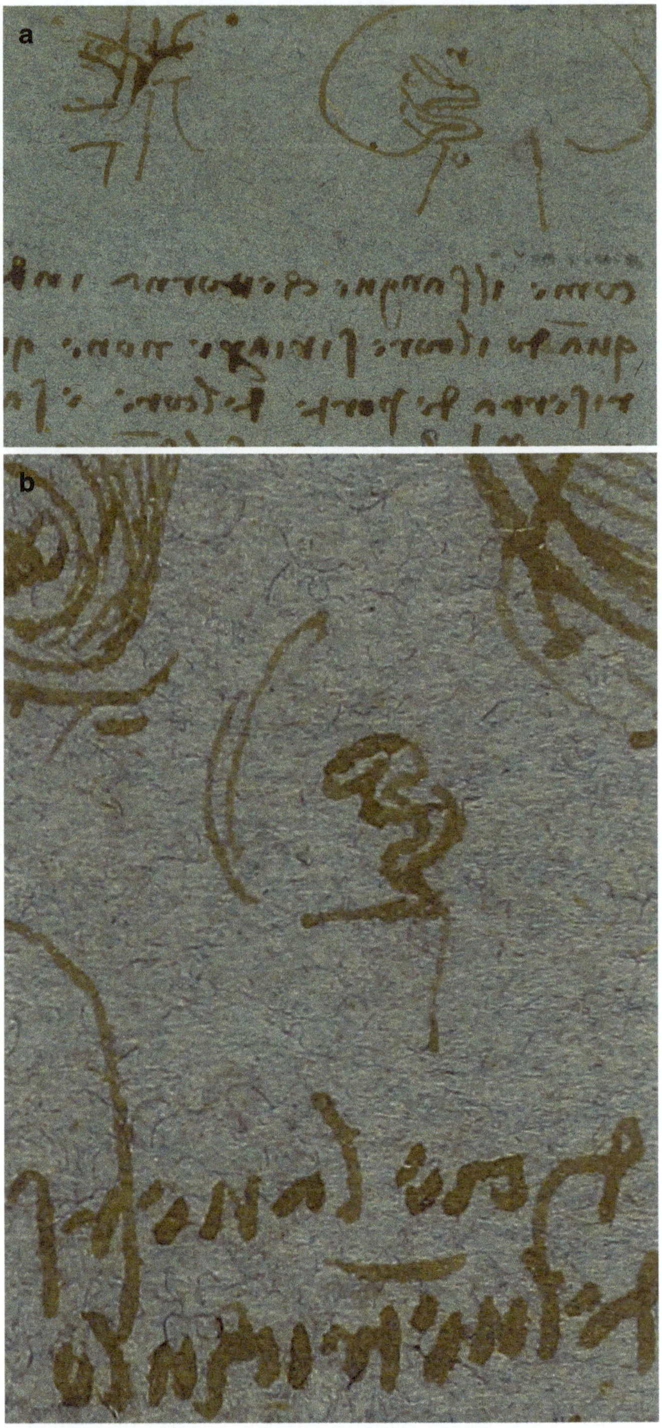

Fig. 4.58 (**a**, **b**) Two small sketches showing the effect on the leaflets if closure were to rely simply upon reflux of the blood. (Part a) Detail from RL 19082 recto, The aortic valve. Leonardo da Vinci, c.1512–13. Pen and ink on blue paper (Lent by Her Majesty The Queen. Royal Collection © Her Majesty Queen Elizabeth II). (Part b) Detail from RL 19083 verso, Blood flow through the aortic valve. Leonardo da Vinci, c.1512–13. Pen and ink on blue paper (Lent by Her Majesty The Queen. Royal Collection © Her Majesty Queen Elizabeth II)

supporting the valve leaflets during the diastolic relaxation phase of the heart. At first, this idea seems unlikely, but there is evidence that this part of the heart is the last to contract and relax, lending support to Leonardo's hypothesis.

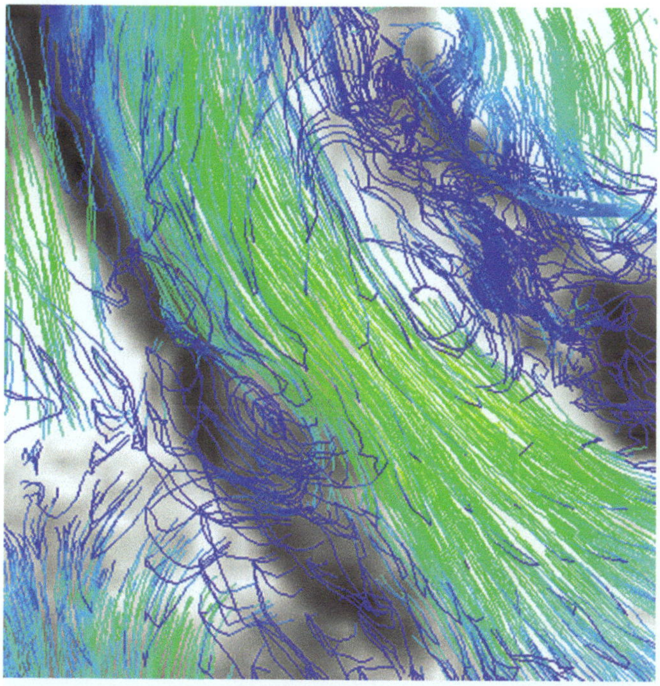

Fig. 4.59 State-of-the-art MR image of vortex formation in the aortic root of a living human heart (Images provided by Dr. Christophe Julie, consultant in Investigational Radiology, Hammersmith Hospital, Ducane Road, London)

In conclusion, this body of work by Leonardo is utterly extraordinary. At the time, it had no relevance other than in the realm of pure investigative science ("blue-sky" research, as we would say now)—Natural Philosophy in its purest form, the pursuit of wisdom and knowledge about the nature of being.

The Bronchial Blood Supply

Although this book is focused on the heart studies, it is important to acknowledge Leonardo's very significant description of the blood supply to the lungs. Once again the specimens under dissection were the organs of the ox. To carry out this work, he must have examined a heart and lung excised en bloc. The most easily identifiable blood supply to the lung is the pulmonary artery. This vessel, which is the same size as the aorta, arises from the right ventricle and then divides into two branches, one for each lung. They provide blood to the lungs for gas exchange (or in Leonardo's terms, to gather the vital spirit, pneuma, and to get rid of heat in the breath). It would seem natural to think that the necessary nutrition for the lungs would be provided in the same way. This is not the case.

In what must have been a very detailed dissection, Leonardo discovered other quite small vessels that followed the branching airways out to their smallest ramifications. This is a significant feat of dissection. These vessels are difficult to find, being no more than 2–3 mm in diameter and very rapidly becoming even smaller. They are intimately related to the surface of the branching airways and supply nutrition to them. Leonardo documents this relationship in a large drawing on RL 19071 recto. This drawing reveals the heart, the trachea, and its branches with an outline of the right lung from behind. A pair of arteries can be identified arising from the posterosuperior aspect of one of the great vessels, the aorta. There is one arterial division for each major airway. These arteries then branch further and extend on to the back of the trachea and its major two divisions into the airway for each lung. It is not clear which of the two major vessels Leonardo had identified as the site of origin of these vessels. His lines demarcating the aorta and the pulmonary artery become unclear at that point. In fact, these vessels arise from the aorta and carry oxygenated blood under arterial pressure.

These vessels are of great significance in heart and lung transplantation today. On excision of the donor lungs, these vessels are inevitably divided. The consequence is subsequent difficulty in the healing of the airways following transplantation because of reduced nourishment to the implanted lung at the bronchial anastomosis. Leonardo distinguished these very small vessels from the pulmonary artery branches by naming them vessels of the first order. He wrote: "The first artery divides at its point of contact with the first division of the trachea…"[193] Under the title "Memorandum" he continued:

> You have to consider the second order of veins and arteries which cover the first [i.e., *bronchial veins and arteries*] minute veins and arteries which nourish and vivify the trachea the substance of which is interposed between the first and second [order] of veins and arteries; and why Nature duplicated artery and vein in such an instrument, one above the other, for the nourishment of one and the same organ. You could say that the trachea and the whole lung had to be nourished and if this had to be done by a single large venous artery this could not be joined to the trachea without great impediment of the movement which the trachea has in its increase and decrease in length as well as breadth. Wherefore this reason she [Nature] gave such veins and arteries to the trachea sufficient for its life and nourishment [*bronchial veins and arteries*] and the other big branches she separated somewhat from the trachea in order to nourish the substance of the lung more conveniently[194] (Fig. 4.60).

Although the reason for their existence, explained by Leonardo as the need to reduce resistance on movement, is probably not right, his description of their purpose is correct. Even today, we have questions about this developing relationship in foetal and embryonic growth. On this page he also correctly described the varying forms of cartilaginous support for the airways as they become smaller.

A final observation on manuscript page RL 19071 recto is the tiny cyst emphasised by Leonardo in his description. This is almost certainly a tuberculous cyst, which may have become colonised with the fungus *Aspergillus*. He describes it in the following way, "Nature prevents ruptures of the ramifications of the trachea (*the terminal bronchi*) by thickening the substance of this trachea and making thereof a crust like a nutshell (*the wall of the cyst*); and it is cartilaginous, and this, with hardness like a callous, repairs the rupture, and the dust and watery humour remain inside." The walls of tuberculous cysts do often calcify, and the remaining cavity may become colo-

[193]RL 19071 recto [XIII].

[194]RL 19071 recto [III].

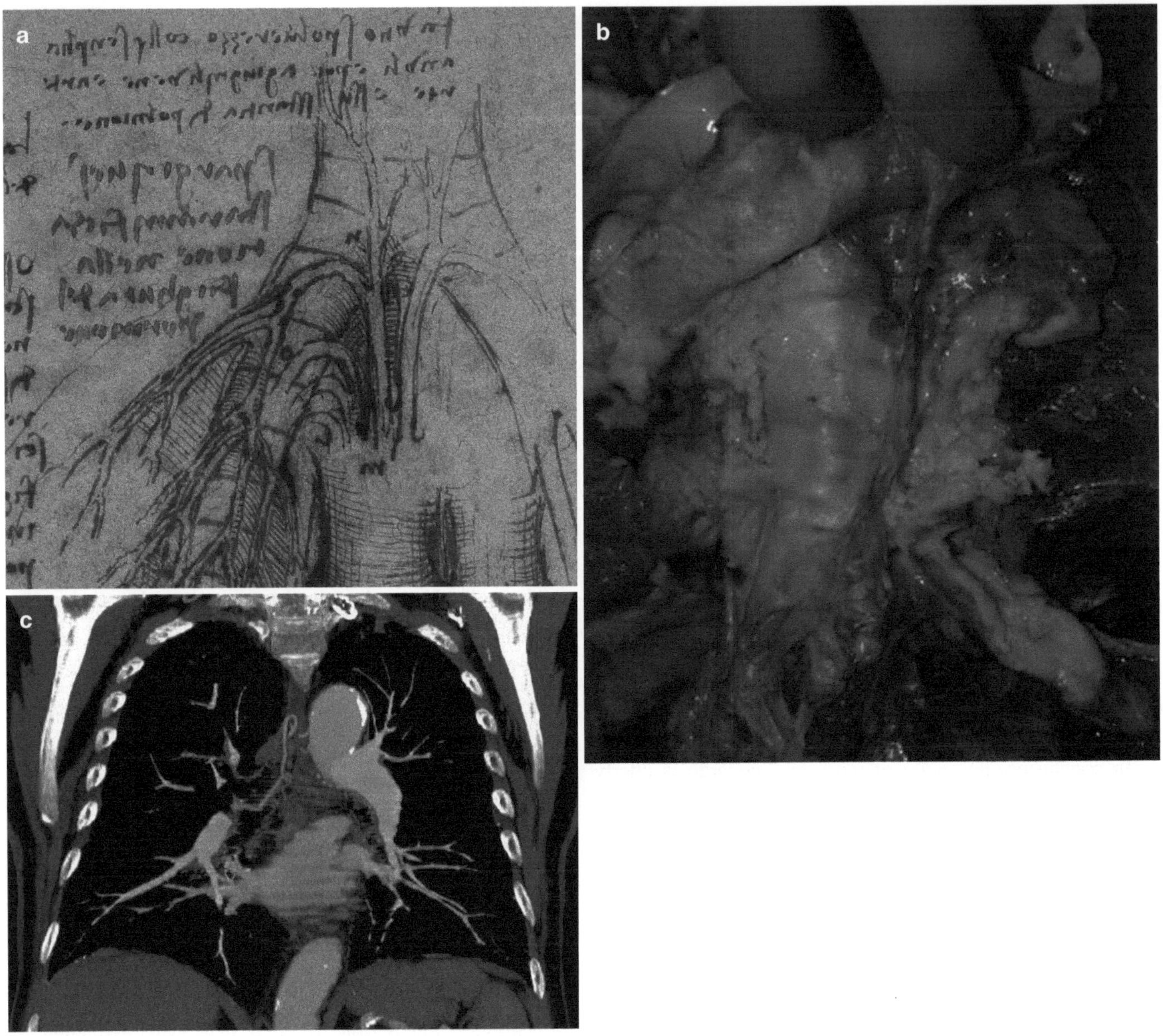

Fig. 4.60 (**a**) Drawing of the bronchial arterial blood supply. Detail from RL 19071 recto, The heart, bronchi and bronchial vessels. Leonardo da Vinci, c. 1511–13. Pen and ink on blue paper (Lent by Her Majesty The Queen. Royal Collection © Her Majesty Queen Elizabeth II). (**b**) Photograph of a dissection of the same part. (**c**) Magnetic resonance image of the bronchial artery in a human

nised with the fungus *Aspergillus*, which often becomes dust-like. Such cysts are quite commonly found in the lungs of human patients with chronic sepsis (Fig. 4.61).

It is on this sheet that Leonardo drew attention to the impossibility that words alone can create a true image in the mind of the student and stressed the importance of illustration.

The Circulation of the Blood

From classical times until the Enlightenment, perceptions of the way in which the body functioned did not command the need for a continuous circulation of the blood. Even when Harvey's description of the continuous circulation was published in *De Motu Cordis* in 1628, it met with a storm in the world of medicine and philosophy.[195] Nowhere in Leonardo's extant notes does he write in terms other than those of the Galenic view of the ebb and flow of the blood.

[195]Breaking with the accepted Galenic description of the distribution of the blood was a slow and difficult process. The appearance of Harvey's book, *De Motu Cordis*, caused something of a storm in the worlds of philosophy and medicine, which reverberated in London, causing his reputation and his practice to falter. Harvey wrote in the book, "...yet when I shall mention them [ideas of a continuous circulation], they are so new and unheard of that not only I fear mischief which may arrive to me from the envy of some persons, but I likewise doubt that every man almost will be my enemy, so much does custom and doctrine once received and deeply rooted (as if it were another Nature) prevail with everyone, and the venerable reverence of antiquity enforces."

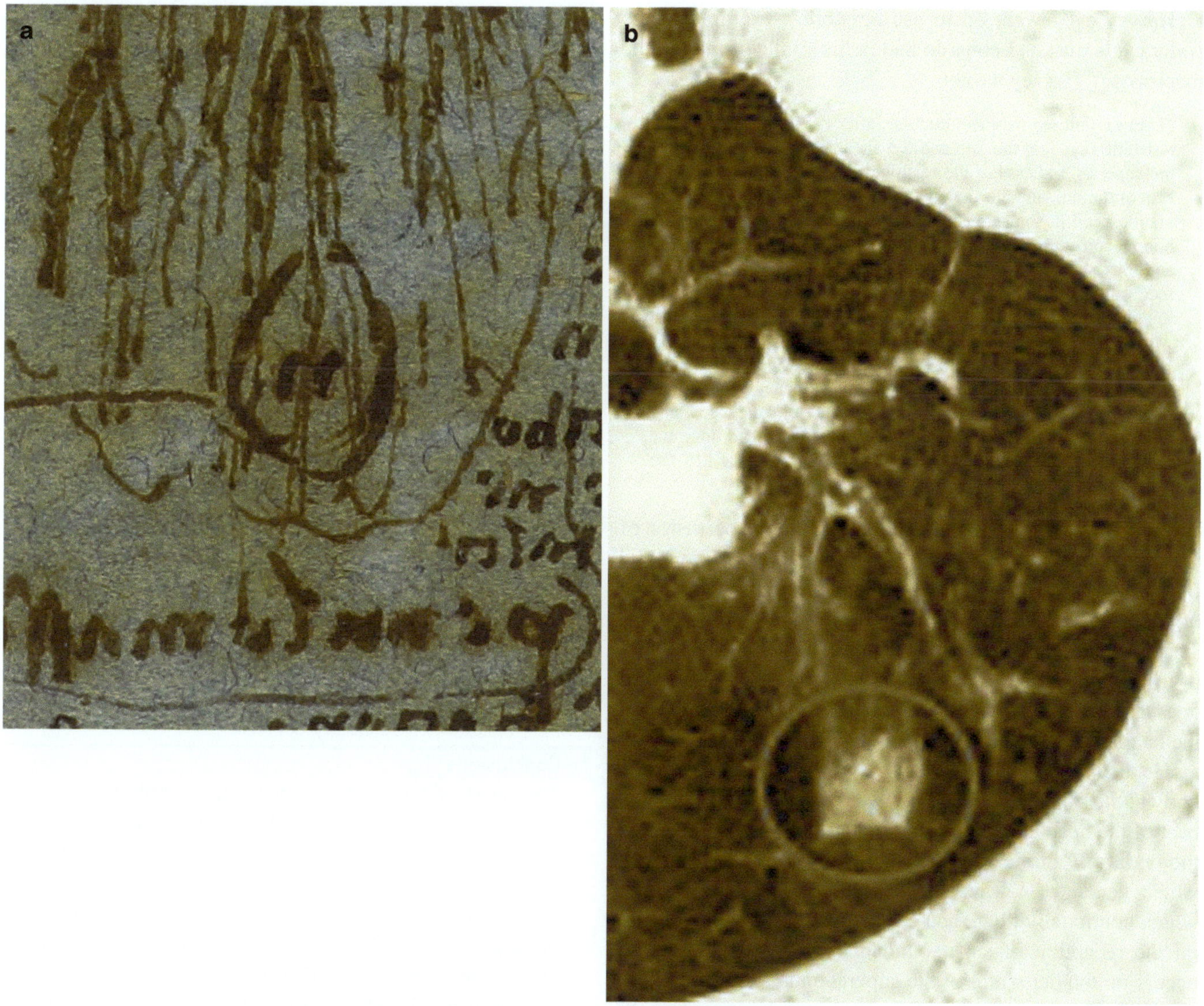

Fig. 4.61 (**a**) A small cyst in the periphery of the lung. Detail from RL 19071 recto, The heart, bronchi and bronchial vessels. Leonardo da Vinci, c. 1511–13. Pen and ink on blue paper (Lent by Her Majesty The Queen. Royal Collection © Her Majesty Queen Elizabeth II). (**b**) Recent CT image of a similar (tuberculous) cyst in the lung of a human

Indeed, as we have shown, much of his discourse is an attempt to define his observations and deductions within Galenic principles. As we have seen, his principal texts were interpretations by Mundinus and de Ketham of Latin translations from Avicenna's Arabic versions of Galen. However, some passages in Leonardo's text do generate an "if only" state of mind in the reader.

For someone as logical in thought as Leonardo, it is surprising that his calculations of the amount of blood that each ventricle would eject per hour did not cause him more concern over the consequences. It is this very problem that largely motivated Harvey's alternative explanation of the movement of the blood around the body. Despite Leonardo's apparently erroneous calculation, which appears to refer more to the amount of blood that the liver must produce to replace that which is consumed (7 oz!), he did note that the amount of blood being expelled was "a great weight"[196] Harvey commented, "At last I perceived that the veins should be quite emptied, and the arteries of the other side be burst with too much intrusion of blood, unless the blood did pass back again by some way out of the veins… and return into the right ventricle of the heart."[197]

[196]RL 19087 verso [VIII].

[197]William Harvey, *The Anatomical Exercises: De Motu Cordis and De Circulatione Sanguinis, in English Translation*. Ed. Geoffrey Keynes. (1653; reprint, New York: Dover Publications, 1995), 58.

Harvey even went on to use Aristotle's example of the water cycle, just as Leonardo had in his simile of man to the macrocosm. Harvey wrote:

> (The) motion we may call circular, after the same manner that Aristotle says that the rain and the air do imitate the motion of the superior bodies. For the earth being wet, evaporates by the heat of the sun, and the vapours being raised aloft are condensed and descend in showers and wet the ground…. So in all likelihood it comes to pass in the body, that all the parts are nourished and cherished and quickened with blood, which is warm, perfect, vaporous, full of spirit, and, that I may so say, alimentative; in the parts the blood is refrigerated, coagulated, and made as it were barren, from thence it returns to the heart, as to the fountain or dwelling house of the body, to recover its perfection, and there again by natural heat, powerful and vehement it is melted and is dispensed again through the body from thence, being fraught with spirits…and that all the things do depend upon the pulsation of the heart. So the heart is the beginning of life, the Sun of the Microcosm."[198]

Leonardo was aware of the importance of the perfusion of the organs (such as the kidneys, bowel, and lungs) with blood for normal function. He had also deduced that, for some organs at least, function was proportionate to blood flow. This is what he had to say about the kidneys: "The quantity of urine shows the quantity of blood which is produced and goes to the kidney; but first it must pass through the valves of the heart *(aortic valve)*."[199]

He also described that the pulse arrived at all parts of the body at the same time and that it was in conjunction with the beat of the heart. He stated: "And the same thing happens in the bodies of animals by means of the beating of the heart which generates a wave of blood through all the vessels, which continually dilate and contract. And dilatation occurs on the reception of superabundant blood, and diminution occurs on the departure of the superabundance of the blood received. This, the beating of the pulse teaches us when we touch the aforesaid vessels with our fingers in any part of the living body."[200]

Elsewhere, Leonardo stated, "The movement of liquid from one direction, proceeds in the original direction as long as the force remains in it which was given to it by its first mover" *(in this case the heart)*.[201] Hence Leonardo had considered all of the important questions that had led Harvey to the idea of a continuous circulation, and yet they apparently did not trigger the same thoughts in him.

Furthermore, an essential part of the Galenic concept was the access of the vital spirits from the air to the heart. At one point in his notes, Leonardo questioned the proposed Galenic solution to this problem:

> Whether air penetrates into the heart or not. To me it seems impossible that any air can penetrate into the heart through the trachea because when if one inflates it, no part of the air escapes from any part of it. And this because of the dense membrane with which the entire ramification of the trachea is clothed. This ramification of the trachea as it goes into the most minute branches together with the most minute ramification of the vessels which accompany them in continuous contact right to the end. It is not here that the enclosed air is breathed out through the fine branches of the trachea and penetrates through the pores of the smallest branches of these vessels. But concerning this I shall not wholly affirm my first statement until I have seen the anatomy which I have in hand.

This is a very important example of Leonardo thinking for himself, with his thoughts being taken directly from his careful observations of the truth revealed by meticulous dissection and careful reasoning.

It is interesting to speculate whether Leonardo did at any time question the Galenic idea of the circulation. It is unlikely that we will ever know, but an interesting aside is the fact that the person who acquired the manuscripts that we now know as the Windsor manuscripts was none other than Thomas Howard, who later accompanied William Harvey on his European tour. Much of Leonardo's work is lost. What might it have contained?

Conclusions

The heart of Leonardo is very recognisable to us today. He began to explain several of the intricate internal mechanisms of the heart, and in so doing he broke free of some Galenic dogma. His observations of the overall structure and the mechanism of the subsystems of the heart were often prescient and described form and its intimate relation to function in terms that would not be out of place in a modern setting. His approach to solving the problem of aortic valve function and his reasoned descriptions of methodology are thoroughly modern.

In the solitude of his own environment, he made several original contributions that did not see the light of day until the nineteenth century. Nevertheless, they are important in their own right. Studying his notes in depth has revealed more to me about the Renaissance use of classical reasoning and this has helped to provoke new ways of thinking about some of the problems that we face today.

This period of intense study has also left a significant sense of likely gaps in the extant work that suggest the scale of the loss of manuscript sheets over time. It is intriguing to wonder what other ideas and questions might have been revealed had all of Leonardo's work been available, but probably we shall never know. What we have is exciting and should encourage us to forever keep an open mind and always to try to see through the questioning but unassuming eyes of a child.

[198] Ibid., 59.
[199] RL 19069 recto [I].
[200] RL 19045 recto [IV].
[201] RL 19045 recto [I].

5 Leonardo: A Personal Legacy

Thou, oh God, dost sell unto us all good things at the price of labour[1]

In this chapter, I will attempt to explain why and how my interest in Leonardo's anatomical studies has influenced me and to some extent my approach to work as a heart surgeon.

The interested reader of Leonardo's notes cannot fail to be impressed with the scope of his work. Perhaps most impressive is the level of detail that he chose to pursue. It is clear that he demanded nothing less than a complete understanding of questions that interested him. As I showed in the preceding chapter, Leonardo was able to derive important and correct conclusions about the function of the heart without any significant prior knowledge, using only observation and reasoning. Two of the most exciting examples are his descriptions of aortic and pulmonary valve closure and the dual circulation to the lungs. The accuracy of this work has only recently been realised, as modern technology has caught up with the vision of Leonardo.

The relevance of these concepts is currently being incorporated into technologies used in modern surgery. Because Leonardo's work was not published until the nineteenth century and still was ahead of its time, the contemporary medical and scientific fraternity remains largely unaware of this seminal work, which is almost 500 years old. It is immensely rewarding to introduce colleagues in our specialist areas of work to this knowledge. The structures that Leonardo chose to study in depth, such as the aortic root sinuses, are now recognised as having a functional importance that cannot be ignored in the reconstruction of these regions. Vascular replacement grafts are now available that have these sinus recesses built into them.

As explained in the last chapter, Leonardo described the elastic nature of the aortic root, the importance of which recently has been recognised with the development of stentless aortic valve replacement prostheses that are more able to expand in a way similar to the normal aortic root. These prostheses can better absorb the shock of the closure of the valve and dissipate this energy into the surrounding tissues, precisely as Leonardo described. The incorporation of this property into the treatment strategy is likely to lead to a longer life for these valves, although conclusive data have not yet been derived.

The importance of the new knowledge Leonardo revealed in these examples has taken us half a millennium to realise. Exciting as this is, perhaps the most interesting thing that I found was Leonardo's intellectual methodology, which began to be revealed only with careful and repeated reading of his notes. It is this that really began to influence my own "style" of thinking. Indeed, it reinforced in me the importance of always being ready to see familiar structures with an open mind, thereby encouraging new ideas.[2] In particular, it has encouraged me to think more expansively about the structural and functional problems encountered in correcting heart valve disease.

Constant critical interrogation of his own ideas is one of the hallmarks of Leonardo's writing, an approach that has informed me of how much can be achieved through a rigorous philosophical approach. Studying this use of rhetoric has led me to try to pursue a more pure understanding of the functional elements of the heart, particularly the mitral valve. Of course Leonardo's rhetorical style is not unique to him but reflects the academic culture of the time. This culture of personal internal discussion, challenging one's own accepted ideas, is something that is rarely taught in the modern medical education system. To find it used to such a powerful effect, revealing scientific truths that are relevant today, has left an indelible impression on me. His reasoning resembles Aristotelian syllogistic logic and other similar deductive processes, and as discussed in an earlier chapter, these logical manoeuvres were reintroduced into the intellectual culture of those times through the translation and dissemination of Arabic versions of ancient Greco-Roman texts.

[1]RL 12642 verso.

[2]F.C. Wells and T. Crowe, "Leonardo da Vinci as a paradigm for modern clinical research," *Journal of Thoracic and Cardiovascular Surgery* 127 (2004): 929–44.

F.C. Wells, *The Heart of Leonardo*,
DOI 10.1007/978-1-4471-4531-8_5, © Springer-Verlag London 2013

In our own time, anatomical knowledge acquired during surgical training is necessarily didactic, and the utilisation of this knowledge is usually perfunctory. We use it in the same way that we use a map to reach a destination. However, as with a simple map, the relevance of the geography encountered on the journey is not considered. For example, a London taxi driver studying for "the Knowledge' (the legendary test that London taxi drivers have to pass before being given their taxi driver's license) learns the interconnectivity of streets and their locations and how to get from "point a" to "point b". The driver does not need to know or understand the reason why a particular street has had to deviate from a direct course because of physical impediments such as a river or an excessively steep incline. And so it is in the use of anatomical knowledge in accomplishing a surgical procedure: It is predominantly used as a route map to be navigated successfully. It informs the surgeon what structures will be encountered and their relationship to one another—very important knowledge for safe completion of the procedure—but it does not encourage a continuing review of the impact of functional requirements upon structural appearance. Whilst establishing a safe practice in most cases, this scenario is also mind-numbing and is likely to inhibit new ideas and further attempts to reinterpret established assumptions.

I found that studious reading of Leonardo's notes led me to a more challenging mind-set. Of course students of the classics and of philosophical thought would remind us that the intellectual tools of logic allied to a rhetorical dialogue are the engines of human development and are the currency of those disciplines. Sadly, the pressures of a surgical life (as in many other high-pressure careers) all too often stifle their use. The modern education system has moved away from the classical form in which these principles were the rules of academic engagement. I have found that through the charisma of Leonardo it is possible to introduce this more reflective form of thought to those who have never before had the luxury of being exposed to it.

To properly understand Leonardo's work, a much wider education about his times is needed. This has led me to a very interesting reading list that is quite different from the reading of the average student of medicine. This personal re-education, allied with the comfort and confidence derived from significant surgical experience, began to encourage a more careful and thoughtful approach to my work. An experience of many procedures allows for a mental space in which the wider issues can be considered. It is this state of mind, coincident with a deepening interest in Leonardo's work, that has led me to find great benefit in these studies.

Clearly Leonardo was well informed about the work of the giants of anatomy from a bygone age, such as Galen, Aristotle, and Hippocrates, but perhaps because he had a limited educational background and was not a physician who had to use his knowledge to intervene in the management of disease, Leonardo was able to pursue his anatomical studies with the innocent but inquisitive eyes of a child, for whom everything has an equal importance. His "looking" was informed by an extremely wide knowledge base in many subjects, however, and his attempts to explain the function of the structures that he found involved the use of basic rather than applied science—what today we would call "blue sky" research. His abiding and often-repeated principle that all forms in nature are forged by the forces that act upon them resulted in some accurate descriptions of complex cardiac function. This teleological style of thought is not endlessly rewarding, however, and many would argue that real advances are possible only if the scientific investigator looks beyond this concept. In trying to understand major biological structural forms, however, this form of logic can be very productive in furthering an understanding of complex structures working in a constantly adapting and stressful environment such as the working heart. We will return to this theme shortly.

It is intriguing to wonder why Leonardo undertook this work. In the context of his times, there was no reason for him to investigate the phenomenon of aortic valve closure. His study preceded the accepted date of the discovery of the circulation and complicated the contemporary explanation of the flux and reflux motion of the blood that Galen's explanation demanded. Thus it is one of the earliest examples of academic research in a form that we instantly recognise today. It is experimental science, and it produced new knowledge for its own sake.

As we have seen, so much of his writing was exercising his understanding of the accepted ideas of the time, particularly the Galenic theory of blood flow. Perhaps his questions about the anatomy of the root of the aorta were fired by its very striking appearance. In the ox heart, this structure is dramatic. The aortic sinuses are particularly large and sit on top of the large muscular outflow tract of the left ventricle. The shape of this structure begs the question of why it is formed like that when the continuation of the aortic tube leading away from the heart is just a simple tube. The cross-sectional shape of this structure clearly brought to Leonardo's mind his work on previous hydrodynamic settings. He was able to transpose the setting from the heart to the more abstract idea of a column of fluid flowing through a channel of this shape.[3] From that point onwards, he would have been in comfortable territory that related to previous work done on rivers and other waterways.[4] His manuscripts contain clear evidence that this recognition of parallel situations led Leonardo to utilise hydrodynamic principles to reach the correct interpretation of the mechanism of valve closure.

Using his hydrodynamic knowledge, Leonardo developed the hypothesis that these bulges (later known as the sinuses of

[3]RL 19116 recto.

[4]RL 12666 recto.

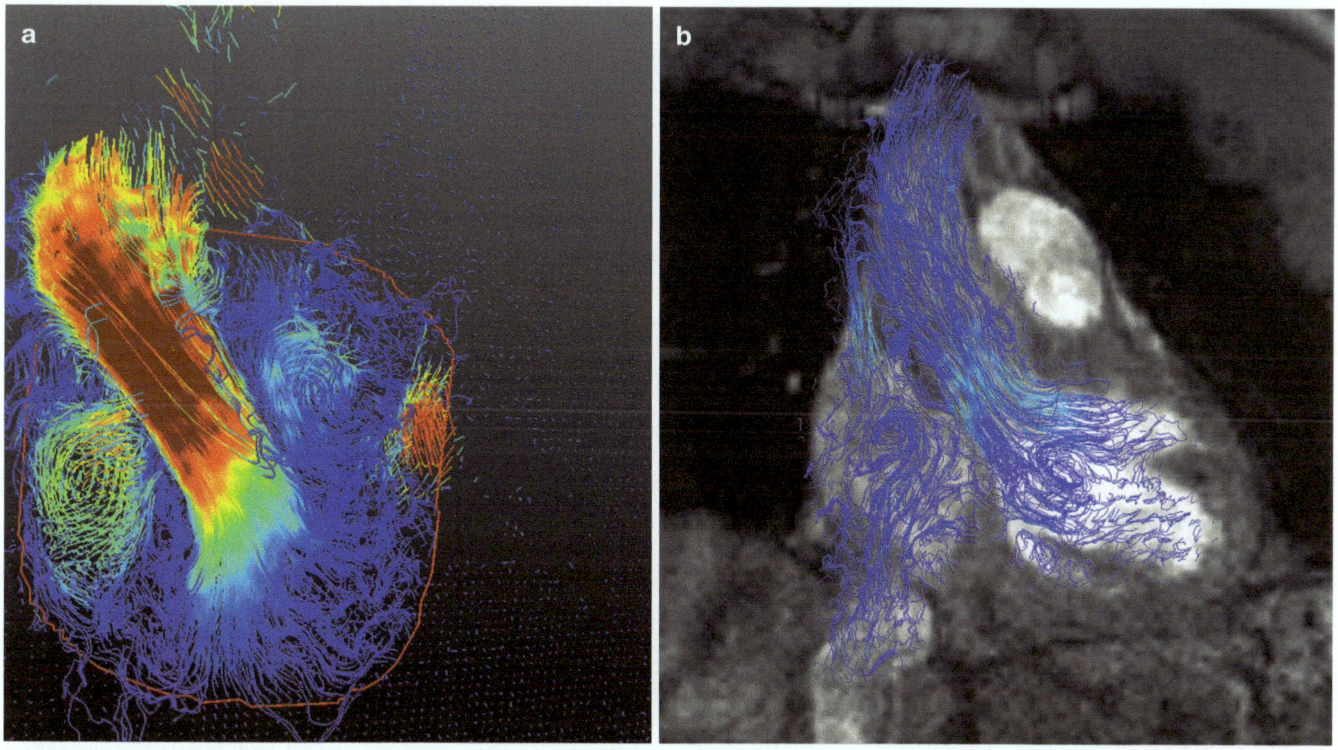

Fig. 5.1 Vortices in the chambers of the human heart, captured with MRI scanning. (**a**) Vortices in the atria during ventricular systole. (**b**) Vortices in the ventricles during systole (Images provided by Dr. Christophe Julie, consultant in Investigational Radiology, Hammersmith Hospital, Ducane Road, London)

Valsalva) give rise to vortex formation in the column of blood escaping from the ventricle, and that these vortices are instrumental in beginning the process of valve closure by unfurling the leaflets and pushing them towards each other. This conclusion is made all the more remarkable by the design of an elegant experiment that proved this hypothesis to his satisfaction.

As a consequence of studying Leonardo's description of these vortices, I set out to try to produce images of them in the intact, functioning human heart. Using new computer software programmes written for advanced magnetic resonance imaging by Dr. Christophe Julie and his colleagues in Switzerland, we were able map the course of the blood flowing through the aortic valve, revealing the vortices as they formed in real time in the functioning human heart. This work has provided tangible evidence that these vortices exist in the intact human heart and not just in an experimental model and that they perform the function predicted by Leonardo about 1513. Furthermore, our studies have revealed that complex vortices form in all chambers of the heart during the normal cardiac cycle (Fig. 5.1).

From these observations, we are now beginning to think about the importance of these vortices in both the normal and the diseased heart. Whilst the heart muscle exerts pressure upon the blood within its cavity as it contracts, the blood exerts an opposing force upon the contracting ventricular wall. It seems as though the normal flow of blood creates multiple vortices in different parts of the heart, and that these vortices are very reproducible. It also appears that they begin to break down as function deteriorates. It seems important to us to try to understand whether this has consequences for the ventricle's ability to function, thereby potentially accelerating a deteriorating situation. At the moment, this would be seen as "blue sky" research, as the applications of such knowledge remain to be seen.

My work as a surgeon is focussed upon the reconstruction of failing heart valves. This type of work depends particularly upon our ability to recreate their natural form. Currently, two surgical solutions are available. The first solution, which is purely functional and somewhat simplistic, is to implant replacement heart valves (an artificial and invented solution). These valves are of two principal types. The first type is mechanical or synthesised-tissue valves, in which the leaflets are made either of ultrasmooth carbon or carefully fashioned pericardium (the fibrous sac that envelopes the heart), usually taken from an ox. The second type is consists of excised and preserved homografts or xenografts[5]. Each of

[5]Homograft: A tissue graft (valve) from a donor of the same species as the recipient. In this case, an aortic valve from a human cadaver is the standard choice. Xenograft: A tissue graft (valve) from a donor of a different species from the recipient; porcine aortic valves are most commonly used.

these valve types will restore the normal, unidirectional flow of blood through the heart. Both can perform satisfactorily for significant (but variable) periods. Mechanical valves are very durable but require lifelong anticoagulation; tissue valves are somewhat less durable but do not require the rigorous use of anticoagulants. None of these replacement valves are able to withstand the forces applied to them for a lifetime with absolute reliability, nor can they endure the harsh chemical and immunological environment of the blood indefinitely.

The second solution (and the favoured one) is reconstruction of the mitral valve in such a way as to mimic its natural form as closely as possible. This reconstructive approach has great attractions: it preserves the normal tissues of the valve, which therefore do not interact adversely with the blood, and it maintains the normal, integrated function of the valve within the overall dynamic structure of the heart. As we have described previously, the valve and the ventricle function as one integrated unit. However, techniques most often used today focus on restoring a fully competent valve on closure by cutting away parts of the valve leaflet and thereby reducing its orifice when the remaining portions are reconstructed. Whilst successful in restoring competence, this method alters the normal form and function of the mitral valve in its opening phase (diastole), restricting its orifice and inhibiting its ability to allow a greater flow, which is needed at the extremes of exercise. The normal mitral valve is able to dilate its orifice as exercise increases in response to an ever-increasing return of blood dictated by the increasing output of the heart. The current techniques may be adequate for people who are not likely to make great demands upon their heart, but not for a younger, more active person. It is surely a goal for the surgeon to achieve a result that matches the demands of the heart through all phases of the cardiac cycle. Any attempt to achieve this goal and return the failing valve to its natural form within the constantly changing environment of the pumping chambers of the heart demands detailed knowledge of its structure and function. This is where a teleological approach with a deep understanding of the intimate relationship of structure to function is required. This more recent approach to the reconstruction of the mitral valve is still evolving and has engendered a greater respect for the complex function of this valve.

Leonardo reminds us that form predicts function and function dictates form in all natural structures. Indeed, it appears to me that an understanding and acceptance of this caveat will refocus our thought on most, if not all, problems of biological structure. I recently heard a dramatic rewording of this principle from Dr. Doris Taylor of the University of Minnesota (USA) Center for Cardiovascular Repair. In describing her team's work on developing a bioartificial heart, in which all of the cells within the organ are first washed away, to be replaced with cardiac stem cells that then repopulate the heart appropriately, she said, "Give nature the tools and then get out of the way."[6] Her point is that science is yet far from being able to construct *de novo* an organ as complex as the heart, but by retaining the natural scaffold of connective tissue, it is likely to be possible to simply reintroduce the progenitor cells (cardiac stem cells) and let them organise themselves in response to the structure that they find themselves in. Moreover, it appears that by placing the neo-heart in the normal loading situation in which it will find itself when working, the organ will respond to those loads and grow into a normally functioning heart. Here we truly find function imposing itself upon developing structure.

This concept has significant resonance with Leonardo's pursuance of Aristotelian philosophy. He states frequently that everything in nature is as it is for a reason, and this tenet sets the tone for his approach to understanding the world around him. It is evidenced in the statement to be found on RL 19063 verso, which we have already quoted: "Nothing is superfluous and nothing is lacking in any species of animals or product of nature unless the defect comes from the means which she produces it."[7] In other words, every observable form and action has both relevance and an effect. Moreover, an observable effect resulting from an action betrays the nature of its cause. Thus through careful and detailed examination allied to intuitive reasoning, we may come to understand the world around us. This evidence-based method of reasoning predicted the essential need for scientific experimentation, the bedrock of modern knowledge-gathering. Much of the genius of Leonardo rests in his simple and yet profound personal resetting of the process that was then accepted as the means of academic education: the assimilation of preconceived ideas, or in other words, the unquestioning study of dogma.

In Codex Atlanticus,[8] Leonardo quotes from a speech by Marius, a Roman military commander, to the Quirites (the earliest citizens of Ancient Rome) saying, "They who adorn themselves in the labours of others will not permit me my own."[9] Leonardo continued, "They will say that because I have no book learning, I cannot properly express what I desire to treat of—but they do not know that my subjects require for their exposition experience rather than the words of others. Experience has been the mistress of whoever has written well; and so as mistress I will cite her in all cases."[10] This exclamation reflects Leonardo's attitude about the importance of independent thought.

---

[6]D.A. Taylor, C. Inaugural, Walton Lillehei Lecture: Bioarticial Heart – "New uses for old cells".

[7]RL 19063 verso [I].

[8]CA 119 verso.

[9]This quote comes from the Roman historian Sallust and is taken from *his Jugurthine War: Bellum Jugurthinum* (chapter 85, paragraph 25). Marius fought against and defeated Jugurtha in 109–104 B.C.

[10]Leonardo da Vinci, *Notebooks*. (Collected by Irma Richter; Oxford: Oxford University Press, 1952; edited with introduction and notes by Thereza Crowe, preface by Martin Kemp, Oxford: Oxford World's Classics, 2008), 3–4.

Leonardo informed his reasoning with intellectual investigation, but just as importantly, he was able to distill the right questions to ask. He also held implacably to the fact that mathematical proof was of fundamental importance, and he took care to record his carefully considered approach to a question in annotated (and frequently illustrated) form. These steps of thought and experience, or what we would refer to today as experiment, allied to mathematical proof and the recording of them, are the fundamental building blocks of what we consider good science. Therefore, despite all of the hagiography that has enveloped him, this form of reasoning makes it impossible to ignore Leonardo's contributions to scientific method, and it behooves the reader to take his writing seriously. His time spent in reflective thought and his detailed examination of natural forms and structures through the eyes of the artist, scientist, and architect brought him to revelations that have been passed over until very recently. In addition, his use of a form of Aristotelian logic afforded him deeper understanding problems.

Perhaps the greatest exponent of this approach in the recent past is D'Arcy Wentworth Thompson. His magnum opus, entitled *On Growth and Form*,[11] although well known in the world of the natural sciences, arts, and classics, is rarely acknowledged in the modern world of medicine, but it has much that is worthy of consideration. He summarises thus for us in his epilogue: "For the harmony of the world is made manifest in Form and Number, and the heart and soul and all the poetry of Natural Philosophy are embodied in the concept of mathematical beauty." The following is a paraphrase of his central tenet, taken from the opening essay in the first chapter of the same book: "The form, then, of any portion of matter, whether it be living or dead, and the changes of form which are apparent in its movements and in its growth, may in all cases alike be described as due to the action of force. In short the form of an object is a diagram of forces, in that sense, at least, that from it we can judge of or deduce the forces that have acted or are acting upon it: in this strict and particular sense, it is a diagram." These sentiments are again a reprise of the ideas expressed by Aristotle and Leonardo, to whom Thompson pays his respects, particularly with reference to the prerequisite for mathematical proof.

As described above, the application of this approach to structures in the heart—particularly the mitral valve—has opened my eyes to important nuances of structure to which we have hitherto paid lip service. By this I mean that we often speak in the abstract of the physiological necessity for the form of the valve, such as the importance of the ability of the orifice of the valve to be flexible, but in our proposed solutions for a leaking valve, we fix the orifice rigidly. We speak of the normal saddle shape of part of the valve circumference but we do not discuss why this shape is naturally imposed upon the valve. We recognise that there are normally clefts in the leaflets but we do not discuss what their normal depth should be. What we can be sure of, however, is that the complex shape that is imposed upon the valve leaflets in the pressure-loaded and volume-loaded ventricle becomes a metaphor for the forces that are imposed upon them. Careful examination of these shapes and forms therefore can reveal the way in which the leaflets have been formed to dissipate the forces that are acting upon them, which in turn may be used to appreciate the normal form and relationships of the leaflets.

Leonardo considered force as one of the four powers of nature and set about defining it in some considerable detail.[12] Of relevance here is his allusion to the way in which forces are dissipated through and across boundary surfaces. First he states, "Force is the same throughout and through all the body where it is generated."[13] Then, in a description of the dissipation of force acquired through an applied weight, he wrote, "It is in the nature of weight to transmit itself of its own accord to the desired place."[14] These observations, in combination with Leonardo's observations upon the ideal surface shape for a semilunar valve within the heart,[15] have been converted into a mathematical form by a bioengineer, Professor Morteza Gharib.[16] He proposed that the commentary by Leonardo on the necessity of a spherical structure for the heart-valve leaflets could be summarised in modern mathematical language in the following way: The shape of the heart valve belongs to the family of minimum-energy surfaces (spherical or parabolic), such as soap bubbles or water drops. This model describes each leaflet as a "minimal energy surface" (MES) that deforms smoothly once exposed to a pressure gradient across the surface area. The tissues comprising the heart valves are among the most thin and delicate in the body, so their shape must be optimal to dissipate the load that falls upon them. The differential equation that Dr. Gharib described can be used to predict the ideal shape for monoleaflet, bileaflet, and trileaflet valves. The solutions derived from these finite element analyses accurately predict the actual shapes that are found in the natural setting. It therefore seems to me that the reverse must be true: If we closely examine and accurately describe the form of failing valves, then we should be able to better understand the reasons for their failure. From this understanding, it should be

[11] Thompson, D'Arcy Wentworth, *On Growth and Form, The complete revised edition.* (Cambridge: Cambridge University Press, 1942; reprint, New York: Dover, 1992).

[12] Codex Atlanticus 302r/299r.

[13] Ibid.

[14] Ibid.

[15] RL 19117 verso.

[16] M. Akdis and M. Gharib, On some peculiar aspects of heart valve structures: a hint by Leonardo (preliminary communication). Bioengineering, California Institute of Technology, Pasadena, California.

possible to derive the most appropriate repair techniques to restore their natural form and function.

The concept of a "minimal energy surface" predicts that the boundary of a cavity (such as the boundary between the left atrium and the left ventricle, and in particular the surfaces of the mitral valve, which forms a boundary between both cavities) will be formed as a result of the forces acting upon it in such a way as to minimise and equalise the forces as much as possible for every part of that surface. Both shape and form will reflect these forces. In this regard, we find that the predictions are true, and the surfaces of the tensioned mitral valve leaflets are very complex. The largest leaflet is domed into the complex shape of a saddle, and both leaflets are in continuation at their junctions to a significant depth. This shape allows the forces to be dissipated across both leaflets. The arrangement of the supporting cords has been reproduced to give the impression of organic form in the wonderful interior of La Sagrada Familia in Barcelona (Fig. 3.7). Recognition of the importance of these arrangements can contribute to the planning of the most natural way possible to reconstruct the failing mitral valve (Fig. 5.2).

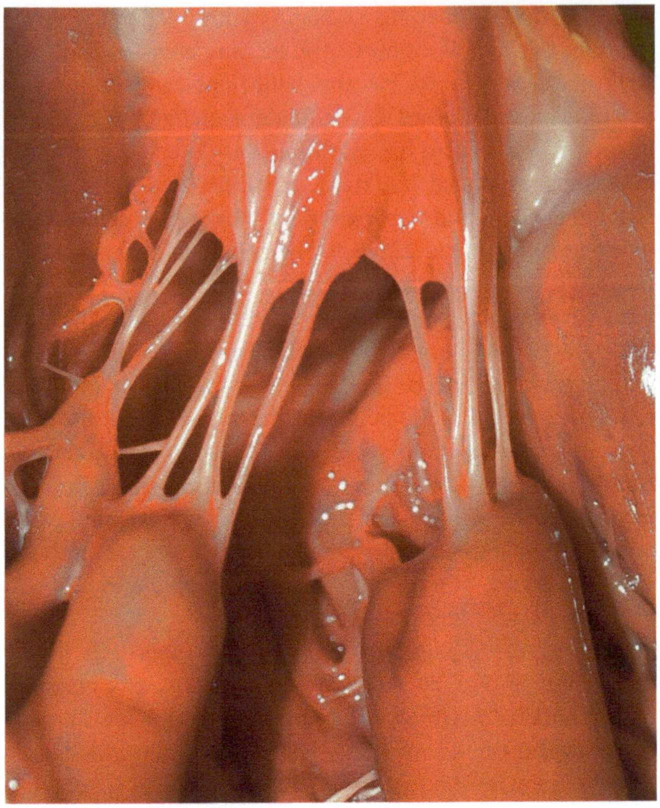

Fig. 5.2 Photograph of the ventricular surface of the human mitral valve

Thus by following the Aristotelian concepts of the formative interaction between cause and effect, for which Leonardo had such respect, closer study of structure at both the macroscopic and the microscopic level of structures such as the valves of the heart is likely to continue to reveal new truths that will enable us to develop more natural solutions to diseased (*dis-eased*!) structures within complex structures such as the heart. Further discussion of these concepts is beyond the scope of this book, but I hope that these considerations have illustrated how this form of reasoning may be of use in understanding problems in structures that are shaped by the forces of nature.

Although it is inspiring to find in Leonardo's ancient manuscripts so many modern descriptions of cardiac structure and function that induce wonder at his genius, his intellectual logic may be the thing from these studies that we can use most productively. To be inspired to look anew at familiar problems and to be encouraged to ask the right questions can only be a positive and invigorating process. It is for this reason that the student of Leonardo must not be blinded by the beauty of his work but rather should be encouraged to look through and beyond his genius with the use of line and to peer into the working of his mind, as revealed in his extensive and detailed notes. It is my hope that by presenting both transcriptions of his notes and our translation, readers will be encouraged to see what they can gain from this powerful repository of human intellect. Perhaps this book will lend some weight to that argument.

6 Leonardo Scholarship and the Cataloguing of the Anatomical Manuscripts

The first academic commentary on Leonardo's anatomical manuscripts was that of the eminent eighteenth century anatomist William Hunter. His comments on the newly discovered manuscripts are the first of the new world of science. The manuscripts and drawings were found in the possession of King George III, whose Royal Librarian, Richard Dalton, rediscovered them and showed them to Hunter. Hunter was deeply impressed. In his "Foundation Lectures on Anatomy", Hunter wrote, "Leonardo must have been the foremost anatomist of that time."[1] He planned to publish the manuscripts with his own commentary, but he died in 1783 before he could undertake the task.

Further scholarship had slow beginnings. The first selection of published drawings was by John Chamberlaine in his "Imitations of Original Designs by Leonardo da Vinci", which contained only a few anatomical drawings. Gaetano Milanesi (1872) and Gustavo Uzielle in the same year began to study them. J.P. Richter undertook the first thorough study of the anatomy.

The first attempts to disseminate the anatomical drawings to a wider audience came about through the efforts of Theodore Sabachnikoff. He published a selection of the drawings in 1898 under the title of *Dell' Anatomia, Fogli A*. The introduction to this volume was written by a professor of anatomy at the École Nationale des Beaux Arts, Mathias Duval. It contained transcriptions of Leonardo's text accompanied by translation into French. In 1901, a second volume (*Fogli B*) was published. This work contains numerous errors in translation and transcription as well as in anatomical understanding and in knowledge of the history of anatomy.

Sabachnikoff continued to photograph the remaining drawings and had deposited the negatives with Édouard Rouveyre, his Paris publisher. Rouveyre betrayed Sabachnikoff, however, and published these remaining drawings before Sabachnikoff had completed the transcriptions and translation of Leonardo's remaining text. It is said that this unethical behaviour brought about Sabachnikoff's premature death. It certainly conspired to undermine the work, which was left in need of completion.

In 1911 and 1916, there appeared the significant publication in Oslo of the *Quaderni d'Anatomia*. This came about through the efforts of C.L. Vangensten, A. Fonahn, and H. Hopstock. It contains a full Italian transcription and translation into English and German. The facsimiles were beautifully made, but the translations are somewhat inaccurate. The various volumes were made up by the grouping of related subjects, so they are not in chronological order.

In 1930, J. Playfair McMurrich (Professor of Anatomy at the University of Toronto) published his book, *Leonardo da Vinci, The Anatomist*. The author of the preface to that book, George Sarton, noted that real scholarship of the anatomy of Leonardo could only be brought to bear by someone academically familiar with the subject, a theoretical knowledge alone not being enough. This observation is certainly true, and it has become ever more obvious to me that those attributes alone may not be enough to deal adequately with the expansive horizons of Leonardo's objectivity and purity of observation. He recorded truthfully his observations and deductions, which very often were ahead of the accepted knowledge of his time. The McMurrich book was the first to begin to systematise the discussion of Leonardo's anatomy.

In 1935, in conjunction with the Royal Library, Kenneth Clark produced what has become the standard reference catalogue of the Windsor drawings. The catalogue numbers for the nonanatomical work were derived from Pompeo Leoni's initial numbering system and begin at 12275. The anatomy pages were bound separately and later, and hence began with the prefix 19, ranging from 19000 to 19152. There are three anatomical manuscripts in this system, *Anatomical manuscripts A, B, and C*. Manuscript A has adopted the Sabachnikoff numbering and ranges from 19000 to 19017. These are principally the musculoskeletal works. Manuscript B also follows Sabachnikoff's second volume and extends

[1]Extract from his lectures on anatomy by William Hunter, given in the Windmill Theatre in 1784. Records kept in the Royal College of Surgeons of Glasgow.

F.C. Wells, *The Heart of Leonardo*,
DOI 10.1007/978-1-4471-4531-8_6, © Springer-Verlag London 2013

from 19018 to 19059; it includes the skull drawings, further musculoskeletal work, and some vascular and visceral anatomy.

All of the drawings not published in Sabachnikoff's A and B volumes, as described earlier, were published in the *Quaderni* and make up the group that Clark called Manuscript C. As mentioned earlier, these are not in chronological order and the definitive work on the archaeology of these drawings remains to be done.[2] Virtually all of the cardiac work is found in this section.

The first attempt to clearly categorise the manuscript sheets into a system of anatomy came about through the work of Charles O'Malley and J.B. de C.M. Saunders of Stanford University and the University of California, published originally by Henry Schuman[3] in 1952. This publication made some attempt at chronology, but it was inadequate.

In 1952, Dr. Kenneth Keele produced his fine review of the cardiovascular system as understood by Leonardo in his book, *Leonardo da Vinci on Movement of the Heart and Blood.*

From 1978 to1980, Dr. Keele and Professor Carlo Pedretti produced a definitive set of facsimiles of the anatomical works with their own updated transcriptions and translations. This remains the definitive work on this section of Leonardo's anatomy.

There have been numerous academic and contextual reviews of Leonardo's anatomy by scholars such as Keele, Kemp, and Pedretti, and reference to much of this work can be found in the text of this book.

The Order of the Drawings

I have kept to the Clark catalogue numbers in the drawings section but wish to add some thoughts of my own about their chronology. It appears to me that there are three distinct periods of effort revealed in the extant cardiac work of Leonardo. The first is the 1480–1495 period, in which we see Leonardo's acceptance of the status quo with regard to his understanding of the Galenic principles of the heart and the lungs and their related vasculature. These are represented by RL 12597, RL 19112 recto, and RL 19104 verso and recto. (The image on the recto side of the manuscript seems to be a reversed tracing of the verso of the drawing, to show the heart and great vessels from the back view.[4]) All of these demonstrate a Galenic heart. However, there are subtle differences within them that deserve further study. Whereas the arterial anatomy displayed in RL 19104 verso is undoubtedly bovine, the same part of the anatomy in RL 19112 recto suggests human form.

RL 12597 recto, although completed early in Leonardo's life, is probably the first in this series of cardiovascular anatomy, and parts of it may date from as early as 1482.[5] When the outline of this drawing is considered in comparison with others of similar design, such as RL 12593, RL 12594, and RL 12595, I am drawn to ask whether the anatomy could have been superimposed later on a drawing taken from what appears to be almost a series of patterns for a full frontal idealised male taken from a model book.

The following group of drawings appear to be related to the time of the dissection of the centenarian in Florence: RL 19023 verso, RL 19027 recto and verso, RL 19028 verso, RL 19049 verso, RL 19050 recto, and RL 19051 recto. These drawings of human arteries and veins reveal elements of Leonardo's descriptions of his findings in the old man. I am inclined to think that RL 19028 recto is from the same period, but it is somewhat different in that its content is of a more philosophical nature: Leonardo argues that the heart (Aristotelian logic) rather than the liver (Galenic logic) is the source of the vasculature of the body. I suspect that RL 19029 is from a similar time, perhaps the later stages of the considered results of the dissection of the centenarian, as it again presents cardiac anatomy with the feel of physiological (functional) interpretation.

The remainder of the drawings fit into a very different style of drawing, which is much more advanced in its incisive questioning of structure as related to function. These drawings reveal Leonardo at his very best as an enquiring anatomical scientist; indeed, he becomes perhaps the first physiologist.

[2]Professor Martin Kemp, Personal communication, 2008.

[3]Charles D. O'Malley and J. B. de C.M. Saunders, *Leonardo da Vinci on the human body*. (New York: Henry Schuman, 1952; reprint, New York, Dover, 1983).

[4]Kenneth Clark, *Leonardo da Vinci. Anatomical Drawings at Windsor Castle*. 2nd edition. (Oxford: Phaidon, 1969), 40–1.

[5]Personal communication from Dr. Domenico Laurenza, whose work suggests a much earlier date than originally projected by Clark. Alcune Precisazioni sul Foglio anatomico di Leonardo RL 12597 recto. Raccolta Vinciana MMV, 2009.

Selected Bibliography

Leonardo's anatomy comprises a relatively small part of his extant work, but the serious nature of his studies in this area is such that they demand rigorous examination. Coming to grips with the significance of his contributions requires extensive contextual knowledge. As a surgeon (and thus artisan) by training and lacking a formal education in the history of art, I have enjoyed a very broad and diverse reading list that has kept me entertained over many years. The following list assembles some examples of that self-education, which has helped me to begin to understand the world that Leonardo inhabited, the forces that shaped his career, and the pantheon of his academic contemporaries, who to a greater or lesser extent informed his revelations.

Ames-Lewis F. Drawing in early renaissance Italy. New Haven/London: Yale University Press; 2000.

Ames-Lewis F, Wright J. Drawing in the Italian Renaissance Workshop. Catalogue to an exhibition of early renaissance drawings from collections in Great Britain held at the University Art Gallery, Nottingham, 12 February–12 March 1983 in association with the Arts Council, UK. London: V&A Publications; 1983.

Belt E. Manuscripts of Leonardo da Vinci. Their history with a description of the manuscript editions in facsimile. Los Angeles: The Elmer Belt Library; 1948.

Belt E. Leonardo the anatomist. Kansas City: University of Kansas Press; 1955.

Brambach C, editor. Leonardo da Vinci, master draftsman. New York/ New Haven & London: The Metropolitan Museum of Art/Yale University Press; 2003.

Clark K. Leonardo da Vinci. Anatomical drawings at Windsor Castle, vol. I–III. 2nd ed. Oxford: Phaidon; 1969.

Clark K. Leonardo da Vinci. Revised and introduced by Martin Kemp. London: Penguin Books; 1988.

Clayton M. Leonardo da Vinci. The anatomy of man. Drawings from the collection of her Majesty Queen Elizabeth II. Boston: Little Brown and Company; 1992.

Covi DA. Andrea del Verrocchio. Life and work. Firenze: Leo S. Olschki; 2005.

Farago C. Leonardo's writings and theory of art. Vol. 4 of Leonardo da Vinci SelectScholarship in English. London: Taylor & Francis; 1999.

Finger FL. Catalogue of the Incunabula in the Elmer Belt Library of Vinciana. Friends of the UCLA Library. Los Angeles: Anderson, Ritchie & Simon; 1971.

Freud S. Leonardo da Vinci and a memory of his childhood, 1910. (Reprint, The Complete Psychological Works of Sigmund Freud, Vol. 11. London: Vintage Books; 2001).

Hall J. Michelangelo and the reinvention of the human body. London: Chatto & Windus; 2005.

Harvey W. The anatomical exercises: De Motu Cordis 1628; De Circulatione Sanguinis 1649: the first English text of 1653 now newly edited by Geoffrey Keynes. (London: Nonesuch Press, 1953); (Reprint, New York: Dover Publications, 1995).

Hibbert C. The rise and fall of the House of Medici. Harmondsworth: Penguin; 1979a.

Jacobs FH. The living image in renaissance art. Cambridge, UK/ NewYork: Cambridge University Press; 2005.

Keele K. Leonardo da Vinci on movement of the heart and blood. London: Harvey and Blythe Ltd.; 1952.

Keele KD. Leonardo da Vinci's elements of the science of man. New York: Academic Press; 1983.

Keele KD, Pedretti C. Corpus of the anatomical studies in the collection of Her Majesty the Queen at Windsor Castle, vol. I and II. London: Harcourt Brace Jovanovich; 1978–80.

Kemp M. Leonardo. Oxford: Oxford University Press; 2004.

Kemp M. Leonardo da Vinci. The marvellous works of nature and man, revised edition. Oxford: Oxford University Press; 2006.

Kemp M. Leonardo da Vinci. Experience, experiment, design. London: V&A Publications; 2007.

MacCurdy E. The notebooks of Leonardo da Vinci. Arranged, rendered into English and introduced by Edward MacCurdy. London: Jonathan Cape; 1945.

McCurdy E. The mind of Leonardo da Vinci. (New York: Dodd, Mead & Co., 1928; reprint, New York: Dover, 2011).

McEwen IK. Vitruvius. Writing the body of architecture. Cambridge: MIT Press; 2003.

McMurrich J. Playfair. Leonardo da Vinci, the anatomist. Baltimore: Williams & Wilkins; 1930.

Nicholl C. Leonardo da Vinci. The flights of the mind. Harmondsworth: Penguin; 2004.

O'Malley CD and Saunders JBCM. Leonardo da Vinci on the human body. (New York, Henry Schuman, 1952; reprint, New York, Dover, 1983).

Pedretti C. Leonardo. A study in chronology and style. London: Thames and Hudson; 1973.

Pedretti C. Il tempio dell'anima [Italian]. Firenze: Cartei & Bianchi Edizioni; 2007.

Popham AE. The drawings of Leonardo da Vinci. London: Jonathan Cape; 1946 (paperback edition, 1964).

Radke GM. (Curator). Verrocchio's David restored. A Renaissance Bronze from the National Museum of the Bargello, Florence. Catalog of an exhibition at the High Museum of Art, Atlanta, 2003.

Reti L, editor. Unknown Leonardo. New York: Abradale/Abrams; 1990.

Richter IA, editor. Notebooks of Leonardo da Vinci. (Oxford: Oxford World's Classics, 1952; edited with an introduction by Thereza Wells and preface by Martin Kemp. New York, Oxford University Press, 2008).

Richter JP. The notebooks of Leonardo da Vinci. Compiled and edited from the original manuscripts. Vols. I and II. New York: Dover Publications; 1970.

Rifkin BA, Ackerman MJ, Folkenberg J. Human anatomy. Depicting the body from the renaissance to today. London: Thames and Hudson; 2006.

Turner A. Richard. Inventing Leonardo. The anatomy of a legend. Berkeley: University of California Press; 1994.

F.C. Wells, *The Heart of Leonardo*,
DOI 10.1007/978-1-4471-4531-8, © Springer-Verlag London 2013

Vasari G. Lives of the artists, vol. I & II. London: Penguin Classics; 1987.
White M. Leonardo: the first scientist. New York: St. Martin's Press; 2000.
Wright A. The Pollaiuolo Brothers: the arts of Florence and Rome. New Haven: Yale University Press; 2005.
Zöllner F. Leonardo da Vinci. The complete paintings and drawings. Köln: TASCHEN; 2005.
Zubov VP. Leonardo da Vinci (Trans. David H. Kraus). New York: MetroBooks; 2002.
Zwijnenberg R. The writings and drawings of Leonardo da Vinci: order and chaos in early modern thought. Cambridge: Cambridge University Press; 1999.

General Reading

Atalay B. Math and the Mona Lisa. The art and science of Leonardo da Vinci. Washington: Smithsonian Books; 2004.
Bramley S. Leonardo: the artist and the man. London: Penguin Books; 1994.
Clayton M. Leonardo da Vinci: the divine and the grotesque. London: The Royal Collection; 2002.
Descartes R. Discourse on method and related writings (trans. Desmond Clarke). London: Penguin Classics; 2000.
Dooley D. The rediscovery of anatomy. Reprinted from the proceedings of the Royal Institution of Great Britain. Vol. 47, 1974. London: Applied Science Publishers; 1974.
Goffen R. Renaissance rivals. Michelangelo, Leonardo, Raphael, Titian. New Haven/London: Yale University Press; 2002.
Gould C. Leonardo the artist and the non-artist. London: Weidenfeld and Nicholson; 1975.
Gould SJ. Leonardo's mountain of clams and the diet of worms. London: Jonathon Cape; 1998.
Henderson J. The Renaissance Hospital. Healing the body and healing the soul. New Haven: Yale University Press; 2006.
Hibbert C. The rise and fall of the House of Medici. London: Penguin; 1979b.
Hibbert C. Florence, the biography of a city. London: Penguin Books; 2004.
Joost-Gaugier CL. Raphael's Stanza della Segnatura. Meaning and invention. Cambridge: Cambridge University Press; 2002.
Keele KD. Leonardo da Vinci and the art of science (Pioneers of Science and Discovery). Hove: Priory Press; 1977.
Kiel H, editor. The Bernard Berenson treasury. A selection from the works, unpublished writings, letters, diaries and journals of the most celebrated humanist and art historian of our times: 1887–1958. New York: Simon and Schuster; 1962.
King R. Brunelleschi's dome: how a renaissance genius reinvented architecture. Harmondsworth: Penguin Books; 2000.
Kraye J. The Cambridge companion to renaissance humanism. Cambridge: Cambridge University Press; 1996.
Martines L. April blood. Florence and the plot against the Medici. Oxford: Oxford University Press; 2003.
Martines L. Scourge and fire. Savonarola and Renaissance Italy. London: Pimlico; 2007.
Masters RD. Fortune is a river. Leonardo da Vinci and Niccolo Machiavelli's magnificent dream to change the course of Florentine History. New York: Plume; 1999.
Nanni R, Testaferrata E. Vinci: the town of Leonardo. Histories and memories. Pacini: Pisa; 2004.
Panofsky E. The life and art of Albrecht Dürer. Princeton/Oxford: Princeton University Press; 2005.
Parks T. Medici money. Banking, metaphysics and art in fifteenth century Florence. New York: Atlas Books & W.W. Norton; 2006.
Porter R. Flesh in the age of reason. How the enlightenment transformed the way we see our bodies and souls. London: Penguin Books; 2003.
Schmitt CB, Skinner Q. The Cambridge history of renaissance philosophy. Cambridge: Cambridge University Press; 1988.
Seward D. The burning of the vanities. Savonarola and the Borgia Pope. Stroud: Sutton Publishing; 2006.
Thomas A. The painter's practice in Renaissance Tuscany. Cambridge: Cambridge University Press; 1995.
Thompson DW. On growth and form, The complete revised edition. Cambridge: Cambridge University Press; 1942; reprint, New York: Dover; 1992.
Van Hensbergen G. Gaudi. London: HarperCollins; 2001.

Index

F.C. Wells, *The Heart of Leonardo*,
DOI 10.1007/978-1-4471-4531-8, © Springer-Verlag London 2013

FSC
www.fsc.org
MIX
Papier | Fördert
gute Waldnutzung
FSC® C083411

Kolibri 360 GmbH
Ferdinand-Jühlke-Straße 7
99095 Erfurt, Deutschland
produktsicherheit@kolibri360.de